PROGRESS IN CLINICAL AND BIOLOGICAL RESEARCH

RECENT TITLES

Vol 286: **Molecular and Cellular Mechanisms of Septic Shock,** Bryan L. Roth, Thor B. Nielsen, Adam E. McKee, *Editors*

Vol 287: **Dietary Restriction and Aging,** David L. Snyder, *Editor*

Vol 288: **Immunity to Cancer. II,** Malcolm S. Mitchell, *Editor*

Vol 289: **Computer-Assisted Modeling of Receptor–Ligand Interactions: Theoretical Aspects and Applications to Drug Design,** Robert Rein, Amram Golombek, *Editors*

Vol 290: **Enzymology and Molecular Biology of Carbonyl Metabolism 2: Aldehyde Dehydrogenase, Alcohol Dehydrogenase, and Aldo-Keto Reductase,** Henry Weiner, T. Geoffrey Flynn, *Editors*

Vol 291: **QSAR: Quantitative Structure-Activity Relationships in Drug Design,** J.L. Fauchère, *Editor*

Vol 292: **Biological and Synthetic Membranes,** D. Allan Butterfield, *Editor*

Vol 293: **Advances in Cancer Control: Innovations and Research,** Paul N. Anderson, Paul F. Engstrom, Lee E. Mortenson, *Editors*

Vol 294: **Development of Preimplantation Embryos and Their Environment,** Koji Yoshinaga, Takahide Mori, *Editors*

Vol 295: **Cells and Tissues: A Three-Dimensional Approach by Modern Techniques in Microscopy,** Pietro M. Motta, *Editor*

Vol 296: **Developments in Ultrastructure of Reproduction,** Pietro M. Motta, *Editor*

Vol 297: **Biochemistry of the Acute Allergic Reactions: Fifth International Symposium,** Alfred I. Tauber, Bruce U. Wintroub, Arlene Stolper Simon, *Editors*

Vol 298: **Skin Carcinogenesis: Mechanisms and Human Relevance,** Thomas J. Slaga, Andre J.P. Klein-Szanto, R.K. Boutwell, Donald E. Stevenson, Hugh L. Spitzer, Bob D'Amato, *Editors*

Vol 299: **Perspectives in Shock Research: Metabolism, Immunology, Mediators, and Models,** John C. Passmore, *Editor,* Sherwood M. Reichard, David G. Reynolds, Daniel L. Traber, *Co-Editors*

Vol 300: **Alpha$_1$-Acid Glycoprotein: Genetics, Biochemistry, Physiological Functions, and Pharmacology,** Pierre Baumann, Chin B. Eap, Walter E. Müller, Jean-Paul Tillement, *Editors*

Vol 301: **Prostaglandins in Clinical Research: Cardiovascular System,** Karsten Schrör, Helmut Sinzinger, *Editors*

Vol 302: **Sperm Measures and Reproductive Success: Institute for Health Policy Analysis Forum on Science, Health, and Environmental Risk Assessment,** Edward J. Burger, Jr., Robert G. Tardiff, Anthony R. Scialli, Harold Zenick, *Editors*

Vol 303: **Therapeutic Progress in Urological Cancers,** Gerald P. Murphy, Saad Khoury, *Editors*

Vol 304: **The Maillard Reaction in Aging, Diabetes, and Nutrition,** John W. Baynes, Vincent M. Monnier, *Editors*

Vol 305: **Genetics of Kidney Disorders,** Christos S. Bartsocas, *Editor*

Vol 306: **Genetics of Neuromuscular Disorders,** Christos S. Bartsocas, *Editor*

Vol 307: **Recent Advances in Avian Immunology Research,** Balbir S. Bhogal, Guus Koch, *Editors*

Vol 308: **Second Vienna Shock Forum,** Günther Schlag, Heinz Redl, *Editors*

Vol 309: **Advances and Controversies in Thalassemia Therapy: Bone Marrow Transplantation and Other Approaches,** C. Dean Buckner, Robert Peter Gale, Guido Lucarelli, *Editors*

Vol 310: **EORTC Genitourinary Group Monograph 6: BCG in Superficial Bladder Cancer,** Frans M.J. Debruyne, Louis Denis, Ad P.M. van der Meijden, *Editors*

Vol 311: **Molecular and Cytogenetic Studies of Non-Disjunction,** Terry J. Hassold, Charles J. Epstein, *Editors*

Vol 312: **The Ocular Effects of Prostaglandins and Other Eicosanoids,** Laszlo Z. Bito, Johan Stjernschantz, *Editors*

Vol 313: **Malaria and the Red Cell: 2,** John W. Eaton, Steven R. Meshnick, George J. Brewer, *Editors*

Vol 314: **Inherited and Environmentally Induced Retinal Degenerations,** Matthew M. LaVail, Robert E. Anderson, Joe G. Hollyfield, *Editors*

Vol 315: **Muscle Energetics,** Richard J. Paul, Gijs Elzinga, Kazuhiro Yamada, *Editors*

Vol 316: **Hemoglobin Switching,** George Stamatoyannopoulos, Arthur W. Nienhuis, *Editors*. Published in two volumes: Part A: *Transcriptional Regulation*. Part B: *Cellular and Molecular Mechanisms*.

Vol 317: **Alzheimer's Disease and Related Disorders,** Khalid Iqbal, Henryk M. Wisniewski, Bengt Winblad, *Editors*

Vol 318: **Mechanisms of Chromosome Distribution and Aneuploidy,** Michael A. Resnick, Baldev K. Vig, *Editors*

Vol 319: **The Red Cell: Seventh Ann Arbor Conference,** George J. Brewer, *Editor*

Vol 320: **Menopause: Evaluation, Treatment, and Health Concerns,** Charles B. Hammond, Florence P. Haseltine, Isaac Schiff, *Editors*

Vol 321: **Fatty Acid Oxidation: Clinical, Biochemical, and Molecular Aspects,** Kay Tanaka, Paul M. Coates, *Editors*

Vol 322: **Molecular Endocrinology and Steroid Hormone Action,** Gordon H. Sato, James L. Stevens, *Editors*

Vol 323: **Current Concepts in Endometriosis,** Dev R. Chadha, Veasy C. Buttram, Jr., *Editors*

Vol 324: **Recent Advances in Hemophilia Care,** Carol K. Kasper, *Editor*

Vol 325: **Alcohol, Immunomodulation, and AIDS,** Daniela Seminara, Ronald Ross Watson, Albert Pawlowski, *Editors*

Vol 326: **Nutrition and Aging,** Derek M. Prinsley, Harold H. Sandstead, *Editors*

Vol 327: **Frontiers in Smooth Muscle Research,** Nicholas Sperelakis, Jackie D. Wood, *Editors*

Vol 328: **The International Narcotics Research Conference (INRC) '89,** Rémi Quirion, Khem Jhamandas, Christina Gianoulakis, *Editors*

Vol 329: **Multipoint Mapping and Linkage Based Upon Affected Pedigree Members: Genetic Analysis Workshop 6,** Robert C. Elston, M. Anne Spence, Susan E. Hodge, Jean W. MacCluer, *Editors*

Vol 330: **Verocytotoxin-Producing *Escherichia coli* Infections,** Martin Petric, Charles R. Smith, Clifford A. Lingwood, James L. Brunton, Mohamed A. Karmali, *Editors*

Vol 331: **Mouse Liver Carcinogenesis: Mechanisms and Species Comparisons,** Donald E. Stevenson, R. Michael McClain, James A. Popp, Thomas J. Slaga, Jerrold M. Ward, Henry C. Pitot, *Editors*

Vol 332: **Molecular and Cellular Regulation of Calcium and Phosphate Metabolism,** Meinrad Peterlik, Felix Bronner, *Editors*

Vol 333: **Bone Marrow Purging and Processing,** Samuel Gross, Adrian P. Gee, Diana A. Worthington-White, *Editors*

Vol 334: **Potassium Channels: Basic Function and Therapeutic Aspects,** Thomas J. Colatsky, *Editor*

Vol 335: **Evolution of Subterranean Mammals at the Organismal and Molecular Levels,** Eviatar Nevo, Osvaldo A. Reig, *Editors*

Vol 336: **Dynamic Interactions of Myelin Proteins,** George A. Hashim, Mario Moscarello, *Editors*

Please contact publisher for information about previous titles in this series.

MOUSE LIVER CARCINOGENESIS

Mechanisms and Species Comparisons

MOUSE LIVER CARCINOGENESIS

Mechanisms and Species Comparisons

Proceedings of the Symposium Held in Austin, Texas, November 30–December 3, 1988

Editors

Donald E. Stevenson
Consultant in Toxicology
Shell Oil Company
Houston, Texas

R. Michael McClain
Department of Toxicology
Hoffmann-La Roche, Inc.
Nutley, New Jersey

James A. Popp
Chemical Industry
Institute of Toxicology
Research Triangle Park, North Carolina

Thomas J. Slaga
Science Park-Research Division
The University of Texas
M.D. Anderson Cancer Center
Smithville, Texas

Jerrold M. Ward
Laboratory of
Comparative Carcinogenesis
National Cancer Institute
Frederick, Maryland

Henry C. Pitot
University of Wisconsin
McArdle Laboratory for
Cancer Research
Madison, Wisconsin

WILEY-LISS • NEW YORK

Address all Inquiries to the Publisher
Alan R. Liss, Inc., 41 East 11th Street, New York, NY 10003

Printed in United States of America

The publication of this volume was facilitated by the authors and editors who submitted the text in a form suitable for direct reproduction without subsequent editing or proofreading by the publisher.

Library of Congress Cataloging-in-Publication Data

Mouse liver carcinogenesis : mechanisms and species comparisons : proceedings of a symposium held in Austin, Texas, November 30-December 3, 1988 / editors, Donald E. Stevenson ... [et al.].
p. cm. -- (Progress in clinical and biological research ; v. 331)
Includes bibliographical references.
Conference hosted by the Science Park-Research Division of the University of Texas M.D. Anderson Cancer Center in Smithville, Tex.
ISBN 0-471-56695-0
1. Liver--Tumors--Congresses. 2. Cancer--Animal models--Congresses. 3. Carcinogenicity testing--Congresses. 4. Carcinogenesis--Congresses. 5. Mice--Diseases--Congresses. I. Stevenson, Donald E. II. M.D. Anderson Cancer Center. Science-Park Research Division. III. Series.
RC280.L5M67 1989
616.99'236--dc20 89-14012
CIP

Contents

Contributors

Alan K. Adams, Department of Carcinogenesis, The University of Texas M.D. Anderson Cancer Center, Science Park-Research Division, Smithville, TX 78957 **[109]**

Diane D. Beal, Oncology Branch, Office of Toxic Substances, U.S. Environmental Protection Agency, Washington, DC 20460 **[5]**

Frederick F. Becker, Section of Experimental Pathology, The University of Texas M.D. Anderson Cancer Center, Houston, TX 77030 **[249]**

David G. Beer, Department of Pharmacology, Toxicology, and Therapeutics, Center for Environmental and Occupational Health, Kansas University Medical Center, Kansas City, KS 66103 **[293]**

Frederick A. Beland, Division of Biochemical Toxicology, National Center for Toxicological Research, Jefferson, AR 72079 **[121]**

Linda Birnbaum, Systemic Toxicology Branch, National Institute of Environmental Health Sciences, Research Triangle Park, NC 27709 **[187]**

Daniel M. Byrd III, Consulting Toxicologist, Falls Church, VA 22044 **[19]**

M. Cadrin, Department of Pathology, University of Ottawa, Ottawa, Ontario, Canada K1H 8M5 **[231]**

C.C. Chang, Department of Pediatrics and Human Development, Michigan State University, East Lansing, MI 48824 **[259]**

George Clarke, Laboratory of Biochemical Risk Analysis, National Institute of Environmental Health Sciences, Research Triangle Park, NC 27709 **[187]**

Katherine E. Cole, Department of Pathology, University of Maryland School of Medicine, Baltimore, MD 21201 and Laboratory of Human Carcinogenesis, DHHS, NIH, NCI, Bethesda, MD 20892 **[395]**

Edmund A.C. Crouch, Meta Systems, Inc., Cambridge, MA 02141-0674 **[19]**

Marcela Del Rio, Department of Carcinogenesis, The University of Texas M.D. Anderson Cancer Center, Science Park-Research Division, Smithville, TX 78957 **[109]**

The number in brackets is the opening page of the contributor's article.

Deborah E. Devor, Tumor Pathology and Pathogenesis Section, Laboratory of Comparative Carcinogenesis, National Cancer Institute, Frederick, MD 21701 **[85]**

Bhalchandra A. Diwan, Chemical Carcinogenesis Section, Biological Carcinogenesis Development Program, Program Resources, Inc., National Cancer Institute-Frederick Cancer Research Facility, Frederick, MD 21701 **[69, 85]**

Norman R. Drinkwater, McArdle Laboratory for Cancer Research, University of Wisconsin, Madison, WI 53706 **[163, 203]**

C.R. Elcombe, I.C.I., Central Toxicology Laboratory, Macclesfield, Cheshire, England SK11 8AH **[385]**

William H. Farland, Office of Health and Environmental Assessment, U.S. Environmental Protection Agency, Washington, DC 20460 **[409]**

Nelson Fausto, Department of Pathology and Laboratory Medicine, Brown University, Providence, RI 02912 **[325]**

S.W. French, Department of Pathology, University of Ottawa, Ottawa, Ontario, Canada K1H 8M5 **[231]**

Nancy F. Fullerton, Division of Biochemical Toxicology, National Center for Toxicological Research, Jefferson, AR 72079 **[121]**

Kevin Gaido, Department of Cellular and Molecular Toxicology, Chemical Industry Institute of Toxicology, Research Triangle Park, NC 27709 **[177]**

Joyce A. Goldstein, Laboratory of Biochemical Risk Analysis, National Institute of Environmental Health Sciences, Research Triangle Park, NC 27709 **[187]**

Jay I. Goodman, Department of Pharmacology and Toxicology, Center for Environmental Toxicology, Michigan State University, East Lansing, MI 48824 **[335]**

Mark Graham, Department of Toxicology, Sterling-Winthrop Research Centre, Avonwick, Northumberland, England NE66 20H **[187]**

William F. Greenlee, Department of Cellular and Molecular Toxicology, Chemical Industry Institute of Toxicology, Research Triangle Park, NC 27709 **[177]**

James A. Hampton, Department of Pathology and Pharmacology, Medical College of Ohio, Toledo, OH 43614 **[277]**

Malgorzata Hanausek-Walasze, Department of Carcinogenesis, The University of Texas M.D. Anderson Cancer Center, Science Park-Research Division, Smithville, TX 78957 **[109]**

Marie H. Hanigan, McArdle Laboratory for Cancer Research, University of Wisconsin, Madison, WI 53706; present address: Department of Anatomy and Cell Biology, University of Virginia Medical Center, Charlottesville, VA 22908 **[163]**

James A. Hartnett, Department of Pathology and Pharmacology, Medical College of Ohio, Toledo, OH 43614 **[277]**

John R. Henneman, Chemical Carcinogenesis Section, Biological Carcinogenesis Development Program, Program Resources, Inc., National Cancer Institute-Frederick Cancer Research Facility, Frederick, MD 21701 **[85]**

Ih-Chang Hsu, Department of Pathology, University of Maryland School of Medicine, Baltimore, MD 21201 **[395]**

Costas Ioannides, Department of Biochemistry, University of Surrey, Guildford, Surrey, England GU2 5XH **[215]**

Thomas W. Jones, Department of Pathology, University of Maryland School of Medicine, Baltimore, MD 21201 **[395]**

Myong W. Kahng, Department of Pathology, University of Maryland School of Medicine, Baltimore, MD 21201 **[395]**

Y. Katsuma, Department of Internal Medicine, Kyoto Prefectural University, Kyoto 602, Japan **[231]**

H. Kawahara, Department of Pathology, University of Ottawa, Ottawa, Ontario, Canada K1H 8M5 **[231]**

M.D. Kelly, I.C.I., Central Toxicology Laboratory, Macclesfield, Cheshire, England SK11 8AH **[385]**

Christopher J. Kemp, Department of Oncology, McArdle Laboratory for Cancer Research, University of Wisconsin, Madison, WI 53706 **[163, 203]**

Takemi Kinouchi, Division of Biochemical Toxicology, National Center for Toxicological Research, Jefferson, AR 72079; present address: Department of Bacteriology, University of Tokushima School of Medicine, Tokushima 770, Japan **[121]**

James E. Klaunig, Department of Pathology and Pharmacology, Medical College of Ohio, Toledo, OH 43614 **[277]**

F.H. Lin, Laboratory of Biochemical Risk Analysis, National Institute of Environmental Health Sciences, Research Triangle Park, NC 27709 **[187]**

Michael M. Lipsky, Department of Pathology, University of Maryland School of Medicine, Baltimore, MD 21201 **[395]**

Ronald A. Lubet, Laboratory of Comparative Carcinogenesis, National Cancer Institute, Frederick, MD 21701 **[85]**

George Lucier, Laboratory of Biochemical Risk Analysis, National Institute of Environmental Health Sciences, Research Triangle Park, NC 27709 **[187]**

B.V. Madhukar, Department of Pediatrics and Human Development, Michigan State University, East Lansing, MI 48824 **[259]**

N. Marceau, Laval University Center for Cancer Research, Quebec, Canada G1R 2J6 **[231]**

R. Michael McClain, Department of Toxicology and Pathology, Hoffmann-La Roche, Inc., Nutley, NJ 07110 **[345]**

Ernest E. McConnell, Science Advisor, International Life Sciences Institute, Raleigh, NC 27613 **[1]**

Jon C. Mirsalis, Cellular and Genetic Toxicology Department, SRI International, Menlo Park, CA 94025 **[149]**

Mark J. Neveu, McArdle Laboratory for Cancer Research, University of Wisconsin, Madison, WI 53706 **[293]**

M. Ohta, Department of Internal Medicine, Kyoto Prefectural University, Kyoto 602, Japan **[231]**

Dennis V. Parke, Department of Biochemistry, University of Surrey, Guildford, Surrey, England GU2 5XH **[215]**

Henry C. Pitot, Department of Oncology and Pathology, McArdle Laboratory, University of Wisconsin, Madison, WI 53706 **[311]**

Miriam C. Poirier, Laboratory of Cellular Carcinogenesis and Tumor Promotion, National Cancer Institute, Bethesda, MD 20892 **[121]**

N.R. Pritchard, I.C.I., Central Toxicology Laboratory, Macclesfield, Cheshire, England SK11 8AH **[385]**

Jerry M. Rice, Laboratory of Comparative Carcinogenesis, National Cancer Institute-Frederick Cancer Research Facility, Frederick, MD 21701 **[69]**

Randall J. Ruch, Department of Pathology and Pharmacology, Medical College of Ohio, Toledo, OH 43614 **[277]**

Thomas R. Skopek, Department of Cellular and Molecular Toxicology, Chemical Industry Institute of Toxicology, Research Triangle Park, NC 27709 **[177]**

Beverly A. Smith, Division of Biochemical Toxicology, National Center for Toxicological Research, Jefferson, AR 72079 **[121]**

Karen L. Steinmetz, Cellular and Genetic Toxicology Department, SRI International, Menlo Park, CA 94025 **[149]**

Donald E. Stevenson, Department of Toxicology, Health, Safety, and Environment, Shell Oil Company, Houston, TX 77210; Scientific Committee, American Industrial Health Council, Washington, DC 20036 **[43, 367]**

Sidney J. Stohs, School of Pharmacy, Creighton University, Omaha, NE 68178 **[187]**

Daniel L. Stout, Section of Experimental Pathology, The University of Texas M.D. Anderson Cancer Center, Houston, TX 77030 **[249]**

J.A. Styles, I.C.I., Central Toxicology Laboratory, Macclesfield, Cheshire, England SK11 8AH **[385]**

J.E. Trosko, Department of Pediatrics and Human Development, Michigan State University, East Lansing, MI 48824 **[259]**

Benjamin F. Trump, Department of Pathology and Maryland Institute for Emergency Medical Services System, University of Maryland School of Medicine, Baltimore, MD 21201 **[395]**

Stan D. Vesselinovitch, Departments of Radiology and Pathology, University of Chicago, Chicago, IL 60637 **[53]**

Roseann L. Vorce, Department of Pharmacology and Toxicology, Center for Environmental Toxicology, Michigan State University, East Lansing, MI 48824; present address: National Institutes of Health, Bethesda, MD 20892 **[335]**

Cheryl Walker, Department of Cellular and Molecular Toxicology, Chemical Industry Institute of Toxicology, Research Triangle Park, NC 27709 **[177]**

Jerrold M. Ward, Tumor Pathology and Pathogenesis Section, Laboratory of Comparative Carcinogenesis, Division of Cancer Etiology, National Cancer Institute, Frederick, MD 21701 **[69, 85]**

Christopher M. Weghorst, Department of Pathology and Pharmacology, Medical College of Ohio, Toledo, OH 43614 **[277]**

Gary M. Williams, Director of Medical Sciences, American Health Foundation, Valhalla, NY 10595 **[131]**

Richard Wilson, Department of Physics and Energy and Environmental Policy Center, Harvard University, Cambridge, MA 02138 **[19]**

Preface

Although at times the target of heated scientific debate, the biologic importance of data from studies of either chemically induced or environmentally caused mouse liver tumors has long been recognized, making this presentation of studies by the world's leading scientists especially relevant. This volume, **Mouse Liver Carcinogenesis: Mechanisms and Species Comparisons,** presents the most recent findings on mouse liver carcinogenesis and will help guide future research, help (we hope) build quantitative and qualitative models for carcinogenicity, and allow the best science possible in the field of carcinogenicity risk assessment.

This book is a compilation of the proceedings of the conference of the same name held in Austin, Texas, November 30 through December 3, 1988. The chapters discuss and evaluate the biologic importance of chemically induced mouse liver tumors to human hazard identification. Representatives of regulatory agencies present their views of the use and interpretation of mouse liver tumor data. One section of the book addresses various models, morphology, and markers and includes, among others, the subjects of liver tumor promotion in the mouse and transplacental and perinatal mouse assay systems. DNA changes, cell proliferation, and cell death also are reviewed, including the role of hyperplasia in liver carcinogenesis and epigenetic mechanisms of liver tumor promotion. The volume also includes sections on genetics and receptors in liver tumor induction and cytochromes, peroxisomes, and free radicals. In the area of cell-to-cell communication, the role of inhibition of gap-junction-mediated intercellular communication in hepatic carcinogenesis is presented, along with a report on oncogene and gap-junction protein expression in rodent liver neoplasms. Another section reviews oncogene and growth factor studies, including an analysis of cell lineages in liver carcinogenesis and the identification of stem cells by oncogene transaction. Studies involving specific chemicals also are presented. These reviews and research studies are drawn together in the chapter "Research Opportunities and Reducing Uncertainty."

It is the general aim of this volume to review and analyze carefully the available information on mouse liver tumors induced by chemicals and environmental factors in order to come to a better understanding of the mechanisms of carcinogenesis in the mouse liver and their relative importance to other species, including man. By compiling these findings, we have provided a valuable and in-depth resource for the scientific investigator.

This volume should be of interest to cancer research scientists, pathologists, toxicologists, pharmacologists, and biochemists, as well as cellular and molecular biologists. Those in governmental administrative and regulatory agencies dealing with environmental health hazards should also find this volume helpful.

Donald E. Stevenson
R. Michael McClain
James A. Popp
Jerrold M. Ward
Thomas J. Slaga
Henry C. Pitot

Introduction

From November 30 through December 3, 1988, the Science Park–Research Division of The University of Texas M. D. Anderson Cancer Center in Smithville, Texas hosted the second conference in a series on Research Directions for Human Relevance in Carcinogenesis. The second meeting, "Mouse Liver Carcinogenesis: Mechanisms and Species Comparisons," included invited lectures by leading scientists from the United States, Canada, and Europe.

This volume represents the proceedings of that conference, including the discussions that followed each lecture. The major objective of this conference was to bring together people from academia, regulatory agencies, industry, and public interest groups to review and analyze the currently available information on mouse liver tumors induced by chemicals and by environmental factors in order to better understand the pertinent mechanisms and their relative importance to other species, including man. Research and interpretation needs relating to the mechanisms involved in the formation of mouse liver tumors were emphasized.

A large part of the conference was devoted to describing and analyzing ongoing mouse liver tumor research and the debate regarding its relevance to cancer research and risk assessment. Presentations addressed mouse liver carcinogenesis on a number of levels. There was extensive discussion about the cellular, biochemical, and molecular events leading to initiation, promotion, and progressions, and studies describing the cellular events leading to changes in cell-to-cell communication were presented. Overall, the conference attempted to evaluate the unifying principles of mouse liver carcinogenesis.

The first part of this book includes detailed overviews of the mouse liver system, use of mouse liver tumor data in risk assessment, and correlations between tumors in different sites and species. This volume also addresses various tumor models, morphological changes and markers, DNA changes, cell proliferation, and cell death, as well as genetic factors and receptors in liver tumor induction. Other chapters are devoted to cytochrome changes, peroxisome and free radical involvement, cell-to-cell communication alterations, altered expression and/or activation of oncogenes, and the involvement of growth factors.

Acknowledgments

The conference "Mouse Liver Carcinogenesis: Mechanisms and Species Comparisons," on which this volume is based, was made possible by generous contributions from both academia and private industry. The conference organizers wish to sincerely thank the Shell Oil Company, ILSI Risk Science, the Motor Vehicle Manufacturers Association, the American Industrial Health Council, E. I. du Pont de Nemours & Company, Procter & Gamble Company, and the Monsanto Company for their generous support. Also, the contributions made by Allied Signal, Inc., the Halogenated Solvents Industrial Alliance, Bristol-Myers Company, Bioproducts for Science, The University of Texas M. D. Anderson Cancer Center, Science Park–Research Division, and the Environmental Protection Agency are sincerely appreciated. Special thanks are extended to the Conference Organizing Committee: Donald Stevenson, Thomas J. Slaga, Henry C. Pitot, James A. Popp, Jerrold M. Ward, William Farland, Frederick F. Becker, R. Michael McClain, Hugh Spitzer, and Carol Henry. The conference organization details were expertly directed by Karen Engel and her staff, Christie Hoy, Mary Lou Fendley, Judy Ing, and LeNel Rice. For help in the preparation of this volume, I would especially like to thank Mary Markell, Edith K. Wilson, and Mary Deiss of the Department of Scientific Publications at The University of Texas M. D. Anderson Cancer Center. Also, special thanks to Mary Slaga, a volunteer at the Science Park–Research Division, for assistance in final manuscript preparation.

Thomas J. Slaga

Mouse Liver Carcinogenesis: Mechanisms
and Species Comparisons, pages 1–3

Mouse Liver Tumors: The Problem

Ernest E. McConnell
International Life Sciences Institute science advisor, Raleigh, NC 27612

The biologic importance of chemically induced mouse liver tumors to human hazard identification has been the subject of considerable debate. There have been many symposia, meetings, workshops, and publications on the subject. I am sure that this book will not be the last.

Some scientists have referred to mouse liver tumor induction as an anomaly. They are convinced that mouse liver tumors are of little or no value for assessing human health hazards because of the normal high background incidence in certain strains of mice, that an increased incidence occurs frequently with so many chemicals that it must be of questionable importance, and/or that the model is inappropriate because liver cancer in humans is uncommon, at least in the United States.

Other scientists are just as convinced that the induction of mouse liver tumors is a clear indication of human hazard because a sensitive model is required as a result of the use of relatively small numbers of animals in typical carcinogenesis studies, that it is only reasonable to expect liver tumors since the liver is a major organ of detoxification and the proximate carcinogen or active metabolite is at higher concentrations at that site, and/or because mechanistic studies such as oncogene activation have shown that the steps involved in mouse liver carcinogenesis are comparable to those observed in other types of tumors and in humans.

There are many inherent problems in evaluating the importance of mouse liver tumor induction. To some, the arguments may appear relatively tedious—such as discussions of pathology nomenclature (e.g., should the lesions be called hepatomas, type A or B nodules, type I, II, III, or IV nodules, adenomas, or carcinomas) and the significance of "foci of cellular alteration" in the development of these lesions. I readily admit that nomenclature and its ramifications are an area of considerable debate in some quarters, but they are not part of this discussion.

Some recent papers may shed light on the reservations that I mentioned about the mouse liver model, or at least put them in a different context. In their reviews of 278 compounds studied by the National Cancer Institute/National Toxicology Program, (NCI/NTP) Haseman et al. (1) and Maronpot et al. (2) found that 141 (51%) were carcinogenic in at least one sex or species. Of those 141, 71 (50%) caused liver tumors in mice. However, only 16 of the 71 (22%) caused liver tumors only and caused them only in mice. This means that of the

278 chemicals studied by the NCI/NTP, only 6% were called carcinogens based on the induction of liver tumors in mice alone.

Nevertheless, it is only fair to point out that some of the chemicals that cause mouse liver tumors only are of extreme commercial importance. For example, the list contains some noteworthy pesticides: aldrin, chlordane, dichlorodiphenyldichloroethylene, dicofol, heptachlor, toxaphene, and tetrachlorvinphos. And some commercially important halogenated aliphatics are on that list: 1,2-dichloropropane, pentachloroethane, 1,1,1,2-tetrachloroethane, 1,1,2,2-tetrachloroethane, trichloroethylene with and without epichlorohydrin, and tetrachloroethylene. Also, on the list are di-(2-ethylhexyl) adipate and tris(2-ethylhexyl) phosphate.

Other recent studies, especially those by Hagiwara and Ward (3) and Hoel et al. (4), affect the relevance of mouse liver tumor induction for human hazard identification. Some investigators have proposed that mouse liver tumors are induced because the inciting chemicals cause a significant amount of liver toxicity. The aforementioned studies clearly show that mouse liver tumors, while they often occur along with liver toxicity, are also frequently seen in its absence. Conversely, severe liver toxicity is often seen in the absence of liver tumors.

The NTP currently has several studies under way to sort out some of the uncertainty in the progression of mouse liver tumors. All current studies of at least 24 months' duration include a 15-month kill time to give an indication of the progression of the liver and other lesions. In addition, several studies have incorporated interim kill points at various times, such as 3, 6, 9, 12, and 18 months, for the same purpose. Finally, several studies have included "stop studies" (exposure to the chemical is stopped at some point, e.g., 12 months, and the animals are observed for the remainder of the study and are compared to animals exposed for the full duration of the experiment) as part of the protocol.

In summary the induction of mouse liver tumors as the result of exposure to chemicals is a fact. They cannot be ignored. Rather than writing the tumors off as being irrelevant, the scientific community should focus on the real question: "What does mouse tumor induction mean in terms of a chemical's potential hazard to humans"? Such a discussion should include whether chemicals that cause tumors in the mouse liver only and not in other organs or species, in one sex of mice only, only in a single strain of mouse, in the presence of concomitant liver toxicity, or only at dose levels that are extremely high compared with human exposure levels should be viewed differently from those chemicals that cause the other previously mentioned effects.

Finally, since we know that tumors, including those in the liver, are probably induced by several completely different mechanisms, wouldn't our efforts be better served by concentrating them on trying to understand the mechanisms underlying the induction of mouse liver tumors as a result of exposure to a given chemical? Then, we could attempt to determine whether the same mechanisms were operative in humans. This is the only scientifically credible way of determining the "real" health hazard of a chemical.

References

1. Haseman JK, Crawford DD, Huff JE, Boorman GA, McConnell EE. Results from 86 two-year carcinogenicity studies conducted by the National Toxicology Program. *J Toxicol Environ Health* 14:621–639, 1984.
2. Maronpot RR, Haseman JK, Boorman GA, Eustis SE, Rao GN, Huff JE. Liver lesions in B6C3F1 mice: The National Toxicology Program, experience and position. *Arch Toxicol [Suppl]* 10:10–26, 1987.
3. Hagiwara A, Ward JM. The chronic hepatotoxic, tumor-promoting, and carcinogenic effects of acetaminophen in male $B6C3F_1$ mice. *Fundam Appl Toxicol* 7:376–386, 1986.
4. Hoel DG, Haseman JK, Hogan MD, Huff J, McConnell EE. The impact of toxicity on carcinogenicity studies: Implications for risk assessment. *Carcinogenesis* 9:2045–2052, 1988.

Mouse Liver Carcinogenesis: Mechanisms
and Species Comparisons, pages 5–18

Use of Mouse Liver Tumor Data in Risk Assessments Performed by the U.S. Environmental Protection Agency[1]

Diane D. Beal
Oncology Branch, Office of Toxic Substances, U.S. Environmental Protection Agency, Washington, DC 20460

The use of mouse liver tumor data in estimating risk to humans has been a controversial issue for some time (1–4). Opinions vary from those of scientists who feel that the induction of such tumors is unique to the mouse and has no relevance to cancer in man to those of others who contend that mouse liver tumors are as relevant a risk model as any other tumor type. Since the U.S. Environmental Protection Agency (EPA) has a major responsibility for estimating cancer risk to humans from a wide variety of chemical substances, it has a keen interest in this controversial issue.

In its continuing effort to improve the quality and consistency of its risk assessments, the EPA formed a group known as the Risk Assessment Forum. This group consists of the EPA's top risk assessment scientists, whose major responsibility is to assure that the agency identifies and addresses its key science/science policy issues. One of the key issues that the forum is addressing is the use of mouse liver tumor data in agency cancer risk assessments.

The analysis presented in this chapter is part of the overall effort of the forum to assess the utility of the mouse liver tumor in predicting potential risks to humans. The main objective of the analysis is to assess the use of mouse liver tumor data by the EPA since 1980. The information used in the analysis was obtained by examining the documents that constitute the formal written record of the agency's risk assessment efforts. This documentation, as of October 1988, consisted of the verified cancer risk assessments for 109 chemicals (5).

The remainder of this chapter is divided into four sections. The first discusses the agency's cancer risk assessment guidelines as background for how mouse liver tumor data has been used by the agency. The next two sections describe the analyses of the 109 cancer risk assessments verified by the EPA. The first identifies the categories of chemicals assessed and classified by the agency and which of these chemicals have produced liver tumors in mice. The second analysis describes how mouse liver tumor data are currently used in EPA cancer risk assessments and how these data have influenced the overall classification

---

[1]Disclaimer: Mention of trade names or commercial products does not constitute endorsement or recommendation for use.

of the weight of evidence for given chemicals. Finally, this chapter discusses the major trends in the agency's use of mouse liver tumor data in cancer risk assessments.

Risk Assessment Guidance

The agency first published cancer risk assessment guidelines in 1976 (6,7). The purpose of those guidelines was twofold: to promote quality and consistency of risk assessments within the EPA and to inform those outside the EPA about the agency's approach to carcinogen risk assessments. These guidelines provided a general framework for developing an analysis of carcinogenic risk and some salient principles for evaluating the quality of data and formulating judgments concerning the nature and magnitude of the cancer hazard from suspect carcinogens.

The most important aspect of the 1976 guidelines from the perspective of the use of mouse liver tumor data was that they provided no salient principles about such use. In essence, mouse liver tumor data were to be used like any other tumor data (i.e., as an indication of potential carcinogenicity in humans).

Ten years later, when the revised EPA guidelines were published (8), they retained the original purposes described for the 1976 guidelines, but they contained new concepts and approaches to cancer risk assessment methodology. Several of these are important to the current analysis. First, the 1986 guidelines incorporated the concept of a classification system for the characterization of the overall weight of the evidence for carcinogenicity from studies in humans and animals. Second, the concept of an overall scheme for categorization of the weight of the evidence of carcinogenicity of a chemical for humans was embodied in the guidelines. Finally, the guidelines contained a modified approach to the use of mouse liver tumor data.

Judgments about the weight of evidence from studies in humans and animals involve consideration of the quality and adequacy of the data. Tables 1 and 2 contain the classification systems for the characterization of the overall weight of evidence from studies in humans and animals, respectively. There are five categories of evidence for each type of data: sufficient, limited, inadequate, and no evidence and no data.

The overall scheme for categorization of the weight of evidence of carcinogenicity of a substance for humans uses a three-step process. First, the weight of evidence in human studies and/or animal studies is summarized as stated above; second, these lines of information are combined to yield a tentative assignment to a category; and third, all relevant support information (e.g., structure-activity relationship, mutagenicity, pharmacokinetics information) is evaluated to determine whether the tentative designation needs to be modified. As indicated in Table 3, the agency places a chemical in one of five groups: group A chemicals are considered carcinogenic in humans; group B chemicals are considered probably carcinogenic in humans; group C chemicals are considered possibly carci-

Table 1. Classification System for the Characterization of the Overall Weight of Evidence for Carcinogenicity from Studies in Humans

Classification	Explanation
Sufficient evidence	Causal relationship between the agent and human cancer
Limited evidence	A causal relationship is credible, but alternative explanations, such as chance, bias, or confounding factors could not be adequately excluded
Inadequate evidence	Few pertinent data are available; or the available studies, while showing evidence of association, did not exclude chance, bias, or confounding factors, and therefore a causal interpretation is not credible
No data	Data are not available
No evidence	No association was found between exposure and increased risk of cancer in well-designed and -conducted independent analytic epidemiologic studies

Table 2. Classification System for the Characterization of the Overall Weight of Evidence for Carcinogenicity from Studies in Animals

Classification	Explanation
Sufficient evidence	Increased incidence of malignant or combined malignant and benign tumors a. In multiple species or strains b. In multiple experiments c. To an unusual degree in a single experiment with regard to high incidence, unusual tumor site or type, or early age at onset
Limited evidence	Data suggest a carcinogenic effect but are limited because a. The studies involve a single species, strain, or experiment and do not meet criteria for sufficient evidence b. The experiments are restricted by inadequate dosage levels, inadequate duration of exposure to the agent, inadequate period of follow-up, or inadequate reporting c. An increase in the incidence of benign tumors only
Inadequate evidence	Studies cannot be interpreted
No data	No data are available
No evidence	There is no increased incidence of neoplasms in at least two well-designed and well-conducted studies in different species

Table 3. Categorization of Overall Weight of Evidence for Carcinogenicity in Humans

Group	Weight of evidence
A. Human carcinogen	Sufficient evidence from epidemiologic studies
B. Probable human carcinogen	
B1	Limited evidence from epidemiologic studies
B2	Sufficient evidence in animal studies
C. Possible human carcinogen	Limited evidence in animal studies and no human data
D. Not classifiable as to human carcinogenicity	Inadequate evidence in human or animal studies or no data are available
E. Evidence of noncarcinogenicity for humans	No evidence in humans and/or animals

nogenic in humans; group D chemicals are substances for which there is no data or insufficient information to determine potential activity in humans; and group E chemicals are considered inactive in humans.

In the 1986 guidelines, the agency recognized that there were widely divergent scientific views about the validity of mouse liver tumors as an indication of potential carcinogenicity in humans. The EPA acknowledged that this is especially true when such tumors occur in strains of mice with a high spontaneous background incidence and when mouse liver tumors are the only tumor response to an agent. Therefore, the agency took the following position in the 1986 guidelines: "...when the only tumor response is in the mouse liver and when the other conditions for a classification of 'sufficient' evidence in animals are met..., the data should be considered as 'sufficient' evidence of carcinogenicity." The guidelines went on to state, "...this classification could be changed on a case-by-case basis to 'limited,' if warranted, when such factors as the following, are observed: an increased incidence of tumors only in the highest dose group and/or only at the end of the study; no substantial dose-related increase in the proportion of tumors that are malignant; the occurrence of tumors that are predominantly benign; no dose-related shortening of time to the appearance of tumors; negative or inconclusive results from a spectrum of short-term tests for mutagenic activity; the occurrence of excess tumors only in a single sex." The guidelines did not specify whether each of these factors were to be given equal weight or whether a combination of factors was needed to decrease the weight of evidence.

Category of Chemical Analysis

As stated earlier, the agency has verified cancer risk assessments for 109 substances. These substances can be grouped into 21 categories (Table 4). Analysis of the information in the table indicates that the vast majority of substances assessed by the EPA to date fall into four major groups (71 substances, 65%).

These groups are chlorinated compounds (37 substances, 34%), metals (15 substances, 14%), nitrosamines (10 substances, 9%), and benzene derivatives (9 substances, 8%).

The chemicals assessed thus far by the EPA are about equally divided among those that produce mouse liver tumors (39 substances, 36%), those that do not appear to produce such tumors (36 substances, 33%), and those that have not been studied in mice (34 substances, 31%). Therefore, mouse liver tumor data have been used in approximately one-third of the agency's cancer risk assessments to date. The 39 substances found to produce liver tumors in mice are presented in Table 5.

Thirty-four of the 39 chemicals (87%) that produce liver tumors in mice were also assessed in rats. Sixteen of these (47%) induced liver tumors in both species; 10 (29%) induced tumors in other tissues of the rat; and 8 (24%) induced only mouse liver tumors. Six of the latter chemicals were chlorinated compounds. Four of the 5 assessed only in mice produced liver tumors in both sexes.

Twenty-six of the 39 (67%) chemicals that produced liver tumors in mice are chlorinated compounds. In fact, 26 of the 31 (84%) chlorinated compounds studied in mice yielded liver tumors. All the chlorinated alkanes/alkenes with diphenyl substitution (four), biphenyls (one), dibenzo-*p*-dioxins (one), cyclohexanes (three), cyclodienes (six), and phenols (two) studied in mice produced liver tumors. Nine of 12 chlorinated alkanes/alkenes and 1 of 3 haloethers also yielded liver tumors in mice. The 2 haloethers that showed no activity in the mouse liver are highly reactive compounds that act as local carcinogens.

Table 4. Categorization of 109 Chemicals (*n*)

Aldehydes/amides/nitriles (1–3 Carbons in Length) (5)
Alkenes (1)
Aromatic amines (6)
Aromatic hydrocarbons (1)
Benzene and derivatives (9)
Carbamates (5)
Chlorinated alkanes/alkenes (14)
Chlorinated alkanes/alkenes with diphenyl substitution (5)
Chlorinated biphenyls (1)
Chlorinated dibenzo-*p*-dioxins (1)
Chlorinated cyclodienes (6)
Chlorinated cyclohexanes (5)
Chlorinated phenols (2)
Ethylene oxides (2)
Haloethers (3)
Hydrazines (2)
Metals (15)
Nitrosamines (10)
Phosphates (6)
Phthalates (5)
Miscellaneous (5)

Table 5. Categorization of 39 Chemicals That Produce Liver Tumors in Mice (*n*)

Category A chemicals (1)	Benzidine
Category B2 chemicals (29)	Aldrin
	bis(2-Chloroethyl) ether
	DDD
	DDE
	DDT
	Carbon tetrachloride
	Chlordane
	Chloroform
	1,2-Dichloroethane
	Dichloromethane
	Dieldrin
	DEHP
	Dimethyl sulfate
	1,4-Dioxane
	1,2-Diphenylhydrazine
	Heptachlor
	Heptachlor epoxide
	Hexachlorocyclohexane, technical grade
	Hexachlorocyclohexane, alpha
	Hexachlorodibenzo-*p*-dioxin mixture
	Hydrazine/hydrazine sulfate
	N-Nitroso-*N*-dibutylamine
	N-Nitroso-*N*-diethylamine
	N-Nitroso-*N*-dimethylamine
	Polychlorinated biphenyls
	Styrene
	Toxaphene
	Trichloroethylene
	2,4,6-Trichlorophenol
Category C chemicals (8)	Acephate
	Dicofol
	Fomesafen
	Hexachlorocyclohexane, beta
	Hexachloroethane
	1,1,2,2-Tetrachloroethane
	1,1,1,2-Tetrachloroethane
	1,1,2-Trichloroethane
Category D chemicals (1)	Assure

Because of their high reactivity, it is highly unlikely that the active moiety could reach the liver.

Eight of the other 13 (61%) chemicals that produced mouse liver tumors are nitrogen-containing compounds. These include 3 aromatic amines, 2 hydrazines, and 3 nitrosamines. None of the 7 metals or 4 carbamates studied in mice induced liver tumors, and only 1 of the 5 benzene derivatives (styrene) studied did so.

Classification of Weight of Evidence

Table 6 provides the overall breakdown of the 109 chemicals assessed thus far by the agency according to the classification of the overall weight of evidence. Forty-eight percent of the chemicals are classified as level B2 carcinogens, indicating that for the vast majority of chemicals, there is sufficient evidence in animals to predict that these chemicals would be active in humans exposed to them but insufficient or no data in humans. Tables 7–11 list these chemicals assessed according to their classification rating.

Of the 12 substances classified as category A chemicals (Table 7), 10 have been studied in mice. However, only 1 of the 10 (benzidine) produced liver tumors. Seven chemicals in this category are metals, and 3 of these are radioactive (radium 226 and 228, radon 222, and uranium). Both the haloethers classified as human carcinogens did not yield liver tumors in mice. As stated earlier, this may be due to the high reactivity of these substances and the inability of the active moiety to reach the liver. Eight of the 12 category A chemicals induced lung tumors in humans.

Of the four substances classified as category B1 chemicals (Table 8), only two have been studied in mice, and neither yielded an increase in mouse liver tumors. One of the four substances is a metal (cadmium). Three of the four substances induced lung tumors in humans.

Of the 52 substances classified as category B2 chemicals (Table 9), 43 have been studied in mice. Twenty-nine of the 43 (67%) chemicals produced liver tumors in mice, while 14 (33%) did not. Therefore, 29 of the 52 (56%) chemicals in this category were classified with the use of mouse liver tumor data. Fifteen of the 29 chemicals that produced liver tumors in mice also produced liver tumors in rats. Eight of these substances are chlorinated compounds, and 5 are nitrogen-containing compounds. Ten of the remaining chemicals induced tumors in other tissues of other species.

Four of the 52 (8%) category B2 chemicals are classified mainly based on production of liver tumors in mice. These compounds are aldrin, *bis*(2-chloroethyl) ether, heptachlor, and hexachlorocyclohexane (technical grade). All, of course, are chlorinated compounds. The overall weight of evidence used to classify these chemicals is given in Table 12.

Table 6. Overall Classification of 109 Chemicals

Category	No. of chemicals	Percentage of total
A Carcinogens	12	11%
B1 Carcinogens	4	4%
B2 Carcinogens	52	48%
C Carcinogens	20	18%
D Carcinogens	21	19%
E Noncarcinogens	0	0%

Of the 20 substances classified as category C chemicals (Table 10), 17 have been studied in mice. Eight of the 17 (47%) produced liver tumors in mice, and 5 of the 8 (62%) are chlorinated compounds. Five of the 8 (62%) chemicals produced liver tumors in both sexes of the strains tested. Five of the eight chemicals have also been assessed in rats; no tumors were observed in this species. Structure-activity relationship evidence was cited as supporting evidence for 7 of the 8 (88%) chemicals that produced liver tumors in mice.

Table 7. Category A Chemicals

Chemicals	Evidence
Arsenic (inorganic)	No studies in animals
Asbestos	No studies in mice
Benzene	No liver tumors in mice
Benzidine	Liver tumors in mice
bis(2-Chloromethyl) ether	No liver tumors in mice
Chloromethyl methyl ether	No liver tumors in mice
Chromium VI	No liver tumors in mice
Nickel refinery dust	No liver tumors in mice
Nickel subsulfide	No liver tumors in mice
Radium 226 and 228	No liver tumors in mice
Radon 222	No liver tumors in mice
Uranium	No liver tumors in mice

Table 8. Category B1 Chemicals

Chemical	Evidence
Acrylonitrile	No studies in mice
Cadmium	No tumors in mice
Creosote	No adequate studies in mice
Formaldehyde	No liver tumors in mice

Table 9. Category B2 Chemicals

Chemical	Evidence
Acetaldehyde	No studies in mice
Acrylamine	No studies in mice
Aldrin	Liver tumors in 3 strains of mice
Aniline	No liver tumors in mice
Azobenzene	No liver tumors in mice
Benzo[*a*]pyrene	No liver tumors in mice
Beryllium	No adequate studies in mice
bis(2-Chloroethyl) ether	Liver tumors in two strains of mice
1,1-*bis*(4-Chlorophenyl)-2,2-dichloroethane (DDD)	Liver tumors in male CF-1 mice

Table 9. Category B2 Chemicals (continued)

Chemical	Evidence
1,1-*bis*(4-Chlorophenyl)-2,2-dichloroethylene (DDE)	Liver tumors in two strains of mice
DDT	Liver tumors in five strains of mice
bis(2-Ethylhexyl) phthalate (DEHP)	Liver tumors in $B6C3F_1$ mice
1,3-Butadiene	No liver tumors in mice
Captan	No liver tumors in mice
Carbon tetrachloride	Liver tumors in six strains of mice
Chlordane	Liver tumors in four strains of mice
Chloroform	Liver tumors in three strains of mice
1,2-Dichloroethane	Liver tumors in male $B6C3F_1$ mice
Dichloromethane (methylene chloride)	Liver tumors in $B6C3F_1$ mice
1,3-Dichloropropene	No liver tumors in mice
Dichlorvos	No liver tumors in mice
Dieldrin	Liver tumors in seven strains of mice
Dimethyl sulfate	No liver tumors in mice
1,4-Dioxane	Liver tumors in mice
1,2-Diphenylhydrazine	Liver tumors in female $B6C3F_1$ mice
Epichlorohydrin	No adequate studies in mice
Ethylene dibromide (EDB)	No liver tumors in mice
Folpet	No liver tumors in mice
Furmecyclor	No liver tumors in mice
Heptachlor	Liver tumors in two strains of mice
Heptachlor epoxide	Liver tumors in two strains of mice
Hexachlorocyclohexane (technical grade)	Liver tumors in four strains of mice
Hexachlorocyclohexane (α)	Liver tumors in five strains of mice
Hexachlorodibenzo-*p*-dioxin (mixture of 1,2,3,6,7,8- and 1,2,3,7,8,9-)	Liver tumors in $B6C3F_1$ mice
Hydrazine/hydrazine sulfate	Liver tumors in four strains of mice
Lead and lead compounds	No liver tumors in mice
Nickel carbonyl	No studies in mice
N-Nitroso-*N*-dibutylamine	Liver tumors in mice
N-Nitroso-*N*-diethanolamine	No studies in mice
N-Nitroso-*N*-diethylamine	Liver tumors in mice
N-Nitroso-*N*-dimethylamine	Liver tumors in mice
N-Nitroso-*N*-diphenylamine	No liver tumors in mice
N-Nitroso-*N*-dipropylamine	No studies in mice
N-Nitroso-*N*-ethylurea	No liver tumors in mice
N-Nitroso-*N*-methylethylamine	No studies in mice
N-Nitroso-*N*-methylurea	No liver tumors in mice
N-Nitrosopyrrolidine	No liver tumors in mice
Polychlorinated biphenyls (PCB)	Liver tumors in two strains of mice
Styrene (vinyl benzene)	No studies in mice
Toxaphene	Liver tumors in $B6C3F_1$ mice
Trichloroethylene	Liver tumors in $B6C3F_1$ mice
2,4,6-Trichlorophenol	Liver tumors in $B6C3F_1$ mice

Table 10. Category C Chemicals

Chemical	Evidence
Acephate	Liver tumors in female CD-1 mice
Acrolein	No studies in mice
Butyl benzyl phthalate	No tumors in mice
1,1-Dichloroethylene	No liver tumors in mice
Dicofol	Liver tumors in male $B6C3F_1$ mice
Dimethipin	No liver tumors in mice
Fomesafen	Liver tumors in CD-1 mice
Fosetyl-A1	No tumors in mice
Hexachlorobutadiene	No studies in mice
Hexachlorocyclohexane (β)	Liver tumors in CD-1 mice
Hexachloroethane	Liver tumors in $B6C3F_1$ mice
Metolachlor	No tumors in mice (two studies)
Oryzalin	No tumors in mice
Paraquat	No tumors in mice
Parathion	No adequate studies in mice
1,1,1,2-Tetrachloroethane	Liver tumors in $B6C3F_1$ mice
Trifluralin	No tumors in mice (two studies)
1,1,1,2-Tetrachloroethane	Liver tumors in both sexes of $B6C3F_1$ mice
2,4,6-Trinitrotoluene	No tumors in mice

Of the 21 substances classified as category D chemicals (Table 11), 11 (52%) had no adequate studies reported on them. Nine of the 21 (43%) chemicals had no data from animal studies. Only 1 chemical in this group (assure) was studied adequately in mice, and it induced liver tumors in male CD-1 mice. This chemical was classified as a category D chemical and not a category C one because the tumors occurred only in the livers of male mice, only at the end of the study, and only in the high-dose group, and the maximum tolerated dose was exceeded.

Summary

In summary, the EPA has begun to look critically at the induction of certain types of tumors in certain species, including liver tumors in mice. The controversy over the use of such tumor data in assessing the cancer risk·for humans has been going on for some time. The present agency policy is to downgrade the weight of evidence for such data under certain conditions.

Review of the cancer risk assessments for the 109 chemicals that the agency has formally verified shows that a variety of chemicals yield liver tumors in mice. However, one group of substances that consistently produced such tumors was chlorinated compounds (84%). Many of these compounds not only

Table 11. Category D Chemicals

Chemical	Evidence
Aldicarb	No adequate studies in aminals
Assure	Liver tumors in male CD-1 mice, MTD exceeded
Copper	No studies in animals
Cyanide	No studies in animals
Dibutyl phthalate	No studies in animals
Diethyl phthalate	No studies in animals
Diisopropylmethylphosphonate	No studies in animals
Dimethyl phthalate	No studies in animals
Ethylbenzene	No studies in animals
Glyphosate	No adequate studies in animals
Δ-Hexachlorocyclohexane	No adequate studies in animals
ε-Hexachlorocyclohexane	No studies in animals
Manganese	No adequate studies in animals
Mercury	No adequate studies in animals
Methoxychlor	No adequate studies in animals
Octahydro-1,3,5,7-tetranitro-1,3,5,7-tetrazocine (HMX)	No studies in animals
Silver	Local sarcomas only
Toluene	No adequate studies in animals
1,1,1-Trichloroethane	No adequate studies in animals
2,4,5-Trichlorophenoxypropionic acid	No adequate studies
Xylenes	No adequate studies in animals

induced liver tumors in mice but also induced liver tumors in rats and/or other types of tumors in mice and rats. However, several of the chlorinated compounds produced only mouse liver tumors. Another group of compounds that often induced liver tumors in mice was nitrogen-containing compounds (aromatic amines, hydrazines, nitrosamines). These latter substances tended to not only induce liver tumors in mice but also a variety of other tumor types in a variety of species.

Mouse liver tumor data have played a major role in the classification of substances in categories B2 and C. Fifty-six percent of the chemicals in category B2 and 40% in category C were classified based at least partially on the use of mouse liver tumor data. In addition, 21 of the 29 category B2 chemicals that produced liver tumors in mice and 5 of the 8 category C chemicals are chlorinated compounds. These two results indicate the importance of chlorinated compounds to the agency, and therefore, the importance of mouse liver tumor data in agency cancer risk assessments.

Table 12. B2 Chemicals with Only Mouse Liver Tumor Data

Chemical	Evidence
Aldrin	Liver tumors in both sexes of C3HeB/Fe mice at 10 ppm
	Liver tumors in both sexes of C3H mice at 10 ppm
	Liver tumors in male $B6C3F_1$ mice at 8 ppm
	No tumors in Carworth and Osborne-Mendel rats
	Causes chromosomal aberrations in mouse, rat, and human cells and unscheduled DNA synthesis (UDS)
	SAR with dieldrin, chlordane, heptachlor, heptachlor epoxide, and chlorendic acid
bis(2-Chloroethyl) ether	Liver tumors in both sexes of two strains of mice
	Injection site tumors in one strain of mice
	Direct acting mutagen
	SAR with two human carcinogens, BCME and CMEE
Heptachlor	Liver tumors in both sexes of C3H and $B6C3F_1$ mice at 10 to 18 ppm
	No tumors in four strains of rats but evidence of liver toxicity
	Only mutagenic activity is increased UDS
	SAR w/ chlordane, aldrin, dieldrin, and chlorendic acid
Hexachlorocyclohexane (technical grade)	Liver tumors in four strains of mice
	No tumors in Wistar rats and Syrian Golden hamsters
	No mutagenicity data available
	SAR with α–hexachlorocyclohexane

Open Forum

Dr. Diane Beal: We are using some of the data I presented here to indicate that there may be reason to doubt the usefulness of the mouse liver tumor data for human risk assessment at this time.

Dr. Gary Williams: You said that the EPA uses structure-activity relationships for categorizing B1 carcinogens and that *bis*-chloromethyl ether was used as the carcinogen to categorize *bis*-chloroethyl ether. To what known human carcinogen did you apply structure-activity relationships to categorize heptachlor as B1?

Dr. Beal: Unfortunately, none of the chlorinated cyclodienes is classified as a known human carcinogen. So we have been using the structure-activity of B2 carcinogens—aldrin, dieldrin, etc.—as the basis for classifying other chemicals in the B2 category.

Dr. Williams: I am not sure I grasp the logic of that. I think there is some circular reasoning here.

Dr. Beal: You are right. They all induce liver tumors. There is some circular reasoning. We use one to support another.

Dr. Williams: One must have been the first, to which all others are related.

Dr. Beal: Probably that was chlorendic acid, which is known to induce liver tumors in mice and rats.

Dr. Robert LeBoeuf: With a number of chemicals, in vitro systems such as the cytogenetics assay will give positive results if pushed long and hard enough. What level of reviewer discrimination is the EPA applying to the genotoxicity data before it is used in the weight-of-evidence assessment?

Dr. Beal: I am still analyzing this area. As you indicated, the more studies that are conducted, the more likely it is that a positive result will be obtained. But so far, the data on these 109 chemicals I have looked at indicate that the EPA does not seem to really use a great deal of the GENETOX information.

Dr. Philip Leder: I have a comment concerning the EPA's classification of carcinogens. Classifications such as "known human carcinogen" or "probable" or "possible human carcinogen" help the public understand the nature of the chemicals. However, classifying a material as a "probable" carcinogen poses a problem from a quantitative standpoint. The public is not sophisticated enough to know that probable carcinogens only pose a health threat if there is heavy exposure. This is a semantic problem that perhaps the agency could address in its next round of cancer risk assessment guidelines. In my opinion, it is overkill to call any of the B2 materials "probable human carcinogens" without some sort of qualification about the kind of massive exposure that would be necessary in most cases to produce tumors in man.

Dr. Beal: My presentation concerned the qualitative part of risk assessment: hazard assessment. The EPA *Carcinogen Risk Assessment Guidelines* contain a whole section on quantitation. The qualitative and quantitative assessments, as well as a discussion of all the uncertainties, go into the overall risk assessment and chemical regulation.

Dr. Leder: That was partly my point. EPA uses a weight of evidence—a qualitative exercise—at this point to deem a material a probable or possible human carcinogen. My point is that the quantitative exposure aspects must be included when something is advertised to the public as a probable human carcinogen.

Dr. Don Stevenson: I think there is a semantic issue here between weight of evidence and sufficiency of evidence. This is one of the things being raised in the EPA's current considerations of whether to revise the cancer risk assessment guidelines. The International Agency for Research on Cancer, for example, uses sufficiency of evidence. They look at whether the positive studies were well done and, if so, that drives the process. In terms of human risk, I would say weight of evidence actually takes many more factors into account, for example, whether human tissues behave the same way as animal tumors. Unfortunately, we often do not have the precise information available to make that kind of comparison. This is one of the things we hope will come out of this meeting: What are the kinds of things we need to look at in terms of weight of evidence to see whether a particular end point is really relevant to the human situation?

References

1. Tomatis L. The value of long-term testing for the implementation of primary prevention. In Hiatt HH, Watson JD, Winstein JA (eds): *Origins of Human Cancer.* Cold Spring Harbor, NY: Cold Spring Harbor Laboratory, 1977, pp 1339–1357.
2. Ward JM, Griesemer RA, Weisburger EK. The mouse liver tumor as an end point in carcinogenesis tests. *Toxicol Appl Pharmacol* 51:389–397, 1979.
3. Nutrition Foundation. *The Relevance of Mouse Liver Hepatoma to Human Carcinogenic Risk: A Report of the International Expert Advisory Committee to the Nutrition Foundation.* Washington, DC, 1983.
4. Office of Science and Technology Policy. Chemical carcinogens: Review of the science and its associated principles. *Federal Register* 50:10372–10442, 1985.
5. U.S. Environmental Protection Agency. *Integrated Risk Information System (IRIS).* Washington, DC: U.S. EPA, 1988
6. U.S. Environmental Protection Agency. Interim procedures and guidelines for health risk and economic impact assessments of suspected carcinogens. *Federal Register* 41:21402–21405, 1976.
7. Albert RE, Train RE, Anderson E. Rationale developed by the Environmental Protection Agency for the assessment of carcinogenic risk. *JNCI* 58:1537–1541, 1977.
8. U.S. Environmental Protection Agency. Guidelines for carcinogen risk assessment. *Federal Register* 51:33992–34003, 1986.

Mouse Liver Carcinogenesis: Mechanisms
and Species Comparisons, pages 19–41

Do Mouse Liver Tumors Predict Rat Tumors? A Study of Concordance Between Tumors Induced at Different Sites in Rats and Mice

Daniel M. Byrd III, Edmund A. C. Crouch,† and Richard Wilson‡*
**Consulting Toxicologist, Falls Church, VA 22044, †Meta Systems, Inc., Cambridge, MA 02141, and ‡Energy and Environmental Policy Center, Harvard University, Cambridge, MA 02138*

Within regulatory and scientific arenas, the use of mouse liver tumors to assess human risk both qualitatively and quantitatively has generated intense controversy (1–5). While liver tumors constitute the most common carcinogenic outcome in some strains of mice (6), liver cancers in humans are rare.

Current knowledge of the potential biologic stages in the progression to cancer contributes to the controversy (7,8). Discrete intermediate steps, such as nodules, may or may not be involved in the stages leading to cancer. In some of these stages, cells can undergo remission, redifferentiation, remodeling, and/or adaptive responses. The similarities and differences between pathologic stages of liver cancer in mouse and man are not well understood. Some scientists also regard the mouse liver as "preinitiated" and thereby as responding differently and perhaps to a different set of substances than those that are of concern for man.

Potential mechanisms of liver carcinogenesis also add to the controversy. Some "nongenotoxic" or "nonmutagenic" substances do induce mouse liver tumors (9). Stimulation of rodent liver DNA synthesis enhances the incidence of liver neoplasia but cannot explain the carcinogenic potency of some substances (10).

It is not likely that DNA synthesis is stimulated in man at environmental exposure levels. According to some theories, stimulation of cell division by concurrent toxicity may also make an essential contribution to the carcinogenic process (11). Acute toxicity correlates well with carcinogenic potency of substances in rodent bioassays (12). However, acute liver toxicity is not likely to have the linear, nonthreshold dose-response relationship that current regulatory risk assessments assume.

The controversy also implies differences of opinion regarding regulatory decision rules. Such rules range from banning the human use of any substance found to induce cancer in an animal to ignoring the results of the bioassays.

Two alternative decision rules are, however, particularly relevant to current risk assessment practice and to the analysis below:

- The induction of tumors at an organ site in animals may imply the induction of tumors at any site in another species. Therefore, evidence of induction at a specific site in a bioassay is used to predict the likelihood and occurrence of cancer in general in a second species.
- The induction of tumors at a specific site in one species may imply the induction of tumors only at the same site in another species. Therefore, in a bioassay evidence of the induction at a specific site is used to predict the likelihood of occurrence of specific cancers in a second species.

The second decision rule has its basis in a classic biomedical paradigm of research on animal models. The first arises from observations by Tomatis and coworkers that animal bioassays for those substances known to cause cancer in man usually cause tumors in experimental animals, but at a variety of sites and not necessarily the analogous organ site seen in man (13,14). This first rule usually forms the basis of regulatory agency risk assessment practices, for example, by the EPA's Carcinogen Assessment Group (15) or by the International Agency for Research on Cancer (IARC). This practice also has the effect of treating tumor induction at any site (including mouse liver) as quantitatively equivalent to tumor induction at any other site.

Crouch and Wilson (16) suggested that rodent data supported this view. They compared the quantitative carcinogenic potency of a number of substances in rats and mice and, using a simple model, showed that potencies in the two species correlated more closely if the site in each species was not specified. To make these estimates, they sometimes had to infer an upper bound on potency in one of the species. Crouch (17,18) also showed that there are few instances in which the carcinogenic potency in mice differs more than 100-fold from the potency in rats. Gaylor and Chen (19) and Chen and Gaylor (20) confirmed the findings of Crouch and Wilson, and they additionally compared estimated safe doses derived from data in mice and rats. Both teams found that body weight, not body surface area, provided the better basis for extrapolation of dose.

An alternative view of the rodent data is to treat evidence of a positive result in one species and a negative result in another as evidence of a qualitative difference in susceptibility. Concordance is a convenient way to view such differences. It refers to the similarity between two objects with respect to one or more characteristics. Concordance, as we are defining it, describes agreement on the statistical significance of tumor induction between two bioassays of the same substance in two rodent species (21). For convenience, we have initially used standard statistical tests of significance.

In effect, we have indirectly inquired about the relevance of mouse liver tumors to man by initiating a retrospective investigation of whether any given substance that induces mouse liver tumors is predictive of tumor induction in rats. While indirect, much better data are available for the rodents than for man. Because the mouse and rat have a much closer phylogenetic relationship to each other than either species has to man, we expect, both on the average and over many substances, that concordance relationships developed in this study will place an upper limit on mouse-to-man or rat-to-man concordance.

Any such investigation has the following important limitations:

- Correlations do not indicate causality or mechanism and may yield spurious results.
- The National Toxicology Program and National Cancer Institute (NTP/NCI) did not choose the chemical substances either randomly or according to a method that gives some estimation of relationship to the chemical universe. The bias introduced into the results by the nonrandom selection of substances is not known.
- The presence or absence of a specific concordance relationship between the mouse and rat does not necessarily imply a similar relationship between either species and man.
- Our statistical criteria have been chosen in a somewhat arbitrary way. We have not yet studied how variation in these criteria affects our conclusions, nor do we know whether intergroup variance is random, as assumed.

Explanation of the Plots

Obtaining the Basic Data

This description is similar to that given at the 1988 Society of Risk Analysis meeting (D. M. Byrd, E. A. C. Crouch, R. Wilson) but is repeated here for clarity.

To obtain some measures of concordance between tumor occurrences in different tissue types and in different strains of animals, a considerable amount of computation was necessary. This computation was carried out in stages, starting from the data in the Carcinogenic Bioassay Database System (CBDS) tapes. At each stage of the computation, various decisions had to be made on methods of analysis.

The CBDS tapes hold data on the individual animals tested in carcinogenesis bioassays. For each animal, types of neoplasms and other lesions found at necropsy and/or histopathologic examination are coded according to a standard classification. The coding is in considerable detail, with more than 4,400 different classifications for primary neoplasms. Such detail was excessive for the comparisons we required, since we followed the practice in the analyses of such bioassays of combining many of the neoplasms that are classified differently (for example, tumors of the same types at nearby sites or tumors of different types at the same site).

A less detailed classification system focused on 102 primary neoplasms, 66 of which were malignant primary neoplasms. These 102 classifications contained all the original, more detailed classifications. All tumors classified as metastatic or as unassignable as to primary or metastatic were excluded. These 102 tumor classifications are listed in Table 1.

Table 1. The Classification of Tumor Sites

No.	Type	No.	Type
1	Skin, breast papilloma, +	52	Lung squamous carcinoma
2	Respiratory, oral + papilloma, +	53	Oral, GI squamous carcinoma
3	GI papilloma, +	54	Urinary, reprod. squamous carc.
4	Urinary, reprod. papilloma +	55	Skin, GI basal cell carcinoma
5	Skin, breast adenoma, +	56	Urinary transitional cell carcinoma
6	Respiratory, oral adenoma, +	57	Skin, breast adenocarcinoma
7	Liver adenoma, +	58	Lung adenocarcinoma
8	GI adenoma, +	59	Oral, GI adenocarcinoma
9	Urinary, reproductive adenoma, +	60	Urinary, reprod. adenocarcinoma
10	Pituitary adenoma, +	61	Endocrine, brain adenocarcinoma
11	Endrocrine + adenoma, +	62	Islet cell carcinoma
12	Skin, urinary + adenomas	63	Bile duct carcinoma
13	Reprod., endocrine + adenomas	64	Hepatocellular carcinoma
14	Tubular cell adenoma +	65	Alveolar, broncheolar carcinoma
15	Follicular, clear cell adenomas	66	Chromaphobe carcinoma
16	Cortical adenoma	67	Tubular cell adenocarcinoma
17	Skin, breast, liver & cystademonas	68	Thyroid follicular cell carcinoma
18	GI, urinary, reprod. cystadenomas	69	Cortical carcinoma
19	Endocrine cystadenomas	70	Clear cell carcinoma
20	Acinar cell adenoma	71	Adnexal, sebaceous, + carcinoma
21	Keratoacanthoma +	72	Thymoma
22	Tubular adenoma +	73	Granulous cell carcinoma
23	Interstitial cell tumor	74	Interstitial cell carcinoma
24	Pheochromocytoma +	75	Pheochromocytoma +
25	Skin, breast fibroma	76	Skin sarcoma +
26	Blood, bone fibroma	77	Other sites sarcoma +
27	Fibroma, other sites	78	Blood, bone sarcoma +
28	Lipoma +	79	Liposarcoma
29	Leiomyoma +	80	Leiomyosarcoma +
30	Endometrial stromal polyp, +	81	Endometrial stromal sarcoma, +
31	Fibroadenoma +	82	Carcinosarcoma +
32	Hemangioma +	83	Mesothelioma, osteosarcoma, +
33	Osteoma +	84	Teratoma +
34	Hamartoma +	85	Hemangiosarcoma +
35	Ganglioneuroma +	86	Granular cell tumor +
36	Chromophobe adenoma	87	Glioma
37	Skin, breast carcinoma	88	Oligodendroglioma +
38	Blood, bone carcinoma	89	Astrocytoma
39	Lung carcinoma	90	Olfactory neuroblastoma +
40	Oral, GI carcinoma	91	Neurofibrosarcoma
41	Urinary carcinoma	92	Lymphoma
42	Reproductive carcinoma	93	Lymphocytic lymphoma
43	Pituitary carcinoma	94	Histiocytic lymphoma
44	Endrocrine carcinoma	95	Mixed lymphoma
45	Brain carcinoma	96	Malignant reticulosis
46	Skin, breast papillary carcinoma	97	Leukemia

continued

Table 1. The Classification of Tumor Sites (continued)

No.	Type	No.	Type
47	Lung papillary carcinoma	98	Myelomonocytic leukemia
48	GI, urinary papillary carcinoma	99	Lymphocytic leukemia
49	Uterus, ovary papillary carcinoma	100	Plasmacytic leukemia +
50	Thyroid papillary carcinoma	101	Granulocytic leukemia
51	Skin + squamous carcinoma	102	Monocytic leukemia

Table 2. Details of the Classifications

Tumor Classification	Site
Classification 7	
Adenoma	Liver NOS
Bile duct adenoma	Liver NOS
Hepatocellular adenoma	Liver NOS
Adenoma, NOS	Intrahepatic bile duct
Adenoma, NOS	Gallbladder NOS
Adenomatous polyp, NOS	Gallbladder NOS
Papillary adenoma	Gallbladder NOS
Classification 63	
Bile duct carcinoma	Liver NOS
Bile duct carcinoma	Bile duct
Bile duct carcinoma, invasive	Pancreas
Bile duct carcinoma, invasive	
Bile duct carcinoma, invasive	Multiple organs NOS
Classification 64	
Hepatocelluar carcinoma, invasive	
Neoplastic nodule	Liver NOS
Hepatocellular carcinoma	Liver NOS
Mixed hepato/cholangiocarcinoma	Liver NOS
Neoplastic nodule	Liver/hepatocytes
Hepatocellular carcinoma, invasive	Pancreas
Hepatocellular carcinoma, invasive	Stomach NOS
Hepatocellular carcinoma, invasive	Omentum NOS
Hepatocellular carcinoma, invasive	Mesentery NOS
Hepatocellular carcinoma, invasive	Adrenal
Hepatocellular carcinoma, invasive	
Mixed hepato/cholangiocarcinoma, invasive	
Hepatocellular carcinoma, invasive	
Hepatocellular carcinoma, invasive	
Hepatocellular carcinoma, invasive	
Hepatocellular carcinoma, invasive	Multiple organs NOS
Mixed hepato/cholangiocarcinoma, invasive	Multiple organs NOS

Abbreviation: NOS, not otherwise specified.

As noted in this chapter, classification 7 contains a number of benign liver tumors, classification 64 contains malignant liver tumors, and classification 63 related malignant tumors (Table 2).

The classifications and chemicals we used are not precisely the same as those used by others. In particular, Ashby and Tennant (22) included in their data set some bioassays from a contractor whose data are explicitly excluded from the CBDS data tapes. Also, Ashby and Tennant used fewer classifications. Therefore, we could not expect a 1:1 correspondence between our analysis and those of others.

All the experiments in the CBDS that met certain criteria were analyzed in a standard way for each of the 102 tumor classifications. Each experiment consisted of groups of animals of the same strain and sex, with every group in the experiment exposed to a single chemical (or a well-defined mixture of chemicals assigned a unique name) at a different dose. Each route of exposure counted as a separate experiment. The different routes of ingestion and different animal strains are also listed in Table 3. There were 1,968 such experiments available for analysis. To be included, an experiment had to meet the following criteria:

• Each experiment had to have at least one control (i.e., undosed) group, and there had to be data available on some of the animals in each group. Of the 1,968 original experiments, 121 lacked either a control group or information on any control group. For some experiments, there was more than one control group and/or there were both gavage and untreated control groups. For gavage experiments, untreated control groups were ignored if there was a control group subjected to sham gavage. All remaining control groups for a single experiment were combined and treated as a single control group.

• The relative magnitudes of the doses administered to the dosed groups had to be evaluated. Chemicals were administered through various routes in different experiments, and the details of dosing are recorded in the CBDS tapes. For the purposes of this analysis, the relative magnitudes of the first dose administered to the different dose groups were taken as representative. However, we were not able to decipher the meaning of the data available for a few of the combinations of dose route and dosing information. A further 137 experiments were excluded for this reason.

• The experiment had to continue long enough for tumors to arise. It was demanded that the longest lifetime recorded for any animal in the experiment exceed 70 weeks. Since the control animals were typically killed as soon as all of the dosed animals had died, or sooner if the experimental protocol demanded it, this criterion ensured that only long-term experiments were included. Seventeen of the 1,968 experiments were excluded for failing to meet this criterion.

Applying the above criteria reduced the available number of experiments to 1,693. The analysis applied to each experiment was designed to obtain standardized measures of the significance of the results and the background rate of tumor occurrence. The same procedure was followed for each of the 102 classifications of neoplasms in the 1,693 experiments—a total of 172,686 times. For each dosed group, two measures of significance were obtained using Fisher's

Table 3. Species, Strains, Sexes, and Routes Included

Species	Hamster	
	Mouse	
	Rat	
Strains	ACI	
	August 28807	
	$B6C3F_1$	
	C57BL/6CR	
	Charles River CD	
	F344	
	Long Evans	
	Marshall	
	Osborne-Mendel	
	Sherman	
	Sprague-Dawley	
	Swiss Albino	
	Swiss CD-1	
	Swiss ICR	
	Swiss NIH	
	Swiss Webster	
	Syrian Golden	
	Wistar	
Sex	Male	
	Female	
Routes of administration	IM	Intramuscular
	SD	Painting
	IP5	Intraperitoneal
	PO1	Food
	PO2	Water
	PO4	Gavage
	RE2	Inhalation

exact tests to compare incidence with that in the control group. The first ("unadjusted") test took no account of early deaths in the experiment, while the second ("adjusted") test made some adjustment for early deaths. In addition, a mortality-adjusted measure of the occurrence of a dose-response relationship was computed. The exact procedure followed was:

(i) The total number of tumors (of the classification under analysis) in the experiment (i.e., in all groups, including the control group) was counted. If this number was less than 4, no further analysis was performed, since even if all the tumors occurred in a single dose group, no measure of significance would reach a critical value. Also if the site/type combinations of tumors included in the classification were not examined in the experiment, no further analysis could be performed.

(ii) The tumors in each dose group were counted and a one-sided Fisher's exact significance value computed to compare the incidence in each group with the incidence in the control group. All the tumor counts were taken at the end of the experiment. The numbers of animals at risk was taken to be either the initial numbers of animals in any given dose group (for the unadjusted analysis) or the number of animals in the given dose group when the first animal in any dose group was found to have a tumor of the classification under analysis (for the adjusted analysis).

(iii) A trend analysis (23) was performed to obtain a dose-response significance level.

(iv) If any of the statistical significance values (*P*) obtained in (ii) and (iii) were less than 0.1, all values were recorded. If all significance values exceeded 0.1, then no data on significance were recorded.

Of the 172,686 products of the number of experiments times the number of classifications, 161,219 had fewer than 4 tumors in any dosed group; 8,385 had all significance values greater than 0.1; and full data on significance values were recorded for 3,082.

In addition to the computations performed on individual experiments, a record was kept of the background incidence of tumors in the control groups. Each experiment was classified according to its combination of species, strain, sex, and route of adminstration of the test material to the dosed groups. For each combination, the number of tumors of each classification seen was totaled, as were the number of control animals in the bioassay and the adjusted number of animals as described in (ii).

Since a few experiments shared control groups, the procedure described in the last paragraph may double count some control groups. However, since only an approximate background incidence was required at this stage, no attempt was made to correct for this. The adjusted numbers of animals were chosen at different times for different classifications and experiments, and are not truly comparable. At this stage, however, these values were not used. The background rate of tumors was estimated by dividing the total observed number of tumors by the total number (unadjusted) of animals in the bioassay control group.

Summary results were obtained for all the distinct combinations of chemical, species, strain, sex, and route of administration in the 1,693 experiments that met the inclusion criteria. For each experiment analyzed, a summary response classification was obtained for each of the 102 tumor classifications. The response was classified as positive (+), negative (–), weak (W), or blank (ø) as follows:

A response was positive if $P < 0.025$ for an adjusted or unadjusted Fisher's exact test for any dosed group, $P < 0.5$ for any two dosed groups, using adjusted or unadjusted tests. Otherwise the response was negative, provided this tumor classification had been examined in the experiment; weak if it was not positive, but there was a dose-response trend at $P < 0.025$; blank if it had not been examined. Any duplications of combinations of chemical, species, strain, sex and route of administration occurring in the experiments were combined by assigning the order + > W > – > ø to the responses and choosing the largest response among duplicates for each of the 102 classifications.

With the procedures described, each chemical has associated with it a list of species/strain/route/sex combinations, each of which has 102 associated results. Any list entry with completely blank results was eliminated, since it contained no information.

Using the Basic Data

The plots of concordance are attempts to summarize some of the information in the files described in the previous section. They illustrate the concordances of tumor responses between different species/strain/ route/sex combinations. The idea is to define such combinations and then write 2 × 2 tables for various groups of tumor site/type classifications. Each 2 × 2 table can be represented as shown in Table 4.

The columns represent a group of tumor site/type classification in one combination of species, strains, routes, and sexes. The rows represent another group of tumor site/type classification in another combination of species, strains, routes, and sexes. The entries A, B, C, and D in the table are the numbers of chemicals that had positive or negative results in the experiments corresponding to the definitions of the columns and rows. By choice of the program user, weak tumor response can be counted as either positive or negative.

Thus A is the number of chemicals that were positive in at least one experiment falling within the definition of the column group and also positive in at least one experiment falling within the definition of the row group; B is the number of chemicals positive in at least one experiment falling within the definition of the row group, but negative in all experiments falling within the definition of the column group; C indicates row negative, column positive values; and D indicates both are negative.

The plots summarize the results of obtaining such 2 × 2 tables for a set of row definitions, with a single column definition. For complete concordance, X = Y = 1.

An example should make this clearer. We may specify that the rows are to be selected for the species mouse; strain $B6C3F_1$; route water, food, or gavage; sex male or female (this gives six species/strain/route/sex combinations). To obtain a plot with more than one point on it, we may define a set of terms for the tumor sites to be included—for example: benign liver tumors (classification 7); malig-

Table 4. Sample Concordance Table

		Columns		
		+	–	
Rows	+	A	B	A + B
	–	C	D	C + D
		A + C	B + D	A + B + C + D

nant liver tumors (classification 64); common tumors (background incidence >20%); (all other tumors). Each entry within parentheses here, in conjunction with the species/strain/route/sex combinations given, makes up a complete definition for the row entries, so that we have here specified four sets of definitions for the row entries.

The column entries may be selected to come from experiments meeting criteria like species, rat; strain, F344 or Osborne-Mendel; route, water, food, or gavage; sex, male or female. The tumor type might be (any tumors).

The definitions in this example lead to computation for four 2 × 2 tables. Each compares mouse against rat (either of two strains) by any of three oral routes of administration for either sex. In addition, the four 2 × 2 tables record the numbers of chemicals that are positive/positive, positive/negative, negative/positive, and negative/negative for liver, rare, and other tumors vs. any tumors in mice vs rats.

For each set of row definitions, any chemical that has an associated experiment matching the row criteria and an associated experiment matching the column criteria will contribute 2 to one of the entries in the corresponding 2 × 2 table. For example, in the 2 × 2 table constructed for row = liver tumors, column = any tumor, A totals all chemicals that had at least one experiment in which liver tumors were positive in mice and also were positive in rats. B equals all chemicals positive for liver tumors in mice, but not in rats. C includes chemicals negative for liver tumors in mice, but positive in rats. And D counts chemicals negative for liver tumors in mice and rats.

The same procedure is used to obtain the 2 × 2 tables for the other row group definitions. From each 2 × 2 table, we obtain coordinates defined by:

$$X = A/(A + B) \text{ and } Y = D/(C + D)$$

and plot the corresponding point on a graph. In addition, the confidence limits for each point may be obtained by treating A/(A + B) or D/(C + D) as binomial samples from an infinite population of chemicals.

X can be called the positive predictive value of the row information for the column outcome or, alternatively, the sensitivity of a column to the row. Y can be called the negative predictive value of the row information for the column outcome or the specificity of the column by the row. Of course, in this chapter we predict nothing; we merely observe patterns in the data. However, predictions can be based on the patterns observed. An extensive group of caveats should accompany such predictions. When we somewhat loosely use the word prediction in what follows, these caveats must be assumed.

Results

We present a selection of results designed to show the power of our procedures in order to illustrate further the concordance plots.

Table 5. Concordance Between Any Tumors in Rats and Mice*

		Any rat tumor		
		+	–	
Any mouse tumor	+	92	53	145
	–	60	85	145
		152	132	290

*Weak significance included as negative.

First we select all routes of ingestion, both for rats and mice, gavage, diet, and water, by selecting the options PO1, PO2, PO4. We select all the mouse strains listed in Table 3 for the row selection and all the rat strains in Table 3 for the column selection. We ask for only one point by selecting "other" for the grouping of sites and select male and female. We count as negative all experiments where there was only weak evidence of carcinogenicity.

Table 5 shows the concordance data for this group. Ninety-two of the 145 chemicals (63.4%) were positive in mice, and the same percentage were positive in rats ($X = 0.634$; $Y = 85/145 = 0.586$). Figure 1 shows a concordance plot of the data. It has one point, and the error bars are selected to be 67% (about one standard deviation). The computer prints just below the origin the list of options selected (left, row; right, column). Complete concordance would be represented on this plot by a point at +1, 1. In some cases the number of strains included in the row is enough to overprint the list of strains in the column.

Table 6 and Figure 2 show the same plot when weak significance is included as positive. The sensitivity (X) has increased. Now 137/180, or 76.1%, of chemicals positive in mice are also positive in rats. On the other hand, the specificity (Y) is reduced to 39/103 (38%).

Tables 7 and 8 set forth the same concordance plots for the benign liver adenoma (classification 7), for the malignant liver carcinoma (classification 64), and for related malignant tumors (classification 63). This is for weak tumors counted as positive. These are all plotted in Figure 3. Also plotted in Figure 3 are the rare tumors; rare is defined as those with an incidence of <5% and classification 63.

The labels on the points in Figure 3 are not the classifications of Table 1 but are in the order of the groups listed on the plot. Label 1 = classification 7; label 2 = classification 63; label 3 = classification 64; and label 4 = rare tumors. Classification 63 had so few cases it was not included. Figure 4 shows a similar plot for weak significance counted as negative and classification 63 omitted.

As we go from a plot of all tumors (Figs. 1,2) to liver tumors only (Fig. 3), we see that there is a slight decrease in predictability—but not much. There is an inappreciable difference between the predictability of benign (classification

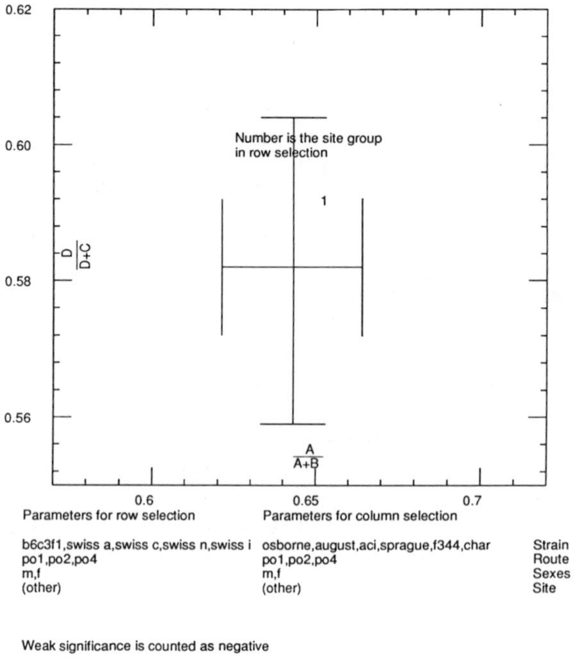

Figure 1. Concordance plots for data in Table 5. In some cases the lists of strains for the rows and columns are overprinted by the computer.

Table 6. Concordance Between Any Tumors in Rats and Mice*

		Any rat tumor		
		+	−	
Any mouse tumor	+	137	50	187
	−	64	39	103
		201	89	290

*Weak significance included as positive.

7), or malignant (classification 64) liver tumors, but there is an improvement when benign and malignant tumors are combined, and they are more predictive than rare tumors.

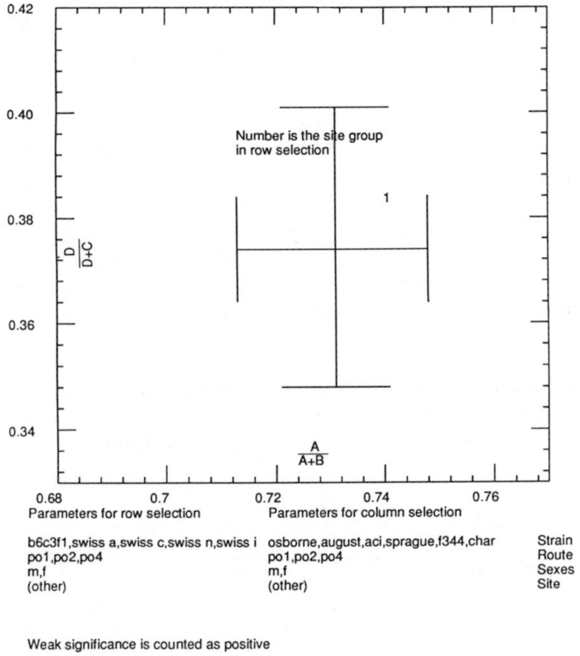

Figure 2. Concordance plots for data in Table 6. Weak significance is counted as positive.

Part of this increase in predictability comes from increased sensitivity resulting from increased numbers. We note that as we go from a plot for all tumors (Figs. 1,2) to any plot for fewer tumors, the error in X increases because the number A falls. But the statistical error in Y stays constant or goes down because the number D increases and the statistical error is dominantly the square root of D. The rest of our results are presented only in the figures. The strains, sexes, and groups are varied as marked.

Figures 5 and 6 show the results for each classification of Table 1. (This is accomplished by setting "all" for the row selection.) In this case each number on this plot corresponds to the classification number in Table 1. Figure 5 is for all rats and mice; Figure 6 is for B6C3F$_1$ mice only.

There is a marked difference in incidence of mouse liver tumors, whether adenoma (classification 7) or carcinoma (classification 64), from that of other tumors. This is evident in both Figures 5 and 6; the Y coordinate is greater than for any other tumor or grouping. The absence of liver tumors in mice is more indicative of an absence of tumors in rats. This may be because of a confirmatory indication of the exquisite sensitivity of the mouse liver for carcinogens. Although this plot is for weak significance included as positive, the difference exists when we exclude weak tumors as well.

Table 7. Concordance Between Mouse Liver Adenoma and Rat Tumors*

		Any rat tumor		
		+	–	
Mouse liver adenoma	+	35	18	53
	–	117	120	237
		152	138	290

*Weak significance is counted as positive.

Table 8. Concordance Between Mouse Liver Carcinoma and Rat Tumors*

		Any rat tumor		
		+	–	
Mouse liver carcinoma	+	46	19	65
	–	106	119	225
		152	138	290

*Weak significance is counted as positive.

Figure 7 is a repeat of Figure 3 but for $B6C3F_1$ mice only. There is little difference, largely because the data base mainly consists of experiments with $B6C3F_1$ mice.

Figures 8 and 9 are like Figure 6 but for male and female $B6C3F_1$ mice, respectively. We note that for female mice there is a slight increase in both positive and negative predictive power.

In Figure 10, we ask whether mouse tumors are predictive of a rat tumor at a specific site (liver) (classifications 7, 64). The sensitivity (X) is not as great as it is for any rat tumor ($X \approx 0.3$), and the uncertainty is greater. There is no particular tendency for mouse liver tumors to predict rat liver tumors (X not different); the points all have $Y \approx 0.9$—a higher value than for any other plot. In other words, the absence of a mouse tumor almost guarantees absence of a rat liver tumor. This is because an individual tumor site in a rat is rare. Therefore B and D are lower than A and C, and Y approaches 1 as X approaches 0. We can still use the results in Figure 10 because the differences are larger than the statistical error.

Figure 11 inverts the question: How well do rat tumors predict for mouse tumors? They certainly predict somewhat better than the reverse, particularly

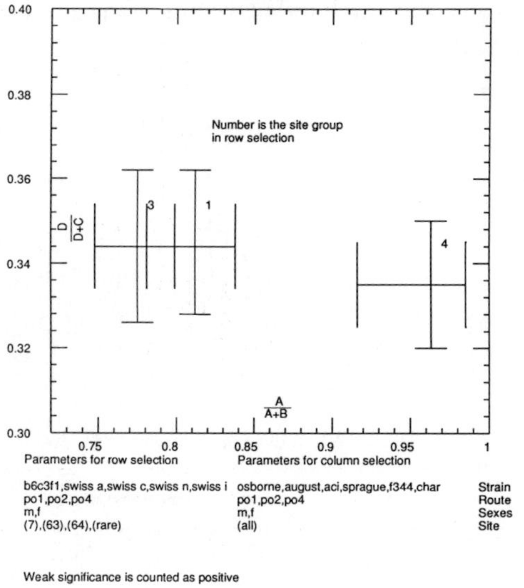

Figure 3. Concordance plots for data in Tables 7 and 8 and data for rare tumors. Point labels indicate order of the groups listed. Weak significance is counted as positive.

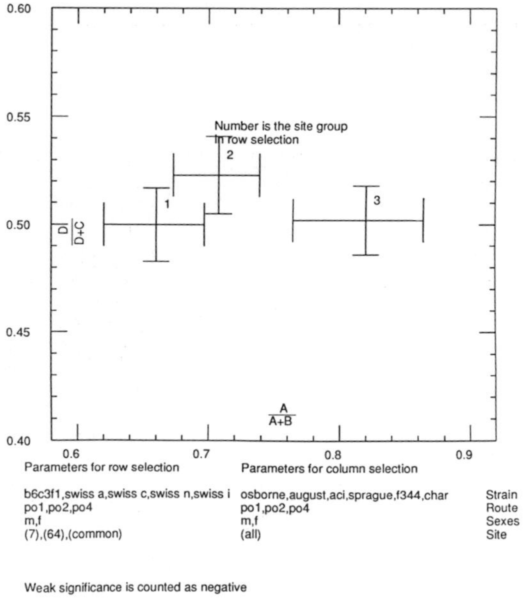

Figure 4. Concordance plot of same data in Figure 3 except that weak significance is counted as negative and classification 63 omitted.

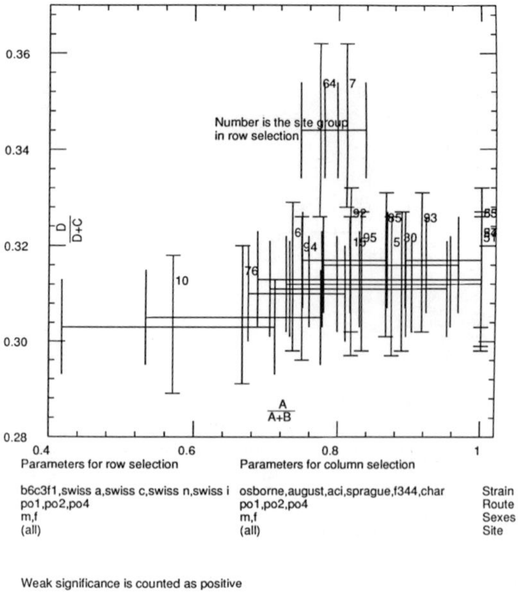

Figure 5. Results for each classification shown in Table 1 for all rats and mice. Weak significance is counted as positive.

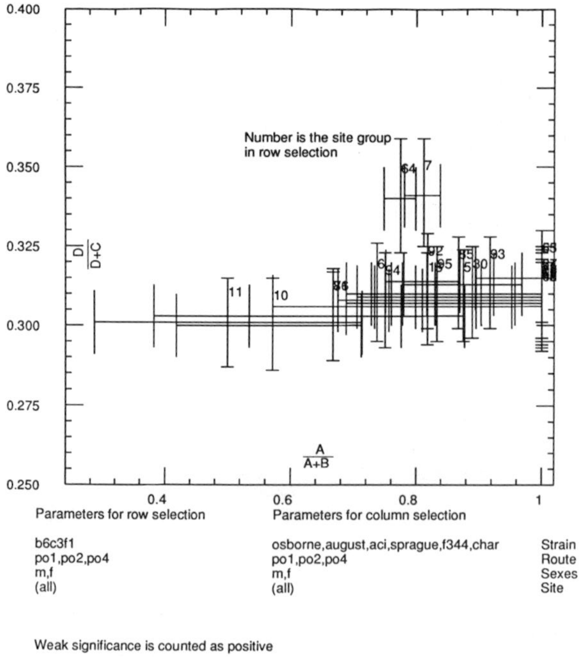

Figure 6. Results for $B6C3F_1$ mice for all Table 1 classifications.

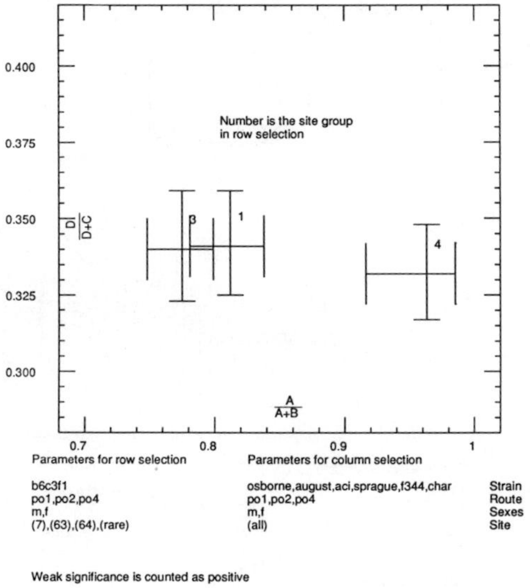

Figure 7. Similar to Figure 3, but for only B6C3F$_1$ mice. Weak significance is counted as positive.

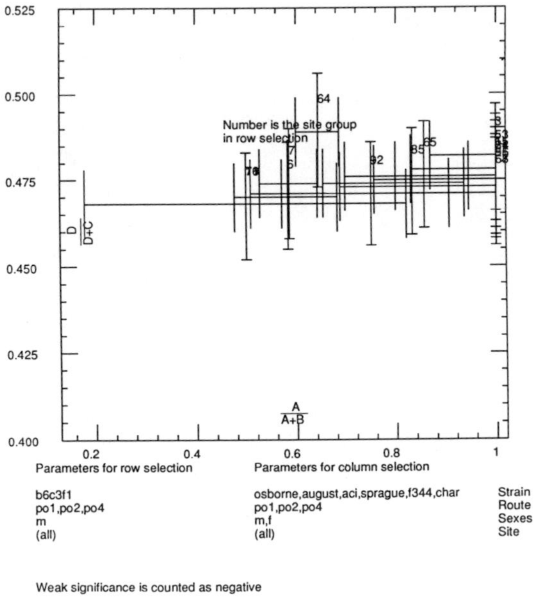

Figure 8. Similar to Figure 6, but for male mice only. Weak significance is counted as negative.

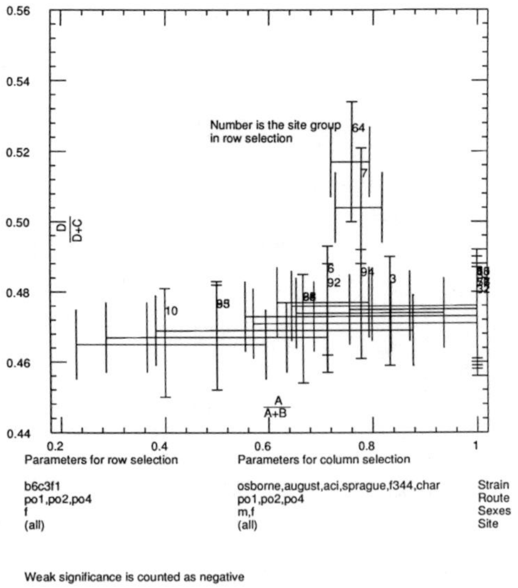

Figure 9. Similar to Figure 6, but for female mice only. Weak significance is counted as negative.

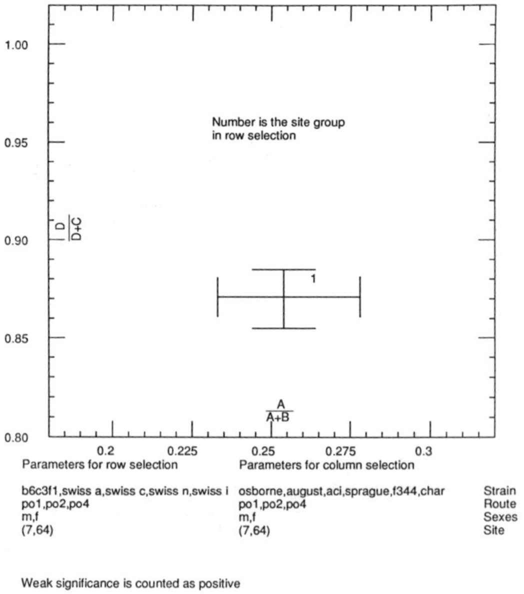

Figure 10. Predictive value of mouse liver tumors for rat liver tumors. Weak significance is counted as positive.

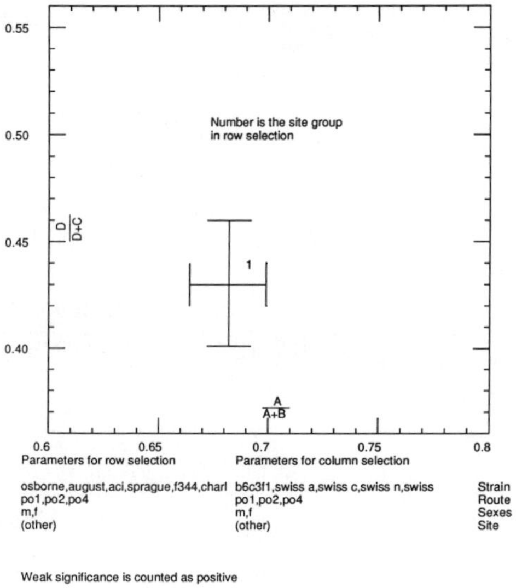

Figure 11. Predictive value of rat tumors for mouse tumors. Weak significance is counted as positive.

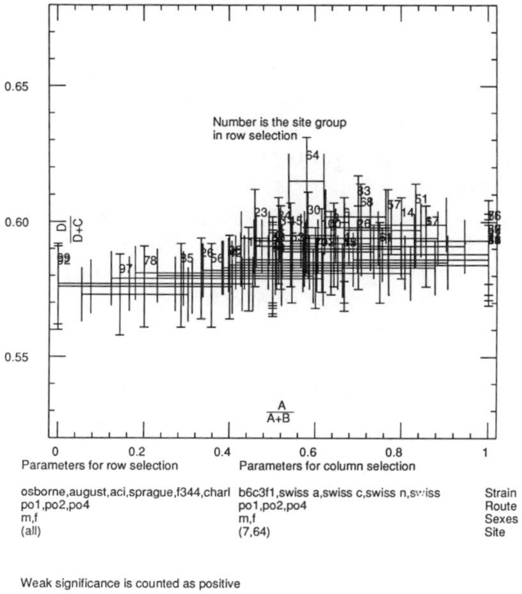

Figure 12. Predictive value of any rat tumor for mouse liver tumors. Weak significance is counted as positive. Note computer overprinting.

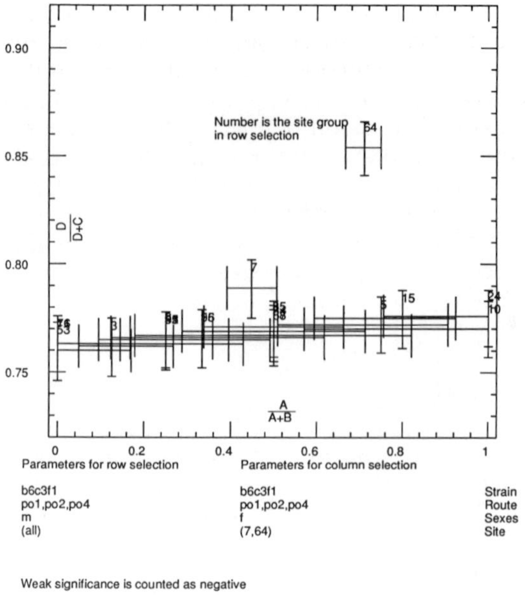

Figure 13. Tumor concordance between male and female $B6C3F_1$ mice. Weak significance is counted as negative.

the rare rat tumors (<5%), label 2, or any rat liver tumor, benign or malignant; label 1 shows slightly more negative predictive power than other tumors.

In Figure 12 we ask how well any rat tumor predicts a mouse liver tumor. We see that malignant rat liver tumors have a slightly higher negative predictive power, but the effect is not as large as for mouse tumors.

The concordance between male and female mice is much better than the concordance between rats and mice (Fig. 13). Again, label 1 shows that absence of tumors in male mice is more predictive of absence of tumors in female mice. Label 2 (rare tumors) are the most predictive of the presence of tumors in the other sex. In Figure 13 we show how well mouse tumors predict rare tumors (normally considered to be most indicative of human carcinogenesis). In this label 1 mouse liver tumors do not predict rare mouse tumors well. Label 2 shows the tautology; there is complete concordance between rare tumors and rare tumors in the same species.

Conclusion

In this paper we give an example of the power of our computerized methods, particularly as applied to mouse liver tumors. Contending views of interpretation hold that mouse liver tumor induction either is irrelevant to human cancer hazard because this organ site is overly sensitive or has the same predictive value as tumor induction at other sites. Our results suggest that both views have some validity with respect to the NCI/NTP bioassay experience. We do not find any particular reason to reject either malignant or benign mouse liver tumors as indicators of carcinogenicity. Indeed we find a small difference in selectivity. However, the carcinogenic potency of a substance in rats, as suggested by mouse liver carcinogenesis, may well be strongly overstated.

We hope that our statistical analyses can give suggestions and leads to those who are studying biologic mechanisms. It is also possible that work such as this may be used to test such mechanisms.

Acknowledgments

It is a pleasure to acknowledge the help of Robert McDonald with computer programming, Tim Bigelow with manuscript editing, and Donna Spiegelman and Dr. John Bailar, who assisted us with the condensation of information into the 102-site classification. Richard Wilson was supported in part by an interagency agreement between Harvard University and the U.S. Food and Drug Administration.

Open Forum

Dr. Michael Gallo: I think it would be particularly intriguing to use this type of program to look for concordance once we start to work out the mechanism of any one particular mouse tumor. With that specificity, we may be able to tease out mechanistic results from the NTP data base. I think many of us in this arena look forward to receiving your PC game.

Dr. Stanley Goldfarb: What do you think the significance is of the fact that the adenomas and the carcinomas plot so closely together in all these graphs?

Dr. Richard Wilson: They do plot very closely, though the adenoma is more different from other mouse tumors than the carcinoma. I think the significance is that the mouse adenomas and carcinomas are probably fairly similar in their ability to predict whether you will see something in the rat. However, that does not mean that they should be combined to calculate carcinogenic potency or risk. That is why I deliberately separated the issues of predictability and assessment. Probably benign liver tumors should not be used to calculate potency. My feeling is that mouse liver tumors probably predict moderately well, but the potency they suggest should be divided by a factor of 20 or so.

References

1. European Chemical Industry Ecology and Toxicology Center, *Hepatocarcinogenesis in Laboratory Rodents: Relevance for Man.* Monograph No. 4. Brussels, 1982, pp 1–77.
2. U.S. Environmental Protection Agency. Summary of a workshop on mouse liver tumors and rat kidney tumors. Washington, DC: Science Advisory Board, U.S. Environmental Protection Agency, 1987, pp 1–12.
3. Maronpot RR, Haseman JJ, Boorman GA, Eustis SE, Rao GN, Huff JE. Liver lesions in B6C3F1 mice: The National Toxicology Program experience and position. *Arch Toxicol [Suppl]* 10:10–26, 1987.
4. Nutrition Foundation. *The Relevance of Mouse Liver Hepatoma to Human Carcinogenic Risk: A Report of the International Expert Advisory Committee to the Nutrition Foundation.* Washington, DC, 1983, pp 1–34.
5. Ward JM, Griesemer RA, Weisburger EK. The mouse liver tumor as an endpoint in carcinogenesis tests. *Toxicol Appl Pharmacol* 51:389–397, 1979.
6. Hottendorf GH, Pachter IJ. An analysis of the carcinogenesis testing experience of the National Cancer Institute. *Toxicol Pathol*10:22–26, 1982.
7. Farber E. Pathogenesis of experimental liver cancer: Comparison with humans. *Arch Toxicol [Suppl]* 10:281–288, 1987.
8. Popp JA (ed). *Mouse Liver Neoplasia.* Washington, DC: Hemisphere, 1984.
9. Waters MD, Bergman HB, Nesnow S. The genetic toxicology of Gene-Tox noncarcinogens. *Mutat Res* 205:129–182, 1988.
10. Michalopoulos GK, Eckl PM, Cruise JL, Novick DL, Jirtle RL. Mechanisms of rodent liver carcinogenesis. *Toxicol Ind Health* 3:119–128, 1987.
11. Stott WT, Reitz RH, Schumann A, Watanabe PG. Genetic and nongenetic events in neoplasia. *Food Cosmet Toxicol* 19:567–576, 1981.
12. Zeise L, Crouch EAC, Wilson R. The use of acute toxicity to estimate carcinogenic risk. *Risk Anal* 4:187–199, 1984.
13. Tomatis L, Partensky C, Montesano R. The predictive value of mouse liver tumor induction in carcinogenicity testing: A literature survey. *Int J Cancer* 12:1–20, 1973.
14. Tomatis L. The contribution of epidemiological and experimental data to the control of environmental carcinogens. *Cancer Lett* 26:5–16, 1985.
15. Anderson EL, and the Carcinogen Assessment Group of the U.S. Environmental Protection Agency. Quantitative approaches in use to assess cancer risk. *Risk Anal* 3:277, 1983.
16. Crouch EAC, Wilson R. Interspecies comparison of carcinogenic potency. *J Toxicol Environ Health* 5:1095–1118, 1979.
17. Crouch EAC. Carcinogenic risk assessment: The consequences of believing models. In Langenbach R, Nesnow S, Rice JM (eds): *Organ and Species Specificity in Chemical Carcinogenesis.* New York: Plenum, 1983, pp 653–665.
18. Crouch EAC. Uncertainties in interspecies extrapolations of carcinogenicity. *Environ Health Perspect* 50:321–327, 1983.
19. Gaylor DW, Chen JJ. Relative potency of chemical carcinogens in rodents. *Risk Anal* 6:283–290, 1986.
20. Chen JJ, Gaylor DW. Carcinogenic risk assessment: Comparison of estimated safe doses for rats and mice. *Environ Health Perspect* 72:305–309, 1987.

21. Haseman JK, Huff JE. Species correlation in long-term carcinogenicity studies. *Cancer Lett* 37:125–132, 1986.
22. Ashby J, Tennant RW. Chemical structure, salmonella mutagenicity and extent of carcinogenicity as indicators of genotoxic carcinogenicity among 222 chemicals tested in rodents by the U.S. NCI/NTP. *Mutat Research* 204: 17–115, 1988.
23. International Agency for Research on Cancer (IARC). Long-term and short-term screening assays for carcinogenesis: A critical appraisal. Appendix 1. *IARC Monogr Eval Carcinog Risk Chem Hum* (Suppl 2):1–426, 1980.

Mouse Liver Carcinogenesis: Mechanisms
and Species Comparisons, pages 43–51

Some Observations on Mouse Liver Tumors

Donald E. Stevenson
Toxicology; Health, Safety and Environment; Shell Oil Company, Houston TX 77210 and Scientific Committee, American Industrial Health Council, Washington, DC 20036

It might be assumed that the end point of experimental toxicology and carcinogenicity is related to the overall reduction in human disease and increased wellness. There are, however, widely divergent views on how to achieve this. For example, recent statements in the *Congressional Record* by Epstein (1) and the National Cancer Institute (2) documented some of the differing perceptions on the major causes of human cancer. These exchanges may be therapeutic to the authors and diverting to the reader, but they also have a more serious aspect because strategies to reduce the disease burden would differ markedly depending upon how the problem has been defined. It is not always recognized that risk reduction strategies also present other risks or that too much emphasis on one risk may distract from the reduction in other risks that may be equally or even more important in terms of lives or years of life lost.

In the arena of risk assessment, the assumption that man should be regarded as being as sensitive as the most susceptible species has long been the topic of vigorous debate. As Daniel Byrd et al. discuss elsewhere in this book, the mouse liver is by far the most common site of tumor response in animal bioassays. In many cases, the mouse liver has provided just such a basis for human health hazard evaluation without our considering fully all the other evidence and end points (weight of evidence) that may be available, such as the tumor responses in other species, metabolism, and pharmacokinetic and other data. The situation becomes more complex when the mouse liver tumor response is accompanied by pathologic changes in other organs and species, such as the male rat kidney or the thyroid. By only utilizing data sets from only the most sensitive response in the most sensitive species, some specific human risks may be vastly overestimated while others are not even considered or compared. Thus, there is a very practical reason for wishing to understand the underlying reasons for the mouse liver response—to know when it provides a good indicator of human hazard.

The mouse, and more particularly the mouse liver, must be viewed as a surrogate for a broader analysis. The discussion should not be whether the mouse liver is relevant to the human situation but rather what we can learn from the mouse that is, or is not, relevant to man. Interestingly, the concept of bioassay has changed—the rabbit and the mouse were first used to test substances already known to be human carcinogens. Since they were quick and sensitive indicators of activity, it was possible to move from complex mixtures to specific chemical ingredients in order to ascertain which were responsible for carcinogenic activity. The concept has now changed from comparative potency to one of absolute potency to man; this change has led to real difficulties in interpretation.

Iversen (3) has suggested that some of the major paradigms in carcinogenesis may focus too much on some overgeneralized conclusions. For instance, the relative importance of environmental and genetic factors has been viewed as an either-or situation rather than a strongly interactive combination of factors, each of which could trigger—or limit—the total process. Roe (4) has discussed the often accepted paradigm that 90%–95% of cancer is caused by environmental factors and that elimination of manmade chemicals would eliminate cancer. Clearly this simplistic paradigm does not withstand scrutiny, either for man or for animals. Information obtained from experiments with mice over the last 50 or more years indicates a far more complex interaction between genetic and other factors. Some of these may be amplified in the mouse, such as the effects of caloric restriction, but the data point the way to the potential for similar effects in man.

Another example of a paradigm that may be too readily accepted for all cases is the classification of carcinogens into genotoxic and nongenotoxic carcinogens and also into complete carcinogens, initiators, promoters, and progressors. This may lead to uneasy attempts to apply a label by examining the reasons why the fit is good, rather than concentrating on all the reasons for and against the fit. This could be the more profitable avenue for advance.

Several compounds have been designated promoters of mouse liver tumors because they are nongenotoxic but increase the incidence of tumors. This implies that the mouse liver cell is already initiated or is initiated by other exogenous means. The evidence that this actually occurs is not clear, although a recent publication by Lee et al. (5) postulates on the basis of cell survival in primary culture that there are immortal or spontaneously initiated cells in the livers of young mice and that they progress to carcinoma by treatment with phenobarbital, which is regarded as a promoter. There may be a correlation within and between strains relating the incidence of these cells and the background liver tumor incidence.

A recent observation in the rat liver suggests a subtle response in that partial hepatectomy may leave a long-lasting promotional imprint for many weeks. Bartsch et al. (6) showed that a dose of nitrosodiethylamine gave a higher tumor response rate after partial hepatectomy—even when this took place 8–10 weeks prior to adminstration of the carcinogen. Recently, Ove et al. (7) found in rats a hepatomitogenic factor that did not stimulate normal hepatocytes in vivo or in vitro without a cofactor or prior partial hepatectomy. This again suggests that increased cell turnover is a factor in liver tumor formation and that the correct combination of factors rather than only a single event is critical.

Consideration of the two-stage Moolgavkar model (8) suggests that an increase in the turnover of normal somatic cells may lead to an increased incidence of tumors. Thus, promotion may include the possibility of increasing the rate of cell division of normal cells, as well as initiated cells. A residual message may be retained after cell division has ceased.

In 1951 Gross (9) reported a transmissible agent that would induce leukemia in mice. Recently, he has reported that the leukemogenic effects of ionizing radiation can be reduced greatly by reduction in food intake (10). In the case of liver tumors, no transmissible factor has been reported, although viral particles have sometimes been identified within liver tumor cells. The effects of nutrition have certainly been documented on many occasions (11). As an example, Roe has used the term "affluent mice" to describe the most tumorigenic situation—ad libitum feeding, single caging, and no sex. Even a relatively small reduction in food intake can produce a major increase in life span and a reduction in liver and lung tumors and leukemia.

Figure 1 summarizes some of the factors shown to increase or decrease the rate of liver tumor formation in mice. Many factors or substances can increase or decrease tumor incidence in this system, including dietary fats, oils, hormones, protein, and total calories (12). The potency of these foodstuffs relative to chemicals per se has been largely ignored in evaluating human risk. In one such comparison, Tennekes et al. (33) concluded that the control incidence of liver tumors in CF1 mice was equivalent to the pressure of about 10 ppm dieldrin in the diet.

The relationship between food, calories, life span, and tumor incidence is intriguing. It has long been recognized that the incidence of cancer increases with age. Aging has been ascribed to an accumulation of genetic damage associated with many generations of cell reproduction (13). However, a more complex explanation is emerging that allows for interactions between mutations and epigenetic effects on gene expression and possibly also interactions

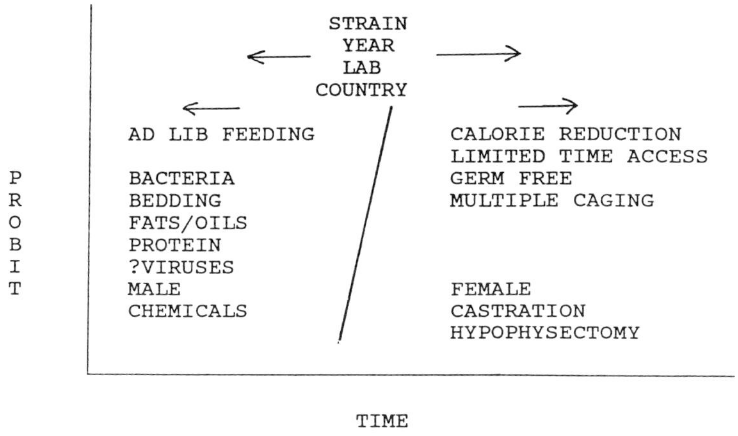

Figure 1. Factors that affect the rate of liver tumor formation in mice.

between epigenetic and molecular events (14). The same interactions may hold for at least some cancers.

Ono et al. (15) have recently investigated age-associated changes in DNA structure and the level of mRNA for the c-*myc* gene in the spleen and liver of mice. In the liver, the mRNA level showed a dramatic decrease with age. Substantial changes in methylation were also described.

It has been pointed out by Cutler (16) that the product of maximum life span and the basal metabolic rate is remarkably similar for all species, i.e., the life-span energy potential (LEP) for most species is 220 kcal/g. For a few species, such as the cat and non-human primates, the values are about 458 kcal/g; for others, such as humans, capuchins, and lemurs, they are about 781 kcal/g. It is possible that one reason for these differences is the ability of different species to deal with free radicals by a variety of defense mechanisms. If this is so, then short-lived animals may be less able to deal with free radical attack. Simic and Cutler (unpublished observations) have found a good correlation between life span and the urinary excretion rates of degradation products of DNA that may result from oxygen or peroxide radical attack.

The mouse and man are at the extremes of the DNA degradation–life span scale. If a disease process is related to free radical formation, then major qualitative or quantitative differences might be anticipated. The high response rate of the mouse liver to a variety of nongenotoxic chemicals and other stimuli suggests that it may sometimes be a nonspecific indicator of the disruption of the physiologic and biochemical processes in an organ already nearing instability, possibly related in part to the high rate of oxygen consumption. It may be conjectured that the mouse, because of its high metabolic rate, is already more prone to oxygen- or other radical-induced damage, particularly if its defense mechanisms are less developed (see ref. 16). The fact that the liver is the primary organ responding by tumor formation may be explained by the observation that there are more mitochondria and cytochromes in liver cells, which are also the primary sites of xenobiotic metabolism. Little attention has been given to the role of mitochondrial DNA in the carcinogenic process (17). However, the near-linear response for some nongenotoxic carcinogens in the liver might be explained by the lack of DNA repair processes in mitochondrial DNA. The mitochondrion might also provide at least a partial answer to the importance of caloric intake to the incidence of cancer.

Saul et al. (18) have reported that the urinary excretion of thymine glycol and thymidine glycol, which result from oxygen damage to DNA, is far higher in mice than in man. There is now an increasing array of techniques that could be used to investigate either the presence of radicals or their effects. These now need to be applied across species. For instance, in vivo radical trapping has shown the presence of carbon tetrachloride–derived radicals in the bile of treated rats (19). Do the mouse and other species respond to a greater or lesser degree?

Diet is important in relation to the bacterial flora of the alimentary tract. Goldman (20) has pointed out that the flora may play an important role in the carcinogenic process in some cases and also that the distribution of bacteria is very different in the rodent and man. For instance, the fore-stomach is an important site of anaerobic bacteria in rodents, but there is no equivalent population in the human stomach. Williams Smith (21) found that ad libitum feeding in rodents induced a large increase in the bacterial population in the fore-stomach, where the pH is relatively high in rodents as compared with man. Reduction in food intake reduced the bacterial populations, the proportions of which also changed according to the type of diet.

The apparent coincidence between the factors that alter mouse liver tumor incidence and those affecting gut bacteria has been an interesting observation, and recently Rowland (22) has summarized data suggesting that this may be a real relationship. He summarized experiments showing that at 1 year of age, germ-free C3H/He mice had less than half the liver tumor incidence of mice with conventional flora. On the other hand, some common gut bacteria either separately or in combination can also increase the incidence of liver tumors. Rowland also reported that the in vivo mutagenicity of aflatoxin in a host-mediated assay was much greater in conventional BALB/c mice than in those with no flora. Thus, mouse bioassays could produce very misleading results if a substance altered the composition of the gut flora indirectly or directly.

Mizutani and Mitsuoka (23) examined the tumorigenic effects of phenobarbital in C3H/He male mice. At 1 year, 30% of the germ-free mice had liver tumors, compared with 75% for those fed a conventional diet. For phenobarbital-fed animals, the figures were 67% and 100%, respectively.

Young (24) has reported an apparent inverse response between leukemia and liver tumor induction in mice. Since leukemia tends to occur earlier than liver tumors, this does not appear to be a question of survival time. This observation raises the issue that risk assessment is carried out on a specific tumor response rather than the total incidence; i.e., substances may alter the spectrum of tumor incidence rather than the total but may be judged only by individual tumor-type increases.

Thus, the mouse data indicate that genetics, viruses, nutrition, and life-style have major impacts on tumor incidence and longevity. In fact, these factors may overwhelm effects due to any chemical exposure per se. Wolff et al. (25) found within a population of C3H × VF hybrid mice a subpopulation that failed to respond to phenobarbital, suggesting a phenotypic difference in gene expression. They also reported that a dominant lethal mutation occurred after exposure to lindane at the A viable yellow locus and concluded that this facilitated the formation of tumors in the mouse lung and liver.

Laboratory strains of mice appear to have a tumor incidence that is very different from wild mice. Gardner and his associates (26) have published a series of papers describing spontaneous tumor rates in wild mice. The overall incidence was very low until 24 months, and the main tumor type was leukemia. Very few liver tumors were found, even in animals surviving more than 3 years.

It has sometimes been assumed that the background incidence of liver tumors in a particular strain is an indication of its susceptibility to hepatocarcinogens. This does not always appear to be the case. For instance, Akamatsu et al. (27) examined the effects of *O*-aminoazotoluene in the C3H and C57BL strains. When the control numbers were subtracted from those for the treated groups, the response was very similar, despite the large differences in the background rates. Ruebner et al. (28) also found with dieldrin that the tumor response did not necessarily depend on the background incidence.

Some interesting aspects of the dose-response relationship have been explored by Tennekes et al. (29) and by Kunz et al. (30). They used the approach of Druckery (31), who found that there is a linear age-specific tumor incidence within dose groups, using a log-probit plot. This implies that the test population is homogeneous. Examination of their plots, which are based on full lifetime data for dieldrin exposure, indicated that if the study had been terminated at 2 years, the control population would have been starting its rise in tumor incidence at a parallel rate to that of the treatment group. Thus, the timing of the end of the experiment was of great importance to the dose-response result if only the total tumor incidence was considered. There is also information for the mouse liver tumor response to phenobarbital (32) and dieldrin (33,34) that allows an evaluation of partial lifetime response. This is of great importance in developing dose-response information, since there is an assumption in modeling that man is exposed continuously from an early age, although this is clearly not always true, either qualitatively or quantitatively.

In essence, it is suggested that the mouse liver tumor response is an example of a toxicologic end point of potential importance to human risk assessment, although constraints on the analysis of the response have reduced our ability to make an estimate of the true risk for man. Further research is needed to determine precisely why the mouse liver tumor responds to such a wide range of chemical and environmental stimuli.

Open Forum

Dr. Ernest McConnell: You mentioned that if you used lifetime studies, you would come up with a different response. Did you make any age adjustment in putting together that graph that you showed?

Dr. Don Stevenson: We did do an actuarial analysis, but I do not have that information with me. We did make an interesting observation in this study. To get the maximum out of the pathologic data, we let the animals live even after they had large palpable liver tumors and the animal technician was convinced the animals only had a few days to live. We discovered that they lived an additional 2–3 months. So the nice gradation in reduction in life span that you saw in that particular slide was really rather artificial. A mouse can live with a massive liver tumor for a very substantial part of its life and may not even have its life expectancy altered that much.

Dr. Gary Williams: I would like to mention another artifact of the animal studies that I think has a great impact on risk assessment and relates to caloric restriction and oxygen utilization. I am referring to voluntary exercise. We have done several experiments in our institute where animals were allowed voluntary exercise in an exercise wheel. The amount of exercise that animals select is remarkable. Rats will run up to 5 miles a day given an exercise wheel. Interestingly, females run more than males. In experiments completed so far on both colon and breast cancer, the cancer incidence has been dramatically reduced despite a higher caloric intake and, obviously, oxygen utilization concomitantly with the exercise. Consequently, I am not particularly enamored of the idea that oxygen utilization is carcinogenic, in contrast to Dr. Bruce Ames' postulate. This is an important element that bears on the quantitative outcome of experiments. Obviously, humans are dramatically different from laboratory animals because we do exercise, albeit generally not enough.

Dr. Stevenson: I do not know of any situation—germ-free, food reduction, or whatever—where you do not finally end up with liver tumors in sensitive strains.

Dr. McConnell: Did your exercised animals weigh less? If they did, some people might say the reduced cancer incidence was strictly because the animals weighed less, and the exercise was simply a method to get there.

Dr. Williams: The animals did not weigh less. The basis is being analyzed. You would think that the exercise would increase muscle mass and reduce body fat. But that does not account for the equilibrium. The main compensation appears to be body water. Exercising animals consume more water.

Dr. McConnell: They essentially weigh the same if they are allowed to eat ad libitum? So the business of caloric restrcition might be viewed in a new light if an animal has plenty of exercise?

Dr. Williams: That's right. Animals eat to calories, to a body weight that is somehow regulated for their condition. Caloric restriction is an artifact where you are actually allowing the animal to eat less than it would choose to if given the opportunity.

References

1. Epstein SS. Losing the war against cancer: Who's to blame and what can be done about it? A position paper on the politics of cancer. *Congressional Record* E3449-E3454, Sept. 9, 1987.
2. National Cancer Institute. National Cancer Institute response to article by Dr. Samuel Epstein. *Congressional Record* E2290–E2297, July 7, 1988.
3. Iversen OH. *Theories of Carcinogenesis*. Washington, DC: Hemisphere, 1988.
4. Roe FJC. Paradigms in cancer research: Biological phenomena in carcinogenesis. In Iversen OH (ed): *Theories of Carcinogenesis*. Washington, DC: Hemisphere, 1988, pp 11–22.
5. Lee G-H, Sawada N, Mochizuki Y, Nomura K, Kitagawa T. Immortal epithelial cells of normal C3H mouse liver cells in tissue culture: Possible precursor populations for spontaneous hepatocellular carcinoma. *Cancer Res* 49:403–409, 1989.
6. Bartsch H, Preat V, Aitio A, Cabral JRP, Roberfroid M. Partial hepatectomy of rats ten weeks before carcinogen administration can enhance liver carcinogenesis: Preliminary observations. *Carcinogenesis* 9:2315–2317, 1988.

7. Ove P, Francavilla A, Coetzee ML, Makowka L, Starzl TE. Response of cultured hepatocytes to a hepatomitogen after initiation by conditioned medium or other factors. *Cancer Res* 49:98–103, 1989.
8. Moolgavkar SH, Dewanji A, Venson DJ. A stochastic two-stage model for cancer risk assessment. I. The hazard function and the probability of tumor. *Risk Anal* 8:383–392, 1988.
9. Gross L. "Spontaneous" leukemia developing in C3H mice following inoculation, in infancy, with Ak-leukemic extracts or Ak-embryos. *Proc Soc Exp Biol Med* 76:27–32, 1951.
10. Gross L. Inhibition of the development of tumors or leukemia in mice and rats after reduction in food intake. *Cancer* 62:1463–1465, 1988.
11. Butler WH, Newberne PM. *Mouse Hepatic Neoplasia.* Amsterdam: Elsevier/North Holland, 1975.
12. Gellatly JBM. The natural history of hepatic parenchymal nodule formation in a colony of C57BL mice with reference to the effect of diet. In Butler WH, Newberne PM (eds): *Mouse Hepatic Neoplasia.* Amsterdam: Elsevier/North Holland, 1975, pp 77–109.
13. Hayflick L. The cellular basis for biological aging. In Finch CE, Schneider EL (eds): *Handbook of the Biology of Aging.* New York: van Nostrand Reinhold, 1977, pp 159–186.
14. Kirkwood TBL. DNA, mutations and aging. *Mutat Res* 219:1–7, 1989.
15. Ono T, Takahashi N, Okada S. Age-associated changes in DNA methylation and m-RNA level of the c-myc gene in spleen and liver of mice. *Mutat Res* 219:39–50, 1989.
16. Cutler RG. Antioxidants, aging and longevity. *Free Radicals in Biology* 6:371–428, 1984.
17. Shay JW, Werbin H. Are mitochondrial DNA mutations involved in the carcinogenic process? *Mutat Res* 186:149–160, 1987.
18. Saul JW, Gee P, Ames BN. *Modern Biological Theories of Aging.* New York: Raven Press, 1987.
19. Knecht KT, Mason RP. In vivo radical trapping and biliary secretion of radical adducts of carbon tetrachloride-derived free radical products. *Drug Metab Dispos* 16:813–817, 1988.
20. Goldman P. Defining the role of anaerobic bacteria in toxicology. In Tashijian AH (ed): Proceedings of Molecular and Cellular Appproaches to Understanding Mechanisms of Toxicity Conference held in Boston, September 1–2, 1983, cosponsored by Harvard University School of Public Health and the United States Environmental Protection Agency, 1983, pp 164–182.
21. Williams Smith H. Observations on the flora of the alimentary tract of animals and factors affecting its composition. *J Pathol Bacteriol* 89:95–122, 1965.
22. Rowland IR. Interactions of the gut microflora and the host in toxicology. *Toxicol Pathol* 16:147–153, 1988.
23. Mizutani T, Mitsuoka T. Effect of dietary phenobarbital on spontaneous hepatic tumorigenesis in germfree C3H male mice. *Cancer Lett* 39:233–237, 1988.
24. Young SS. Exploration of the negative correlation between proliferative hepatocellular lesions and lymphoma in rats and mice: Establishment and implications. *Fundam Appl Toxicol* 4:632–640, 1984.
25. Wolff GL, Roberts DW, Morrissey RL, Greenman DL, Allen RR, Campbell WL, Bergman H, Nesnow S, Frith CH. Tumorigenic responses to lindane: Potentiation by a dominant mutation. *Carcinogenesis* 8:1889–1897, 1987.

26. Gardner MB, Henderson BE, Rongey RW, Estes JD, Huebner RJ. Spontaneous tumors of aging wild house mice: Incidence, pathology, and C-type virus expression. *JNCI* 50:719–726, 1973.
27. Akamatsu Y, Takemura T, Ikegami R, Takahashi A, Miyajiama H. Growth behavior of hepatomas in O-aminoazotoluene treated mice in comparison with spontaneous hepatomas. *Gann* 58:323–330, 1967.
28. Ruebner BH, Gershwin ME, French SW, Meierhenry E, Dunn P, Hsieh LS. Mouse hepatic neoplasia: Differences among strains and carcinogens. In Popp JA (ed): *Mouse Liver Neoplasia*. Washington, DC: Hemisphere, 1984, pp 115–144.
29. Tennekes H, van Ravenzwaay B, Kunz HW. Quantitative aspects of enhanced liver tumor formation in CF1 mice by dieldrin. *Carcinogenesis* 6:1457–1462, 1985.
30. Kunz HW, Tennekes H, Port RE, Schwartz M, Lorke D, Schaude G. Quantitative aspects of chemical carcinogenesis and tumor promotion in liver. *Environ Health Perspect* 50:113–122, 1983.
31. Druckery H. Quantitative aspects of chemical carcinogenesis. *UICC Monograph Series* 7:60–78, 1967.
32. Ward JM, Lynch P, Riggs C. Rapid development of hepatocellular neoplasms in aging male C3H/HeNCr mice given phenobarbital. *Cancer Lett* 39:9–18, 1988.
33. Walker AIT, Thorpe E, Stevenson DE. The toxicology of dieldrin (HEOD). I. Long-term oral studies in mice. *Food Cosmet Toxicol* 11:415–432, 1972.
34. van Ravenzwaay B. Mechanistic and quantitative aspects of liver tumor formation in mice. Doctoral Thesis. Landbouwuniversitatt of Wagengingen, the Netherlands, 1988.

Mouse Liver Carcinogenesis: Mechanisms
and Species Comparisons, pages 53–68

Perinatal Mouse Liver Carcinogenesis as a Sensitive Carcinogenesis Model and the Role of the Sex Hormonal Environment in Tumor Development

Stan D. Vesselinovitch
Departments of Radiology and Pathology, University of Chicago, Chicago, IL 60637

The induction and development of hepatocellular tumors depends upon the nature of the chemical carcinogen used (1), the requirement for its activation to an ultimate carcinogenic moiety (2), the enzymatic competence of tissues to activate the procarcinogens (3–5), the degree of interaction of ultimate carcinogenic moiety with specific macromolecular sites (6,7), the rate of removal of the formed adducts (8), the rate of DNA (cell) replication at carcinogenic treatment, the rate and the degree of macromolecular repair (9,10), and the caloric intake (11–13), immune competence, genetic makeup regarding carcinogen susceptibility (14–17), and sex hormonal environment of the host (18–21). The interplay of all these factors contributes in varying degrees to the inception, development, expression, and progression of hepatocarcinogenesis.

The use of neonatal and infant mice for chemical carcinogenesis as a highly sensitive carcinogenesis system in general and hepatocarcinogenesis in particular began shortly after the high susceptibility of newborn mice to viral leukemogenesis was demonstrated. The early perinatal studies on chemical carcinogenesis showed that high susceptibility to hepatocarcinogenesis by single or limited numbers of carcinogenic treatments was restricted to a relatively short perinatal age period (22–25). Thus the intrauterine exposure to carcinogens was effective if implemented following organogenesis on the 12th day of gestation. The incidence of liver tumors correlated directly with the gestational age at the time of the administration of the carcinogen (26). However, when hepatocarcinogens were administered within 24 h of birth, the carcinogenic response of liver was further increased (23–25). The high susceptibility to hepatocarcinogenesis persisted at least for the first 3 weeks of postnatal life, depending directly upon the degree of macromolecular replication and the enzymatic competence of the liver to activate procarcinogens to their ultimate carcinogenic moieties. The susceptibility to hepatocarcinogenesis decreased subsequently in direct relationship to the decrease in macromolecular replication, reaching the lowest tumor response at 6 weeks of age, when macromolecular and cellular replication reached their minimum. This low susceptibility to hepatocarcinogenesis persisted throughout the animal's life unless cell replication was increased by mitogenic stimuli. During the first 3 postnatal weeks, the circadian DNA replication shifted continuously so that the level of DNA replication at 9:00 a.m. (usual time at carcinogenic treatment) varied daily from high to intermediate to low and then back to high, resulting in a characteristic ultradian DNA cycle (27). The high replicating levels of DNA were identified on the 1st, 5th, 10th, 15th, and 20th postnatal days and the low levels were observed on days 3, 8, 13, and 18. During the first 3 postnatal weeks, however, high DNA replication also occurred later in the day on the days when the low rates were observed at 9:00 a.m. Correlative studies between the ultradian DNA replication and the neoplastic liver response to single carcinogenic treatment were then carried out (28). These pointed out the existence of a high susceptibility to hepatocarcinogenesis, even when carcinogen administration was timed for the low levels of DNA replication (at 9:00 a.m.), as long as hepatocytes entered high DNA replication shortly after carcinogen treatment (27). The absence of DNA replication at and following carcinogen treatment (at 42 days of age and later) did not result in hepatocarcinogenesis, even when the carcinogenic dose was increased 40-fold (50.0 μg instead of 1.25 μg of DEN/g). Thus the high macromolecular replication that occurred during the neonatal-infant period was an essential factor contributing to the fixation and amplification of the primary macromolecular lesions leading to effective hepatocarcinogenesis. The absence of the high DNA replication rate at a later age was compatible with the repair of macromolecular lesions and consequent lack of carcinogenesis.

The species, strain (29), age at carcinogen treatment (30), and sex (30) of the animals may influence the degree of an agent's activation. Recently, Drinkwater and his associates (14,15) investigated genetic control of hepatocarcinogenesis in C57BL/6J, C3H/HeJ, and their F_1 hybrids ($B6C3F_1$). The authors also investigated the segregation of genes controlling susceptibility to hepatocarcinogenesis and analyzed recombinant inbred mouse strains by analyzing the outcome of the inbreeding of the F_2 mice. These investigators concluded that although at least two genes controlled sensitivity to liver tumor induction, a single locus, denoted *Hcs* (hepatocarcinogen sensitivity), controlled the higher susceptibility of C3H/HeJ mice relative to C57BL/6J in about 85%. The C3H/HeJ and C57BL/6J alleles at this locus were semidominant, yielding intermediate sensitivity in the heterozygous $B6C3F_1$ hybrids (14). From subsequent studies (15), the authors concluded that this gene controls susceptibility to hepatocarcinogenesis by acting at the level of postinitiation events or promotion of carcinogenesis, affecting the rate of the emergence rather than the final multiplicity of the developing lesions. Similarly, Dragani et al. (16) studied the growth rate of liver nodules in DEN-treated C57BL/6J × C3H/HeJ F_1 ($B6C3F_1$) and C57BL/6J × BALB/c F_1 mice. They concluded that $B6C3F_1$ mice were more susceptible to hepatocarcinogenesis than C57BL/6J × BALB/c F_1 mice because of a higher growth rate of the initiated cells in the $B6C3F_1$ animals. The above studies showed that under the experimental conditions, the genetic background governed the rate of tumor promotion without affecting initiation.

The sex of the animal may influence the outcome of hepatocarcinogenesis in two ways. First, the sex-dependent difference in the metabolic competence between males and females may influence the level of the ultimate carcinogenic moiety and consequently the degree of the primary macromolecular events leading to initiation and hepatocarcinogenesis (31). Second, the sex hormonal environment may influence the rate of promotion and morphologic expression of the induced hepatocytes.

The promoting role of sex hormones in hepatocarcinogenesis was investigated in the infant mouse liver model, which was developed and defined in our laboratory. This model possesses the following characteristics: (i) a single carcinogenic dose results in the development of both early and late hepatocellular lesions, culminating in the emergence of hepatocellular carcinomas, (ii) there is an equal degree of metabolic activation of DEN and a similar ethylation of hepatic DNA (*N*-6-ethylguanine or *O*-6-ethylguanine) in both sexes following DEN treatment (15), and (iii) a similar pattern of morphologic response occurs in males and females. Thus, B6C3F$_1$ mice, following low dose treatment with DEN at 15 days of age, developed sequentially tinctorial foci, hepatocellular nodules, hepatocellular adenomas, and hepatocellular carcinomas (32). The stereomorphometric quantitation of the above lesions was achieved by translating two-dimensional images of the lesions into three-dimensional space (33–35). Such quantitation resulted in the establishment of dose-time-response kinetics regarding the development of each of the above lesions (36).

Materials and Methods

To define the effect of the sex hormonal environment in general and androgens in particular upon the postinitiation phase of hepatocarcinogenesis, the process was studied in hormonally intact (sham-gonadectomized) and sex-hormone deficient (gonadectomized) B6C3F$_1$ mice. Gonadectomies were performed at various intervals following carcinogenic treatment. The modifying effect of sex hormones was evaluated by comparing kinetics of hepatocarcinogenesis between (i) intact male and female mice, (ii) the animals that were gonadectomized at different intervals following DEN treatment, and (iii) 4-week sham-orchidectomized (hormonally intact or competent) and 4-week orchidectomized (hormonally deficient) males following the induction of hepatocarcinogenesis by 1.25, 2.5, and 5.0 μg DEN/g. DEN was administered intraperitoneally once at 15 days of age. Details of each experiment are presented, in turn, in the Results section. Liver tissues were stained with hematoxylin and eosin, and the liver lesions were classified diagnostically and quantitated as reported (36).

Results and Comments

The original studies utilizing the incidence of tumor-bearing animals as the end point showed that B6C3F$_1$ female mice responded similarly to male animals, although after a longer period of time (36). This observation was substantiated more recently by a study in which the average numbers of adenomas per liver (rather than the incidence of tumor-bearing animals) were used as quantitative end points (37). The latter data showed that the expression of the hepatocellular adenomas occurred by an average of 16 weeks later in females than in males. This indicated that in spite of the same degree of initiation of hepatocellular adenomas in both sexes, the rate of development of adenomas was influenced by the sex hormonal environment of the host.

In subsequent studies (37), 15-day-old mice were administered a single dose of DEN (males 2.5 μg and females 10.0 μg/g). Groups of mice were sham-gonadectomized or gonadectomized at various time intervals, and the respective subgroups were killed subsequently at 4-week intervals. Orchidectomies were performed at 20, 30, or 38 weeks of age. Data showed that the timing of orchidectomy delayed differentially the emergence of hepatocellular adenomas. The intact males reached their maximal tumor response by 44 weeks, plateauing afterward. The animals that were orchidectomized at 20 weeks showed a significantly lower multiplicity of expressed adenomas through the 54th week of age (18 adenomas per liver in orchidectomized vs. 51 adenomas in hormonally intact males). The average difference in the temporal distribution of the development of the hepatocellular adenomas between the hormonally competent and the 20-week orchidectomized males was 19 weeks. Delaying orchidectomy from 20 to 30 and 38 weeks, respectively, was increasingly less effective in delaying the emergence of the hepatocellular adenomas. Thus, approximately 18 adenomas per liver were observed at 54, 50, and 46 weeks of age in mice that were orchidectomized at 20, 30, and 38 weeks, respectively. However, by 58 weeks of age, regardless of age at orchidectomy, all orchidectomized groups developed a similar number of hepatocellular adenomas per liver to the number of tumors observed in the sham-orchidectomized mice.

In contrast to the male series, the time of ovariectomy (4, 28, and 40 weeks of age) accelerated differentially the emergence of hepatocellular adenomas in female mice. In the hormonally intact females, the adenomas were first observed at 48 weeks, reaching the multiplicity of 56 adenomas per liver by 56 weeks of age. However, the females ovariectomized at 4 weeks of age showed first adenomas 16 weeks earlier than the intact females (32 vs. 48 weeks of age, respectively) and continued to express neoplasia at a relatively slow rate, reaching 55 adenomas per liver at 52 weeks. The acceleration of the development of hepatocellular adenomas was somewhat less manifest in the females ovariectomized at 28 and 40 weeks, respectively. However, all three ovariectomized groups reached 55 adenomas per liver by 52 weeks of age, only 4 weeks earlier than observed for the intact females. The mean of 28 adenomas per liver was observed by 43 weeks in 4-week ovariectomized females and by 54 weeks in hormonally intact females, representing an acceleration of 11 weeks in tumor development in the absence of ovarian hormones.

The recently terminated studies showed that 4-week orchidectomy was even more effective in delaying and in partly preventing hepatocarcinogenesis. These data are illustrated in Tables 1–3. Table 1 compares the multiplicities of tinctorial foci, hepatocellular (or hyperplastic) nodules, hepatocellular adenomas, and hepatocellular carcinomas per liver between the sham-orchidecto-

Table 1. The Average Numbers of Tinctorial Foci, Hyperplastic Nodules, Hepatocellular Adenomas, and Hepatocellular Carcinomas per Liver following Administration of 1.25 μg DEN/g*

	Tinctorial foci		Hyperplastic nodules		Hepatocellular adenomas		Hepatocellular carcinomas		All hepatocellular lesions	
Weeks	A†	B	A	B	A	B	A	B	A	B
36	63.7	28.8	14.1	1.1	9.2	—	—	—	87.0	29.9
44	83.9	42.4	27.2	5.3	24.7	—	0.1	—	135.9	47.7
52	35.0	38.0	22.4	7.6	29.7	3.8	1.1	—	88.2	49.4
60	22.6	44.8	12.8	13.5	38.8	4.0	6.0	—	80.2	62.3
68	22.4	75.3	11.4	16.0	63.2	10.3	9.2	0.3	106.2	101.9

*$B6C3F_1$ males treated at 15 days of age.
†A, sham-orchidectomized males at 4 weeks of age; B, orchidectomized males at 4 weeks of age.

Table 2. The Average Number of Tinctorial Foci, Hyperplastic Nodules, Hepatocellular Adenomas, and Hepatocellular Carcinomas per Liver following Administration of 2.5 μg DEN/g*

	Tinctorial foci		Hyperplastic nodules		Hepatocellular adenomas		Hepatocellular carcinomas		All hepatocellular lesions	
Weeks	A†	B	A	B	A	B	A	B	A	B
28	172.9	71.0	17.8	—	7.1	—	—	—	197.8	71.0
36	140.3	85.9	37.2	15.0	27.5	—	—	—	105.0	100.9
44	201.2	253.8	57.0	29.8	32.5	4.4	1.3	—	292.0	288.0
52	64.2	141.2	27.3	38.4	59.7	7.9	1.1	0.3	152.3	187.8
60	22.2	155.0	35.3	48.9	80.7	39.9	5.4	0.7	143.6	244.5
68	47.3	57.1	31.9	46.9	99.2	46.2	13.4	2.1	191.8	152.3

*$B6C3F_1$ males treated at 15 days of age.
†A, sham-orchidectomized males at 4 weeks of age; B, orchidectomized males at 4 weeks of age.

mized and orchidectomized males in animals exposed to 1.25 μg DEN/g. The subgroups of animals were killed at 8-week intervals between the 36th and 68th week. Hepatocellular adenomas occurred earlier and reached a higher number in the hormonally competent than in the hormonally deprived animals. Thus, while a multiplicity of 9.2 adenomas was seen in hormonally competent males at 36 weeks, a similar multiplicity (10.3) was observed at 68 weeks (32 weeks later) in the hormonally deprived males. Similarly the development of hepato-

Table 3. The Average Numbers of Tinctorial Foci, Hyperplastic Nodules, Hepatocellular Adenomas, and Hepatocellular Carcinomas per Liver following Administration of 5.0 μg DEN/g*

	Tinctorial foci		Hyper-plastic nodules		Hepato-cellular adenomas		Hepato-cellular carcinomas		All hepatocellular lesions	
Weeks	A†	B	A	B	A	B	A	B	A	B
20	118.8	46.8	4.0	—	—	—	—	—	122.8	46.8
28	181.2	143.7	39.4	7.0	4.7	—	—	—	225.3	150.7
36	308.4	131.4	73.0	18.4	44.3	1.9	—	—	425.7	151.7
44	322.1	276.6	111.1	51.4	130.2	7.8	0.4	—	563.8	335.8
52	111.2	298.5	81.7	75.5	126.2	16.4	8.0	0.5	327.1	390.9

*$B6C3F_1$ males treated at 15 days of age.
†A, sham-orchidectomized males at 4 weeks of age; B, orchidectomized males at 4 weeks of age.

cellular carcinomas was first detected at 44 and 68 weeks in animals with the two hormonal environments, respectively (a 24-week time difference). The last two columns of Table 1, which refer to the number of combined focal and nodular liver lesions, reveal that the overall number of lesions was at a plateau between 36 and 68 weeks in the hormonally competent males. In contrast, in the hormonally deprived animals, the overall multiplicity of lesions for the same period of time increased from 29.9 at 36 weeks to 101.9 lesions at 68 weeks. The enumeration of the diagnostically differentiated lesions listed in the preceding columns of Table 1 revealed, however, that the latter phenomenon was due to a significant delay in the expression and progression of liver tumors.

The same phenomenon was observed in the animals exposed to 2.5 μg DEN/g (Table 2). The relatively smaller difference, however, between the hormonally competent and deprived animals was due to the greater propensity for hepatocarcinogenesis with the administration of higher doses of DEN. The temporal difference in development of similar frequencies per liver of hepatocellular adenomas (7.1 vs. 7.9) was 24 weeks between the intact and orchidectomized animals (28 vs. 52 weeks, respectively). The difference in the development of carcinomas was 16 weeks for the same two groups (1.3 at 44 weeks in intact and 0.7 in the hormonally deprived animals at 60 weeks).

At the highest dose level of DEN (5.0 μg/g) (Table 3), the difference between the two hormonal environments regarding the emergence of hepatocellular adenomas was 16 weeks and regarding the development of hepatocellular carcinomas was 8 weeks. Thus the degree of the modulating effect of androgens on hepatocarcinogenesis decreased further with the increase in the carcinogenic dose of DEN.

The temporal effect of orchidectomy upon development of adenomas was similar to the effects resulting from three- to fourfold changes in the administered DEN dose. Thus, the time to development of 50% incidence of tumors induced by 5.0 μg DEN/g in the orchidectomized males was similar to the time to 50% incidence following the administration of 1.25 μg DEN/g to the hormonally intact males.

Tables 4–6 present the average diameters of and the percentages of liver occupied by focal and nodular liver lesions in animals exposed to 1.25, 2.5, and 5.0 μg DEN/g, respectively. The data show that, irrespective of the DEN dose used, the overall growth of lesions in the hormonally intact males was consistently more evident than in the orchidectomized mice. Thus, the presence of

Table 4. Diameters and Percentages of Livers Occupied by All Focal and Nodular Liver Lesions following Administration of 1.25 μg DEN/g*

	Average diameter of all focal and nodular liver lesions		Percentage of liver occupied by all focal and nodular liver lesions	
Weeks	A†	B	A	B
36	999.0	394.6	3.79	0.07
44	1311.0	563.3	16.56	0.28
52	2019.1	695.0	26.05	1.02
60	2682.3	879.2	47.84	2.32
68	3574.4	930.4	62.99	6.50

*$B6C3F_1$ males treated at 15 days of age.
†A, sham-orchidectomized males at 4 weeks of age; B, orchidectomized males at 4 weeks of age.

Table 5. Diameters and Percentages of Livers Occupied by All Focal and Nodular Liver Lesions following Administration of 2.5 μg DEN/g*

	Average diameter of all focal and nodular liver lesions		Percentage of liver occupied by all focal and nodular liver lesions	
Weeks	A†	B	A	B
28	553.3	406.4	1.53	0.12
36	877.3	504.8	9.30	0.46
44	1158.4	477.2	27.85	4.24
52	1820.6	752.0	31.62	9.07
60	4051.0	822.8	68.36	11.20
68	3673.0	1679.8	70.52	28.37

*$B6C3F_1$ males treated at 15 days of age.
†A, sham-orchidectomized males at 4 weeks of age; B, orchidectomized males at 4 weeks of age.

Table 6. Diameters and Percentages of Livers Occupied by All Focal and Nodular Liver Lesions following Administration of 5.0 μg DEN/g*

Weeks	Average diameter of all focal and nodular liver lesions		Percentage of liver occupied by all focal and nodular liver lesions	
	A†	B	A	B
20	429.2	306.5	0.32	0.04
28	577.7	505.3	2.03	3.98
36	776.0	549.8	15.43	5.40
44	1098.9	598.9	48.71	8.90
52	1931.2	627.4	73.26	9.75

*$B6C3F_1$ males treated at 15 days of age.
†A, sham-orchidectomized males at 4 weeks of age; B, orchidectomized males at 4 weeks of age.

Table 7. Effect of Orchidectomy on the Growth of DEN-Induced Liver Neoplasia*

Age in weeks	Untreated controls	DEN dose (μg/g)		
		1.25	2.5	5.0
Sham orchidectomy at 4 weeks				
36	1492 ± 54	2024 ± 38	1994 ± 67	2221 ± 51
44	1702 ± 31	2287 ± 91	2602 ± 239	3405 ± 225
52	1660 ± 83	2791 ± 217	3063 ± 325	5776 ± 178
Average		2367	2553	3800
Orchidectomy at 4 weeks				
36	1492 ± 54	1370 ± 36	1406 ± 43	1744 ± 159
44	1702 ± 31	1380 ± 32	1275 ± 41	1485 ± 113
52	1660 ± 83	1577 ± 66	1491 ± 75	1982 ± 288
Average		1442	1390	1737

*The average weights of the livers include weights of the preneoplastic and neoplastic nodular lesions; numbers represent mean liver weight ± SE.

endogenous androgens early in the postinitiation period was essential for earlier inception and more pronounced tumor growth in comparison with the growth of liver tumors in the orchidectomized males. However, the comparison of the average diameters and the average percentages of liver lesions between the series exposed to 1.25, 2.5, and 5.0 μg of DEN/g (Tables 4–6) shows that the average growth of the induced liver lesions was independent of DEN dose both within the hormonally competent and hormonally deprived animal groups. Since the DEN dose affected the number of induced liver lesions, the carcinogenic dose influenced the overall neoplastic growth by this mechanism, as shown in Table 7. This table shows that the increase in liver weights due to overall neoplastic growth was dependent upon the DEN dose, the time at kill, and the levels of the endogenous androgens. Thus, regardless of the age at which the animals were killed, the weights of livers, inclusive of neoplastic lesions, were 2,367 mg, 2,553 mg, and 3,800 mg in the hormonally competent males following the administration of 1.25, 2.5, and 5.0 µg of DEN/g, respectively. The corresponding weights of the livers with tumors in the hormonally deprived males were 1,442 mg, 1,390 mg, and 1,737 mg, respectively.

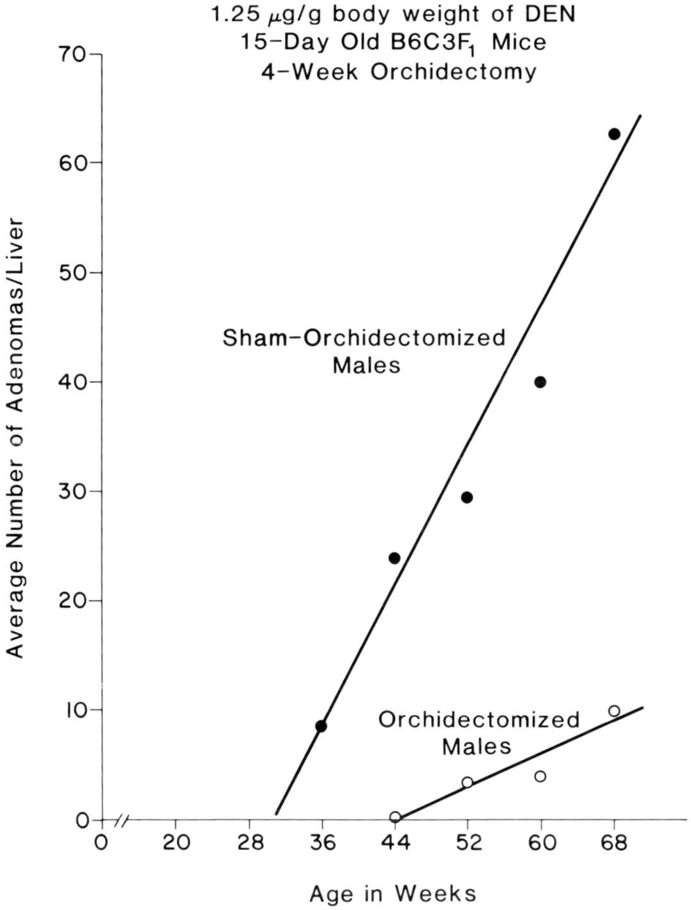

Figure 1. The effect of 4-week orchidectomy on the average number of adenomas/liver following administration of 1.25 µg/g DEN to 15-day-old $B6C3F_1$ mice.

Figures 1–4 illustrate the development of hepatocellular adenomas (number of lesions/liver and percentage of liver occupied by adenomas) under two hormonal conditions for two DEN dose levels: 1.25 μg/g (Figs. 1,2) and 2.5 μg/g (Figs. 3,4). The figures show that the multiplicity of the hepatocellular adenomas and the percentage of liver occupied by adenomas were consistently lower in the orchidectomized than in the hormonally intact males throughout the whole observational period (up to the 68th week of age). It is apparent, therefore, that the level of endogenous androgens controls the multiplicity and the rate of growth of the neoplastic lesions. Consequently, the dependence of growth rate upon the sex hormonal environment may be considered the underlying mechanism that controls the rate at which nodular liver lesions emerge.

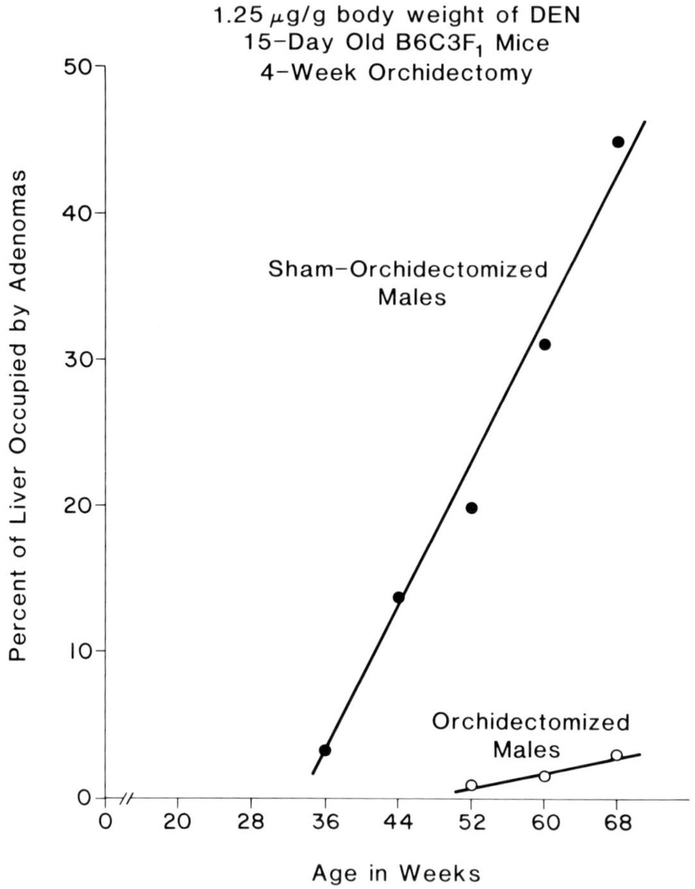

Figure 2. The effect of 4-week orchidectomy on the percentage of the liver occupied by adenomas following administration of 1.25 μg/g DEN to 15-day-old $B6C3F_1$ mice.

Discussion

The degree of the induction and development of tumors in general and of the liver in particular depend upon the carcinogenic dose and the age at carcinogenic treatment. Perinatal age contributes to the most effective hepatocarcinogenesis due to the physiologic presence of a high macromolecular rate, which favors fixation of the primary macromolecular lesions and the amplification of the initiated cell population. In addition to other contributing factors, the sex hormonal environment of animals and humans can influence the rate and the degree of development of liver tumors (38–40). From the operational point of view, the sex difference in hepatocarcinogenesis may be due to modulation of the initiation and/or promotion of the neoplastic process. Thus the degree of the metabolic activation of the procarcinogens and the fixation-repair ratio of the primary molecular lesions could be causally related to different degrees of initiation and tumor induction. However, in the experimental model

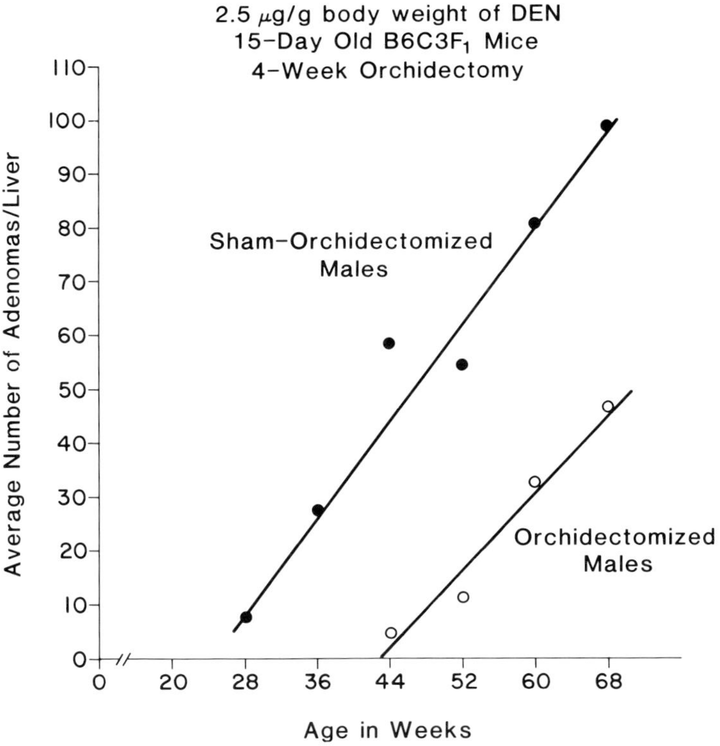

Figure 3. The effect of 4-week orchidectomy on the average number of adenomas/liver following administration of 2.5 μg/g DEN to 15-day-old $B6C3F_1$ mice.

in which the degree of initiation is equivalent in males and females (30), the sex hormonal environment of the animals may influence the promoting stage of hepatocarcinogenesis. In the latter instance, the hormonal environment could influence one or several of the following: the number of initiated cells that would be expressed, the time of onset and the rates of development of the early tinctorial foci, and the rates of neoplastic expression and progression.

The early- (tinctorial foci) and late-emerging (hepatocellular adenomas and carcinomas) lesions were associated with cell replication, as indicated by an increase in the labeling indices within tinctorial foci (41), an increase in the mitotic figures in the neoplastic lesions, and an increase in tumor weights, respectively. Thus, one can assume that the sex hormonal environment in general and androgens in particular could be considered regulators of preneoplastic and neoplastic cell growth. The androgenic environment (sham-orchidectomized males) accelerated and the estrogenic environment (sham-ovariecto-

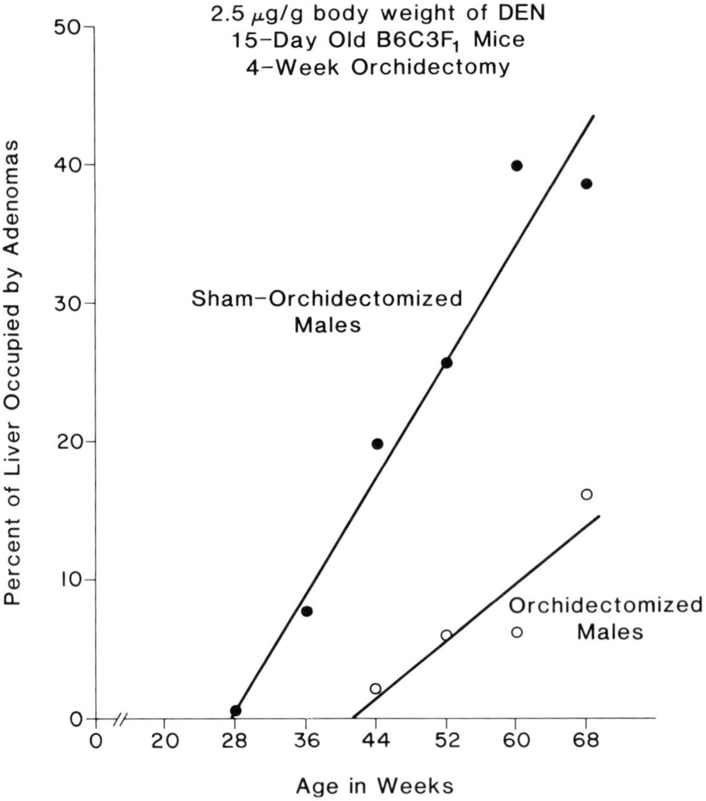

Figure 4. The effect of 4-week orchidectomy on the percentage of liver occupied by adenomas following administration of 2.5 μg/g DEN to 15-day-old $B6C3F_1$ mice.

mized females) delayed neoplastic growth and progression. The shift in the hormonal environment of the host away from the hormonally competent state delayed hepatocarcinogenesis following orchidectomy and accelerated hepatocarcinogenesis following ovariectomy.

The studies presented were undertaken to elucidate the effect of the endogenous sex hormones upon the developmental phase of hepatocarcinogenesis. The data obtained from the orchidectomized series pointed out that as long as orchidectomy was performed at 20 weeks of age or later, only the rate of neoplastic expression and progression was affected. In contrast, when orchidectomies were carried out 2 weeks after initiation (DEN treatment) not only was the rate of neoplastic expression and progression delayed but the number of tumors that emerged was reduced as well. The latter was due apparently to the fact that a certain fraction of the initiated cells failed to undergo neoplastic expression and progression. Thus we speculate that the presence of endogenous androgens early in carcinogenesis not only might be essential for the timely expression of the neoplastic process but also might be crucial for a certain fraction of initiated cells to undergo changes necessary for the eventual emergence of liver tumors. Hypothetically, it was assumed that DEN treatment at infancy conferred differing initiating potential upon the affected cells. The majority of the affected cells became conditioned by DEN treatment in such a way as to respond to the concurrent high DNA replication by being converted from the initiated into the induced neoplastic state. The balance of the initiated cells, however, apparently required the androgenic environment shortly after initiation for their conversion to the induced neoplastic state. Thus in the hormonally intact males, all the initiated hepatocytes were apparently converted in the presence of endogenous androgens during the concurrent cell replication into the induced cell state (first phase of promotion). These cells, in the continuous presence of hormones, emerged as neoplasia within a relatively short latency period (second phase of promotion). Therefore, the endogenous androgenic hormones can be considered as modifiers of hepatocarcinogenesis by affecting the level of conversion of initiated into induced neoplastic cells and by influencing their growth and the rate of the events leading to the development, expression, and progression of liver tumors.

In conclusion, the literature review, the presented data, and the discussion showed that perinatal age represents a sensitive period regarding the induction of liver tumors, which is due primarily to high DNA and cellular replication. The outcome of the hepatocellular neoplastic response, however, can be modulated by the sex hormonal environment of the host, which operates mainly by affecting the rate of tumor development (latent period) and to a lesser degree by influencing the level of tumor induction (tumor multiplicity).

Acknowledgment

This investigation has been supported in part by grant CA-25522 from the National Cancer Institute.

References

1. Vesselinovitch SD. Liver tumor induction. *Toxicol Pathol* 10:110–120, 1982.
2. Remmer H. Metabolism of carcinogens: Its significance for the initiation of liver tumors. In Remmer H, Bolt HM, Bannasch P, Popper H (eds): *Primary Liver Tumors*. Lancaster: MTP Press, 1978, pp 31–49.
3. Miller EC, Miller JA. Mechanisms of chemical carcinogenesis. *Cancer* 47:1055–1064, 1981.
4. Miller EC, Miller JA. Searches for ultimate chemical carcinogens and their reactions with cellular macromolecules. *Cancer* 47:2327–2345, 1981.
5. Gelboin HV, Wiebel FJ, Kinoshita N. Microsomal aryl hydrocarbon hydroxylases: On their role in polycyclic hydrocarbon carcinogenesis and toxicity and the mechanism of enzyme induction. *Biochem Soc Symp* 34:103–133, 1972.
6. Becker RA, Shank RC. Kinetics of formation and persistence of ethylguanine in DNA of rats and hamsters treated with diethylnitrosamine. *Cancer Res* 45:2076–2084, 1985.
7. Brambilla G, Martelli A, Pino A, Robbiano L. Sequential analysis of DNA damage and repair during the development of carcinogen-induced rat liver hyperplastic lesions. *Cancer Res* 46:3475–3481, 1986.
8. Scherer E, Steward AP, Emmelot P. Kinetics of formation of O^6-ethylguanine in, and its removal from liver DNA of rats receiving diethylnitrosamine. *Chem Biol Interact* 19:1–11, 1977.
9. Stich HF, Laishes BA. DNA repair and chemical carcinogens. *Pathobiology Annual* 3:341–376, 1973.
10. Kleihues P, Cooper HK, Buecheler J. Involvement of DNA repair in the organ-specific carcinogenicity of alkylating agents. In Remmer H, Bolt HM, Bannasch P, Popper H (eds): *Primary Liver Tumors*. Lancaster: MTP Press, 1978, pp 319–326.
11. Tucker MJ. The effect of long-term food restriction on tumors in rodents. *Int J Cancer* 23:803–807, 1979.
12. Conybeare G. Effect of quality and quantity of diet on survival and tumor incidence in outbred Swiss mice. *Food Cosmet Toxicol* 18:65–75, 1980.
13. Rogers AE, Newberne PM. Dietary effects on chemical carcinogenesis in animal models for colon and liver tumors. *Cancer Res* 35:3427–3431, 1975.
14. Drinkwater NR, Ginsler JJ. Genetic control of hepatocarcinogenesis in C57BL/6J and C3H/HeJ inbred mice. *Carcinogenesis* 7:1701–1707, 1986.
15. Hanigan MH, Kemp CG, Ginsler JJ, Drinkwater NR. Rapid growth of preneoplastic lesions in hepatocarcinogen-sensitive C3H/HeJ male mice relative to C57BL/6J male mice. *Carcinogenesis* 9:885–890, 1988.
16. Dragani TA, Manenti G, Della Porta G. Genetic susceptibility to murine hepatocarcinogenesis is associated with high growth rate of NDEA-initiated hepatocytes. *J Cancer Res Clin Oncol* 113:223–229, 1987.
17. Vesselinovitch SD, Rao KVN, Mihailovich N, Rice JM, Lombard LS. Development of broad spectrum of tumors by ethylnitrosourea in mice: Modifying role of age, sex, and strain. *Cancer Res* 34:2530–2538, 1974.
18. Vesselinovitch SD, Mihailovich N. The effect of gonadectomy on the development of hepatomas induced by urethan. *Cancer Res* 27:1788–1791, 1967.
19. Vesselinovitch SD, Itze L, Mihailovich N, Rao KVN. Modifying role of partial hepatectomy and gonadectomy in ethylnitrosourea-induced hepatocarcinogenesis. *Cancer Res* 40:1538–1542, 1980.

20. Vesselinovitch SD, Mihailovich N. Modifying effect of the sex hormonal environment on the growth of basophilic foci and hepatocellular carcinomas. In Davis W, Maltoni C, Tanneberger ST (eds): *The Control of Tumor Growth and Its Biological Bases*. Berlin: Akademie-Verlag, 1983, pp 282–286.
21. Vesselinovitch SD, Mihailovich N, Rao KVN, Goldfarb S. Relevance of basophilic foci to promoting effect of sex hormones on hepatocarcinogenesis. In Hecker E, Rusenig N, Marks F, Kunz E (eds): *Carcinogenesis*, Vol 7. New York: Raven Press, 1982, pp 127–131.
22. Vesselinovitch SD, Mihailovich N. The neonatal and infant age periods as biologic factors which modify multicarcinogenesis by urethan. *Cancer Res* 27:1422–1429, 1967.
23. Vesselinovitch SD, Mihailovich N, Pietra G. The prenatal exposure of mice to urethan and the consequent development of tumors in various tissues. *Cancer Res* 27:2333–2337, 1967.
24. Vesselinovitch SD, Mihailovich N, Rao KVN, Itze L. Perinatal carcinogenesis by urethan. *Cancer Res* 31:2143–2147, 1971.
25. Vesselinovitch SD, Rao KVN, Mihailovich N. Neoplastic response of mouse tissues during perinatal age periods and its significance in chemical carcinogenesis. *Natl Cancer Inst Monogr* 51:239–250, 1979.
26. Vesselinovitch SD, Koka M, Rao KVN, Mihailovich N, Rice JM. Prenatal multicarcinogenesis by ethylnitrosourea in mice. *Cancer Res* 37:1822–1828, 1977.
27. Itze L, Vesselinovitch SD, Rao KVN. Inter and intra diurnal variations. DNA, RNA, and protein synthetic activity in newborn, infant and young adult mouse livers. *Physiol Bohem* 25:289–293, 1976.
28. Vesselinovitch SD. Factors modulating response to carcinogeneic mutagens. In Alacevic M (ed): *Progress in Environmental Mutagenesis*. New York, Amsterdam: Elsevier/North Holland Biomedical Press, 1980, pp 281–296.
29. Vesselinovitch SD, Koka M, Mihailovich N, Rao KVN. Carcinogenicity of diethylnitrosamine in newborn, infant and adult mice. *J Cancer Res Clin Oncol* 108:60–65, 1984.
30. Rao KVN, Vesselinovitch SD. Age- and sex-associated diethylnitrosamine dealkylation activity of the mouse liver and hepatocarcinogenesis. *Cancer Res* 33:1625–1627, 1973.
31. Toh YC. Effect of neonatal castration on liver tumor induction by N-2-fluorenylacetamide in suckling BALB/c mice carcinogenesis. *Carcinogenesis* 2:1219–1221, 1981.
32. Vesselinovitch SD, Mihailovich N, Rao KVN. Morphology and metastatic nature of induced hepatic nodular lesions in C57BL × C3H F_1 mice. *Cancer Res* 38:2003–2010, 1978.
33. Fulmann RL. Measurement of particle sizes on opaque bodies. *Trans AIME* 197: 447–452, 1953.
34. Campbell HA, Pitot HC, Potter VR, Laishes BA. Application of quantitative stereology to the evaluation of enzyme-altered foci in rat liver. *Cancer Res* 42:465–472, 1982.
35. Moore MR, Drinkwater NR, Miller EC, Miller JA, Pitot HC. Quantitative analysis of the time-dependent development of glucose-6-phosphatase-deficient foci in the livers of mice treated neonatally with diethylnitrosamine. *Cancer Res* 41:1585–1593, 1981.
36. Vesselinovitch SD, Mihailovich N. Kinetics of diethylnitrosamine hepatocarcinogenesis in the infant mouse. *Cancer Res* 43:4253–4259, 1983.

37. Vesselinovitch SD, Mihailovich N, Negri S. Sex hormones as modulators of liver tumor development. In Roberfroid MB, Preat V (eds): *Experimental Hepatocarcinogenesis*. New York: Plenum Publishing, 1988, pp 51–62.
38. Higginson J. The epidemiology of primary carcinoma of the liver. In Pock GT, Islami AH (eds): *Tumors of the Liver*. Berlin: Springer-Verlag, 1970, pp 38–52.
39. Higginson J. Geographical pathology of primary liver cancer. *Cancer Res* 23:1624–1633, 1963.
40. Toh YD. Physiological and biochemical review of sex differences and carcinogenesis with particular references to the liver. *Adv Cancer Res* 18:155–209, 1973.
41. Vesselinovitch SD, Hacker HF, Bannasch P. Histochemical characterization of focal hepatic lesions induced by single diethylnitrosamine treatment in infant mice. *Cancer Res* 44:2774–2780, 1985.

Mouse Liver Carcinogenesis: Mechanisms
and Species Comparisons, pages 69–83
Published 1990 by Alan R. Liss, Inc.

Strain-Dependent Effects of Phenobarbital on Liver Tumor Promotion in Inbred Mice[1]

Bhalchandra A. Diwan, Jerry M. Rice,† and Jerrold M. Ward**
**Biological Carcinogenesis and Development Program, Program Resources, Inc., and †Laboratory of Comparative Carcinogenesis, National Cancer Institute–Frederick Cancer Research Facility, Frederick, MD 21701*

Evidence that chemical carcinogenesis is a two-stage process came from the pioneering work of Berenblum (1) and Rous and Kidd (2), mainly on mouse skin. Since then the two-stage process of carcinogenesis, initiation-promotion, has been experimentally demonstrated in various tissues and organs, including the liver, thyroid, kidney, and urinary bladder (3–9). Peraino and his coworkers (3) presented the first definitive evidence for two-stage liver carcinogenesis in rats—initiation by short-term feeding of a carcinogen, 2-acetylaminofluorene, followed by long-term oral dosing with phenobarbital (PB), which promotes tumorigenesis in liver parenchyma. This basic observation was later confirmed by several other investigators using different experimental protocols (10–12).

The phenomenon of two-stage carcinogenesis has also been demonstrated in mouse liver (13–15). PB has been shown to enhance both the development of naturally occurring hepatocarcinogenesis in genetically susceptible strains of mice such as C3H (16) and hepatocarcinogenesis initiated by chemical carcinogens in this and other strains of mice (14–16). In our laboratory, consistent enhancement of hepatocarcinogenesis was observed only when PB was given to mature mice after *N*-nitrosodiethylamine (DEN) initiation (13,17).

Several studies have shown that mouse strains differ significantly in their susceptibility to hepatocarcinogenesis induced by DEN alone (18–20). For example, RF mice were found to be susceptible to the development of hepatocellular tumors after oral administration of DEN, while no such tumors developed in BALB/c mice after similar exposure (19). In an earlier study (20), we reported that the hepatocarcinogenic effect of a single dose of DEN varied greatly with the strain of mouse used (Table 1). No hepatocellular tumors were found in AKR/J and SWR/J mice, which developed a high incidence of leukemia and/or lung tumors. On the other hand, 5/10 male and 1/10 female mice of the C57BL/6J strain developed hepatocellular adenomas. Thus, this strain of mouse, known for its resistance to spontaneous liver tumor development, is at least moderately susceptible to hepatocarcinogenesis by DEN.

Table 1. Hepatocellular Tumors* in DEN-Treated Mice of Three Inbred Strains†

	Total no. of mice		Tumors/mouse	
Strain	Males	Females	Males	Females
AKR/J	12	13	0	0
SWR/J	12	14	0	0
C57BL/6J	10	11	5(1.2)	2(1.0)

*All hepatocellular tumors were adenomas.
†Mice, 8 weeks old, were given a single intraperitoneal injection of 90 mg DEN/kg.
Modified from Diwan et al. *Cancer Lett* 1:249–253, 1976.

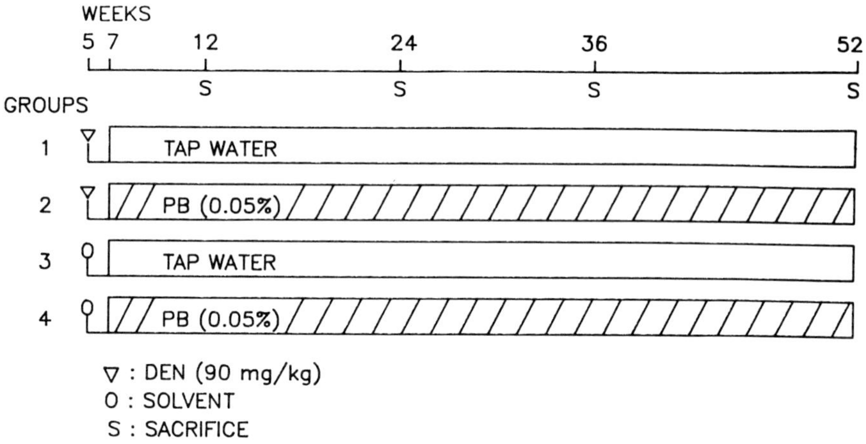

Figure 1. Schematic representation of experimental design.

Interstrain Differences in Susceptibility to Two-Stage Hepatocarcinogenesis

Until recently, no information was available on interstrain differences in sensitivity of the liver to two-stage carcinogenesis. In a recent study, we investigated the relative sensitivities of a number of strains of mice to hepatocarcinogenesis initiated by DEN and promoted by phenobarbital (21). Figure 1 shows our experimental design. Five-week-old male mice of strains C57BL/6NCr (C57), C3H/HeNCr^{MTV-} (C3H), DBA/2NCr (DBA), and NSF/NCr (NSF) were given a single intraperitoneal injection of 90 mg DEN/kg (groups 1 and 2) or the solvent alone (groups 3 and 4). Beginning 2 weeks later, mice in groups 2 and 4 were given drinking water containing 0.05% PB. Ten mice from each treatment group were killed at 12, 24, 36, and 52 weeks of age (5, 17, 29, and 45 weeks' exposure

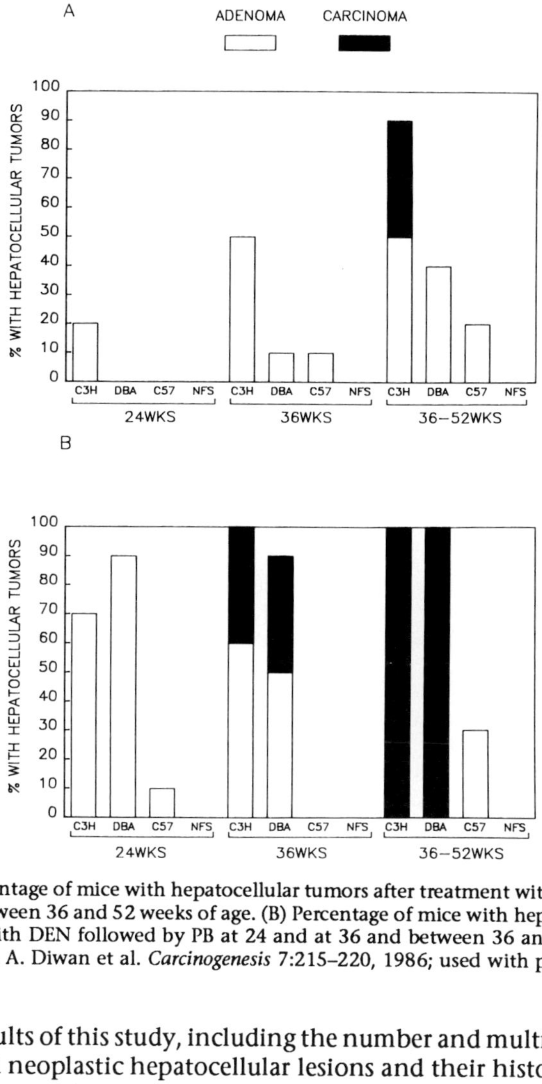

Figure 2. (A) Percentage of mice with hepatocellular tumors after treatment with DEN alone, at 24 and at 36 and between 36 and 52 weeks of age. (B) Percentage of mice with hepatocellular tumors after treatment with DEN followed by PB at 24 and at 36 and between 36 and 52 weeks of age. Reprinted from B. A. Diwan et al. *Carcinogenesis* 7:215–220, 1986; used with permission.

to PB). The results of this study, including the number and multiplicities of preneoplastic and neoplastic hepatocellular lesions and their histological characteristics have been published (21) and are briefly summarized here. The incidences of tumors in different strains of mice at different time points after treatment with DEN alone or DEN followed by PB are shown in Figure 2A and B, respectively. PB significantly increased the incidence of hepatocellular tumors after 17 weeks of treatment in 24-week-old DEN-initiated C3H (20% vs. 70%) and DBA mice (0 vs. 90%). At 36 weeks of age, the incidence of liver cell tumors in mice given DEN but not PB was 10% for DBA, 10% for C57, and 50% for C3H mice; the incidence was increased by PB to 90% in DBA and 100% in C3H mice, but there was no increase for C57 mice. Even at 52 weeks, the low incidence of hepatocellular tumors in C57 mice given DEN only (20%) was not significantly increased by subsequent exposure to PB (Fig. 2).

Although both DBA and C3H mice were susceptible, the pattern of susceptibility was different in these two strains. DBA mice were sensitive to DEN with promotion but relatively insensitive in the absence of promotion. For example, by 36 weeks, 90% of DBA mice given PB following DEN initiation had developed hepatocellular adenomas and carcinomas, whereas only one animal exposed to DEN alone had a hepatocellular adenoma, and that mouse had only one adenoma. C3H mice were susceptible to DEN either with or without promotion with PB. Thus, the incidence of hepatocellular tumors in DEN-treated C3H mice was 20% at 24 weeks, 50% at 36 weeks, and 90% at 52 weeks (Fig. 2). Although subsequent administration of PB significantly enhanced this incidence in C3H mice, the increase was not as dramatic as in DBA mice.

Our results clearly showed that inbred strains of mice differ in susceptibility to two-stage liver carcinogenesis. DBA and C3H mice were susceptible, while C57 mice were resistant; NSF mice were totally refractory to two-stage liver carcinogenesis.

Differences in Susceptibility to Promotion as a Basis for Differences in Susceptibility to Hepatocarcinogenesis

It is clear from our results that there were no significant differences in tumor incidence between C57 and DBA mice that received DEN alone (Fig. 2A). Thus, the significant differences in two-stage liver carcinogenesis between these two strains of mice appear to be due to differences in the promotion phase of carcinogenesis. Our earlier study also showed that C57BL/6J mice were moderately susceptible to hepatocarcinogenesis induced by DEN (20). Drinkwater and Ginsler (22) recently showed that differences in susceptibility to liver tumor induction by DEN between strains C57BL/6J and C3H/HeJ are consistent with allelic differences at two genetic loci, one of which accounts for about 90% of the difference in responsiveness. Their results further suggested that the major locus, designated as *Hcs*, controlling susceptibility to hepatocarcinogenesis in C3H/HeJ male mice, affects the promotion phase of carcinogenesis. The level of initiation appeared to be the same in the two strains. No differences were observed in the levels of ethylation of hepatic DNA or proportions of ethylated bases, i.e., *N*7-ethylguanine or *O*6-ethylguanine, between C3H/HeJ and C57BL mice following administration of [^{14}C]DEN. They recently showed (23) that the greater susceptibility of C3H/HeJ male mice to hepatocarcinogenesis relative to C57BL/6J mice resulted from a strain-dependent difference in the rate of growth of preneoplastic lesions. Similar results were recently reported by Della Porta et al. (24), who quantified the development of hepatocellular nodules in the livers of DEN-treated B6C3F$_1$ and C57BL/6 × BALB/c (B6CF$_1$) mice. The different susceptibility to hepatocarcinogenesis in B6C3F$_1$ and B6CF$_1$ mice appeared to be

Table 2. Incidence of Preneoplastic and Neoplastic Hepatic Lesions in $D2B6F_1$ and Reciprocal F_1 Hybrid Mice by Treatment Group at 33 Weeks of Age

Mice	Treatment groups (10 mice/ group)	THL*/cm² (mean ± SD)	Mean volume of THL (mm³) (mean ± SE)	Hepatocellular adenomas		Mice with hepato- blastoma
				Incidence†	Multiplicity‡	
$B6D2F_1$	1 (DEN)	0.38 ± 0.3	0.3 ± 0.26	3/10	2.0 ± 1.0	0
	2 (DEN,PB)	3.6 ± 0.7§	0.94 ± 0.06	10/10	5.2 ± 0.96§	0
$D2B6F_1$	1 (DEN)	0.5 ± 0.1	0.1 ± 0.03	2/10	1.0 ± 0.0	0
	2 (DEN,PB)	3.7 ± 0.4§	0.95 ± 0.26	9/10	3.8 ± 1.0§	1

*THL indicates total hepatic lesions including hepatocellular foci, adenomas, and hepatoblastomas.
†Mice with one or more hepatocellular adenomas/mice at risk.
‡Total hepatocellular adenomas/number of mice with adenomas.
§Statistically significant compared with group 1 ($P < 0.05$).

related to a higher susceptibility of the $B6C3F_1$ initiated liver cells to growth stimulation.

Recently, Becker (25) also reported that a higher proportion of hepatocellular tumors, both spontaneous and chemically induced, progressed to a more malignant phenotype in $B6C3F_1$ mice than in C57BL/6J mice. Thus, our own findings and those of others (23–25) support the argument that interstrain differences in susceptibility to hepatocarcinogenesis are probably due to differences in the postinitiation phase of carcinogenesis.

Genetic Basis for Susceptibility to Liver Tumor Promotion by PB

To analyze the genetics of susceptibility to tumor promotion by PB, the reciprocal F_1 hybrids between C57 and DBA mice were given DEN and PB using the protocol described above (Fig. 1). Thus, 5-week-old male mice of $B6D2F_1$ and $D2B6F_1$ crosses were given 90 mg DEN/kg intraperitoneally, followed by 0.05% PB in drinking water. Ten mice from each group were examined at 33 weeks of age (28 weeks after DEN injection), and the rest were killed when moribund or at 47 weeks (42 weeks after DEN injection).

At 33 weeks, $D2B6F_1$ mice given DEN followed by PB had significantly more ($3.7 \pm 0.4/cm^2$) and larger ($1.5 \pm 0.6/mm^3$) hepatocellular lesions than mice given DEN alone and 9/10 had hepatocellular adenomas, an average of 3.8 ± 1.0 tumors/mouse (Table 2). Administration of PB also significantly increased the incidence (30% vs. 100%, $P < 0.05$) and the number of tumors (5.2 ± 0.9 per mouse) initiated by DEN in the reciprocal F_1 cross $B6D2F_1$. By 47 weeks (Table 3), all mice of reciprocal F_1 hybrids $D2B6F_1$ and $B6D2F_1$ that received PB follow-

Table 3. Incidence of Preneoplastic and Neoplastic Hepatic Lesions in D2B6F_1 and Reciprocal F_1 Hybrid Offspring Observed between 33 and 47 Weeks of Age

Mice	Treatment groups (10 mice/ group)	No. of mice with hepatocellular tumors	Hepatocellular tumors per mouse with tumors		No. of mice with hepato-blastomas
			Adenomas (mean ± SE)	Carcinomas (mean ± SE)	
B6D2F_1	1 (DEN)	6	2.2 ± 0.53	0	0
	2 (DEN,PB)	10	15.0 ± 1.36*	2.7 ± 0.41	1
D2B6F_1	1 (DEN)	7	2.7 ± 0.22	1.0 ± 0.0	0
	2 (DEN,PB)	10	12.5 ± 2.55*	2.6 ± 0.3	8 (1.75 ± 0.37/ mouse)

*Statistically significant compared with group 1 ($P < 0.01$).

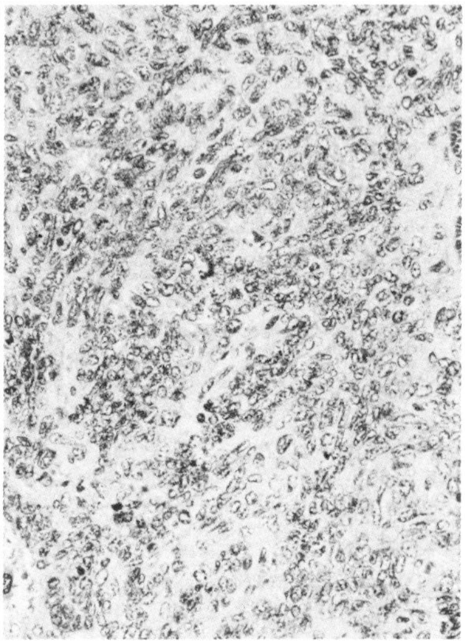

Figure 3. Hepatoblastoma in D2B6F_1 male mouse after DEN initiation and PB promotion. Undifferentiated cells and rosette-like structures can be seen. Hematoxylin and eosin, ×250.

ing DEN administration had multiple hepatocellular tumors, both adenomas (12.5 ± 2.6 and 15 ± 1.36/mouse, respectively) and carcinomas (1.6 ± 0.3 and 2.7 ± 0.46/mouse, respectively). It is clear from these results that susceptibility to promotion of hepatocarcinogenesis by PB was a dominant trait in crosses between DBA and C57 mice, and both reciprocal F_1 hybrids responded similarly to PB promotion.

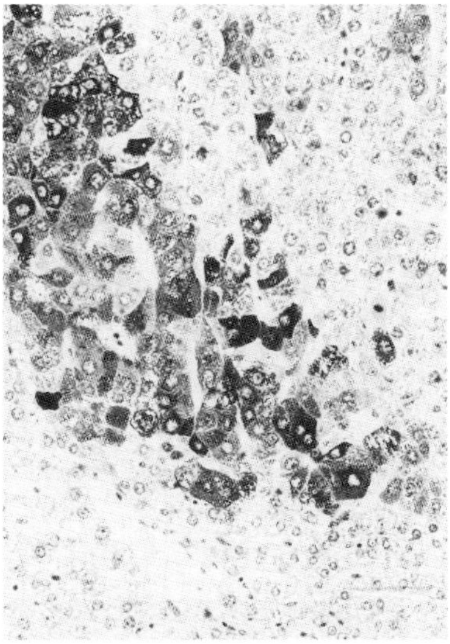

Figure 4. Alpha-fetoprotein in hepatocytes in hepatocellular adenoma. ABC immunohistochemistry, hematoxylin, ×250.

The most interesting finding was the occurrence of an unexpected high incidence of hepatoblastomas in $D2B6F_1$ mice that had been initiated with DEN and promoted with PB. These distinctive and highly malignant tumors were mainly found within or adjacent to hepatocellular carcinomas or adenomas. They were intensely basophilic when stained with hematoxylin and eosin (Fig. 3) and were similar histologically to those described earlier (26–29). The tumor cells were arranged radially or concentrically around branches of the hepatic vein, although in some cases they formed columns, rosettes, or rows. In some animals, there were multicentric masses of hepatoblastomas, many of them hemorrhagic and some cystic in appearance. Alpha-fetoprotein was consistently absent in hepatoblastomas, while focal areas in most of these tumors and the surrounding normal bile duct epithelial cells and hyperplastic oval cells were positive for keratin by avidin-biotin complex (ABC) immunohistochemistry (Figs. 4,5). Some of these tumors (4/22, total from two experiments) metastasized to the lung (Fig. 6). The incidence of hepatoblastomas was low at 33 weeks (10%) but increased to 80% between 33 and 47 weeks of age. These results showed for the first time that PB, a well-known promoter of hepatocellular carcinogenesis in rodents, also promotes the development of hepatoblastomas in DEN-initiated mice. The histogenesis of hepatoblastomas is not yet established, and it is not clear whether the effect of PB is to stimulate evolution of preexisting hepatocellular neoplasms into hepatoblastomas as a third stage in a possible adenoma-carcinoma-hepatoblastoma sequence or represents promotion of a previously unrecognized target cell population related to oval or bile duct cells.

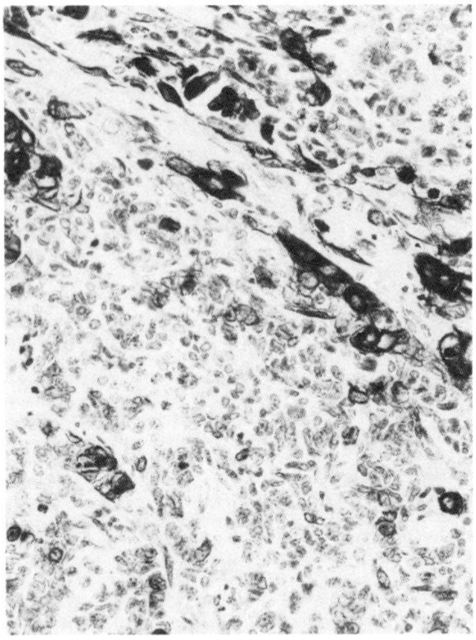

Figure 5. Keratin-immunoreactive hepatoblastoma cells in D2B6F$_1$ mouse. ABC immunohistochemisty, hematoxylin, ×250.

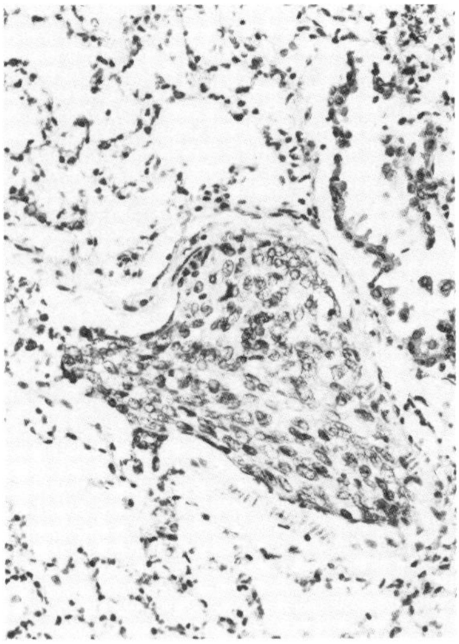

Figure 6. Pulmonary metastasis of hepatoblastoma in D2B6F$_1$ mouse. Hematoxylin and eosin, ×250.

In contrast to the high incidence of hepatoblastomas in D2B6F$_1$ males, only one reciprocal F$_1$ hybrid (B6D2F$_1$) mouse developed such a tumor (Table 3). To investigate the possible influence of sex on the development of hepatoblastomas, in a separate experiment, we gave both male and female D2B6F$_1$ mice DEN and PB under identical protocols. PB promoted the development of hepatocellular tumors initiated by DEN in female mice (Table 4). PB, as in the previous study, significantly enhanced the development of hepatoblastomas in D2B6F$_1$ males but failed to enhance such tumors in D2B6F$_1$ female mice. Furthermore, in this experiment, too, hepatoblastomas were not seen in either male or female hybrids given DEN alone. These results suggest a sex-linked differential response to the development of hepatoblastomas in reciprocal F$_1$ hybrid mice that are genetically identical except for the reverse origin of their X and Y chromosomes.

Biochemical and Pharmacokinetic Factors in Interstrain Differences in Susceptibility to PB Promotion

How PB promotes liver carcinogenesis is not known. PB belongs to a group of liver tumor promoters that cause little if any hepatocyte injury or loss at dose levels that are effective for promotion. According to Schulte-Hermann (30,31),

Table 4. Incidence of Neoplastic Hepatic Lesions in Male and Female Mice of $D2B6F_1$ Cross Given a Single Dose of DEN and Subsequently Fed PB for 53 Weeks

Treatment groups	No. of mice	Mice with hepatocellular tumors (%)	Hepatocellular tumors per mouse with tumors: Adenomas (mean ± SE)	Hepatocellular tumors per mouse with tumors: Carcinomas (mean ± SE)	Hepatoblastomas/mouse with hepatoblastomas (mean ± SE)
Females					
DEN	13	5 (39)	1.5 ± 0.23	0	0
DEN,PB	13	10 (77)	5.1 ± 0.3*	0	0
PB	13	0	0	0	0
Males					
DEN	20	12 (60)	3.0 ± 0.4	1 ± 0	0
DEN,PB	20	20 (100)	9.4 ± 1.0*	2.4 ± 0.4	13 (2.2 ± 0.4)*
PB	20	4 (20)	1 ± 0	0	0

*Statistically significant compared with group exposed to DEN alone ($P < 0.01$).

the common denominator of many nontoxic liver tumor promoters is their ability to induce adaptive responses in normal hepatocytes, i.e., stimulation of organ growth and increases in the drug-metabolizing enzymes in normal liver. We have found a good correlation between activities of different barbiturates (32,33) and hydantoin derivatives as promoters of hepatocarcinogenesis and their abilities to induce hepatic cytochrome P450 IIB1–mediated pentoxy- and benzyloxyresorufin *O*-dealkylase activities. To further elucidate the relationship between cytochrome P450 induction and tumor-promoting activity, we investigated the ability of PB to induce these hepatic cytochrome P450-dependent *O*-dealkylase activities in PB-sensitive DBA and PB-resistant C57 mice. It is clear from Table 5 that PB induced hepatic cytochrome P450 and also increased liver weights in mice of both these strains. Thus, PB-induced adaptive responses in normal hepatocytes were similar in both strains. However, it is possible that the adaptive responses to PB in the hepatocytes from preneoplastic lesions may be different in different strains of mice; such possibilities are being considered in our laboratory.

PB is metabolized in the liver, mostly to 5-ethyl-5(4-hydroxyphenyl) barbituric acid (34). This compound lacks anticonvulsant activity and is rapidly removed from the blood by metabolic conjugation and renal filtration (35). Thus, this appears to be purely an elimination pathway rather than a mechanism for activation of promoting activity. Also, there is no unequivocal evidence in the literature to suggest that PB requires metabolic activation for any pharmacologic activity, including tumor promotion. As a working hypothesis it is reasonable to assume that PB itself, rather than any metabolite, is responsible for the promotion of hepatocellular neoplasms in rodents to whom this drug is given for prolonged periods.

Table 5. Effects of Phenobarbital Administration in the Diet on Cytochrome P-450 Activities in the Mouse*

Treatment	Mean liver weight (g)	*O*-Dealkylation of phenoxazone ethers† Ethoxy-	Pentoxy-	Benzyloxy-
C57BL/6				
Control	0.92 ± 0.09	10.7 ± 1.6	20.2 ± 1.6	36.4 ± 2.5
Phenobarbital	2.09 ± 0.41‡	12.4 ± 2.0	164 ± 40‡	369 ± 139‡
DBA/2				
Control	0.93 ± 0.07	9.3 ± 2.2	22.3 ± 3.5	30.7 ± 5.1
Phenobarbital	1.23 ± 0.13‡	15.3 ± 1.8	171 ± 33‡	309 ± 105‡

*Mice received control diet or diet containing 500 ppm PB for 15 days. Enzyme assays were done in duplicate; values given are mean ± standard deviation for five male mice per treatment.
†Values are in units of picomoles resorufin formed/min/mg S-9 protein.
‡Significantly different from appropriate control group, $P \leq 0.05$.

Table 6. Serum PB Levels in Different Strains of Mice at 12, 24, and 36 Weeks of Age*

Strain	Serum PB levels (μg/ml) 12 weeks	24 weeks	36 weeks
C3H/HeNCr^{MTV-}	9.6 ± 0.4	8.5 ± 0.12	8.3 ± 0.6
DBA/2NCr	28.7 ± 0.9†	29.0 ± 0.17†	27.3 ± 0.9†
C57BL/6NCr	9.4 ± 1.35	6.8 ± 1.5	5.3 ± 0.9

*All mice were from group 4, originally receiving tricaprylin but no carcinogen at 7 weeks of age and then given PB in drinking water thereafter.
†Statistically significant compared to values for C57BL/6NCr or C3H/HeNCr^{MTV-} ($P < 0.05$).
Reprinted with permission from Diwan et al. *Carcinogenesis* 7:215–220, 1986.

It is, therefore, possible that differences in the rates at which PB is detoxified and/or eliminated could account for the observed strain differences in tumor promotion by PB. Our results (Table 6) on serum PB levels in different strains of mice at different time points suggest that mouse strains used in this study differed significantly in their abilities to metabolize, detoxify, and eliminate PB. Serum PB levels in DBA mice were consistently higher than in either C3H or C57 mice. In a separate study, male mice of strains C3H, DBA, and C57 were given PB continuously in drinking water beginning at 7 weeks of age and until they were killed at different times—for a PB administration period of up to 20 days (Fig. 7). Maximum PB levels in blood were observed in mice of all strains at 24 h. PB levels thereafter decreased rapidly; the lowest levels were

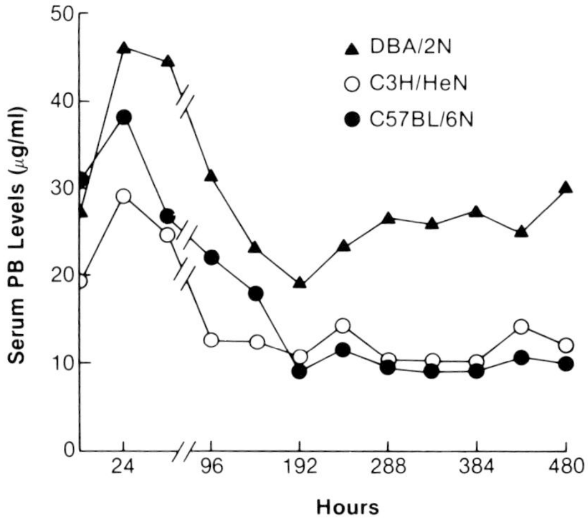

Figure 7. Comparisons of serum PB levels in different strains of mice. Male mice were given PB (0.05%) continuously in drinking water beginning at 7 weeks of age and continuing until they were killed at different times during this 20-day study. Each point is the mean of six samples. Reprinted from B. A. Diwan et al. *Carcinogenesis* 7:215–220, 1986; used with permission.

Table 7. Urinary Excretion of 5-Ethyl-5(4-hydroxyphenyl) Barbituric Acid (4-OH.PB) and PB in Mice

Strain	Ratio of 4-OH.PB to PB (μg/ml) (mean ± SD)
C57BL/6N	2.5 ± 0.4[†]
DBA/2N	1.04 ± 0.2

*Five mice of each strain were given 0.05% PB in drinking water for 20 days. Urine was collected on days 18–20.
[†]$P < 0.01$ compared with the value for DBA/2N mice.

observed after 192 h of PB administration. Note that in DBA mice, serum PB levels were consistently the highest, gradually increasing between 192 and 480 h of treatment to 30 mg/ml—three times that seen in C3H or C57 mice.

DBA mice appear to be unable to metabolize PB or other barbiturates and sleep longer than C57 mice given this compound or other barbiturates (36). C57 mice effectively metabolize and detoxify these compounds (37). We measured the levels of PB and 4-OH.PB in the urine of C57 and DBA mice that were given 0.05% PB in drinking water for 20 days (Table 7). The ratio of 4-OH.PB to PB was significantly higher in C57 mice than in DBA mice. Thus, to some extent, strain differences in the promoting effects of PB are correlated with differences in capacity to metabolize and detoxify this drug.

However, detoxification and elimination of PB do not appear solely responsible for the observed strain differences in PB promotion since C3H mice, which we have found maintain blood levels of PB comparable to those of C57 mice, are relatively susceptible to liver tumor promotion by this agent. If interstrain differences in the promoting effects of PB result from differences in the metabolism/detoxification of this drug, then use of promoting agents that are not metabolized might abolish the differential response between the susceptible and resistant strains. Barbital (BB) and 1,4-*bis*-[2-(3,5-dichloropyridyloxy)]benzene (TCPOBOP) are such agents. BB, a liver tumor promoter in rats (38,39) is not metabolized by the liver and is excreted unchanged by the kidney (40), while TCPOBOP, a liver tumor promoter in mice (41), has a long half-life and is stored in the fat primarily as the parent compound (42). These two compounds appear to be excellent agents for exploring further the role of metabolism/detoxification in liver tumor promotion in mice; studies using these compounds are in progress in our laboratory.

Summary and Conclusions

Inbred strains of mice were found to differ significantly in their susceptibility to liver tumor promotion by PB. The susceptibility to promotion of hepatocarcinogenesis by this drug was a dominant trait in crosses between the sensitive DBA and resistant C57 mice. The reciprocal F_1 hybrids responded similarly to tumor promotion by PB. PB promoted the development of hepatoblastomas in $D2B6F_1$ males but not in $B6D2F_1$ mice. Female mice of the $D2B6F_1$ cross, similarly exposed to PB, failed to develop hepatoblastomas. These results suggested a sex-linked differential response to the development of hepatoblastomas in reciprocal F_1 hybrids that are genetically identical except for the reverse origin of their X and Y chromosomes. Differences in the promoting effects of PB between C57 and DBA mice appeared to correlate with differences in the metabolism/detoxification of this drug.

In conclusion, mouse strains that differ in susceptibility to two-stage liver carcinogenesis provide excellent opportunities to investigate the genetic and/or biochemical mechanisms responsible for these differences. Understanding these mechanisms may lead to the identification of factors that affect liver tumor development not only in mice but in other species as well.

Acknowledgments

This project has been funded at least in part with Federal funds from the Department of Health and Human Services under contract number N01-CO-74102. The content of this publication does not necessarily reflect the views or policies of the Department of Health and Human Services, nor does mention of trade names, commercial products, or organizations imply endorsement by the

U.S. Government. We also would like to acknowledge the skilled technical assistance of Debbie Devor and John Henneman.

References

1. Berenblum I. The cocarcinogenic action of croton oil. *Cancer Res* 1:44–48, 1941.
2. Rous P, Kidd JG. Conditional neoplasms and subthreshold neoplastic states: A study of the tar tumors of rabbits. *J Exp Med* 73:365–390, 1941.
3. Peraino C, Fry RJM, Staffeldt E. Reduction and enhancement by phenobarbital of hepatocarcinogenesis induced in the rat by 2-acetylaminofluorene. *Cancer Res* 31: 1506–1512, 1971.
4. Peraino C, Richards WL, Stevens FJ. Multistage carcinogenesis. In Slaga TJ (ed): *Mechanisms of Tumor Promotion*, Vol. 1. *Tumor Promotion in Internal Organs*. Boca Raton, FL: CRC Press, 1983, pp 1–53.
5. Hiasa Y, Kitahori Y, Ohshima M, Fujita Y, Yuasa T, Konishi N, Miyashiro A. Promoting effects of phenobarbital and barbital on the development of thyroid tumors in rats treated with N-bis(2-hydroxypropyl)nitrosamine. *Carcinogenesis* 3:1187–1190, 1982.
6. Diwan BA, Palmer AE, Ohshima M, Rice JM. N-Nitroso-N-methylurea initiation in multiple tissues for organ-specific tumor promotion in rats by phenobarbital. *JNCI* 75:1099–1105, 1985.
7. Diwan BA, Ohshima M, Rice JM. Promotion by sodium barbital of renal cortical and transitional cell tumors, but not intestinal tumors, in F344 rats given methyl-(acetoxymethyl)nitrosamine, and lack of effect of phenobarbital, amobarbital, or barbituric acid on development of either renal or intestinal tumors. *Carcinogenesis* 10:183–188, 1989.
8. Shirai T, Ohshima M, Masuda A, Tamano S, Ito N. Promotion of 2-(ethylnitrosamino)ethanol-induced renal carcinogenesis in rats by nephrotoxic compounds: Positive responses with folic acid, basic lead acetate, and N-(3,5-dichlorophenyl)succinimide but not with 2,3-dibromo-1-propanol phosphate. *JNCI* 72:477–482, 1984.
9. Cohen SM, Murasaki G, Ellwein LB, Greenfield RE. Tumor promotion in bladder carcinogenesis. In Slaga TJ (ed): *Mechanisms of Tumor Promotion*, Vol. 1. *Tumor Promotion in Internal Organs*. Boca Raton, FL: CRC Press, 1983, pp 131–149.
10. Pitot HC, Sirica AE. The stages of initiation and promotion of hepatocarcinogenesis. *Biochim Biophys Acta* 605:191–215, 1980.
11. Emmelot P, Scherer E. The first relevant cell stage in rat liver carcinogenesis: A quantitative approach. *Biochim Biophys Acta* 605:247–304, 1980.
12. Ito N, Tatematsu M, Nakanishi K, Hasegawa R, Takano R, Imaida IC, Ogiso T. The effects of various chemicals on the development of hyperplastic liver nodules in hepatectomized rats treated with N-nitrosodiethylamine or N-2-fluorenylacetamide. *Gann* 71:832–842, 1980.
13. Ward JM, Rice JM, Creasia P, Lynch P, Riggs C. Dissimilar patterns of promotion by di(2-ethylhexyl)phthalate and phenobarbital of hepatocellular neoplasia initiated by diethylnitrosamine in $B6C3F_1$ mice. *Carcinogenesis* 4:1020–1029, 1983.
14. Dragani TA, Manenti G, Galliani G, Della Porta G. Promoting effects of 1,4-bis[2-(3,5-dichloropyridyloxy)]benzene in mouse hepatocarcinogenesis. *Carcinogenesis* 6:225–228, 1985.

15. Uchida E, Hirono I. Effect of phenobarbital on induction of liver and lung tumors by dimethylnitrosamine in newborn mice. *Gann* 70:639–640, 1979.
16. Peraino C, Fry RJ, Staffeldt E. Enhancement of spontaneous hepatic tumorigenesis in C3H mice by dietary phenobarbital. *JNCI* 51:1349–1950, 1973.
17. Diwan BA, Rice JM, Ward JM, Ohshima M, Lynch PH. Inhibition by phenobarbital and lack of effect of amobarbital on the development of liver tumors induced by N-nitrosodiethylamine in juvenile B6C3F$_1$ mice. *Cancer Lett* 23:223–234, 1984.
18. Takayama S, Oota K. Induction of malignant tumors in various strains of mice by oral administration of N-nitrosodimethylamine and N-nitrosodiethylamine. *Gann* 56:189–199, 1965.
19. Clapp NK, Tyndall RL, Otter JA. Differences in tumor types and organ susceptibility in BALB/cJ and RF mice following dimethylnitrosamine and diethylnitrosamine. *Cancer Res* 31:196–198, 1971.
20. Diwan BA, Meier M. Carcinogenic effects of a single dose of diethylnitrosamine in three unrelated strains of mice: Genetic dependence of induced tumor types and incidence. *Cancer Lett* 1:249–253, 1976.
21. Diwan BA, Rice JM, Ohshima M, Ward JM. Interstrain differences in susceptibility to liver carcinogenesis initiated by N-nitrosodiethylamine and its promotion by phenobarbital in C57BL/6NCr, C3H/HeNCr^{MTV-} and DBA/2NCr mice. *Carcinogenesis* 7:215–220, 1986.
22. Drinkwater NR, Ginsler JJ. Genetic control of hepatocarcinogenesis in C57BL/6J and C3H/HeJ inbred mice. *Carcinogenesis* 7:1701–1707, 1986.
23. Hanigan MH, Kemp CJ, Ginsler JJ, Drinkwater NR. Rapid growth of preneoplastic lesions in hepatocarcinogen-sensitive C3H/HeJ male mice relative to C57BL/6J male mice. *Carcinogenesis* 9:885–890, 1988.
24. Della Porta G, Dragani TA, Manenti G. Two-stage liver carcinogenesis in the mouse. *Toxicol Pathol* 15:229–233, 1987.
25. Becker FF. Morphological classification of liver tumors based on biological characteristics. *Cancer Res* 42:3918–3923, 1982.
26. Turusov VS, Deringer MK, Dunn TB, Stewart HL. Malignant mouse liver tumors resembling human hepatoblastomas. *JNCI* 51:1589–1595, 1973.
27. Turusov VS, Takayama S. Tumors of the liver. *IARC Sci Publ* 23:193–232, 1973.
28. Turusov VS, Day NE, Tomatis L, Gati E, Charles RT. Tumors in CF1 mice exposed for six consecutive generations to DDT. *JNCI* 51:983–997, 1973.
29. Nonoyama T, Reznik G, Bucci TJ, Fullerton F. Hepatoblastoma with squamous differentiation in a B6C3F$_1$ mouse. *Vet Pathol* 23:619–622, 1986.
30. Schulte-Hermann R, Timmermann-Trosiener I, Schuppler J. Promotion of spontaneous preneoplastic cells in rat liver as a possible explanation of tumor production by nonmutagenic compounds. *Cancer Res* 43:839–844, 1983.
31. Schulte-Hermann R, Timmermann-Trosiener I, Schuppler J. Facilitated expression of adaptive responses to phenobarbital in putative prestages of liver cancer. *Carcinogenesis* 7:1651–1655, 1986.
32. Nims RW, Devor DE, Henneman JR, Lubet RA. Induction of alkoxyresorufin O-dealkylase, epoxide hydrolase and liver weight gain: Correlation with liver tumor-promoting potential in a series of barbiturates. *Carcinogenesis* 8:67–71, 1987.
33. Diwan BA, Rice JM, Nims RW, Lubet RA, Hu H, Ward JM. P-450 enzyme induction by 5-ethyl-5-phenylhydantoin and 5,5-diethylhydantoin analogues of barbiturate tumor promoters phenobarbital and barbital, and promotion of liver and thyroid carcinogenesis initiated by N-nitrosodiethylamine in rats. *Cancer Res* 48: 2492–2497, 1988.

34. Butler TC. The metabolic hydroxylation of phenobarbital. *J Pharmacol Exp Ther* 116:326–336, 1956.
35. Craig CR, Schideman FE. Metabolism and anticonvulsant properties of mephobarbital and phenobarbital in rats. *J Pharmacol Exp Ther* 176:35–41, 1971.
36. Belknap JK, Waddingham S, Ondrusek G. Barbiturate dependence in mice induced by a simple short-term oral procedure. *Physiol Psychol* 1:394–396, 1973.
37. Nabeshita T, Ho IK. Pharmacological responses to phenobarbital in different strains of mice. *J Pharmacol Exp Ther* 216:198–204, 1981.
38. Mori H, Tanaka T, Nishikawa A, Takahashi M, Williams GM. Enhancing effect of barbital on N-2-fluorenylacetamide-induced rat liver lesions. *Gann* 72:798–801, 1981.
39. Diwan BA, Rice JM, Ohshima M, Ward JM, Dove LF. Comparative tumor-promoting activities of phenobarbital, amobarbital, barbital sodium, and barbituric acid on livers and other organs of male F344/NCr rats following initiation with N-nitrosodiethylamine. *JNCI* 74:509–516, 1985.
40. Conney AM, Burns JJ. Factors influencing drug metabolism. *Adv Pharmacol* 1:31–58, 1962.
41. Polland A, Mak I, Glover E, Boatman RJ, Ebetino F, Kende AS. 1,4-bis[2-(3,5-Dichloropyridyloxy)]benzene, a potent phenobarbital-like inducer of microsomal monooxygenase activity. *Mol Pharmacol* 18:571–580, 1980.
42. Polland A, Mak I, Glover E. Species differences in responsiveness to 1,4-bis[2-(3,5-dichloropyridyloxy)]benzene, a potent phenobarbital-like inducer of microsomal monooxygenase activity. *Mol Pharmacol* 20:442–450, 1981.

Mouse Liver Carcinogenesis: Mechanisms
and Species Comparisons, pages 85–108

Liver Tumor Promoters and Other Mouse Liver Carcinogens

Jerrold M. Ward, Bhalchandra A. Diwan,† Ronald A. Lubet,* John R. Henneman,† and Deborah E. Devor**
**Tumor Pathology and Pathogenesis Section, Laboratory of Comparative Carcinogenesis, Division of Cancer Etiology, National Cancer Institute, and †Chemical Carcinogenesis Section, Biological Carcinogenesis Development Program, Program Resources, Inc., National Cancer Institute–Frederick Cancer Research Facility, Frederick, MD 21701*

Background

The Mouse Liver Tumor as an End Point in Carcinogenicity Assays for the Safety Assessment of Chemicals

Over the past 20 years, the mouse liver has been the major target of chemicals tested for safety in routine carcinogenicity assays. In large-scale carcinogenesis testing programs, approximately 50% of the chemicals tested have been shown to increase the incidence of some tumor type in at least one assay, and 50% of these chemicals were carcinogenic in mouse liver (1–3). The potency (measured as TD_{50}) of these carcinogens varied greatly, from very potent [e.g., 87 ng/kg/day for 2,3,7,8-tetrachlorodibenzo-para-dioxin (TCDD)] to weak (e.g., 5,230 mg/kg/day for chloramben) (Fig. 1) (4). Although in the majority of these assays, other target organs of mice or rats are also involved, the mouse liver has been the sole target site for 20%–36% of the chemicals (2,3). The numerous mouse hepatocarcinogens are derived from many chemical classes and use categories including food additives, drugs, pesticides, industrial intermediates, and other environmental chemicals (1,2,5). At least eight human carcinogens also have induced liver tumors in mice, suggesting the validity of this target site for human risk assessment (6). These chemicals, however, usually also cause other tumors in mice and/or rats, as well.

Carcinogenicity and Mutagenicity of the Mouse Liver Carcinogens

The exact mechanisms of cell transformation and multistage carcinogenesis are not known. It has been postulated that carcinogens are mutagens

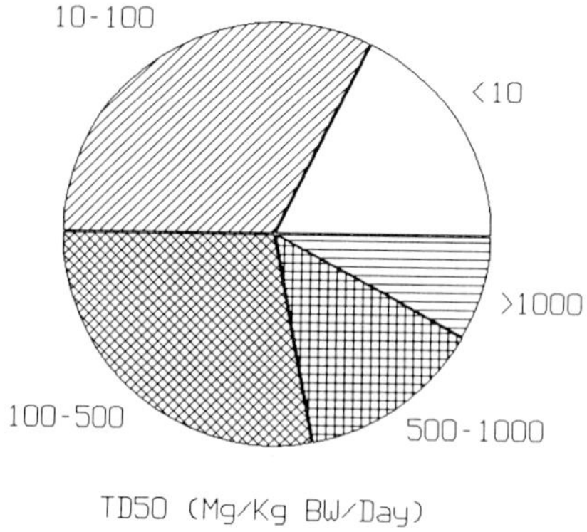

Figure 1. Carcinogenic potency (TD_{50}) of 50 mouse liver carcinogens. Adapted from Gold et al. (1).

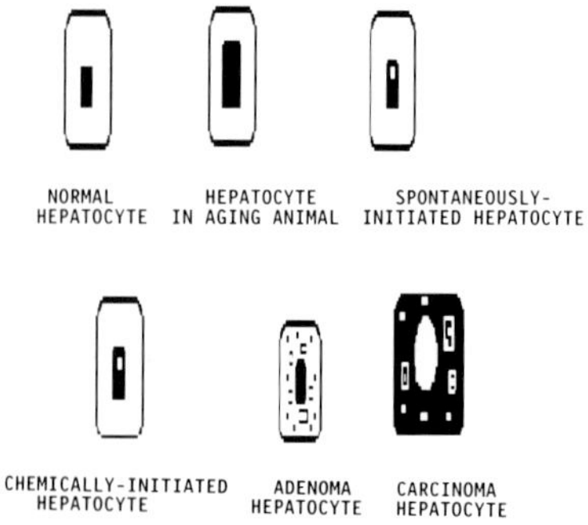

Figure 2. Potential target cells for liver tumor promoters.

and that their ability to cause tumors occurs through a genotoxic mechanism. The majority of human carcinogens have been shown to be mutagenic or genotoxic in some relevant assay system (6). Although a majority of chemicals that cause cancer in laboratory animals are mutagenic or genotoxic in some assay system, a significant number of carcinogens (15%–30%) possess little or no genotoxic activity. They include promoters, cytotoxic agents, hormones, immunosuppressors, solid-state chemicals, and peroxisomal proliferators and have been suggested to act via epigenetic mechanisms (7), perhaps by modifying gene expression. The mouse liver has been a target of chemicals that do not possess classic genotoxic activities. For example, 65% of a series of 61 mouse hepatocarcinogens also causing other tumors in mice and/or rats were mutagenic, usually in *Salmonella* spp., while only 30%–35% of those causing mouse liver tumors alone were mutagenic (2,8). In a large series of 222 chemicals reviewed, 74%–85% of chemicals causing tumors at two or more sites in rats or mice were mutagenic in *Salmonella* spp. (9). In contrast, 45%–50% of those causing tumors at a single site other than mouse liver were mutagenic, and 30%–35% carcinogenic only in mouse liver were mutagenic. Noncarcinogens were mutagenic in 30%–32% of the cases. Thus, if one were to postulate that chemicals causing only mouse liver tumors act via some nongenotoxic (epigenetic) mechanism, one would have to explain how 30%–35% of these mouse liver carcinogens that are genotoxic also cause these tumors. In addition, many chemicals have been shown to increase or promote the natural occurrence, development, and/or progression of liver tumors by as-yet-undefined processes. It is also possible that the mutagenic chemicals causing liver tumors act via epigenetic mechanisms.

Tumor Promotion and Multistage Carcinogenesis

Tumor promotion has been defined by several investigators (7,10–17) but remains a complex array of biologic processes. In reality, all, or almost all, tumor promoters are carcinogenic for the cells of the target organ, as detected in 2-year rodent carcinogenicity assays. Additionally, they are often target organ toxins. Chemicals characterized as carcinogens by other criteria, including their genotoxic activity, often have tumor-promoting activity. Frequently the carcinogenic activity of tumor promoters is characterized as weak, i.e., they induce small or moderate increases in often benign tumors only after the complete 2-year assay (18–26), yet their promoting activity is characterized as strong. Tumor promotion has been defined in various ways. Originally, Berenblum defined promotion simply as the phenomenon of increasing papilloma incidence and multiplicity in skin initiated by 9,10-dimethyl-1,2-benzanthracene (DMBA) or 3,4-benzpyrene and promoted by croton oil (12–14). After much research on such two-stage skin and liver carcinogenesis systems over the past 20 years, tumor promotion has shown to be a complex process or set of processes, defined in theoretical terms of multistage carcinogenesis (7,10–12). Tumor promoters enhance carcinogenesis, perhaps through altering gene expression, after a single exposure to an initiating agent, which is usually a complete and very potent carcinogen.

Several carcinogens have been shown to induce mouse liver tumors with specific point mutations at codon 61 of the H-*ras* oncogene, a possible mechanism of cell transformation or initiation (M. Anderson, at the symposium that this volume reports). The enhancement of carcinogenesis may occur at one or at several steps of the theoretical processes of carcinogenesis (Fig. 2). After the initiator causes initial DNA damage followed by DNA repair and/or DNA mutation, it may facilitate the appearance of hitherto-uncharacterized and -unrecognized initiated cells. Initiated cells are poorly defined in vivo and have perhaps not yet been characterized or seen (27). Theoretically, they are putative precursors of preneoplastic, precancerous, or neoplastic cells. They may contain oncogene site-specific mutations induced previously by exposure to a carcinogen, or they may be primed for transformation and clonal expansion, by promoters or other cofactors. Presumably, these cells are susceptible to the mitogenic effects of tumor promoters, a process that may progress through clonal expansion, but are resistant to the cytotoxic effects of drugs and carcinogens.

In the liver, the first visible evidence of these proposed initiated cells occurs with the appearance of hyperplastic putative preneoplastic foci. In the rat, single cells, immunoreactive for the placental form of glutathione S-transferase (GSTP), have been suggested to be the first visible evidence of these cells (27). In mice, histologically visible liver foci are evident (28,29). Promoters may act on these foci and individual cells by enhancing or inducing mitogenesis or encouraging a selective growth advantage (10–12,30). If tumor promoters act on initiated cells by causing clonal expansion, we might expect more visible foci and adenomas to be found in promoted animals. Thus, we have seen increased

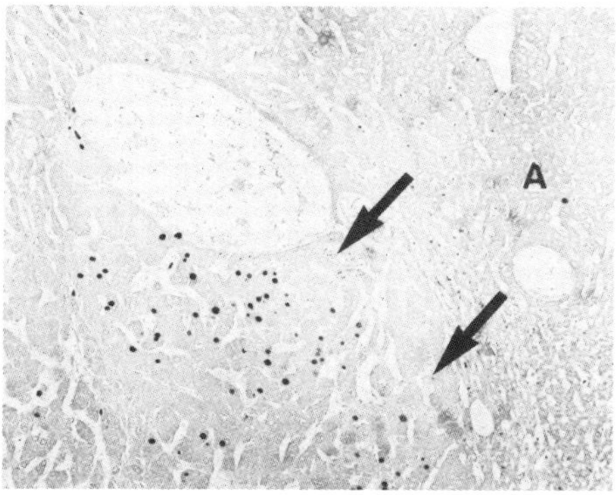

Figure 3. A trabecular hepatocellular carcinoma with a high BrdU labeling index (arrows), within an adenoma (A) of an untreated male C3H/HeNCr mouse. BrdU immunohistochemistry, hematoxylin, ×100.

numbers of foci in *N*-nitrosodiethylamine (DEN)-initiated mice after phenobarbital (PB) exposure (29). Hepatocellular adenomas and carcinomas may also be directly affected by promoters. Di(2-ethylhexyl) phthalate (DEHP) was shown to increase the size of foci and adenomas in mice after DEN initiation (18,29). PB has been shown to induce γ-glutamyl transpeptidase (GGT) in spontaneous liver nodules of C3H mice (31), and focus size was significantly enhanced in female $B6C3F_1$ mice initiated as infants with DEN, both phenomena suggesting the direct effects of PB on these lesions. Increased rates of metastasis in tumor promotion experiments would suggest effects on tumor progression, assuming that metastatic capability is associated with progression and earlier appearance of the lesions (17).

Promoters may also enhance the natural progression of carcinomas from adenomas, since carcinomas have been shown to arise within hepatocellular adenomas of the mouse (21) (Fig. 3). As the methods of evaluation improve and molecular biology techniques become available, they may provide additional insight into the specific mechanisms involved.

Methods of Evaluation

The methods of accumulating and evaluating data are extremely important to the proper formulation of valid conclusions. Various evaluation methods have been used in liver carcinogenesis studies, including determination of the incidence of tumors or preneoplastic lesions. These end points, however, may be a relatively crude estimate of the biologic effects of tumor promoters. Alternatively, it has been suggested that appropriate stereologic methods could be used for evaluation of liver foci in mice (28,32). We have found that the final

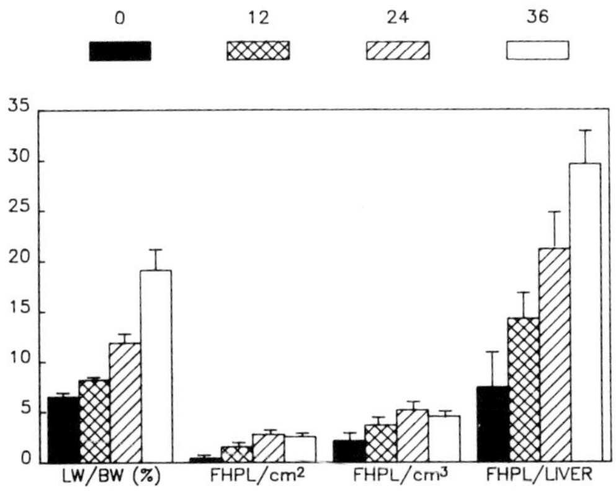

Figure 4. Comparison of parameters for measuring focal hepatocellular proliferative lesions (FHPL) including all hepatocellular hyperplastic foci, adenomas, and carcinomas. The FHPL/liver and body weights were significantly increased. Fifty-two-week-old male retired breeder C3H/HeNCr mice were given water containing phenobarbital for up to 36 weeks (24). Mean ± SEM.

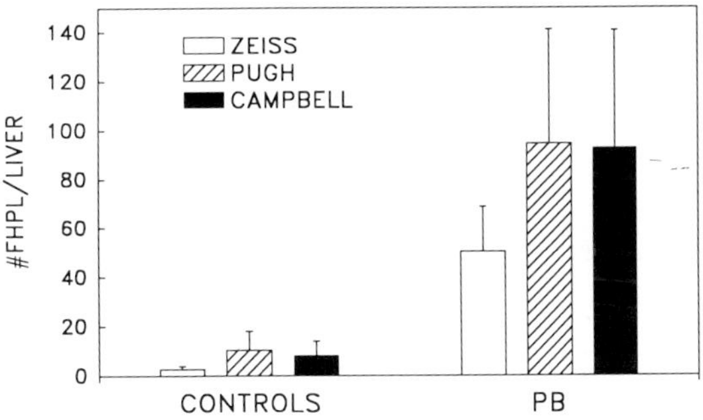

Figure 5. Comparison of numbers of FHPL/liver using methods of Zeiss Videoplan software (29), Pugh et al. (28), or Campbell et al. (32). Mean ± SEM.

conclusions from a study are dependent, in part, on the evaluation methods used. If the number of focal hepatocellular proliferative lesions (FHPL), including hyperplastic foci, hepatocellular adenomas, and carcinomas, per unit area (cm^2) is measured, conclusions can be different than if the number of FHPL per unit volume of liver is used (Figs. 4,5). In our case, the numbers/cm^2 in two groups were identical, while numbers/liver were greatly enhanced after PB exposure. Older stereologic methods may provide lower estimates without affecting the study's final conclusion. Our early stereologic methods (29) used with software supplied with our early image analysis system (Videoplan, Zeiss, Inc.) provided mouse liver foci/liver values that were 50%–100% less than with newer stereologic methods (28,32) (Fig. 5), but there were no significant differences

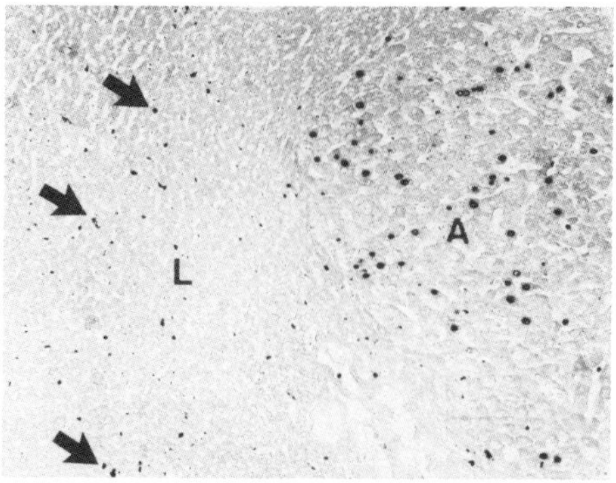

Figure 6. High BrdU labeling index in an eosinophilic adenoma (A) of an 88-week-old male C3H/HeNCr mouse after receiving PB from 52 weeks of age. Note immunoreactive BrdU in Kupffer cells (arrows) in nonneoplastic liver (L). BrdU immunohistochemistry, ×100.

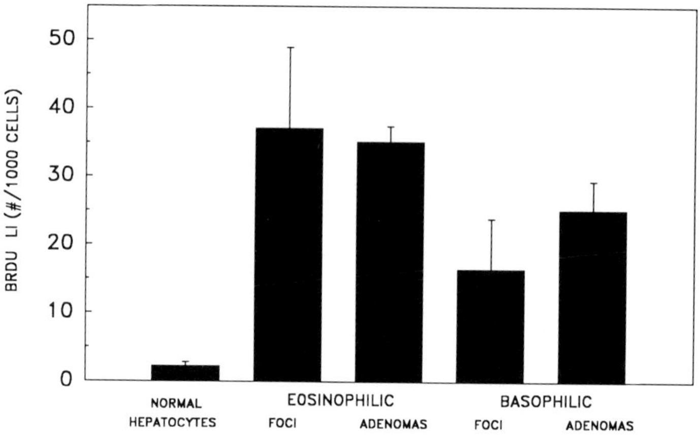

Figure 7. Single BrdU pulse labeling indices of hepatic lesions in male C3H/HeNCr mice, 88 weeks of age, after 36 weeks of drinking water with PB, 500 ppm. Mean ± SEM.

among these three methods in their powers to distinguish active from inactive compounds. Using any of the three methods, the tumor promoter, PB, would have significantly enhanced the numbers of FHPL per liver.

Gross evaluations include measuring liver weight and visible nodules or lesions per liver. The rough size of gross lesions may be surveyed with a caliper or ruler using surface lesions only, or the entire liver may be step sectioned to identify all gross nodules found within the liver parenchyma. The number of nodules, usually defined as nodular lesions greater than 1 mm in diameter, is only an approximation of the actual number of tumors in the liver, including both hepatocellular adenomas and carcinomas. As the nodules get larger and more numerous, however, coalescing of adjacent nodules leads to difficulties in

Table 1. Postulated Outcome of Tumor Promotion, Dependent on Hepatocyte Target for Promoter

Hepatocyte target for promoter	Major parameter measured: Increased no. of foci	Increased size of lesion	Increased progression to carcinoma	Increased metastases
Initiated cell	+	–	–	–
Focus cell	–	+	–	–
Adenoma cell	–	+	+	–
Carcinoma cell	–	+	–	+

accurately assessing the actual numbers of gross nodules. The same problem also occurs histologically.

More accurate measurements are made histologically using image analysis and stereologic computer programs (24,28,29,32). Area-perimeter measurements can be made via a digitizer tablet and appropriate computer programs. We have recently defined FHPL to include all microscopic (hyperplastic) foci, hepatocellular adenomas, and hepatocellular carcinomas (24,33), since all these lesions have similar greatly enhanced levels of DNA synthesis as compared with normal hepatocytes (Figs. 3,6,7). All microscopically identifiable lesions can be measured and evaluated (18,24,29,33), or only lesions of a defined size can be measured (34). In the latter instance, foci are often measured, while adenomas and carcinomas are not. It would be an artificially selective method of evaluation if all lesions were not included.

Using stereologic methods, a more accurate estimation of the numbers of lesions per unit volume (mm^3 or cm^3) and per liver can be determined (24,32). Additional evaluation provides the mean volumes of the average lesion and the percentage of the volume of the liver occupied by all lesions. It is important to determine and compare all these parameters. We have found that they often do not agree; this depends, in part, on the mode of action of the tumor promoter. For example, if the growth, clonal expansion, and progression of the initiated cell was not affected by the promoter, we would not expect to find an increase in absolute numbers of lesions in promoted mice compared with those that received only the initiator (29) (Table 1). In contrast, promotion of the growth rate of foci, adenomas, or carcinomas may lead to larger foci (as measured by area, volume, percentage of volume), larger or more numerous adenomas or carcinomas, a decreased latency period, increased tumor metastases, or shortened life spans of tumor-bearing animals (17,18,29). On the other hand, if the promoter acted primarily or solely on initiated cells, we might expect to find greatly increased numbers of small foci early in the experiment (24,29). These differential effects of two different promoters were, in fact, seen in DEN-initiated $B6C3F_1$ mice. PB increased the numbers of foci only, while DEHP significantly enhanced the size (volume) of the FHPL (29).

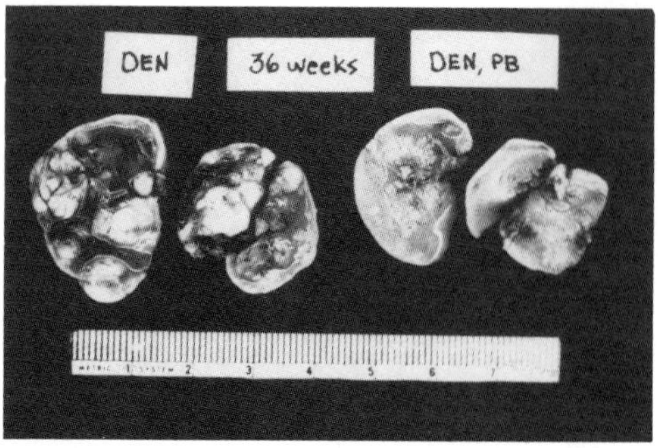

Figure 8. Gross liver lesions (adenomas and carcinomas) in male B6C3F$_1$ mice 36 weeks after receiving DEN (1.5 mg/kg) at 15 days of age or DEN followed by PB (500 ppm) in water from 6 weeks of age. Note that there are fewer and smaller lesions in the DEN/PB-treated mice's livers.

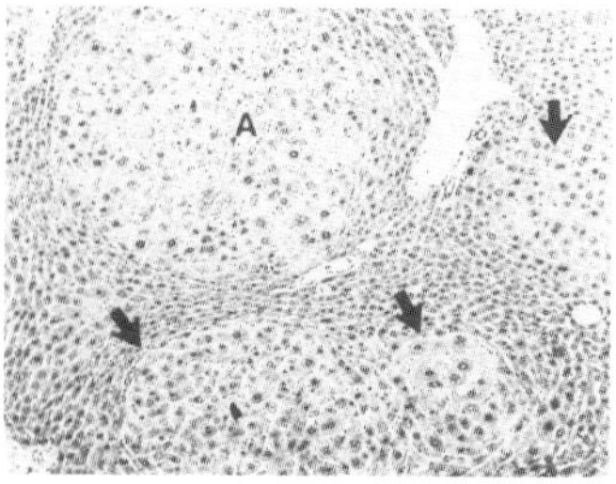

Figure 9. Eosinophilic foci (arrows) and a small adenoma (A) in an 88-week-old male C3H/HeNCr mouse after PB promotion for 36 weeks; phenotypic lesions typical of PB promotion after chemical or spontaneous initiation; H & E, ×100.

Experimental Models

Tumor Initiation in the Fetal, Neonatal, and Infant Mouse

Relatively few investigators have used transplacental initiation for a study of two-stage carcinogenesis. In such systems, however, acetylaminofluorene (35) and *N*-nitrosoethylurea (36,37) have been used as initiators with PB (36), barbital (37), or DEHP (J. Ward, unpublished observation) as promoters with some success.

Neonatal mice have been initiated with dimethylnitrosamine (DMN) (38, 39–41), 2-acetylaminofluorene (AAF) (42), methyl-4-aminoazobenzene (MAB) (42), *N*-nitrosoethylurea (42), or DEN (44). Promoters or modifiers have included PB (40,41,43,44), polychlorinated biphenyls (39), phorbol (38), or food restriction (44). Findings with these regimens have been mixed. While PB promoted liver lesions in DDD and Swiss-Webster mice, it failed to do so in A/He mice after ethylnitrosourea (ENU) initiation (43), in Swiss mice after DEN initiation (44), or in B6C3F$_1$ mice after AAF or MAB initiation (42) when effects were measured with nonstereologic methods.

Infant mice, 7–15 days of age, with high levels of hepatocyte DNA synthesis and cell replication (similar to that seen after partial hepatectomy [PH]), were initiated with DEN (15,16,30,45–52), ENU (54), or DMN (38). Promoters included PB (15,16,45–49,51–53), chloroform (54), 1,4-*bis*[2-(3,5-dichloropyridyloxy)]benzene (TCPOBOP) (16,51), amobarbital (52), or phorbol (38). While PB promoted liver tumors in male BALB/c mice after DEN initiation (48), it did not promote liver tumors in male B6C3F$_1$ (15,49–52) or male C57BL/6N mice (46), as measured by nonstereologic methods (Fig. 8). PB did promote liver tumors,

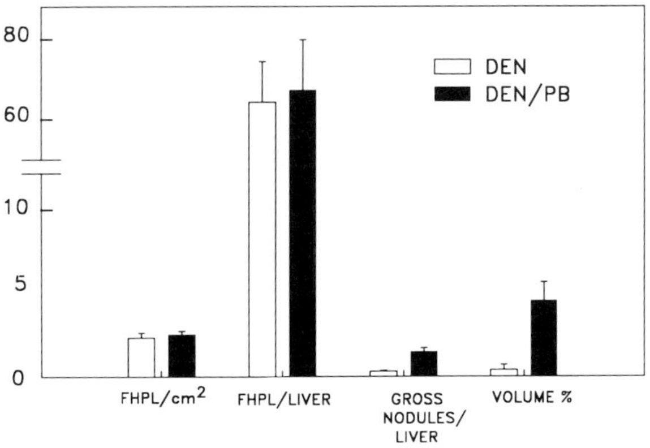

Figure 10. Increase in volume of FHPL, which were primarily of the eosinophilic phenotype and evidenced by increases in gross nodules/liver, in female B6C3F$_1$ mice 30 weeks after DEN (1.5 mg/kg) at 15 days of age and PB from 6 weeks of age.

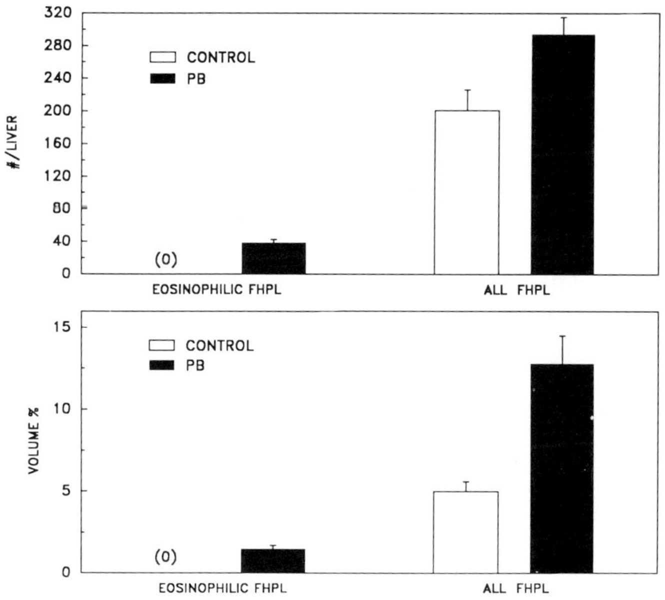

Figure 11. Increase in numbers of eosinophilic FHPL (hyperplastic foci and adenomas) in male B6C3F$_1$ mice 30 weeks after receiving DEN (1.5 mg/kg) and PB from 6 weeks of age. No eosinophilic FHPL were observed in controls. Mean ± SEM.

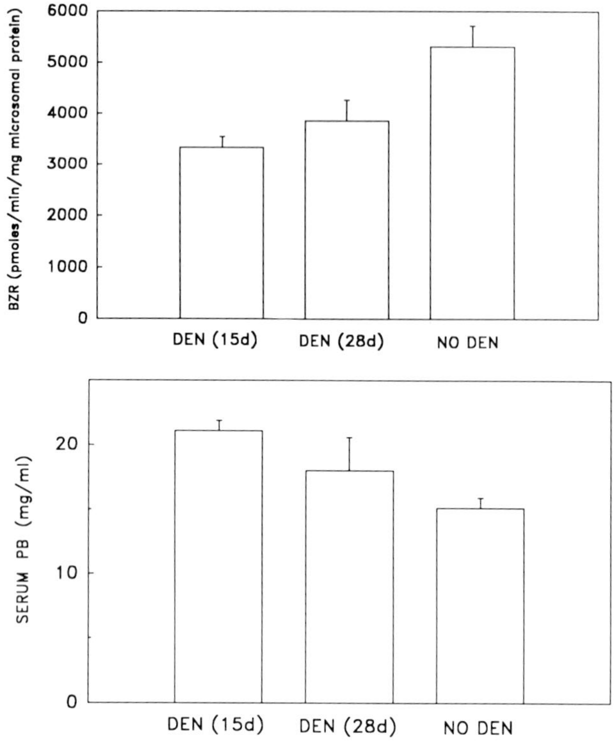

Figure 12. (A) P450 induction as measured by dealkylation of benzyloxyresorufin. Male $B6C3F_1$ mice receiving DEN (5 mg/kg) at 15 or 28 days of age and PB (500 ppm) in water from 35–49 days of age. Mean ± SEM. (B) Serum PB levels in mice noted in A. Mean ± SEM.

however, in infant CD-1 mice after initiation with ENU (54). The lack of obvious promotion of liver foci or tumors in some mouse strains by PB after initiation as infants was in marked contrast to potent promoting effects of PB after initiation in young adults (49,55). It was suggested that the effect of PB on testicular maturation may play a role. Although PB does not usually produce testicular atrophy in adult mice, it was suggested that PB may affect sexual maturation in young mice (49). After several studies, however, it could not be demonstrated that PB effectively castrated or feminized male mice, regardless of age at administration (47). We could not detect decreased testicular weights after exposure of young weanling mice to PB. We and others (30) did find, however, that PB has promoting activity in female $B6C3F_1$ mouse liver after initiation with DEN at 15 days of age. The promoting activity was most evident as increased lesion volume and development of the phenotypic eosinophilic foci (30) (Fig. 9), in marked contrast to increased numbers of lesions after DEN initiation of young adult male mice (29). With careful evaluation, we could find evidence of weak tumor-promoting activity in male mouse liver, as evidenced by increased lesion volume and, more important, by the development of the eosinophilic FHPL, the characteristic phenotypic lesions promoted by PB (Figs. 10,11).

To study possible mechanisms for the resistance of infant male B6C3F$_1$ mice to an initiating-promoting regimen in liver, we directly determined blood levels of PB and indirectly measured levels of the major PB-inducible forms of cytochrome P450 by measuring the levels of benzyloxyresorufin *O*-dealkylase activities (56) after DEN initiation. This study was performed because in rats we have shown, for a variety of pharmacologic agents including barbiturates, hydantoins, and alkylacetylureas, that there is a strong correlation between the ability of a xenobiotic to induce the major PB-inducible form of cytochrome P450 and to serve as a liver tumor promoter (56). We found that DEN initiation at 15 days significantly increased serum PB levels after 14 days of PB exposure and partially inhibited P450 induction by PB (Fig. 12A,B). Despite the partially inhibitory effects of DEN on PB induction of cytochrome P450, the levels of benzyloxyresorufin *O*-dealkylase were still 30–40 times higher in DEN/PB–treated animals relative to control mice exposed to neither of these xenobiotics. These results would appear to argue against differences in P450 induction being the primary determinant of promotion in the neonatal mouse model. Our findings and those of others demonstrating the ineffectiveness of PB promotion in neonatal mice argues against utilizing an infant mouse model for tumor initiation-promotion studies except with mouse strains specifically selected to study the effects of neonatal initiation-promotion.

Tumor Initiation in the Young Adult

DEN has been typically employed for initiation in young adult mice, usually at 4–6 weeks of age (18,29,33,47–49,55,57–60). Promoters have included PB (18,29,33,47–49,55), DEHP (18,29,33,61), carbon tetrachloride (57), dichlorodiphenyltrichloroethane (DDT) (58), butylated hydroxyanisole (61), acetaminophen (62), chlordane and heptachlor (58), diazepam and oxazepam (59), and choline-deficient diets (60). Some strains of mice, notably C57BL and NFS, were resistant to tumor promotion by PB (46,55), while others, i.e., C3H, DBA/2, and B6C3F$_1$, were highly susceptible (18,29,46,55). Choline-deficient diets did not enhance liver carcinogenesis after DEN initiation (60,63) but did promote spontaneous tumors (63).

Other initiators used in young adult mice are dimethylhydrazine (64), 2-amino-5-azotoluene (65), hydrazine sulfate (22), and DMN (66). Butylated hydroxytoluene (64), PB (18,29,33,47–49,55), carbon tetrachloride (66), and diazepam (59) had promoting activity after initiation with one of the previous agents, while PB did not promote tumorigenesis in mice initiated with hydrazine (22), and ethyl alcohol did not enhance DMN hepatocarcinogenesis (67).

Role of Androgens

In most strains of mice thus far used in studying susceptiblility to liver tumor induction, males almost invariably demonstrate a higher incidence, greater yield, and/or quicker generation of hepatic tumors than females of the same strain (3,68). This finding, and the fact that ovariectomy and castration can modify both spontaneous and chemically induced hepatocarcinogenesis (68), strongly suggests that sex hormones play some role in carcinogen metabolism (69,70), tumor promotion, and/or tumor progression. Strains or hybrids showing a high natural occurrence of liver tumors, such as C3H and B6C3F_1, illustrate the effect of gonadectomy most dramatically; ovariectomy of females increases the tumor incidence toward male levels, while castration reduces the male incidence down to that of females (68). This phenomenon is repeated in assays using chemical initiation. Gonadectomy is especially effective when done prior to puberty and/or carcinogen administration, but it is interesting that regardless of surgical intervention or chemical induction, the response of males is always more pronounced, i.e., females either intact or ovariectomized will approach but never reach male liver tumor levels, while castrated males do attain normal, low female levels.

Enhancement of Carcinogenesis by Partial Hepatectomy (PH)

PH enhanced liver carcinogenesis when performed prior to or after carcinogen administration (63,68,71–73), and such enhancement was usually found when the carcinogen was administered from 17–48 h after PH (68,71,73,74), at peak levels of DNA synthesis. Recently, we attempted a modification of the Solt-Farber-Ito procedure used in rats (34). Male B6C3F_1 mice, 4 weeks of age, were injected intraperitoneally with a single dose of DEN (80 mg/kg) and held for 2 weeks. At 6 weeks of age, they were placed on diets containing PB or DEHP. One week later a 2/3 PH was performed by the Higgins-Anderson procedure (73), which removes the gallbladder with the median lobe. Mice were killed 5, 11, and 22 weeks after diets were started. At 5 and 11 weeks, few foci were found. After 22 weeks of dietary promotion, however, both PB and DEHP had enhanced the numbers of eosinophilic FHPL, with PB being the more effective (Figs. 13,14). The volumes of the lesions in partially hepatectomized mice receiving PB or DEHP were similar, in contrast to when young intact adult mice were initiated (29). Our findings with this PH system, however, did not demonstrate any advantage over that of initiation of young adults without PH. One major disadvantage was the development of biliary hyperplasia and inflammation in 20%–40% of the mice after PH because of removal of the gallbladder.

PH can also modify hepatocarcinogenesis in some strains of mice, even when carcinogens are not given at peaks of DNA synthesis. Thus, PH 1 week after urethan administration or 4 days beforehand enhanced liver carcinogenesis (72). Interestingly, PH alone enhanced spontaneous hepatocarcinogenesis in male B6C3F_1 mice (63) but not in BALB/c mice (71).

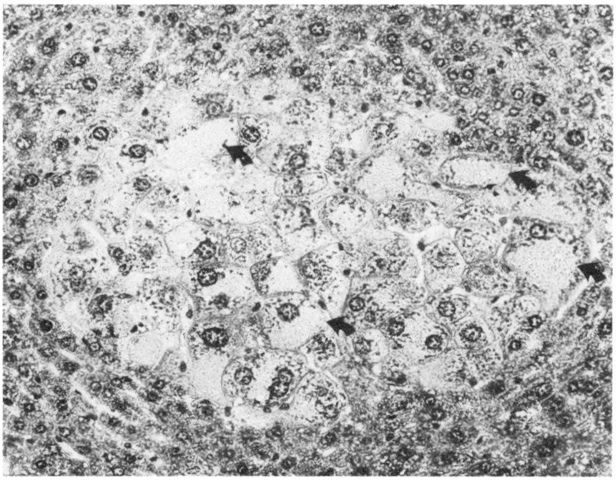

Figure 13. Eosinophilic hyperplastic focus in liver of male $B6C3F_1$ mouse 24 weeks after DEN (80 mg/kg) at 4 weeks of age, partial hepatectomy at 7 weeks, and PB from 6 weeks of age. Note pale eosinophilic granular cytoplasm of focus hepatocytes (arrows), evidence of proliferation of smooth endoplastic reticulum (18); hematoxylin and eosin, ×250.

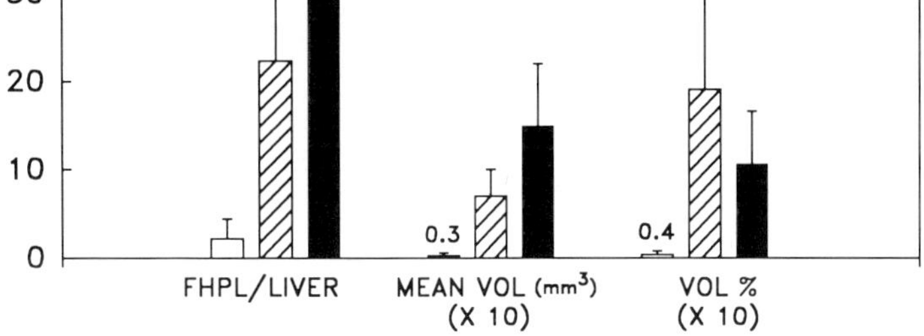

Figure 14. FHPL in male $B6C3F_1$ mice 24 weeks after DEN at 4 weeks of age, PB or DEHP from week 6, and partial hepatectomy (PH) at week 7.

High Susceptibility of Aging Male C3H/HeNCr Mice to PB Carcinogenesis: Promotion of Spontaneously Initiated Hepatocytes?

Hepatocytes in aging rats and mice have been shown to possess altered abilities to metabolize xenobiotics, repair DNA damage, and replicate in response to mitogens and to respond differently to a variety of other stimuli than do those in young animals (reviewed in refs. 24,25). These differences may account, in part, for differences in susceptibility to carcinogens or tumor promoters. Since $B6C3F_1$ mice, a hybrid strain with a high incidence of spontaneous liver tumors in males, has been used extensively in carcinogenesis assays, it has been suggested that nongenotoxic chemicals that increase the incidence of liver tumors in these mice may actually be enhancing the development of naturally occurring liver preneoplastic foci or tumors rather than inducing tumors de novo (24,63).

In F344 rats, the incidence of spontaneous liver tumors is low, but almost all aged rats have multiple basophilic foci in the liver (23,25,75). We were unsuccessful at promoting the growth and development of these foci by PB (25) or Aroclor 1254 (23) in F344 rats, although PB and Aroclor seemed to stimulate the growth and progression of a unique phenotypic focus composed of eosinophilic hepatocytes.

To further investigate the mechanisms of carcinogenesis by nongenotoxic carcinogens, we developed a model assay system in mice to study the promoting effects of nongenotoxic tumor promoters using retired breeder male C3H/HeNCr mice, a strain that demonstrates a high natural liver tumor incidence (24). In a more recent experiment, we killed untreated mice at 52 weeks of age and observed a 20%–30% incidence of liver foci or tumors. Additional mice of the same age were placed on diets containing PB or DEHP for periods of 12–36 weeks (64–88 weeks of age) and killed in groups of 10. Quantitative image analysis and stereologic evaluation of livers from these animals showed PB markedly increased the numbers of FHPL with a primarily eosinophilic phenotype (Fig. 15) without enhancing tumor size. DEHP had no effect on these lesions. The more common FHPL of the basophilic phenotype were not significantly enhanced in number or size (volume) by either of the chemicals. These findings suggest that PB caused clonal expansion of putative spontaneously initiated hepatocytes or induced the transformation of susceptible hepatocytes rather than clonal expansion of already existing FHPL (visible lesions). These results are identical to those found with PB administration after DEN initiation in mice but in marked contrast to that seen in rats (17) after PB promotion, or, interestingly, in younger mice after DEN initiation and DEHP promotion. In the latter model, PB or DEHP appears to cause faster progression of foci to adenomas or carcinomas. Also, when young, 6-week-old C3H/HeNCr mice are exposed to PB or DEHP without DEN initiation, no foci were found by 36 weeks (24). Thus, the aging liver of the C3H mouse is highly susceptible to the carcinogenic activity of PB. This model system has implications for identifying the causes of suscepti-

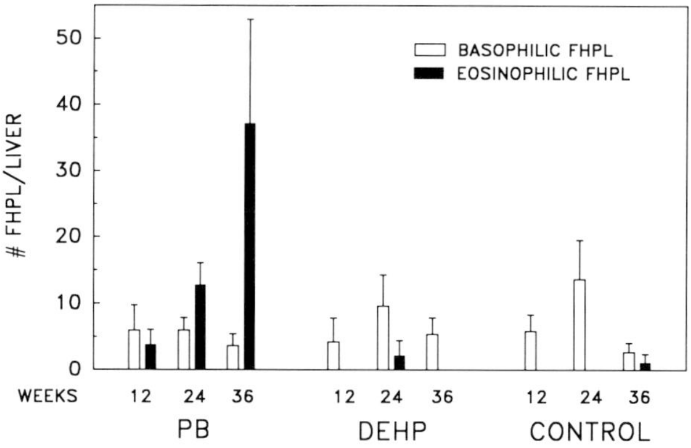

Figure 15. FHPL in 64- to 88-week-old male C3H/HeNCr mice after exposure to PB (500 ppm in water) or DEHP (6,000 ppm in diet) from 52 weeks of age.

bility of the aging mouse liver to carcinogens. In preliminary studies, we did not see differences in P450 induction or serum PB levels between young (6 week) and aging (52 week) female C3H/HeNCr mice. This finding suggests that differences in susceptibility to PB carcinogenesis for males are not due to differences in metabolism of PB or induction of cytochrome P450.

Role of Increased Cell Replication (Hyperplasia) in Tumor Promotion and Carcinogenesis by Chemicals without Significant Genotoxic Activities

It has been hypothesized that cytotoxicity may lead to transient or persistent increases in levels of DNA synthesis in target cells and, subsequently, to carcinogenesis itself or to tumor promotion or progression (13,62,63,76–80). Some associations evidently exist. Chronic *Schistosoma japonica* infection associated with liver granulomas has been found to enhance liver carcinogenesis in ddY mice (81) and CBA mice (65). Carbon tetrachloride, given 24 or 48 h prior to DEN or given 5 weeks after DEN for 4 weeks, enhanced carcinogenesis (57,79). Conversely, there is also some evidence that chronic toxicity and hyperplasia may not always be associated with carcinogenesis or tumor promotion (62,77,80). In our studies with acetaminophen, a human and rodent liver toxin, we found chronic, persistent increases in levels of DNA synthesis in hepatocytes at dose levels that were not carcinogenic in B6C3F$_1$ mice (62,80). The higher dose, 1%, was tumor promoting in mouse liver but not carcinogenic. A large review of the carcinogenicity of 99 chemicals conducted by the National Toxicology Program revealed no general correlation between toxicity and carcinogenesis (77). For 25 chemicals that were liver carcinogens in mice, there were 42 positive (carcinogenic) sex/species experiments. Of these 42 experiments, 21 produced liver toxicity, confirmed only by histopathologic changes, in 2-year studies. Of 26 nonmutagens tested in *Salmonella* spp., only 12 produced liver lesions, while 5/13 mutagens produced toxic lesions. Although chronic toxicity may play a role in carcinogenesis and tumor promotion, especially by chemicals without significant genotoxic activities, it is frequently not associated with carcinogenesis (77).

Summary and Conclusions

Mouse liver tumors are important end points for safety assessment of chemicals in rodent carcinogenicity assays. The mechanisms of induction of these tumors, whether by chemicals with significant genotoxic activities or by chemicals without such activities, remain unknown. If test mice have spontaneous tumors of high or low background incidence, the promotion of these tumors or of spontaneously initiated or susceptible cells may provide a basis for carcinogenesis induced by these compounds. Likewise, chronic toxicity and its associated target cell hyperplasia may also lead to promotion of these tumors. Experimental systems have been developed to study the mechanisms of tumor promotion in mouse liver. These models should significantly advance research into mechanisms of hepatocarcinogenesis and tumor promotion in mouse liver and may subsequently be used for human risk assessment.

Acknowledgments

This project has been funded at least in part with federal funds from the Department of Health and Human Services under contract number NO1-CO-74102. The content of this publication does not necessarily reflect the views or policies of the Department of Health and Human Services, nor does mention of trade names, commercial products, or organizations imply endorsement by the U.S. government. We are grateful for the editorial and secretarial assistance of Kathy Breeze, Jennifer Sipes, Deborah Shores, and Debora Brandenburg and technical assistance of Mark Shrader, Dan Logsdon, Areitha Smith, Charles Riggs, Dane Miller, Craig Driver, Gwenn Magruder, Shirley Hale, Lynda Rickerd, and Cindy Harris.

Open Forum

This exchange relates to the chapters by B. A. Diwan et al., J. M. Ward et al., and a presentation by S. Goldfarb at the symposium this book is based on.

Dr. Robert LeBoeuf: I think we need to be very careful about the term nongenotoxic. Currently, point mutation and cytogenetic damage are the two categories assessed, but evidence suggests that other phenomena such as aneuploidy and translocation may be important in the cancer process. So I think we need to be careful how we define nongenotoxic.

Dr. Jerrold Ward: I think that point should be emphasized to the U.S. regulatory agencies because there is a stepwise evaluation procedure that takes that into account. I think, though, most of the human carcinogens studied are genotoxic. Also, the human carcinogens are carcinogenic at very high doses—basically maximally tolerated doses.

Dr. LeBoeuf: You showed a table of human carcinogens. Most were genotoxic, but for some, the genotoxicity was not determined. Until those determinations are made with in vivo cytogenetic assays, I think we should be careful in interpreting that data.

Dr. Dennis Parke: Twenty years ago Rudolf Peters showed that the enzyme induction produced by phenobarbitone and its analogues was due to futile cycling. In other words, it was due to the production of oxygen radicals rather than the metabolism of the phenobarbitone. For example, it was shown that 4-chloro-phenobarbitone and 4,4-dichloro-phenobarbitone produced considerably more furylcyclin oxygen radicals than did phenobarbitone, and the enzyme induction paralleled the amount of futile cycling. That fits beautifully with your findings, Dr. Diwan. My question is, Have either you or Dr. Ward looked at glutathione? As Don Stevenson has mentioned, the mouse is unique in having this very high oxygen turnover and high oxygen tension in the liver, which indicates a very high rate of production of these oxygen radicals. These are not only cytotoxic but also mutagenic and carcinogenic. Glutathione is the protective agent.

Dr. Bhalchandra Diwan: We have not looked at glutathione. Shinozuka has shown that the promotion by phenobarbital is not inhibited by antioxidants, whereas promotion by hypolipidemic drugs, which probably act through free radical formation to promote peroxisome proliferation, is inhibited by antioxidants like BHT and BHA.

Dr. Michael Gallo: This hypothesis of an oxygen radical mechanism was also made for the TCDD class, and at high doses it may very well be correct. I think the work on TCDD by Stohs and others has shown that you can increase oxygen radicals at extremely high doses, but the activation of the putative receptor protein may be playing a role, although I recognize that that is a different enzyme system. Dr. Diwan, I wanted to ask you about the striking difference in susceptibility that you showed between the C57BL/6 and the DBA strains. The susceptibility for both induction of tumorigenesis and toxicity is exactly reversed from what happens with TCDD in those two strains.

Dr. Diwan: Is this true for TCDD in the liver, too? Are there strain differences in susceptibility to tumor promotion by TCDD?

Dr. Gallo: Yes. It is exactly reversed in susceptibility.

Dr. Diwan: C57BL/6 mice are more susceptible to liver promotion by TCDD?

Dr. Gallo: Yes.

Dr. James Barrett: One of the important issues of this conference is the progression from the preneoplastic lesions to malignant lesions, and when that commitment to progression really occurs. I was interested in Dr. Goldfarb's observations that there are differences in the rat and the mouse, with the mouse showing a more continuous progression, while the rat, in his view, has more discrete lesions. Typically at this time, we tend to combine the benign and the malignant lesions for both species when we evaluate chronic bioassays. Dr. Goldfarb, do you have any thoughts on the appropriateness of doing that, given that there might be some difference in the commitment to progression?

Dr. Stanley Goldfarb: That depends on which model you are talking about. For example, with the Solt-Farber model, 80%–90% of the hyperplastic nodules actually regress. Obviously, you could not use nodules as a measure of carcinogenic effect in that model. In contrast, in the single-dose model that I showed, there appears to be a continuum. This distinction of adenomas from carcinomas is very arbitrary. It is really a growth-rate-dependent phenomenon. With the single-dose model, I would prefer that we just analyze by size. That is why I suggested using the micro- and macrohepatocellular approach to categorize nodules as either microscopic (smaller than 1 mm) or macroscopic. This size-dependent classification makes a nice, clean distinction.

Dr. Gabriel Altmann: Dr. Goldfarb, I wonder whether those cells that accumulate the eosinophilic globules are the original cells of the foci, which would thus have an adaptive response, or some new cells that form a new type of focus?

Dr. Goldfarb: The replication rate of the globule-containing cells is less than that of the surrounding nonglobular-containing cells, and many of the cells contain large numbers of globules. Therefore, I have to assume that this phenomenon represents a time-dependent change in which existing cells gradually accumulate these globules and become manifest, not that they are a new population.

Dr. Jerry Styles: Dr. Diwan, could you comment on the differences in the unpromoted tumor incidences that you found with DEN? How much of that represents inherent differences, and how much is due to differences in the metabolic capabilities of those particular mouse strains?

Dr. Diwan: There are no differences in the metabolic activation of DEN between these two strains. Also, no differences were observed in the formation and persistence of O^6-guanine or N^7-ethylguanine and total levels of ethylation of hepatic DNA between these two strains of mice.

Dr. Styles: But does that also apply to the refractory strain that you showed?

Dr. Diwan: We did not use the NFS/NCr strain in that study, but used only the C57BL/6NCr and DBA/2NCr, which had shown remarkable differences in tumor promotion by phenobarbitol.

Dr. Frederick Becker: Some years ago, Dan Stout and I conducted an experiment in which we exposed both spontaneous tumors of C3H and DEN-induced tumors of C57BL mice to phenobarbital after the tumors had been established. Both types of tumors showed a very substantial induction of P-450 and associated enzymes. In fact, percentagewise, those tumors showed a greater capacity for induction than did the background liver, but they never achieved the absolute levels of the background liver. Dr. Diwan, let us accept for the moment that these tumors were derived from the initiated cells, and we are talking about the responsivity of these cells plus or minus the phenobarbital. Here we have a C57 DEN-induced tumor that responds to phenobarbital but is not promoted by phenobarbital. I think that is what was suggested by induction of enzyme systems.

Dr. Diwan: What we showed is that in normal hepatocytes of these two strains, there were no differences in enzyme induction. But Shulte-Hermann showed that there are differences in preneoplastic hepatocytes. In other words, there is more proliferation in preneoplastic hepatocytes exposed to phenobarbital than in the normal hepatocytes. Also after phenobarbital treatment there are more phenobarbital-specific enzyme-positive foci.

Dr. Becker: Of course we have to be wary because Shulte-Hermann's work was done in the rat. Dr. Goldfarb, I am concerned about the concept of defining a tumor by size. Dr. Ward was correct in stating that we often see huge tumors—2–3 cm in diameter—quite late in the life of the mouse, and the animals seem to live merrily with them. This type of tumor is not associated with any of the normal connotations of malignancy. It does not even show local invasion. It shows no metastasis, etc. I appreciate the fact that you are restricting your proposal to the Vesselinovitch system.

Dr. Goldfarb: Let me clarify again. You understand that I feel rather strongly that this is a continuum, that the basic difference is in growth rate. I was just proposing separating micro- and macrocancers as an easy way of defining two categories that are rate-dependent. The smaller ones are growing more slowly than the larger ones. Now Stan Vesselinovitch has actually proposed a lesion classification system based on size. Under this system, size is one of the criteria defining a hyperplastic nodule, an adenoma, and a cancer. I do not think that is correct. If we want to study promotion and progression in terms of its biologic significance, then we have to appreciate that there are differences in growth rates. An easy way to address this biologic problem is to divide the lesions into small and large ones. But I certainly feel that early lesions are a usable parameter because they predict what will happen in the later development of the large ones. They do give rise to carcinomas.

Dr. Becker: I think Dr. Vesselinovitch was the first to suggest the qualitative difference in the hits. But I think that some of the studies that Dr. Ward and I have presented suggest that different populations of initiated cells may indeed result from a single dose of carcinogen in a single strain. When I give a single dose of a carcinogen followed by phenobarbital, I see not just one but several types of tumors arising. Within the same liver (and possibly even within the same field, based on what Dr. Ward showed) one can see these eosinophilic tumors that are seen only in the presence of phenobarbital plus initiation. We see the basophilic foci, and I frequently see de novo foci of malignancy where I cannot see a preceding nodule. I am simply saying that some of these differences in rates and kinetics may be due to different populations of initiated cells that look like a continuum when added up. From the human pathology standpoint, Dr. Morson has demonstrated in ulcerative colitis patients that foci of colonic carcinoma arise de novo in the crypts of the colon without a preceding benign or preneoplastic lesion. So that possibility still pertains in the liver of a mouse.

Dr. Mark Walter: I, too, have some problems with classifying lesions based on their size, but if you find it useful I would not say that you should not do it. Dr. Goldfarb, would you suggest that this classification be used in the evaluation of the large bioassays?

Dr. Goldfarb: No. I recognize that the regulators have a different problem. All I am suggesting is that the regulators should start thinking that there is this continuum and that it is not a sharply defined separation. I was simply saying that separation into small and large lesions is useful for biologists studying this problem as a way of looking at two categories of low-growth-rate vs. more-rapid-growth-rate tumors.

References

1. Gold LS, Bernstein L, Magaw R, Slone TH. Interspecies extrapolation in carcinogenesis: Prediction between rats and mice. *Environ Health Perspect*, in press.
2. Ward JM, Griesemer RA, Weisburger EK. The mouse liver tumor as an endpoint in carcinogenesis tests. *Toxicol Appl Pharmacol* 51:389-397, 1979.
3. Maronpot RR, Haseman JK, Boorman GA, Eustis SE, Rao GN, Huff JE. Liver lesions in B6C3F1 mice: The National Toxicology Program, experience and position. *Arch Toxicol [Suppl]* 10:10–26, 1987.
4. Gold LS, Ward JM, Bernstein L, Stern B. Association between carcinogenic potency and tumor pathology in rodent carcinogenesis bioassays. *Fundam Appl Toxicol* 6: 677–690, 1986.
5. Tomatis L, Partensky C, Montesano R. The predictive value of mouse liver tumour induction in carcinogenicity testing: A literature survey. *Int J Cancer* 12:1–20, 1973.
6. International Agency for Research on Cancer. Overall evaluations of carcinogenicity: An updating of IARC Monographs Volumes 1 to 42. *IARC Monogr Eval Carcinog Risk Chem Hum* 7(Suppl):1–440, 1987.
7. Williams GM, Weisburger JH. Chemical carcinogens. In Klaassen CD, Andur MO, Doull J (eds): *Casarett and Doull's Toxicology*. New York: MacMillan, 1986, pp 99–173.
8. Newberne PM, Suphakarn V, Punyarit P, De Camargo J. Nongenotoxic mouse liver carcinogens. In Butterworth BE, Slaga TJ (eds): *Nongenotoxic Mechanisms in Carcinogenesis*. Banbury Report 25. Cold Spring Harbor, NY: Cold Spring Harbor Laboratory, 1987, pp 165–176.
9. Ashby J, Tennant RW. Chemical structure, Salmonella mutagenicity and extent of carcinogenicity as indicators of genotoxic carcinogenesis among 222 chemicals tested in rodents by the U.S. NCI/NTP. *Mutat Res* 204:17–115, 1988.
10. Pitot HC, Sirica AE. The stages of initiation and promotion in hepatocarcinogenesis. *Biochim Biophys Acta* 605:191–215, 1980.
11. Slaga TJ. Overview of tumor promotion in animals. *Environ Health Perspect* 50:3–14, 1983.

12. Shubik P. Progression and promotion. *JNCI* 73:1005–1011, 1984.
13. Mottram JC. A developing factor in experimental blastogenesis. *J Pathol Bacteriol* 56:181–187, 1944.
14. Berenblum I, Shubik P. The role of croton oil applications, associated with a single painting of a carcinogen, in tumour induction of the mouse's skin. *Br J Cancer* 1:379–382, 1947.
15. Becker FF. The direct and indirect effects of promoters may depend upon the nature of the initiated cell. In Fujiki H, Hecker E, Moore RE, Sugimura T, Weinstein IB (eds): *Cellular Interactions by Environmental Tumor Promoters.* Tokyo: VNU Science Press, 1984, pp 349–359.
16. Della Porta G, Dragani TA, Manenti G. Two-stage liver carcinogenesis in the mouse. *Toxicol Pathol* 15:229–233, 1987.
17. Kaufmann WK, MacKenzie SA, Kaufman DG. Quantitative relationship between hepatocytic neoplasms and islands of cellular alteration during hepatocarcinogenesis in the male F344 rat. *Am J Pathol* 119:171–174, 1985.
18. Ward JM, Diwan BA, Ohshima M, Hu H, Schuller HM, Rice JM. Tumor-initiating and promoting activities of di(2-ethylhexyl) phthalate in vivo and in vitro. *Environ Health Perspect* 65:279–291, 1986.
19. Wolff GL, Morrissey RL, Chen JJ. Amplified response to phenobarbital promotion of hepatotumorigenesis in obese yellow A^{vy}/A (C3H × VY) F-1 hybrid mice. *Carcinogenesis* 7:1895–1898, 1986.
20. Wolff GL, Morrissey RL, Chen JJ. Susceptible and resistant subgroups in genetically identical populations: Response of mouse liver neoplasia and body weight to phenobarbital. *Carcinogenesis* 7:1935–1937, 1986.
21. Ward JM. Morphology of potential preneoplastic hepatocyte lesions and liver tumors in mice and a comparison with other species. In Popp JA (ed): *Mouse Liver Neoplasia.* Washington, DC: Hemisphere, 1984, pp 1–26.
22. Cavaliere A, Bufalari A, Vitali R. Carcinogenicity and co-carcinogenicity test of phenobarbital sodium in adult BALB/c mice. *Tumori* 72:125–128, 1986.
23. Ward JM. Proliferative lesions of the glandular stomach and liver in F344 rats fed diets containing Aroclor 1254. *Environ Health Perspect* 60:89–95, 1985.
24. Ward JM, Lynch P, Riggs C. Rapid development of hepatocellular neoplasms in aging male C3H/HeNCr mice given phenobarbital. *Cancer Lett* 39:9–18, 1988.
25. Ward JM, Ohshima M. Evidence for lack of promotion of the growth of the common naturally occurring basophilic focal hepatocellular proliferative lesions in aged F344/NCr rats by phenobarbital. *Carcinogenesis* 6:1255–1259, 1985.
26. Ashby J, Purchase IFH. Reflections on the declining ability of the Salmonella assay to detect rodent carcinogens as positive. *Mutat Res* 205:51–58, 1988.
27. Moore MA, Nakagawa K, Satoh K, Ishikawa T, Sato K. Single GST-P positive liver cells: Putative initiated hepatocytes. *Carcinogenesis* 8:483–486, 1987.
28. Pugh TD, King JH, Koen H, Nychka D, Chover J, Wahba G, He Y-H, Goldfarb S. Reliable stereological method for estimating the number of microscopic hepatocellular foci from their transections. *Cancer Res* 43:1261–1268, 1983.
29. Ward JM, Rice JM, Creasia D, Lynch P, Riggs C. Dissimilar patterns of promotion by di(2-ethylhexyl) phthalate and phenobarbital of hepatocellular neoplasia initiated by diethylnitrosamine in $B6C3F_1$ mice. *Carcinogenesis* 4:1021–1029, 1983.
30. Weghorst CM, Klaunig JE. Phenobarbital promotion in diethylnitrosamine-initiated infant $B6C3F_1$ mice: Influence of gender. *Carcinogenesis* 10:609–612, 1989.
31. Williams GM, Ohmori T, Katayama S, Rice JM. Alteration by phenobarbital of

membrane-association enzymes including gamma-glutamyl transpeptidase in mouse liver neoplasms. *Carcinogenesis* 1:813–818, 1980.
32. Campbell HA, Xu Y-D, Hanigan MH, Pitot HC. Application of quantitative stereology to the evaluation of phenotypically heterogeneous enzyme altered foci in rat liver. *JNCI* 76:751–767, 1986.
33. Ward JM, Ohshima M, Lynch P, Riggs C. Di(2-ethylhexyl)phthalate but not phenobarbital promotes N-nitrosodiethylamine-initiated hepatocellular proliferative lesions after short-term exposure in male B6C3F$_1$ mice. *Cancer Lett* 24:49–55, 1984.
34. Ito N, Tsuda H, Tatematsu M, Inoue T, Tagawa Y, Aoki T, Uwagawa S, Kagawa M, Ogiso T, Masui T, Imaida K, Fukushima S, Asamoto M. Enhancing effect of various hepatocarcinogens on induction of preneoplastic glutathione S-transferase placental form positive foci in rats: An approach for a new medium-term bioassay system. *Carcinogenesis* 9:387–394, 1988.
35. Armuth V, Berenblum I. Possible two-stage transplacental liver carcinogenesis in C57BL/6 mice. *Int J Cancer* 20:292–295, 1977.
36. Tsuchiya E, Concetti HF, Sugano H, Kitagawa T. Lack of promotion effect of phenobarbital on pulmonary tumorigenesis in ddY mice initiated by transplacental exposure to 1-ethyl-1-nitrosourea. *Gann* 75:305–310, 1984.
37. Diwan BA, Rice JM, Ohshima M, Wenk ML. Transplacental carcinogenesis by N-nitrosoethylurea in B10.A mice: Effects of postnatal administration of different barbiturates on the incidence and types of tumors (Abstract). *Proc Am Assoc Cancer Res* 26:137, 1985.
38. Armuth V, Berenblum I. Systemic promoting action of phorbol in liver and lung carcinogenesis in AKR mice. *Cancer Res* 32:2259–2262, 1972.
39. Anderson LM, Ward JM, Fox SD, Isaaq HJ, Riggs CW. Effects of a single dose of polychlorinated biphenyls to infant mice on N-nitrosodimethylamine-initiated lung and liver tumors. *Int J Cancer* 38:109–116, 1986.
40. Cater KC, Gandolfi AJ, Sipes IG. Characterization of dimethylnitrosamine-induced focal and nodular lesions in the livers of newborn mice. *Toxicol Pathol* 13:3–9, 1985.
41. Uchida E, Hirono I. Effect of phenobarbital on induction of liver and lung tumors by dimethylnitrosamine in newborn mice. *Gann* 70:639–644, 1979.
42. Dooley KL, Reznik GK, Sloane TW, Kadlubar FF. Phenobarbital and saccharin as inhibitors of hepatic tumorigenesis in the B6C3F$_1$ mouse (Abstract). *Proc Am Assoc Cancer Res* 25:151, 1984.
43. Naito M, Naito Y, Ito A. Effect of phenobarbital on the development of tumors in mice treated neonatally with N-ethyl-N-nitrosourea. *Gann* 73:111–114, 1982.
44. Lagopoulos L, Stalder R. The influence of food intake on the development of diethylnitrosamine-induced liver tumours in mice. *Carcinogenesis* 8:33–37, 1987.
45. Becker FF. Characterization of spontaneous and chemically induced mouse liver tumors. In Popp JA (ed): *Mouse Liver Neoplasia*. Washington, DC: Hemisphere, 1984, pp 145–159.
46. Becker FF. Tumor phenotype and susceptibility to progression as an expression of subpopulations of initiated murine cells. *Cancer Res* 45:768–773, 1985.
47. Klaunig JE, Weghorst CM, Pereira MA. Effect of the age of B6C3F$_1$ mice on phenobarbital promotion of diethylnitrosamine-initiated liver tumors. *Toxicol Appl Pharmacol* 90:79–85, 1987.
48. Klaunig JE, Pereira MA, Ruch RJ, Weghorst CM. Dose-response relationship of

diethylnitrosamine-initiated tumors in neonatal BALB/c mice: Effect of phenobarbital promotion. *Toxicol Pathol* 16:381–385, 1988.

49. Pereira MA, Klaunig JE, Herren-Freund SL, Ruch RJ. Effect of phenobarbital on the development of liver tumors in juvenile and adult mice. *JNCI* 77:449–452, 1986.
50. Dragani TA, Manenti G, Della Porta G. Genetic susceptibility to murine hepatocarcinogenesis is associated with high growth rate of NDEA-initiated hepatocytes. *J Cancer Res Clin Oncol* 113:223–229, 1987.
51. Dragani TA, Manenti G, Galliani G, Della Porta G. Promoting effects of 1,4-bis[2-(3,5-dichloropyridyloxy)]benzene in mouse hepatocarcinogenesis. *Carcinogenesis* 6:225–228, 1985.
52. Diwan BA, Rice JM, Ward JM, Ohshima M, Lynch PH. Inhibition of phenobarbital and lack of effect of amobarbital on the development of liver tumors induced by N-nitrosodiethylamine in juvenile $B6C3F_1$ mice. *Cancer Lett* 23:223–234, 1984.
53. Pereira MA, Knutsen GL, Herren-Freund SL. Effect of subsequent treatment of chloroform or phenobarbital on the incidence of liver and lung tumors initiated by ethylnitrosourea in 15-day-old mice. *Carcinogenesis* 6:203–207, 1985.
54. Tsuji S, Ogawa K, Takasaka H, Sonoda T, Mori M. Clonal origin of γ-glutamyl transpeptidase-positive hepatic lesions induced by initiation-promotion in ornithine carbamoyl transferase mosaic mice. *Jpn J Cancer Res* 79:148–151, 1988.
55. Diwan BA, Rice JM, Ohshima M, Ward JM. Interstrain differences in susceptibility to liver carcinogenesis initiated by N-nitrosodiethylamine and its promotion by phenobarbital in C57BL/6NCr, $C3H/HeNCr^{MTV-}$ and DBA/2NCr mice. *Carcinogenesis* 7:215–220, 1986.
56. Nims RW, Devor DE, Henneman JR, Lubet RA. Induction of alkoxyresorufin O-dealkylases, epoxide hydrolase, and liver weight gain: Correlation with liver tumor-promoting potential in a series of barbiturates. *Carcinogenesis* 8:67–71, 1987.
57. Pound AW. Influence of carbon tetrachloride on induction of tumours of the liver and kidneys in mice by nitrosamines. *Br J Cancer* 37:67–75, 1978.
58. Williams GM, Numoto S. Promotion of mouse liver neoplasms by the organochlorine pesticides chlordane and heptachlor in comparison to dichlorodiphenyltrichloroethane. *Carcinogenesis* 5:1689–1696, 1984.
59. Diwan BA, Rice JM, Ward JM. Tumor-promoting activity of benzodiazepine tranquilizers, diazepam and oxazepam, in mouse liver. *Carcinogenesis* 7:789–794, 1986.
60. Shivapurkar N, Wilson MJ, Hoover KL, Mikol YB, Creasia D, Poirier LA. Hepatic DNA methylation and liver tumor formation in male C3H mice fed methionine- and choline-deficient diets. *JNCI* 77:213–217, 1986.
61. Hagiwara A, Diwan BA, Ward JM. Modifying effects of butylated hydroxyanisole, di(2-ethylhexyl)phthalate or indomethacin on mouse hepatocarcinogenesis initiated by N-nitrosodiethylamine. *Jpn J Cancer Res* 77:1215–1221, 1986.
62. Hagiwara A, Ward JM. The chronic hepatotoxic, tumor-promoting, and carcinogenic effects of acetaminophen in male $B6C3F_1$ mice. *Fundam Appl Toxicol* 7:376–396, 1986.
63. Newberne PM, De Camargo JLV, Clark AJ. Choline deficiency, partial hepatectomy, and liver tumors in rats and mice. *Toxicol Pathol* 10:95–109, 1982.
64. Witschi HP. Enhanced tumour development by butylated hydroxytoluene (BHT) in the liver, lung and gastrointestinal tract. *Food Chem Toxicol* 24:1127–1130, 1986.

65. Domingo EP, Warren KS, Steiner RJ. Increased incidence of hepatoma in mice with chronic schistosomiasis treated with a carcinogen. *Am J Pathol* 51:307–321, 1967.
66. Cole LJ, Nowell PC. Accelerated induction of hepatomas in fast neutron irradiated mice with carbon tetrachloride. *Ann NY Acad Sci* 114:259–267, 1964.
67. Griciute L, Castegnaro M, Bereziat J-C. Influence of ethyl alcohol on carcinogenesis induced with N-nitrosodiethylamine. *IARC Sci Publ* 56:413–417, 1984.
68. Vesselinovitch SD, Itze L, Mihailovich N, Rao KVN. Modifying role of partial hepatectomy and gonadectomy in ethylnitrosourea-induced hepatocarcinogenesis. *Cancer Res* 40:1538–1542, 1980.
69. Hashimoto Y, Degawa M, Kojima M, Hishinuma T. Induction of carcinogen activation enzyme(s) by feeding of a carcinogenic tryptophan pyrolysate correlates to sex difference in the carcinogenesis of the mouse. *Gann* 73:508–510, 1982.
70. Williams GM. Sex hormones and liver cancer. *Lab Invest* 46:352–354, 1982.
71. Lane M, Liebelt A, Calvert J, Liebelt RA. Effect of partial hepatectomy on tumor incidence in BALB/c mice treated with urethane. *Cancer Res* 30:1812–1816, 1970.
72. Hollander CF, Bentvelzen P. Enhancement of urethan induction of hepatomas in mice by prior partial hepatectomy. *JNCI* 41:1303–1306, 1968.
73. Tsuda H, Sarma DSR, Rajalakshmi S, Zubroff J, Farber E, Batzinger RP, Cha Y-N, Bueding E. Induction of hepatic neoplastic lesions in mice with a single dose of hycanthone methanesulfonate after partial hepatectomy. *Cancer Res* 39:4491–4496, 1979.
74. Chernozemski N, Warwick GP. Liver regeneration and induction of hepatomas in $B6AF_1$ mice by urethan. *Cancer Res* 30:2685–2690, 1970.
75. Schulte-Hermann R, Timmermann-Trosiener I, Schuppler J. Promotion of spontaneous preneoplastic cells in rat liver as a possible explanation of tumor production by non-mutagenic compounds. *Cancer Res* 42:839–844, 1983.
76. Grasso P. Persistent organ damage and cancer production in rats and mice. *Arch Toxicol [Suppl]* 11:75–83, 1987.
77. Hoel DG, Haseman JK, Hogan MD, Huff J, McConnell EE. The impact of toxicity on carcinogenicity studies: Implications for risk assessment. *Carcinogenesis* 9:2045–2052, 1988.
78. Newberne PM, Punyarit P, De Camargo J, Suphakarn V. The role of necrosis in hepatocellular proliferation and liver tumors. *Arch Toxicol [Suppl]* 10:54–67, 1987.
79. Pound AW, McGuire LJ. Influence of repeated liver regeneration on hepatic carcinogenesis by diethylnitrosamine in mice. *Br J Cancer* 37:595–602, 1978.
80. Ward JM, Hagiwara A, Anderson LM, Lindsey K, Diwan BA. The chronic hepatic or renal toxicity of di(2-ethylhexyl)phthalate, acetaminophen, sodium barbital, and phenobarbital in male $B6C3F_1$ mice: Autoradiographic, immunohistochemical, and biochemical evidence for levels of DNA synthesis not associated with carcinogenesis or tumor promotion. *Toxicol Appl Pharmacol* 96:494–506, 1988.
81. Amano T, Oshima T. Hepatoma formation in ddY mice with chronic Schistosomiasis japonica. *Jpn J Cancer Res* 79:173–180, 1988.

Mouse Liver Carcinogenesis: Mechanisms
and Species Comparisons, pages 109–120

Structural and Immunological Identity of p65 Tumor-Associated Factors from Rat and Mouse Hepatocarcinomas

Malgorzata Hanausek-Walaszek, Marcela Del Rio, and Alan K. Adams
Department of Carcinogenesis, Science Park, The University of Texas M. D. Anderson Cancer Center, Smithville, TX 78957

An oncofetal protein found to stimulate the nucleocytoplasmic transport of mRNA (1) and therefore referred to as oncofetal mRNA–transport protein exhibits certain properties that strongly favor its candidacy as a general marker of premalignant and malignant alterations (2). Polyacrylamide gel electrophoresis (PAGE) under denaturing conditions followed by antibody probing of western blots of blood plasma from tumor-bearing rats revealed that native species of this oncofetal protein exhibit a relative molecular weight of 65 kDa (3). At least two proteins have been identified (see ref. 4) that promote nucleocytoplasmic mRNA transport in a cell-free system: the 35-kDa protein in normal adult cells and the 65-kDa protein (p65) coexisting with the former in fetal and tumor cells (1,5). Small amounts of these factors are released to the circulatory system (1–3).

The oncofetal mRNA–transport protein p65 was present in blood plasma and tumors of more than 200 patients with confirmed diagnoses of cancer at 31 different sites, as well as in tumor-bearing animals (2,6). These data suggest that this protein may be produced by all tumors. It appears to have an M_r of 65 kDa in humans, rats, and mice (3). p65 has been found in both human and rat amniotic fluid and in fetal rats at 18 days of gestation (1). However, it has not been detected in maternal blood (5,7). p65 is also produced by and released into cell culture medium by transformed cells, but not by untransformed cells (8). This factor is not induced by cellular proliferation associated with nonneoplastic diseases and conditions and is consistently absent from the plasma of normal, healthy control subjects (2,7). It has not been detected in the rat during liver regeneration (hyperplasia) in response to a 2/3 partial hepatectomy (8).

Data accumulated to date indicate that all of 18 known chemical carcinogens, but none of 12 known noncarcinogens tested, induce the p65 factor in the rat (8,9). The absence of inducing activity of selected promoting agents was confirmed in the C3H/10T1/2 cells/12-*O*-tetradecanoylphorbol-13-acetate system in vitro (8), and with 2,3,7,8-tetrachlorodibenzo-*p*-dioxin and phenobarbital in vivo (8). Induction appears to require an initiating agent (10). Although it is convenient to measure p65 activity in blood plasma, it was also present in the tumor cytosols examined to date (1,3,5,11) and in the target organ (liver) within a relatively short period of time (i.e., 40–48 h) after carcinogen treatment of the experimental animals (10,12). Specifically, significant amounts of p65 are detectable in rat liver cytosol within 40 h of treatment with the hepatocarcinogen 3′-methyl-4-dimethylaminoazobenzene (12).

The 35-kDa factor is present in all adult cells and coexists with the 65-kDa factor in tumor cells (1,3,5,11), fetal cells (1), and carcinogen-treated adult cells (10,12) of the rat. Evidence indicates that both the 65-kDa (1) and the 35-kDa factors (13) are phosphoproteins. Neither exhibits exo- or endonuclease, protease, or protein kinase activities (1,13). p65 induces the release of some RNA sequences, i.e., non-intronic repetitive sequences not released from the nuclei in response to the 35-kDa factor (12). p65 produced by transformed rat cells or rat fetal cells and induced in normal adult cells of the rat by carcinogens (early phase) appear to be immunologically identical (1,9,11). The antibody against rat p65 does not crossreact with the rat 35-kDa factor (11). Both the 35-kDa (13) and 65-kDa (3) protein can be purified by nucleic acid affinity chromatography. p65 may be identical with one of the two cytoplasmic, poly (A)–binding proteins that have recently been characterized (4). This poly (A)–binding protein was shown (4,14) to stimulate mRNA transport in a cell-free system similar to the system we have been using to measure the mRNA transport activity of p65 (11).

The primary objective of this study was to characterize the p65 factor isolated from mouse and rat hepatocellular carcinomas and to show that antibodies to p65 are useful markers of new cell populations that appear in the livers of rats subjected to a chemical hepatocarcinogenesis protocol.

Materials and Methods

Rat and Mouse Liver Tumors

Rat hepatoma cell lines McA-RH7777 and McA-RH8994 were purchased from the American Type Culture Collection (Rockville, MD) and carried as cell cultures in Swim's S77 medium with 4 mM L-glutamine and supplemented with 5% fetal calf serum and 20% horse serum (Gibco, Grand Island, NY). After several passages in cell culture, rat hepatoma cells (1 × 106/0.2 ml phosphate-buffered saline [PBS]) were inoculated subcutaneously into a hind leg of male Buffalo rats (120–150 g) (Harlan Labs, Indianapolis, IN) and were carried as solid tumors according to the standard procedure (see refs. 2,9). Transplantable hepatocellular carcinoma (THC) 1682C was derived from primary hepatic tumors in ACI rats maintained for 4 months on the choline-deficient diet of Shinozuka (15) containing 0.2% ethionine and 8 months on a choline-supplemented (0.8%) diet without ethionine. THC T52 was established by intraperitoneal transplantation of growing tumors induced in ACI rats by 2-acetylaminofluorene (AAF) (16). THC cultures were maintained in vitro (17). THC were also carried as solid (1682C) and ascitic (T52) tumors in male ACI rats (The University of Texas M. D. Anderson Cancer Center, Science Park–Veterinary Resources Division, Bastrop, TX). The Reuber hepatoma cell culture H35 was provided by Dr. Andrew P. Butler, UT M. D. Anderson Cancer Center Carcinogenesis Department, Science Park, Smithville, TX). Mouse liver carcinoma CRL 6421 (formerly NBL #MM45T.Li) was obtained from American Type Culture Collection and carried as cell cultures as described above for rat hepatoma cell lines.

Purification of p65 and Preparation of Antibodies

Rat p65 was purified to apparent homogeneity (see ref. 3) as follows. The proteins precipitating at 90% saturation with ammonium sulfate from the culture medium in which THC 1682C cells were grown were dissolved in TMK buffer (50 mM Tris-HCl, pH 7.5; 25 mM KCl; 2.5 mM $MgCl_2$), dialyzed against the same buffer, then fractionated on a 0.8 × 30 cm TSK 3000 SW column in an LKB high-performance liquid chromatography system (LKB Instruments, Gaithersburg, MD). p65 was further purified on an LKB TSK Phenyl-5PW hydrophobic interactions column. A final purification was performed using affinity chromatography on a single-strand DNA-cellulose column (3). Mouse p65 was isolated from the mouse liver carcinoma cell line CRL 6421 (MM45) and purified as described above for the rat factor. Polyclonal antibodies to the purified rat p65 preparation were raised in rabbits, essentially as previously described (1,11), in collaboration with Dr. Lezlee Coghlan (M. D. Anderson Cancer Center, Science Park–Veterinary Resources Division, Bastrop, TX).

CNBr Cleavage of p65

Cyanogen bromide (CNBr) cleavage maps of p65 purified from cell culture medium of rat transplantable hepatocellular carcinoma cell line 1682C or mouse liver carcinoma cell line CRL 6421 (MM45) were obtained as follows. p65 preparations (20 μg) purified from cell culture medium were subjected to 12.5% sodium dodecyl sulfate (SDS)-PAGE. The p65 band located on slab gels by Coomassie blue staining of parallel gel tracks was cut from appropriate gel tracks and incubated at room temperature for 16 h in 1 ml of 88% formic acid containing 20 mg/ml of CNBr (18). The gel slices were then rinsed five times with 1 ml of water and washed several times for 10 min with 1 ml 120 mM Tris-HCl, pH 7.0, 20% glycerol (vol/vol), 2% SDS until the pH of the slices reached 7.0. CNBr-treated gel slices were placed onto 15% SDS-PAGE slabs, electrophoresed, and then stained with silver (19). Slices of gel containing p65 treated with 88% formic acid served as controls.

SDS-PAGE and Immunoblotting Analyses

Carefully dissected hepatomas and liver fragments from normal rats and mice were rinsed with an ice-cold saline solution and processed for immunochemical determination of p65. Liver tumor cells grown in culture were separated from medium by centrifugation, rinsed with the cold saline solution, and processed for immunoassay of the p65 tumor-associated factor. Briefly, small pieces of tissues or tumor cell pellets were homogenized in TMK-sucrose buffer, pH 7.2, and samples of the total homogenate containing the same amount of protein were mixed with electrophoresis sample buffer, boiled for 3 min, and subjected to 10% SDS-PAGE. The proteins precipitated at 90% saturation of ammonium sulfate from the conditioned medium of hepatocarcinoma cells were dissolved in TMK buffer and electrophoresed in a similar manner. Subsequently the gel slabs were prepared for transfer by equilibration for 1 h in 0.025 M Tris, 0.192 M glycine, 20% methanol (vol/vol), pH 8.3, and then transblotted onto nitrocellulose sheets. For immunoassay, the nitrocellulose sheets were treated with the appropriate blocking solution to block nonspecific binding sites and incubated overnight either with preimmune serum (controls) or antibody against rat p65 diluted 1:200. Secondary biotinylated antibody was applied next, and the color was developed using an ABC Elite Kit (Vector Laboratories, Burlingame, CA). The highly specific polyclonal antibody against p65 was obtained by immunizing rabbits with the rat p65 preparation purified as described above.

Induction of Altered Hepatic Foci

Altered hepatic foci (AHF) were induced in the course of 2-acetylaminofluorene (AAF)-initiated, phenobarbital (PB)-promoted hepatocarcinogenesis in the rat (20). Rats (male weanling albinos, Sprague-Dawley strain) were purchased from Harlan Labs. The 0.06% (wt/wt) AAF diet and 0.05% PB diet were prepared and pelleted by Altromin (Lage, FRG) and by Dyets, Inc. (Bethlehem, PA), respectively. Each of two experimental groups of rats and the control group consisted of 40 rats. All rats were 22 days old when first used in the experiment. One group received an AAF diet for 18 days and then an AIN-76A diet. The second group received an AAF diet for 18 days, AIN-76A for 24 days, and then an AIN-76A diet supplemented with 0.05% phenobarbital. The third group (controls) received an AIN-76A diet for 42 days, then an AIN-76A diet supplemented with 0.05% phenobarbital. The livers of rats (4 animals per each time point) killed at different times, up to 48 weeks, in the course of AAF and/or PB feeding were used for immunochemical studies.

Immunohistochemical Staining for p65

Paraffin-embedded liver sections were prepared and stained for p65 with specific antibodies raised in rabbits by immunization with purified rat p65. p65 was visualized in the liver sections using the avidin-biotin-peroxidase complex system (Vectastain ABC Elite Kit, Vector Laboratories). Antiserum to p65 was diluted 1:200 in PBS with 1% goat serum for use in the staining protocol. Appropriate controls with nonimmune serum were performed routinely.

Immunochemical Staining for Glutathione S-Transferase P (GST-P)

GST-P was visualized in paraffin liver sections by an immunohistochemical procedure (21). An antibody to rat GST-P was kindly provided by Dr. E. Farber (University of Toronto, Canada). The anti-GST-P antiserum was diluted 1:400 in PBS with 1% normal goat serum.

Results and Discussion

Immunological Identity of p65 Released by Rat and Mouse Liver Carcinoma Cells

PAGE under denaturing conditions, followed by anti-rat p65 antisera probing of western blots of the conditioned media, was used to confirm the secretion of p65 by hepatocellular carcinoma cells. As shown in Figure 1 a single

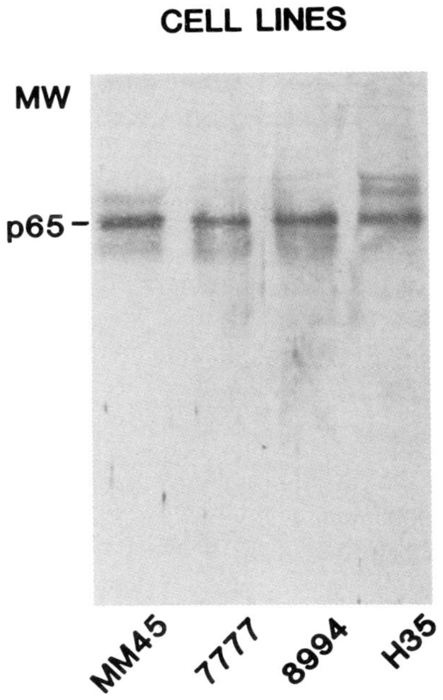

Figure 1. Analysis by SDS-PAGE and immunoblotting (western blot) of total protein samples (20 μg) from the conditioned medium of the following cell lines: mouse liver carcinoma MM45, Morris hepatoma 7777, Morris hepatoma 8994, Reuber hepatoma H35. After electrophoresis, the proteins were transferred to nitrocellulose paper and probed with antibodies to rat p65 (lanes MM45, 7777, 8994, H35) or preimmune serum (data not shown), as described in Materials and Methods.

band, in the region of 65 kDa, was detected by polyclonal antibodies to p65. In fact, p65 was released to culture medium by all transformed liver cells tested. Some weak bands in the blots resulted from nonspecific staining. No band was detected in the 65-kDa region when unconditioned medium was tested under the same conditions (data not shown).

It was previously demonstrated (8) with a biochemical assay (induced transport of mRNA from isolated nuclei) that p65 was secreted to and accumulated in the conditioned medium of cultured transformed cells in vitro. A combination of the biochemical assay and enzyme-linked immunosorbent assay (ELISA) was then used (11) to show the immunological identity of p65 induced in rats by chemical carcinogens and released by transformed cells in vitro. In the current study, immunoblot analysis of the cell culture media was successfully used for the first time to reveal the presence of p65 production and secretion by three transplantable rat hepatoma cell lines and a mouse liver carcinoma cell line. Strong crossreactivity of the antibodies to rat p65 with mouse p65 suggests that rat and mouse factors may be immunologically related, if not identical.

Presence of p65 in Rat and Mouse Liver Tumors and Its Absence from Normal Mouse and Rat Liver

When total protein samples from rat and mouse liver carcinomas grown in vitro were analyzed by SDS-PAGE and then anti-rat p65 antibody probing of western blots, a single, prominent band was detected in the 65-kDa region of the blots shown in Figure 2. Some weak bands seen in the 60–64 kDa region may represent degradation products of the native species of p65 (3). Other minor

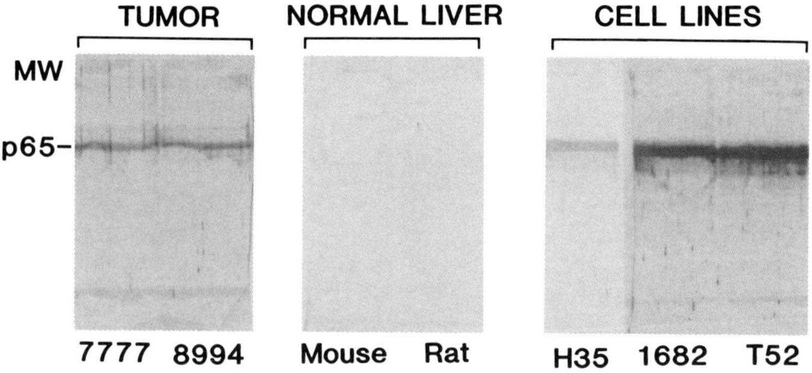

Figure 2. SDS-PAGE and immunoblot analysis of total protein samples from rat hepatomas, normal rat and mouse liver, and liver tumor cell lines of rat and mouse origin. Twenty micrograms of protein from Morris hepatoma 7777 and 8994 and the same amount of protein from normal mouse and rat liver and three different hepatocellular carcinoma cell lines were electrophoresed by 12.5% SDS-PAGE, transblotted to nitrocellulose sheets, and probed with anti-rat p65 antibody.

bands resulted from nonspecific staining. There is no band characteristic of p65 in immunoblots representing normal adult rat or mouse liver.

All hepatocarcinoma cells grown in vitro as cell cultures and tumor tissues from Morris hepatomas carried as solid tumors in vivo were positive for the p65 antigen. There was no reaction with normal liver cells of either rat or mouse origin. Thus, immunoblotting analysis has demonstrated that p65 was specifically produced by liver cancer cells but not by the cells of normal adult rat or mouse liver.

p65 as a Marker of New Cell Populations Induced During Chemical Hepatocarcinogenesis

Since p65 synthesis had not been previously studied at the cellular and subcellular level, it appeared to be essential to determine the sites of p65 synthesis in the cell and assess the value of this protein as a marker for preneoplastic lesions during chemical hepatocarcinogenesis. Altered hepatic foci were induced by feeding male weanling Sprague-Dawley rats AAF and then a diet containing phenobarbital (20).

As shown in Figure 3, p65 was predominantly seen in the cells of putative preneoplastic foci found at 24 weeks in the liver of a rat fed the AAF and PB diets. p65 was highly concentrated in the foci, but little or none was detected in the surrounding cells (Fig. 3, panels A, C). The p65-positive cells were negative when preimmune rabbit sera were used instead of immune sera. No positive staining was found in control livers from rats fed only the PB diet (data not shown). Most of the p65 activity appeared to be associated with the nuclei of p65-positive hepatocytes and, more precisely, with nuclear envelopes; relatively less was detected in the cytoplasm (panel D). These results affirm the putative role of p65 in the nucleocytoplasmic transport of mRNA.

GST-P has recently been shown (see refs. 22,23) to be useful in the phenotypic analysis of some hepatocyte populations, including altered hepatic foci or nodules that appear during cancer development. Therefore we have included GST-P in our comparative study of the expression of p65 during chemical hepatocarcinogenesis. Panels A and B show consecutive sections stained for p65 and GST-P. Immunohistochemical staining for p65 appears to detect foci within the foci positive for GST-P. A detailed paper on the kinetics of p65 production during AAF-initiated, PB-promoted carcinogenesis is now being prepared for publication. Differential expression of p65 and other markers of AHF, including γ-glutamyltranspeptidase (GGT) and GST-P, in foci of different zones of rat liver is also under study.

Structural Identity of Rat and Mouse p65

Rat and mouse p65 were purified to apparent homogeneity as described in Materials and Methods. CNBr cleavage maps shown in Figure 4 were obtained as the first step toward the final characterization of the amino acid composition and sequence of p65. The cleavage of p65 with CNBr resulted in six major peptides identifiable by silver staining of the SDS-PAGE gels. The peptides have molecular weights of 8 to 45 kDa. Identical cleavage maps were obtained for rat and mouse p65.

Thus, rat and mouse p65 may be structurally related—if not identical. A study of the amino acid composition and sequence of p65 and the peptides yielded by CNBr cleavage is in progress. Our ultimate goal is to produce cDNA for p65 to study how chemical carcinogens activate its gene.

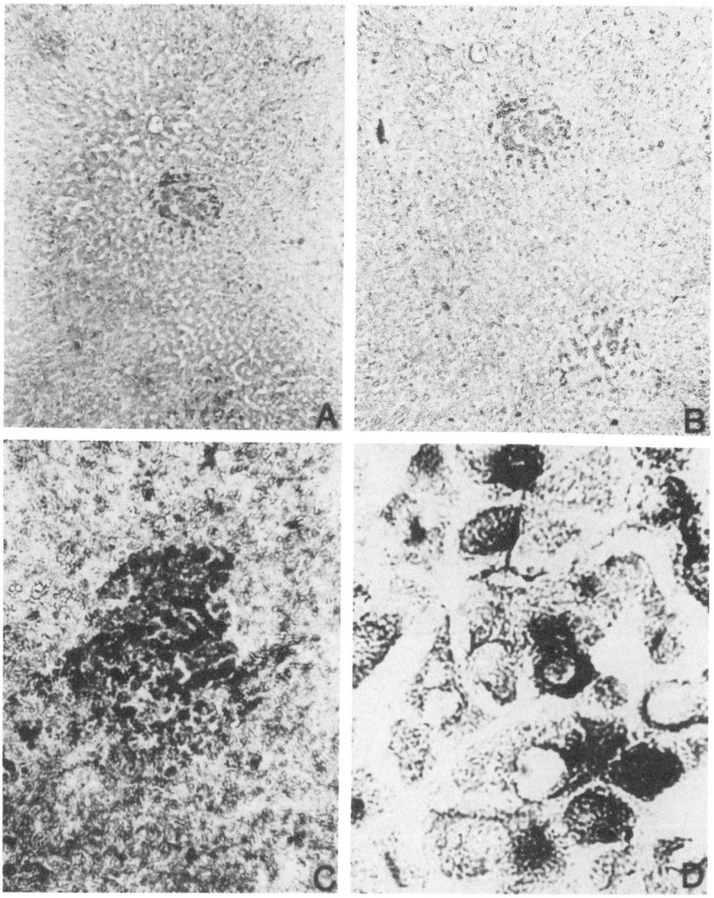

Figure 3. Immunohistochemical staining for p65 (A,C,D) and GST-P (B) of paraffin-embedded sections of livers taken at 24 weeks from rats fed AAF and PB diets (see Materials and Methods). Consecutive sections stained for p65 and GST-P are shown in panels A and B, respectively. Magnification × 100 (A,B); × 400 (C); × 1000 (D).

Conclusions

We conclude that p65 tumor-associated factors from rat and mouse liver carcinomas appear immunologically and structurally related, if not identical. In fact, polyclonal antibodies produced in rabbits against rat p65 detected the p65 factor in both rat and mouse liver carcinomas but not in normal rat or mouse liver. Furthermore, p65 was found in the cells of putative preneoplastic lesions induced in livers of rats by the AAF/PB diets; little or none was found in the surrounding cells. The cellular origin of the early, rapid rise of the 65-kDa protein after administration of a chemical carcinogen (6,8,9,12) remains to be determined.

The number of 65-kDa protein-positive foci detected at 12–48 weeks represents less than 10% of the total number of altered hepatic foci positive for GST-P (M. Hanausek-Walaszek, unpublished data). Whether antibodies to the 65-kDa protein are specific for persistent foci and nodules induced by this and by other hepatocarcinogenesis protocols remains to be determined. Studies from a number of laboratories have demonstrated that the majority of focal and

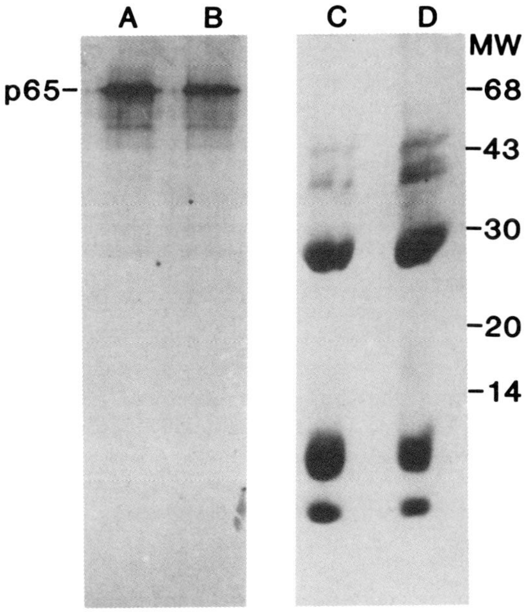

Figure 4. CNBr cleavage of rat and mouse p65 factor. p65 proteins from medium of rat hepatocellular carcinoma cell line 1682C or mouse liver carcinoma cell line MM45 were purified to apparent homogeneity (see Materials and Methods) and electrophoresed by 12.5% SDS-PAGE. Lanes A and B, rat and mouse p65 factors, respectively, treated with 88% formic acid. Lanes C and D, CNBr-digested rat and mouse p65 factors, respectively. p65 was digested with 20 mg/ml CNBr in 88% formic acid at room temperature for 16 h, as described in Materials and Methods.

nodular lesions that occur during the early phase of multistage hepatocarcinogenesis in the rodents disappear or remodel upon removal of the carcinogenic stimulus (see refs. 24–26). The alterations that persist upon cessation of a carcinogenic regimen may be of great importance in the delineation of changes critical to the development of malignant tumors (27). If 65-kDa protein production during chemical hepatocarcinogenesis is associated with persistent preneoplastic lesions, antibodies to the 65-kDa protein may prove useful not only as a tumor marker but also as a marker of these phenotypes on the pathway to transformed cells, both in rats and mice.

Acknowledgments

The authors wish to thank Drs. H.C. Pitot and H. Shinozuka for comments on this manuscript. The gifts of antisera to GST-P and the AAF diet obtained from Drs. E. Farber and F. F. Becker, respectively, are greatly appreciated. The expert photographic assistance of J. Ing is also appreciated. This work was aided, in part, by NIH grant RR-5511-23 and by grant SIG 14 from the American Cancer Society.

References

1. Hanausek-Walaszek M, Walaszek Z, Lang RW, Webb TE. Characterization of a 60,000-dalton oncofetal protein from the plasma of tumor-bearing rats. *Cancer Invest* 2:433–441, 1984.
2. Hanausek-Walaszek M, Walaszek Z, Webb TE. A 60 kilodalton oncofetal protein as tumor marker. *J Med* 17:13–24, 1986.
3. Hanausek-Walaszek M, Adams AK, Del Rio-Larcher M, Kriewaldt SD, Walaszek Z. Plasma oncofetal mRNA-transport protein is proportional to tumor burden in rats bearing transplantable hepatocarcinomas (Abstract). *Proc Am Assoc Cancer Res* 29:167, 1988.
4. Schroeder HC, Rottman M, Bachmann M, Mueller WEG, McDonald AR, Agutter PS. Proteins from rat liver cytosol which stimulate mRNA transport: Purification and interactions with the nuclear envelope mRNA translocation system. *Eur J Biochem* 159:15–59, 1986.
5. Walaszek Z, Hanausek-Walaszek M, Schumm DE, Webb TE. An oncofetal 60 kilodalton protein in the plasma of tumor-bearing and carcinogen-treated rats. *Cancer Lett* 20:277–282, 1983.
6. Schumm DE, Webb TE. Putative transformation-dependent proteins in the blood plasma of tumor-bearing rats and cancer patients. *Cancer Res* 42:4964–4969, 1982.
7. Schumm DE, Hanausek-Walaszek M, Walaszek Z, Webb TE. Absence of the cancer-associated factors with a molecular weight of 60,000 from the plasma of patients with a spectrum of non-neoplastic conditions. *Cancer Res* 44:401–406, 1984.
8. Hanausek-Walaszek M, Walaszek Z, Webb TE. Carcinogens as specific inducers of a 60-kilodalton protein in rats. *Carcinogenesis* 6:1725–1730, 1985.

9. Hanausek-Walaszek M, Montgomery N, Walaszek Z, Lang RW, Webb TE. Correspondence between biochemical and antigenic activity of a 60 kilodalton oncofetal protein during carcinogenesis and tumorigenesis. *Cancer Lett* 33:55–61, 1986.
10. Hanausek-Walaszek M, Walaszek Z, Webb TE. Use of the 60 kilodalton oncofetal protein for monitoring chemical hepatocarcinogenesis. *Fed Proc* 45:698, 1986.
11. Hanausek-Walaszek M, Lang R, Walaszek Z, Webb TE. Immunological identity of a 60 kd oncofetal protein induced in rats by chemical carcinogens and released by transformed cells. *Biochem Biophys Res Commun* 127:779–785, 1985.
12. French BT, Hanausek-Walaszek M, Walaszek Z, Schumm DE, Webb TE. Nucleocytoplasmic release of repetitive DNA transcripts in carcinogenesis correlates with a 60 kilodalton cytoplasmic protein. *Cancer Lett* 23:45–52, 1986.
13. Moffett RB, Webb TE. Characterization of a messenger RNA-transport protein. *Biochim Biophys Acta* 740:231–242, 1983.
14. Schroeder HC, Becker R, Bachmann M, Gramzow M, Seve AP, Monsigmy M, Muller WEG. Differential changes of nuclear-envelope-associated enzyme activities involved in nucleocytoplasmic mRNA transport in the developing rat brain and liver. *Biochim Biophys Acta* 868:108, 1986.
15. Shinozuka H, Lombardi B, Sell S, Immarino RM. Early histological and functional alterations of ethionine liver carcinogenesis in rats fed a choline-deficient diet. *Cancer Res* 38:1092–1098, 1978.
16. Becker FF, Klein KH, Wolman SR, Asofky R, Sell S. Characterization of primary hepatocellular carcinomas and initial transplant generations. *Cancer Res* 33:3330–3338, 1973.
17. Hixon DC, McEntire KD, Obrink B. Alterations in the expression of a hepatocyte cell adhesion molecule by transplantable rat hepatocellular carcinomas. *Cancer Res* 45:3742–3749, 1985.
18. Zingde SM, Shirsat NV, Gothoska BP. Peptide mapping of proteins in gel band after partial cleavage with acidic cyanogene bromide vapors. *Anal Biochem* 155:10–13, 1986.
19. Wray W, Boulikas T, Wray VP, Hancock R. Silver staining of proteins on polyacrylamide gels. *Anal Biochem* 118:197–203, 1981.
20. Peraino C, Fry RJM, Staffeldt E, Kisieleski WE. Effects of varying the exposure to phenobarbital on its enhancement of 2-acetylaminofluorene-induced hepatic tumorigenesis in the rat. *Cancer Res* 33:2701–2705, 1973.
21. Rushmore TH, Harris L, Nagai M, Sharma RN, Hayes MA, Cameron RG, Murray RK, Farber E. Purification and characterization of P-52 (glutathione S-transferase-P or 7-7) from normal liver and putative preneoplastic liver nodules. *Cancer Res* 48: 2805–2812, 1988.
22. Rushmore TH, Sharma RN, Roomi MW, Harris L, Satoh K, Sato K, Murray RK. Identification of characteristic cytosolic peptide of rat preneoplastic hepatocyte nodules as placental gluthathione S-transferase. *Biochem Biophys Res Commun* 143:98–103, 1987.
23. Pitot HC, Campbell HA. An approach to the determination of the relative potencies of chemical agents during the stages of initiation and promotion in multistage hepatocarcinogenesis. *Environ Health Perspect* 76:49–56, 1987.
24. Farber E, Sarma DSR. Chemical carcinogenesis: The liver as a model. *Pathol Immunopathol Res* 5:1–28, 1986.
25. Goldfarb S, Pugh TD. Multistage rodent hepatocarcinogenesis. *Prog Liver Dis* 8: 597–620, 1986.

26. Goldsworthy TL, Hanigan MH, Pitot HC. Models of hepatocarcinogenesis in the rat: Contrasts and comparisons. *CRC Crit Rev Toxicol* 9:61–89, 1985.
27. Teebor GW, Becker FF. Regression and persistence of hyperplastic hepatic nodules induced by N-2-fluorenylacetamide and their relationship to hepatocarcinogenesis. *Cancer Res* 31:1–30, 1971.

Mouse Liver Carcinogenesis: Mechanisms
and Species Comparisons, pages 121–129

DNA Adduct Formation in Relation to Tumorigenesis in Mice Chronically Fed 2-Acetylaminofluorene

Frederick A. Beland, Nancy F. Fullerton,* Takemi Kinouchi,[1] Beverly A. Smith,* and Miriam C. Poirier†*
**National Center for Toxicological Research, Jefferson, AR 72079 and †National Cancer Institute, Bethesda, MD 20892*

A dominant theme in cancer research is that most tumors arise from the covalent interaction of chemical carcinogens with DNA (1). However, the relationship between the concentration of carcinogen-DNA adducts in a target tissue and the resultant tumorigenic response is by no means clear. Efforts to elucidate this relationship have been impaired for two reasons: first, few studies have been conducted in which tumors have been induced over a wide range of carcinogen concentrations (2–6); and second, methods that permit the measurement of DNA adducts after the administration of carcinogens over the entire dose range used in the tumor protocols have been developed only recently (1,7).

One of the largest dose-response tumor studies so far conducted involved the feeding of 2-acetylaminofluorene (AAF) to 20,880 female BALB/c mice at seven different doses for up to 33 months (2). Treatment-related tumors were found in two organs, the liver and bladder; however, there was a difference in response between these two organs. In the liver, a linear relationship was observed between the administered dose and the tumor incidence (Fig. 1). In the bladder, this was not the case; at doses below 75 mg AAF/kg diet the tumor incidence did not differ from that of controls, whereupon it rose very steeply (Fig. 1). The difference in tumor response between these two tissues suggested that this might be an interesting model with which to examine the relationship between the concentration of DNA adducts and tumor incidence.

Materials and Methods

Weanling female BALB/c mice (four to five per group; obtained from the breeding colony at the National Center for Toxicological Research) were used in

---

[1]Permanent address: Department of Bacteriology, The University of Tokushima School of Medicine, Tokushima, Japan

each of the experiments. In the first experiment, the mice were fed either 30 or 150 mg [^{3}H]AAF (982 mCi/mmol)/kg diet for 21 days; in the second, they were fed 30 or 150 mg AAF/kg diet for 0, 3, 7, 14, 21, 28, 56, or 112 days; and in the third, the mice were fed 0, 5, 10, 15, 30, 45, 75, 100, or 150 mg AAF/kg diet for 28 days. After the mice were killed, hepatic nuclei were prepared by the method of Basler et al. (8). DNA was then isolated from whole bladders and hepatic nuclei as outlined in Beland et al. (9). Adduct levels in liver and bladder DNA from the mice administered [^{3}H]AAF were quantified by liquid scintillation counting. Hepatic DNA from these mice was then hydrolyzed enzymatically to nucleosides, and the adducts were analyzed by high-performance liquid chromatography (HPLC) as previously reported (10). The amount of radioactivity associated with the bladder DNA was insufficient for adduct analysis by HPLC. Adduct levels in liver and bladder DNA from mice fed nonradioactive AAF were determined by radioimmunoassay using a polyclonal rabbit antibody specific for *N*-(deoxyguanosin-8-yl)-2-aminofluorene (dG-C8-AF) (11). The identity of this adduct was confirmed by ^{32}P-postlabeling (12,13) through comparison with DNA modified with dG-C8-AF (14).

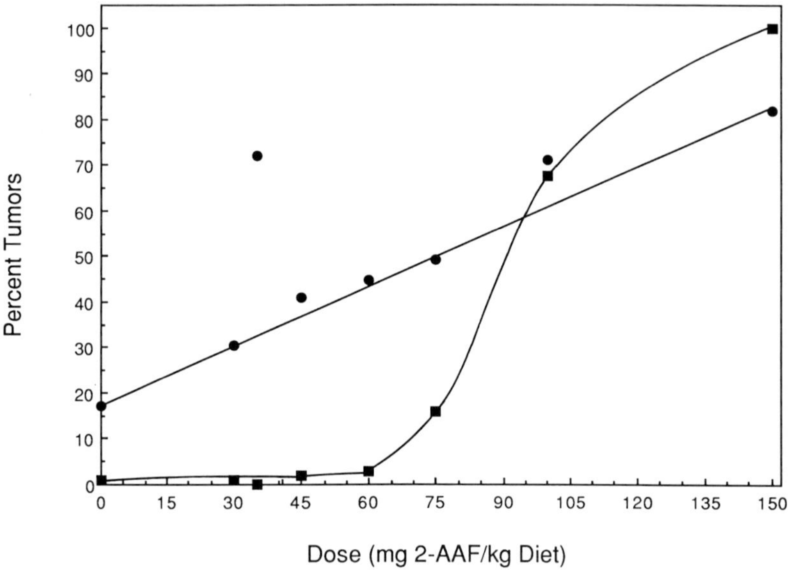

Figure 1. Relationship between the dose of AAF fed to female BALB/c mice for 33 months and the percentage of animals having liver (circles) and bladder (squares) tumors. Data are adapted from Farmer et al. (2).

Results

To determine the spectrum of carcinogen-DNA adducts formed during the chronic feeding of AAF, hepatic DNA adducts were analyzed by HPLC in mice fed 30 or 150 mg [^{3}H]AAF/kg diet for 21 days. As has been reported previously for mice treated with *N*-hydroxy-AAF or *N*-hydroxy-2-aminofluorene (15–17), dG-C8-AF was the major adduct detected in the liver DNA (data not shown). Adduct analyses were also conducted by ^{32}P-postlabeling of liver and bladder DNA from mice fed 30 or 150 mg nonradioactive AAF/kg diet for 28 days. With both tissues, one adduct was observed that had chromatographic characteristics identical to dG-C8-AF (Fig. 2). Although adduct analyses were conducted with DNA samples obtained from the entire bladder rather than just the epithelium, which is presumably the target cell, immunohistochemical analyses have indicated that the adducts are associated with the epithelial cells to a much greater extent than the underlying stroma (Huitfeldt et al., unpublished data).

In mice fed 30 or 150 mg AAF/kg diet for up to 112 days, there was a rapid increase in DNA adduct levels in liver and bladder for approximately 21 days of feeding (data not shown). The rate of increase then became much slower; therefore, a 28-day feeding period was selected to examine the relationship between DNA adducts and tumorigenicity over the entire dose range used in the tumor study (2). The relationship between mice fed 30, 45, 75, 100, or 150 mg AAF/kg diet for 28 days and the concentration of dG-C8-AF in liver and bladder DNA is shown in Figure 3. A linear correlation was observed in both tissues, and at each dose the adduct concentration was two- to fourfold higher in the bladder as compared to the liver. In this experiment, AAF was also fed at 5, 10, and 15 mg/kg diet, which were doses lower than those used in the tumor study (2). With these groups, the same linear correlation was observed between the adminis-

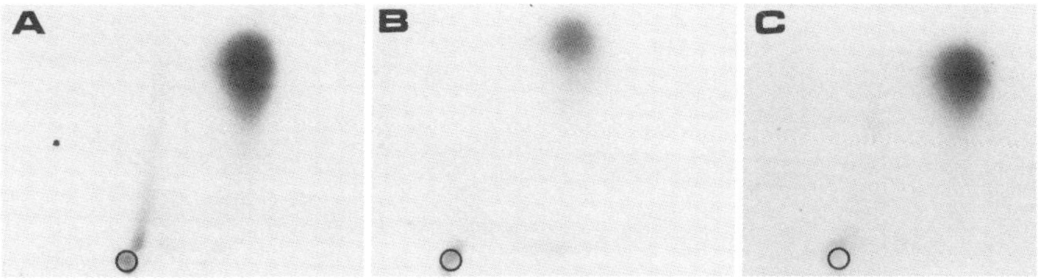

Figure 2. Autoradiograms of adducts obtained from: (A) reacting *N*-hydroxy-2-aminofluorene with calf thymus DNA at pH 5 to give dG-C8-AF; (B) liver DNA of a female BALB/c mouse fed 30 mg AAF/kg diet for 28 days; (C) bladder DNA of the same mouse. ^{32}P-Postlabeling was conducted as described in Gupta (12). Adducts were resolved by the contact transfer method of Lu et al. (13) using the following solvents: D1 and D5, 0.9 M sodium phosphate, pH 6.8; D3, 3.6 M lithium formate, 8.5 M urea, pH 3.5; D4, 1.2 M lithium chloride, 0.5 M Tris-HCl, 8 M urea, pH 8.0. Adducts were visualized by autoradiography at –80°C for 14.5 h (A) or 4 h (B and C). The adduct levels are: (A) 9 fmol/µg DNA; (B) 29 fmol/µg DNA; (C) 38 fmol/µg DNA.

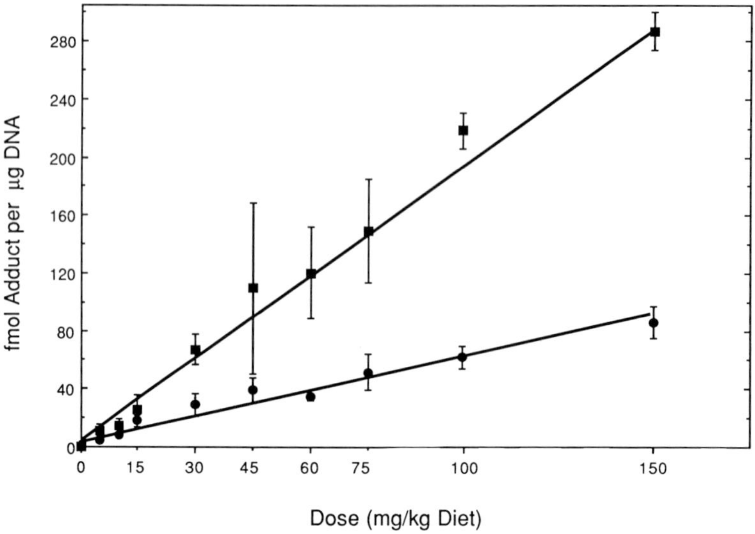

Figure 3. Relationship between the dose of AAF fed to female BALB/c mice for 28 days and the concentration of dG-C8-AF in liver (circles) and bladder (squares) DNA as determined by radioimmunoassay. The data are expressed as the mean ± standard deviation for four or five mice.

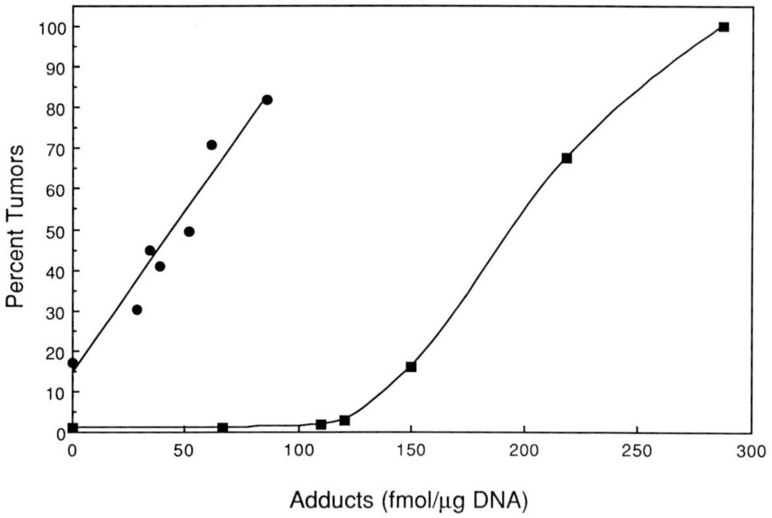

Figure 4. Relationship between the concentration of dG-C8-AF in liver (circles) and bladder (squares) DNA of female BALB/c mice fed 0 to 150 mg AAF/kg diet for 28 days and the tumor incidence in the same tissues after feeding the same concentrations of AAF for 33 months.

tered dose and the concentration of dG-C8-AF as was found with the higher dose levels (Fig. 3).

The relationship between the levels of dG-C8-AF, as determined after 28 days of AAF feeding, and the tumor incidence in mice fed AAF for 33 months is illustrated in Figure 4. In the liver, there was a linear correlation between the adduct concentration and the tumor incidence. In the bladder, however, a linear correlation was not found because at adduct levels <150 fmol/μg DNA, the bladder tumor incidence did not exceed that of mice on the control diet. It is also clear from this figure that much higher adduct concentrations are necessary in the bladder to induce the same tumor incidence observed in the liver.

Discussion

These data show that tissue-specific differences play a significant role in AAF-induced tumorigenesis. In the liver, there is a high spontaneous tumor incidence among mice killed at 33 months (Fig. 1), which suggests the presence of predisposing factors for tumors in this tissue. In addition, a dose of 150 mg AAF/kg diet, which resulted in an adduct level in the liver of ~100 fmol dG-C8-AF/μg DNA, produced a hepatic tumor incidence of ~80%. By comparison, at dose levels that gave an adduct concentration of ~100 fmol dG-C8-AF/μg DNA in the bladder (~50 mg AAF/kg diet), the tumor incidence did not exceed the ~1% level observed in unexposed control mice. Thus, the presence of particular quantities of DNA adducts is only one factor contributing to the tumorigenic process, and it is possible to have significant adduct quantities with no tumorigenesis. In fact, if one assumes that predisposing factors in the liver are responsible for the high spontaneous tumor incidence and contribute to the neoplastic conversion, then fewer adducts should be needed to complete the process. Conversely, it can be postulated that in the absence of predisposing factors, an increased number of DNA adduct-related events would be required for tumorigenesis.

The number of DNA adduct-related events required for tumorigenesis can be derived from the following equation (3,4):

$$-\ln(1-P) = A^n,$$

where P is the fraction of mice with tumors, A is the adduct concentration, and n is the number of DNA adduct-related events, or "hits." As shown in Figure 5, the number of hits is determined from the slope of the line obtained from plotting log (−ln(1−P)) vs. log A. For the liver of female BALB/c mice fed AAF, the slope of the line is 1.4, while the slope of the line for the bladder is 4. This suggests that for each adduct-related event required to induce a liver tumor, three adduct-related events are required in the bladder of female BALB/c mice fed AAF. For example, a G→T transversion mutation in codon 61 of the *ras* proto-oncogene has been associated with the induction of hepatic tumors in

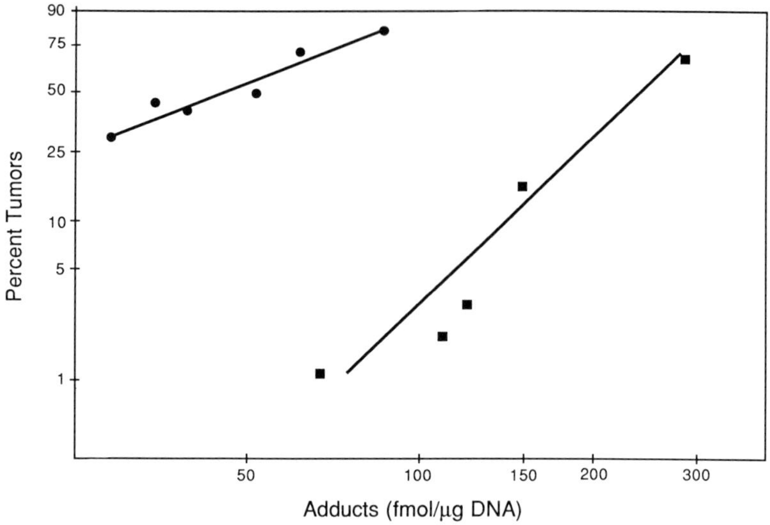

Figure 5. Number of adduct-related hits necessary to induce liver (circles) and bladder (squares) tumors in female BALB/c mice fed AAF. The number of adduct-related hits within a single cell required before a tumor occurs can be estimated from the equation $-\ln(1-P) = A^n$, where P is the fraction of mice with tumors after feeding AAF for 33 months, A is the dG-C8-AF level after feeding the same concentrations of AAF for 28 days, and n is the number of hits. The number of hits can be determined from the slope of the line obtained from plotting log $(-\ln(1-P))$ vs. log A.

mice administered *N*-hydroxy-AAF (18). This is one way a single adduct-related mutation may be sufficient for the induction of hepatic tumors in mice. Furthermore, if only one hit is required for the induction of liver tumors while three are required for bladder tumors in BALB/c mice, a higher spontaneous tumor incidence should be observed for liver tumors, which is the case (Fig. 1). Finally, if more than one hit is required in the bladder, a higher concentration of adducts should be necessary to induce bladder tumors than hepatic tumors, as we found (Fig. 4).

Open Forum

Dr. James Trosko: Dr. Poirier, your data raise some implicit assumptions that many of us tend to be afraid to talk about when we discuss carcinogenesis. These fall under the category, Is carcinogenesis more than mutagenesis? The first rhetorical question is, Are all cells in any target organ, in fact, susceptible to the presumptive carcinogen? For example in the liver, is the oval cell, which is a presumptive stem cell, really the target cell? And are the biochemical and molecular studies done on the bulk of the cells of the liver (e.g., hepatocytes) to be considered, as coined by Dr. Becker, "nonresults" because the hepatocytes might not be the target cells?

Dr. Miriam Poirier: We have very strong evidence that there is not nearly as much binding in the oval cells as in the hepatocytes. Another thing that is interesting and confounding is that when we do this immunohistochemistry in the liver, we can show by double staining liver nuclei with Hoechst dye, for example, that there are quite a few hepatocytes that also do not have adducts. So it is a heterogeneous phenomenon. The more we get into it, the more confusing it is.

Dr. Trosko: I think this is going to be important when you are trying to explain the disconcordance between adducts and tumors. My second question is, Do these presumptive carcinogens, such as the AAF carcinogen you used, have other effects besides those associated with DNA damage adducts and mutations? I believe the National Center for Toxicology Research has shown that AAF has strong promoting activity. My third point was that we have two types of organs here. The liver is a contained organ; in other words, there must be strong mitogenic control to keep the volume in a homeostatic state. The bladder, on the other hand, is an open-ended system. Consequently, the controls for regulating cell growth will be very different in these two organs, and the number of hits may be reflected by the type of data you are generating.

Dr. Poirier: You would think that a replicating tissue like the bladder would be a more sensitive target tissue than a nonreplicating tissue like the liver. I think we have a lot to learn about tissue-specific effects when it comes to the relationship between DNA adducts and tumorigenesis.

Dr. Frank Richardson: I had two questions. One concerned dose-response with the adducts. In studies we did in Jim Swenberg's lab, we also found that adducts were basically linear as the dose went down, but tumor response was not. However, cell replication showed a break point, after which there was no increase in cell replication as the dose went down. Has anybody looked at the response between AAF dose and cell replication?

Dr. Poirier: Cell replication during chronic administration has been investigated in the rat; I do not know about the mouse. A study done 10 or 15 years ago looked at the mitotic index, but I believe they used only one or two doses, 0.02% and, possibly, 0.04% AAF. My recollection is that after 28 days of AAF feeding, the mitotic index had not increased much. It may have doubled, which is pretty low. Today, we can pulse rats with BrdU in vivo and use anti-BrdU immunohistochemically. This is a much better way of doing these studies, so these are things that really should be done over again.

Dr. Richardson: Did you look at *mdr* expression in the foci?

Dr. Poirier: Henrik Huitfeldt has looked at *mdr* immunohistochemically. The results do not make much sense. An animal that has not been exposed to AAF obviously has lower levels of *mdr* than an animal that has been fed AAF for a few days. Within preneoplastic foci, an *mdr* pattern different from that in induced liver foci is observed. *mdr* is clearly present in surrounding liver in quite large quantities, but the foci can be distinguished from the rest of the liver because of the pattern. We had anticipated that *mdr* would be much higher within the foci, and that is not the case.

Dr. Don Stevenson: I would suspect that liver cells have many more mitochondria than bladder cells. Do you have any evidence that there is any adduct formation on mitochondrial DNA?

Dr. Poirier: We have never done this.

Dr. Stringner Sue Yang: Part of the research program in my laboratory focuses on adduct formation by aflatoxin B_1 epoxide with dG. I was carrying out nearest-neighbor analysis on the binding of dG by aflatoxin B_1 epoxide and I found it specifically dependent on the nearest neighbors, i.e., the vicinal nucleotide sequence. I wonder if the AAF-dG adducts exhibit the same kind of specificity. And, if so, could it be that you were getting a point mutation within a regulatory codon rather than creating a mutational change within the structure gene. For instance, in aflatoxin B_1-activated hhc^M (a human hepatocellular carcinoma–transforming DNA sequence of 3.1 kb)–induced transformation of NIH3T3 cells, instead of a mutational change in the protein, I found that one single dG change to T resulted in a sequence resembling an RNA polymerase II promoter consensus sequence and an increase in the mRNA expression specific for hhc^M. I wonder if anyone has done a walk through for a gene with AAF adduct-induced mutation(s) and has had similar observations.

Dr. Poirier: People are doing those kinds of studies, but I am not aware of any particular sequence or confirmation yet that is particularly associated with this.

Acknowledgment

We thank Cindy Hartwick for helping prepare this manuscript.

References

1. Beland FA, Poirier MC. DNA adducts and carcinogenesis. In Sirica AE (ed): *The Pathobiology of Neoplasia*. New York: Plenum Publishing, 1989, pp 57–80.
2. Farmer JH, Kodell RL, Greenman DL, Shaw GW. Dose and time response models for the incidence of bladder and liver neoplasms in mice fed 2-acetylaminofluorene continuously. *J Environ Pathol Toxicol* 3:55–68, 1979.
3. Frei JV, Lawley PD. Thymomas induced by simple alkylating agents in C57BL/Cbi mice: Kinetics of the dose response. *JNCI* 64:845–856, 1980.
4. Frei JV. Methylnitrosourea induction of thymomas in AKR mice requires one or two "hits" only. *Carcinogenesis* 1:721–723, 1980.
5. Peto R, Gray R, Brantom P, Grasso P. Nitrosamine carcinogenesis in 5120 rodents: Chronic administration of sixteen different concentrations of NDEA, NDMA, NPYR and NPIP in the water of 4440 inbred rats, with parallel studies on NDEA

alone of the effect of age of starting (3, 6 or 20 weeks) and of species (rats, mice or hamsters). *IARC Sci Publ* 57:627–665, 1984.

6. Schieferstein GJ, Littlefield NA, Gaylor DW, Sheldon WG, Burger GT. Carcinogenesis of 4-aminobiphenyl in BALB/cStCrlfC3Hf/Nctr mice. *Eur J Cancer Clin Oncol* 21:865–873, 1985.
7. Boucheron JA, Richardson FC, Morgan PH, Swenberg JA. Molecular dosimetry of O^4-ethyldeoxythymidine in rats continuously exposed to diethylnitrosamine. *Cancer Res* 47:1577-1581, 1987.
8. Basler J, Hastie ND, Pietras D, Matsui S-I, Sandberg AA, Berezney R. Hybridization of nuclear matrix attached deoxyribonucleic acid fragments. *Biochemistry* 20:6921–6929, 1981.
9. Beland FA, Fullerton NF, Heflich RH. Rapid isolation, hydrolysis and chromatography of formaldehyde-modified DNA. *J Chromatogr* 308:121–131, 1984.
10. Heflich RH, Morris SM, Beranek DT, McGarrity LJ, Chen JJ, Beland FA. Relationships between the DNA adducts and the mutations and sister-chromatid exchanges produced in Chinese hamster ovary cells by *N*-hydroxy-2-aminofluorene, *N*-hydroxy-*N'*-acetylbenzidine and 1-nitrosopyrene. *Mutagenesis* 1:201–206, 1986.
11. Poirier MC, Hunt JM, True B, Laishes BA, Young JF, Beland FA. DNA adduct formation, removal and persistence in rat liver during one month of feeding 2-acetylaminofluorene. *Carcinogenesis* 5:1591–1596, 1984.
12. Gupta RC. Enhanced sensitivity of ^{32}P-postlabeling analysis of aromatic carcinogen:DNA adducts. *Cancer Res* 45:5656–5662, 1985.
13. Lu L-JW, Disher RM, Reddy MV, Randerath K. ^{32}P-Postlabeling assay in mice of transplacental DNA damage induced by the environmental carcinogens safrole, 4-aminobiphenyl, and benzo(*a*)pyrene. *Cancer Res* 46:3046–3054, 1986.
14. Beland FA, Allaben WT, Evans FE. Acyltransferase-mediated binding of *N*-hydroxyarylamides to nucleic acids. *Cancer Res* 40:834–840, 1980.
15. Lai C-C, Miller JA, Miller EC, Liem A. N-Sulfoöxy-2-aminofluorene is the major ultimate electrophilic and carcinogenic metabolite of N-hydroxy-2-acetylaminofluorene in the livers of infant male C57BL/6J × C3H/HeJ F_1 (B6C3F_1) mice. *Carcinogenesis* 6:1037–1045, 1985.
16. Lai C-C, Miller EC, Miller JA, Liem A. Initiation of hepatocarcinogenesis in infant male B6C3F_1 mice by *N*-hydroxy-2-aminofluorene or *N*-hydroxy-2-acetylaminofluorene depends primarily on metabolism to *N*-sulfooxy-2-aminofluorene and formation of DNA-(deoxyguanosin-8-yl)-2-aminofluorene adducts. *Carcinogenesis* 8:471–478, 1987.
17. Lai C-C, Miller EC, Miller JA, Liem A. The essential role of microsomal deacetylase activity in the metabolic activation, DNA-(deoxyguanosin-8-yl)-2-aminofluorene adduct formation and initiation of liver tumors by *N*-hydroxy-2-acetylaminofluorene in the livers of infant male B6C3F_1 mice. *Carcinogenesis* 9:1295–1302, 1988.
18. Wiseman RW, Stowers SJ, Miller EC, Anderson MW, Miller JA. Activating mutations of the c-Ha-*ras* protooncogene in chemically induced hepatomas of the male B6C3F_1 mouse. *Proc Natl Acad Sci USA* 83:5825–5829, 1986.

Mouse Liver Carcinogenesis: Mechanisms
and Species Comparisons, pages 131–145

Epigenetic Mechanisms of Liver Tumor Promotion

Gary M. Williams
American Health Foundation, Valhalla, NY 10595

Several types of enhancement of carcinogenesis have been recognized (1) and delineated in rodent liver (2). Among these, promotion was originally defined by Berenblum (3) as the enhancement of latent neoplastic cells to develop into tumors. By this definition, evidence does not exist for promotion in the human liver by most of the agents known to promote rodent liver carcinogenesis, with the possible exception of oral contraceptives (2). Nevertheless, promotion is an important phenomenon in rodent liver, especially the mouse. One reason is that many chemicals exert a promoting effect in rodent liver carcinogenesis, particularly drugs (4) or candidate drugs (5), and this promoting effect appears to be the basis for the liver carcinogenesis caused by some chemicals (6).

Liver Neoplasm Promotion

Often, promotion is assumed to be the underlying mechanism for any enhancement of carcinogenesis by sequential exposure; this assumption has been extended to liver carcinogenesis (7–9). However, the importance of distinguishing promotion from the other types of enhancement—cocarcinogenesis and syncarcinogenesis—has been pointed out (1). With the discovery by Kitagawa and Sugano (10), Pitot et al. (11), and Watanabe and Williams (12) that preneoplastic liver lesions (altered foci) were enhanced by phenobarbital, an approach became available to extend to liver carcinogenesis Berenblum's concept (3) of promotion as the facilitation of the growth of latent neoplastic cells. Foci are an abnormal population that includes the progenitors of adenomas and carcinomas. The exogenous enhancement of foci therefore represents facilitation of the growth of abnormal cells, a small number of which will develop into neoplasms. A trophic effect on foci (Table 1) is produced by numerous chemicals that enhance liver tumor development when given after a carcinogen of the type that reacts with DNA. Accordingly, in the liver, promotion may be defined as the enhancement of altered foci, leading to an increase in liver neoplasms (13). Such enhancement also occurs with exposure to a second genotoxic carcinogen (14) and may represent either promotion or syncarcinogenesis (1,2). To distinguish promotion from cocarcinogenesis, it is necessary to separate the two exposures by a sufficient interval for the acute effects of the initiating carcinogen to be eliminated. To distinguish promotion from syncarcinogenesis, it is

Table 1. Effect of Chemicals Studied for Enhancement of Altered Foci in Rodent Liver

Chemical	Species	Effect	Reference
Phenobarbital	Rat	+	12
	Hamster	−	18
Barbital	Rat	+	60
Amobarbital	Rat	+	61
Pentobarbital	Rat	+	61
TCDD	Rat	+	62
Hexachlorocyclohexane	Rat	+	63
DDT	Mouse	+	17
	Hamster	−	19
Polychlorinated biphenyls	Rat	+	64
Methapyrilene	Rat	+	18
Butylated hydroxytoluene	Rat	+	65
Nafenopin	Rat	−	66
Clofibrate	Rat	−	67
Di(2-ethylhexyl)phthalate	Rat	−	67
Diazepam	Rat	−	69
Clorazepate	Rat	−	69
Oxazepam	Rat	−	69
Lorazepam	Rat	−	69

necessary to conduct reverse-sequence exposures. In the rat, only phenobarbital has been adequately demonstrated to be a promoter (15,16), while in the mouse, dichlorodiphenyltrichloroethane (DDT), chlordane, and heptachlor have been appropriately studied (17). In the hamster, phenobarbital and DDT were inactive as promoters (18,19). The importance of distinguishing different types of enhancement is illustrated by the fact that methapyrilene was initially suspected to be a promoter (20,21) but was later found to act as a syncarcinogen (22).

Mechanisms of Liver Neoplasm Promotion

A number of mechanisms have been postulated to underlie the process of promotion (1). These fall into two general categories—promoters specifically affect partially transformed cells or fully neoplastic cells rendering them capable of development into tumors; or promoters act generally on a tissue to disturb homeostasis, thereby allowing neoplastic cells to proliferate into overt neoplasms. Versions of these hypotheses are discussed as they apply to the trophic effects of liver promoters on foci, an essential feature of promotion.

One example of the first proposed mechanism is the hypothesis that promoters complete or somehow render permanent neoplastic conversion in carcinogen-altered cells. This concept has been supported by reports of promoter effects on cell differentiation (23–25) and on the genome (26). Demonstration of such effects, however, has been limited so far to the skin-tumor promoters, phorbol esters and teleocidins. These agents bind to cellular receptors; this led Weinstein et al. (27) to suggest that this class of promoters may act by usurping the action of endogenous hormone or growth-regulating substances. These studies on the mechanism of action and promoting effect of phorbol esters have so far been restricted to skin and cell cultures. Although some liver tumor promoters induce enzyme activities, reflecting a modulation of gene expression (28,29), they have not been reported to produce the permanent effects on differentiation and chromosomes reported for phorbol esters. In fact, the reported genetic effects of promoters have not been confirmed in some studies. Thus, no substantial evidence exists for any action that might be construed as completing the process of neoplastic conversion by liver tumor promoters at the level of genetic organization.

Another formulation of the hypothesis of a specific effect of promoters on cells in foci is that of Schulte-Hermann (28), which postulates that promoters provide the signal that triggers expression of a specific gene program in initiated cells that are committed for overexpression of this program. This concept is somewhat similar to that of Boutwell (30), who suggested that skin-tumor promoters induce the tumor phenotype in all epidermal cells and that this phenotype cannot be reversed in initiated cells. These hypotheses are consistent with current information on the effects of liver promoters and warrant further study.

The second broad means by which promoters may function is by disturbing homeostasis and thereby allowing already neoplastic cells to proliferate and form tumors. This concept corresponds to the original postulate of Berenblum (3) on the mechanism of promotion. Liver neoplasm promoters produce a number of effects in cells throughout the liver, such as enzyme induction (28,29,31) and increased cell replication (28) leading to hyperplasia. The role of cytotoxicity and resultant increased cell proliferation is under much study (32,33). A recent report, however, found that hepatic hyperplasia was not always associated with tumor-promoting activity (34). In fact, this lack of association was demonstrated nearly 50 years ago in skin tumor promotion (35). This is not to say that proliferation is unimportant. To the contrary, it is essential to the development of tumors.

The effects of promoters on cell membranes have been studied for many years (36–38). Yotti et al. (39) and Murray and Fitzgerald (40) first delineated an important effect of promoters that could underlie the augmented growth of restrained neoplastic cells into tumors. These groups described the ability of tumor promoters to inhibit intercellular communication in fibroblast cell cultures. This action was confirmed in liver cell cultures for a number of liver neoplasm promoters by Williams and coworkers (Table 2) and Klaunig and coworkers (in another chapter of this book). The significance of this effect is discussed in detail in the following section.

Table 2. Results in the Hepatocyte–Liver Epithelial Cell Assay for Inhibition of Intercellular Molecular Exchange

Compound	Result	Reference
Phenobarbital	+	59
DDT	+	70
Chlordane	+	71
Heptachlor	+	71
Polybrominated biphenyls	+	72
Butylated hydroxyanisole	+	73
Butylated hydroxytoluene	+	Unpublished
1,3-*bis*[2-Cyano-5-(trifluoromethyl)-phenyl]triazene	+	5
Benzo[*a*]pyrene	–	71
2-Acetylaminofluorene	+	Unpublished

Inhibition of Intercellular Communication as a Mechanism of Liver Neoplasm Promotion

The inhibition of intercellular communication between liver cells demonstrated in cell culture has been suggested to reflect an action that could be the basis for in vivo promotion of liver tumors (6). In all tissues, cell proliferation is a precisely controlled process in which the production of new cells balances those lost through differentiation or death. The regulation of this complex process is incompletely understood but apparently involves some kind of cell-to-cell communication. One type could be transfer of factors that regulate growth, including chalones and other substances that induce differentiation and, thereby, loss of proliferative capability. Factors of this type have been described for the liver (41). Several lines of evidence indicate that growth regulation is exerted on neoplastic cells. Most significantly, an interval of up to 1 year can be allowed between carcinogen exposure and the beginning of promotion, with essentially the same incidence of tumors as when promotion immediately follows initiation (42). It seems certain that during the prolonged interval neoplastic cells continue to turn over in the tissue along with normal cells and are kept under control by the regulatory factors that operate on the normal cells. In rat liver carcinogenesis, it has been demonstrated that foci of cellular alteration can undergo phenotypic reversion after cessation of carcinogen exposure (43,44), and this may also reflect homeostatic control. This is illustrated in Figure 1. If the control of preneoplastic or neoplastic cells in foci can be mediated by intercellular communication, as seems likely, then inhibition of communication could release such cells for proliferation as directed by their abnormal genome. Once proliferation was sufficient to achieve a mass of cells that could isolate themselves from the surrounding parenchyma, the action of the promoter would no longer be necessary. The inhibition of intercellular communication as the basis for liver neoplasm promotion accounts for several features of the process. It is consistent

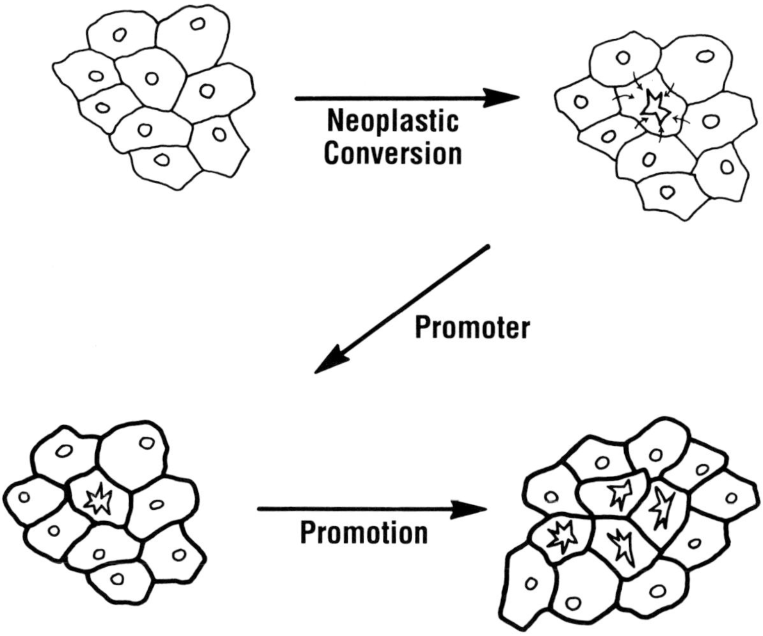

Figure 1. Schematic of generalized membrane effects of promoters leading to clonal expansion of neoplastic cells.

with the generalized effects of promoters on the entire tissue, while accounting for the trophic effect on foci cells. It explains the often high dose requirement as the need to block cell-to-cell communication over the entire cell membrane. It is compatible with reversibility, since the postulated membrane effects (as opposed to genetic effects or alterations of differentiation) are reversible. Inhibition of intercellular communication, however, does not perfectly correlate with promoting activity, although it has been demonstrated for a wide variety of tumor promoters (45).

The inhibition of intercellular communication by liver tumor promoters could be produced in several ways. DDT is highly lipophilic and is known to accumulate in cell membranes, where it could interfere directly with the function of gap junctions. The means of inhibition by phenobarbital, which is not highly lipophilic, may be more complex. Williams et al. (46) reported that phenobarbital affects the activity of several membrane-associated enzymes in mouse liver tumors. This suggests that phenobarbital modulates gene expression, giving rise to alterations in the composition of the cell membrane that impair intercellular communication. Of particular importance, Sugie et al. (47) have reported that gap junctions are diminished between hepatocytes in rats exposed to phenobarbital or DDT. In addition to these biochemical changes, surface-active substances may cause other effects. Regardless of the mechanism underlying inhibition of cell-to-cell communication, the outcome is to release abnormal cells in foci from growth control, allowing a clonal expansion of cells in foci.

A Hypothetical Sequence for Neoplastic Development in Rodent Liver

I have previously proposed a sequence of cellular events in the development of experimental hepatocellular carcinoma (6,48). The essential features of this proposal were that genetic alterations in hepatocytes rendered them capable of forming altered foci and that a very small fraction of altered foci (less than 1 in 1,000) could progress to form either adenomas or hepatocellular carcinomas. This series of events could come about in several ways (Fig. 2). As a function of the extent of genetic alteration, which increasingly appears to involve activation of oncogenes or inactivation of tumor suppressor genes (49,50) by either cryptogenic sources or carcinogens that modify DNA, several types of altered populations of liver cells might arise, including: those with alterations insufficient for the neoplastic state, those with alterations sufficient for the neoplastic state but insufficient for progressive growth; those with alterations sufficient for the neoplastic state but only for limited abnormal growth, and those with alterations sufficient both for the neoplastic state and for progressive growth into neoplasms. The postulated genetic abnormalities in foci are supported by the finding that foci contain abnormal amounts of DNA (51). These different theoretical populations also may be reflected in the heterogeneity of phenotypes in foci in rats (11,52) and mice (53).

In the first condition (i.e., alterations insufficient for the neoplastic state), the altered cell could remain dormant as a partially transformed cell (designated a in Fig. 2). Additional genetic alterations would be required to fully convert such a cell to a neoplastic cell (Fig. 2, b). Such additional genetic alterations could be the basis for the cumulative effect of repeated carcinogen exposures. The finding that cells from foci display a lesser DNA repair response to DNA-damaging carcinogens (54) may indicate the acquisition of a greater susceptibility to genetic effects of carcinogen exposure. As a result of the genetic alterations or with the assistance of a promoter, the partially transformed cell could proliferate to form an altered focus (c in Fig. 2), which would be highly susceptible to phenotypic reversion and which, without additional genetic alterations, would not progress to form neoplastic lesions.

In the second and third conditions, the cell is fully converted to a neoplastic cell as a consequence of genetic alterations but is limited in its growth capability. With alterations insufficient for progressive growth, the neoplastic cell would remain inapparent (Fig. 2, b), whereas with changes sufficient for

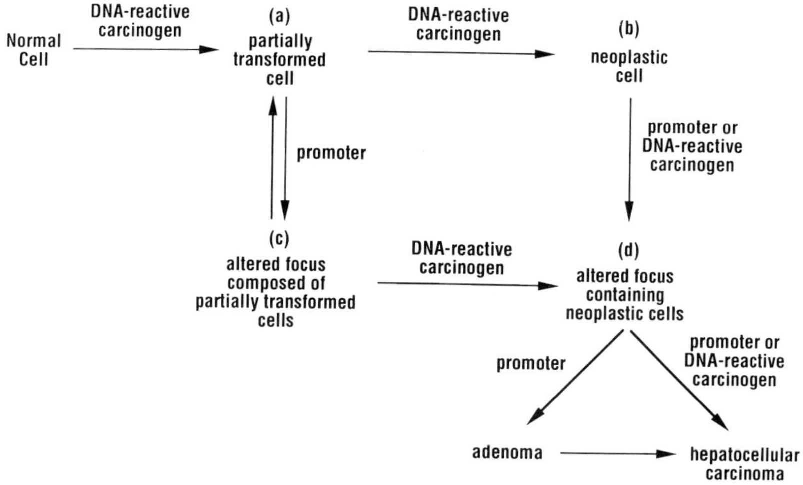

Figure 2. Hypothetical sequences for liver neoplastic development.

only limited growth it could undergo clonal expansion to form an altered focus, indicated by (d) in Figure 2. A major factor restricting the growth of the neoplastic cells at these stages could be imposition of growth inhibition/differentiation on them by the intercellular transmission of regulatory factors. Thus, for populations (a), (b), and (d) in Figure 2, reduction of intercellular communication could accomplish a promoting action by releasing the cells from growth restraint and allowing them to undergo clonal expansion according to their altered genetic constitution. Continuous maintenance of interrupted intercellular communication would be required for repressed neoplastic cells (b and d in Fig. 2) to progress to form tumors. Once a sufficient mass of neoplastic cells was achieved, the majority of neoplastic cells would be effectively isolated from interaction with surrounding normal cells and thus would be unable to receive regulatory signals.

In the fourth condition, in which the genetic alterations in the cell are sufficient for neoplastic conversion and progressive growth, one basis for the latter could be cell membrane alterations that decrease the recognition or uptake of normal tissue constituents regulating growth and differentiation. The progeny of such a cell would be able to grow in a progressive manner without assistance from promoters. Nevertheless, promoters or further carcinogen action could enhance neoplastic development. This state would presumably be the rarest cellular event in carcinogenesis.

Table 3. Chemicals That Are Hepatocarcinogenic after Prolonged Administration to Rodents

	Hepatocarcinogenicity*	
Chemical	Mouse	Rat
Phenobarbital	+	+/–
Clofibrate	ND	+
Chlordane	+	+
DDT	+	+
Dieldrin	+	–
Heptachlor	+	–
Lindane	+	+/–
Mirex	+	+
Polychlorinated biphenyls	+	+
Polybrominated biphenyls	ND	+

*Evaluation of hepatocarcinogenic findings after long-term administration: +, active hepatocarcinogen; –, inactive; +/–, studies mostly negative; ND, not determined.

Role of Promotion in the Carcinogenicity of Epigenetic Hepatocarcinogens

A number of drugs and chemicals produce tumors primarily or exclusively in the rodent liver, especially in mice, when they are administered for prolonged periods at high dose levels (Table 3). The properties of these substances differ considerably from those of the more thoroughly studied carcinogens described by Miller and Miller (55) as giving rise to reactive electrophiles. The latter are sometimes effective with a single exposure, are often active at low doses, and generally produce malignant neoplasms at multiple sites, usually after short or moderate expression times. Partly because of such differences in the carcinogenic effects of diverse carcinogens, Weisburger and Williams (56) proposed the classification of two major categories of carcinogens, DNA-reactive (genotoxic) and epigenetic (Table 4).

DNA-reactive carcinogens are defined as those that are capable of producing DNA damage through covalent reactions and thus correspond to carcinogens that act as electrophilic reactants. Epigenetic carcinogens are defined as those that do not damage DNA but act by other biologic means, leading to tumor formation. The DNA-reactive capacity of carcinogens can be evaluated by determining chemical modification of DNA or by short-term testing for genetic damage (57). Substantial evidence from short-term tests for genetic damage carried out on the hepatocarcinogenic chemicals shown in Table 3 indicates a lack of

Table 4. Classification of Carcinogenic Chemicals

Category and class	Example
DNA-reactive carcinogens	
Activation-independent	Nitrosomethylurea
Activation-dependent	2-Acetylaminofluorene, diethylnitrosamine
Epigenetic carcinogens	
Promoter	Phenobarbital
Cytotoxic	Nitrilotriacetic acid
Hormone-modifying	Amitrole
Immunosuppressor	Azothioprine
Peroxisome proliferator	Clofibrate
Unclassified	
Miscellaneous	Dioxane
Inorganic*	Hydrazine

*Some metals appear to be genotoxic.

genotoxic action. Moreover, several of these carcinogens, such as phenobarbital and DDT have been shown to enhance foci development and increase liver tumor yield (12,17). Therefore, it has been concluded that such chemicals are epigenetic carcinogens of the promoter class (6,58,59). The understanding of the mechanisms of liver neoplasm promotion, reviewed above, help clarify how epigenetic hepatocarcinogens of the promoter class operate.

This concept provides an explanation for some of the dose-response characteristics of carcinogens of this type. Such carcinogens would have to be administered in a sufficiently high dose to alter the cell membrane extensively, either through accumulation in the lipid layer, as postulated for DDT, or through alteration of the membrane composition, as seems to be the case for phenobarbital, to inhibit intercellular exchange at the many sites of communication between cells. In addition, such chemicals would have to be administered for a period long enough to enable the altered cells either to achieve a mass sufficient to insulate themselves from regulatory signals transmitted by the normal tissue or to acquire during proliferation additional abnormalities that would enable them to become resistant to intercellular communication.

Conclusions

The mechanisms of liver neoplasm promotion are not yet established, although several hypotheses have much to commend them. Regardless, it seems likely that promotion is an important element of the ability of nongenotoxic chemicals to increase the incidence of liver neoplasms in rodents.

Open Forum

Dr. Don Stevenson: Many of the organochlorines do not stimulate their own metabolism in mice but do so in other species. I think Ralph Gingell did some work with DDT and found that it stimulated its own metabolism in the hamster but not in the mouse. My laboratory in England did the same thing with dieldrin and found that predosing could stimulate the metabolism of dieldrin in the rat but not in the mouse.

Dr. Williams: I was talking with Ralph before this presentation and we did agree that another important thing this work showed was differences in metabolism of DDT in the hamster as compared to the rat.

Dr. Ralph Gingell: I do remember from related studies that the basal level of hamster mixed-function oxidase was much higher than in the rat. Tumors in the hamster could be induced by phenobarbital, Arochlor, and some other inducers, but only two- or threefold, while rat tumors were induced to a much greater extent. So rat tumors seemed to be much more inducible than hamster tumors.

Regarding the species differences in DDT metabolism, the excreted metabolites were similar in both the mouse and hamster after a single administration, but after dietary adminstration of 250 ppm DDT for several months, the mouse excreted increasing amounts of the metabolite DDE in the urine, whereas the hamster did not excrete DDE. This was reflected in the tissue residues, with greater amounts of DDE being present in the liver and fat of the mouse than the hamster.

Dr. Dennis Parke: Some years ago we showed that the first malignant event we could detect in human gastric carcinoma was a change in the cell surface sugars. Several other scientists at the meeting we organized to promulgate this had similar findings. In London, Dr. Filipe subsequently diagnosed her patients with colon cancer by undertaking histochemistry for the cell surface sugars. These sugars alter the tight junctions between cells. The glycoprotein sugars are regulated by the glycosyltransferases and the glycosidases, but it appears that it is the glycosyltransferases of the endoplasmic reticulum that undergo change. We have not said anything here yet about these sugars and glycoproteins in cell surfaces. Do you think this could be an epigenetic mechanism by which the cell-to-cell transfer of nutrients and the cell-to-cell tight junctions are altered?

Dr. Williams: I am intrigued by any possibility relating to cell-to-cell communication as an important event in carcinogenesis. We recently used immunohistochemistry to look at fibronectin surrounding the cells in the foci nodules and cancers vs. normal hepatocytes. We found a subpopulation of foci that show reduced fibronectin staining, whereas most tumors were clearly deficient in fibronectin. Perhaps cell membrane abnormality indicates that something is fundamentally wrong with the cell membrane that allows those cells to escape the growth control mechanisms.

Dr. Parke: But does the chemical nature of that fibronectin change? I think the terminal sugars seem to change, and that changes the polarity and the adhesion of one cell with another.

Dr. Williams: That could be. I even wonder if it would change the antigenicity.

Dr. Parke: It definitely would.

Dr. Miriam Poirier: Dr. Williams, for your study in which glutamine synthetase-positive cells were essentially destroyed by administration of 2-acetylaminofluorene (AAF), what route and dose of administration did you use?

Dr. Williams: AAF was a fed at 0.02% in the diet for 8 weeks. The post-exposure phase was 24 weeks.

Dr. Poirier: When we feed AAF, we generally find the concentration of adducts is periportal, and there are fewer adducts in the centrilobular region. People have done immunochemistry on DEN in the liver, and it would be interesting to see if there is a centrilobular association.

Dr. Williams: Did you look at the adducts at 1 month?

Dr. Poirier: We have done them at various times, and we have found that the concentration of adducts is periportal from about 5 days through 1 month.

Dr. Williams: I was hoping you were going to say that you had not looked at it for 1 month, because I was going to suggest that by then the centrilobular cells would have been obliterated and lost along with their adducts. The only way I could reconcile what you are describing and what we found is that the cells in the centrilobular region may require very few adducts to initiate the process.

References

1. Williams GM. Modulation of chemical carcinogenesis by xenobiotics. *Fundam Appl Toxicol* 4:325–344, 1984.
2. Williams GM. Interactive carcinogenesis in the liver. In Bannasch P, Keppler D, Weber G (eds): *Liver Carcinogenesis.* Falk Symposium 51, in press.
3. Berenblum I. *Carcinogenesis as a Biological Problem.* New York: Elsevier/North Holland, 1974, p 376.
4. Peraino C, Fry RJM, Grube DD. Drug-induced enhancement of hepatic tumorigenesis. In Slaga TJ, Sivak A, Boutwell RK (eds): *Carcinogenesis: A Comprehensive Survey,* Vol. 2. *Mechanisms of Tumor Promotion and Cocarcinogenesis.* New York: Raven Press, 1978, pp 421–432.
5. Maruyama H, Tanaka T, Williams GM. Activity of the anorectic agent 1,3-bis[2-cyano-5-(trifluoromethyl)phenyl]-triazene in in vitro and in vivo liver promoting assays. *Cancer Lett* 40:13–21, 1988.
6. Williams GM. Liver carcinogenesis: The role for some chemicals of an epigenetic mechanism of liver-tumour promotion involving modification of the cell membrane. *Food Cosmet Toxicol* 19:577–583, 1981.
7. Leonard TB, Dent JG, Graichen ME, Lyght O, Popp JA. Comparison of hepatic carcinogen initiation-promotion systems. *Carcinogenesis* 3:851–856, 1982.

8. Solt DB, Cayama E, Tsuda H, Enomoto K, Lee G, Farber E. Promotion of liver cancer development by brief exposure to dietary 2-acetylaminofluorene plus partial hepatectomy or carbon tetrachloride. *Cancer Res* 43:188–191, 1983.
9. Saeter G, Schwarze PE, Nesland JM, Seglen PO. 2-Acetylaminofluorene promotion of liver carcinogenesis by a non-cytotoxic mechanism. *Carcinogenesis* 9:581–587, 1988.
10. Kitagawa T, Sugano H. Enhancing effect of phenobarbital on the development of enzyme-altered islands and hepatocellular carcinomas initiated by 3′-methyl-4-(dimethylamino)azobenzene or diethylnitrosamine. *Gann* 69:679–687, 1978.
11. Pitot HC, Barsness L, Goldsworthy T, Kitagawa T. Biochemical characterization of stages of hepatocarcinogenesis after a single dose of diethylnitrosamine. *Nature* 271:456–458, 1978.
12. Watanabe KI, Williams GM. The enhancement of rat hepatocellular altered foci by the liver tumor promoter phenobarbital: Evidence that foci are precursors of neoplasms and that the promoter acts on carcinogen-induced lesions. *JNCI* 61: 1311–1314, 1978.
13. Williams GM. Phenotypic properties of preneoplastic rat liver lesions and applications to detection of carcinogens and tumor promoters. *Toxicol Pathol* 10:3–10, 1982.
14. Williams GM, Katayama S, Ohmori T. Enhancement of hepatocarcinogenesis by sequential administration of chemicals: Summation versus promotion effects. *Carcinogenesis* 2:1111–1117, 1981.
15. Schwarz M, Bannasch P, Kunz W. The effect of pre- and post-treatment with phenobarbital on the extent of γ-glutamyl transpeptidase positive foci induced in rat liver by N-nitrosomorpholine. *Cancer Lett* 21:17–21, 1983.
16. Williams GM, Furuya K. Distinction between liver neoplasm promoting and syncarcinogenic effects demonstrated by exposure to phenobarbital and diethylnitrosamine either before or after N-2-fluorenylacetamide. *Carcinogenesis* 5:171–174, 1984.
17. Williams GM, Numoto S. Promotion of mouse liver neoplasms by the organochlorine pesticides chlordane and heptachlor in comparison to dichlorodiphenyltrichloroethane. *Carcinogenesis* 5:1689–1696, 1984.
18. Stenbäck F, Mori H, Furuya K, Williams GM. The pathogenesis of dimethylnitrosamine-induced hepatocellular cancer in hamster liver and lack of enhancement by phenobarbital. *JNCI* 76:327–333, 1986.
19. Tanaka T, Mori H, Williams GM. Enhancement of dimethylnitrosamine-initiated hepatocarcinogenesis in hamsters by subsequent administration of carbon tetrachloride but not phenobarbital or p,p′-dichlorodiphenyltrichloroethane. *Carcinogenesis* 8:1171–1178, 1987.
20. Couri D, Wilt SR, Milks MM. Methapyrilene effects on initiation and promotion of γ-glutamyl-transpeptidase positive foci in rat liver. *Res Commun Chem Pathol Pharmacol* 35:51–61, 1982.
21. Furuya K, Mori H, Williams GM. An enhancing effect of the antihistaminic drug methapyrilene on rat liver carcinogenesis by previously administered N-2-fluorenylacetamide. *Toxicol Appl Pharmacol* 70:49–56, 1983.
22. Furuya K, Williams GM. Neoplastic conversion in rat liver by the antihistamine methapyrilene demonstrated by a sequential syncarcinogenic effect with N-2-fluorenylacetamide. *Toxicol Appl Pharmacol* 74:63–69, 1984.
23. O'Brien TG. The induction of ornithine decarboxylase as an early, possibly obligatory, event in mouse skin carcinogenesis. *Cancer Res* 36:2644–2653, 1976.

24. Yamasaki H, Fibach E, Nudel U, Weinstein IB, Rifkind RA, Marks PA. Tumor promoters inhibit spontaneous and induced differentiation of murine erythroleukemia cells in culture. *Proc Natl Acad Sci USA* 74:3451–3455, 1977.
25. Diamond L, O'Brien TG, Rovera G. Tumor promoters: Effects on proliferation and differentiation of cells in culture. *Life Sci* 23:1979–1988, 1979.
26. Kinsella AR, Radman M. Tumor promoter induces sister chromatid exchanges: Relevance to mechanisms of carcinogenesis. *Proc Natl Acad Sci USA* 75:6149–6153, 1978.
27. Weinstein IB, Nigler M, Fisher PB, Sisskin E, Pietropaolo C. Cell culture studies on the biologic effects of tumor promoters. In Slaga TJ, Sivak A, Boutwell RK (eds): *Carcinogenesis: A Comprehensive Survey*, Vol. 2. *Mechanisms of Tumor Promotion and Cocarcinogenesis*. New York: Raven Press, 1978, pp 313–333.
28. Schulte-Hermann R. Tumor promotion in the liver. *Arch Toxicol* 57:147–158, 1985.
29. Kitchin KT, Brown JL. Biochemical effects of two promoters of hepatocarcinogenesis in rats. *Food Chem Toxicol* 25:603–607, 1987.
30. Boutwell RK. Biochemical mechanism of tumor promotion. In Slaga TJ, Sivak A, Boutwell RK (eds): *Carcinogenesis: A Comprehensive Survey*, Vol. 2. *Mechanisms of Tumor Promotion and Cocarcinogenesis*. New York: Raven Press, 1978, pp 49–58.
31. Furukawa K, Maeura Y, Furukawa NT, Williams GM. Induction by butylated hydroxytoluene of rat liver gamma-glutamyl transpeptidase activity in comparison to expression in carcinogen-induced altered lesions. *Chem Biol Interact* 48:43–58, 1984.
32. Mirsalis JC. In vivo measurement of unscheduled DNA synthesis and S-phase synthesis as an indicator of hepatocarcinogenesis in rodents. *Cell Biol Genet Toxicol* 3:165–173, 1987.
33. Busser MT, Lutz WK. Stimulation of DNA synthesis in rat and mouse liver by various tumor promoters. *Carcinogenesis* 8:1433–1437, 1987.
34. Ward JM, Hagiwara A, Anderson LM, Lindsey K, Diwan BA. The chronic hepatic or renal toxicity of di(2-ethylhexyl) phthalate, acetaminophen, sodium barbital, and phenobarbital in male B6C3F$_1$ mice: Autoradiographic, immunohistochemical, and biochemical evidence for levels of DNA synthesis not associated with carcinogenesis or tumor promotion. *Toxicol Appl Pharmacol* 96:494–506, 1988.
35. Berenblum I. The cocarcinogenic action of croton resin. *Cancer Res* 1:44–48, 1941.
36. Rohrschneider LR, O'Brien DH, Boutwell RK. The stimulation of phospholipid metabolism in mouse skin following phorbol ester treatment. *Biochim Biophys Acta* 280:57–70, 1972.
37. Sivak A. Induction of cell division: Role of cell membrane sites. *J Cell Physiol* 80: 167–173, 1972.
38. Driedger PE, Blumberg PM. The effect of phorbol diesters on chicken embryo fibroblasts. *Cancer Res* 37:3257–3265, 1977.
39. Yotti LP, Chang CC, Trosko JE. Elimination of metabolic cooperation in Chinese hamster cells by a tumor promoter. *Science* 206:1089–1091, 1979.
40. Murray AW, Fitzgerald DJ. Tumor promoters inhibit metabolic cooperation in cocultures of epidermal and 3T3 cells. *Biochem Biophys Res Commun* 91:395–401, 1979.
41. Mashima K, Kimura T, Huang W, Yano K, Ashida Y, Yamagata Y, Miyazaki K, Yamashita J, Horio T. Multiple forms of growth inhibitors secreted from cultured rat liver cells: Purification and characterization. *J Biochem* 103:1020–1026, 1988.
42. Van Duuren BL, Sivak A, Katz C, Seidman I, Melchionne S. The effect of aging and

interval between primary and secondary treatment in two-stage carcinogenesis on mouse skin. *Cancer Res* 35:502–505, 1975.
43. Williams GM, Watanabe K. Quantitative kinetics of development of N-2-fluorenylacetamide-induced, altered (hyperplastic) hepatocellular foci resistant to iron accumulation and of their reversion or persistence following removal of carcinogen. *JNCI* 61:113–121, 1978.
44. Furuya K, Maeura Y, Williams GM. Abnormalities in liver iron accumulation during N-2-fluorenylacetamide hepatocarcinogenesis that are dependent or independent of continued carcinogen action. *Toxicol Pathol* 12:136–142, 1984.
45. Barrett JC, Kakunaga T, Kuroki T, Neubert D, Trosko JE, Vasiliev JM, Williams GM, Yamasaki H. Short-term assays to predict carcinogenicity: In-vitro assays that may be predictive of tumour-promoting agents. *IARC Sci Publ* 83:287–302, 1986.
46. Williams GM, Ohmori T, Katayama S, Rice JM. Alteration by phenobarbital of membrane-associated enzymes including gamma glutamyl transpeptidase in mouse liver neoplasms. *Carcinogenesis* 1:813–818, 1980.
47. Sugie S, Mori H, Takahashi M. Effect of in vivo exposure to the liver tumor promoters phenobarbital or DDT on the gap junctions of rat hepatocytes: A quantitative freeze-fracture analysis. *Carcinogenesis* 8:45–51, 1987.
48. Williams GM. The pathogenesis of rat liver cancer caused by chemical carcinogens. *Biochim Biophys Acta* 605:167–189, 1980.
49. Bishop JM. Viral oncogenes. *Cell* 42:23–38, 1985.
50. Harris H. The analysis of malignancy by cell fusion: The position in 1988. *Cancer Res* 48:3302–3306, 1988.
51. Mori H, Tanaka T, Sugie S, Takahashi M, Williams GM. DNA content of liver cell nuclei of N-2-fluorenylacetamide-induced altered foci and neoplasms in rats and human hyperplastic foci. *JNCI* 69:1277–1282, 1982.
52. Hirota N, Williams GM. The sensitivity and heterogeneity of histochemical markers for altered foci involved in liver carcinogenesis. *Am J Pathol* 95:317–328, 1979.
53. Lipsky MM, Hinton DE, Klaunig JE, Goldblatt PJ, Trump BF. Biology of hepatocellular neoplasia in the mouse. II. Sequential enzyme histochemical analysis of BALB/c mouse liver during safrole-induced carcinogenesis. *JNCI* 67:377–392, 1981.
54. Mori H, Sugie S, Ohbayashi F, Shima H, Yoshimi N, Takahashi M, Williams GM. Reduced DNA repair response of cells from carcinogen-induced rat liver altered foci exposed to chemical carcinogens in culture. *Carcinogenesis* 6:1087–1090, 1985.
55. Miller EC, Miller JA. The metabolism of chemical carcinogens to reactive electrophiles and their possible mechanisms of action in carcinogenesis. In Searle CE (ed): *Chemical Carcinogens*. Washington, DC: *American Chemical Society Monographs* 173: 737–762, 1976.
56. Weisburger JH, Williams GM. Carcinogen testing: Current problems and new approaches. *Science* 214:401–407, 1981.
57. Williams GM. Methods of evaluating chemical genotoxicity. *Annu Rev Pharmacol Toxicol* 29:189–211, 1989.
58. Williams GM. Liver cell culture systems for the study of hepatocarcinogenesis. In Margison GP (ed): *Advances in Medical Oncology Research and Education.* Proceedings of the XIIth International Cancer Congress. Vol. 1. Carcinogenesis. New York: Pergamon Press, 1979, pp 273–280.
59. Williams GM. Classification of genotoxic and epigenetic hepatocarcinogens using liver culture assays. *Ann NY Acad Sci* 349:273–282, 1980.

60. Mori H, Tanaka T, Nishikawa A, Takahashi M, Williams GM. Enhancing effect of barbital on N-2-fluorenylacetamide-induced rat liver lesions. *Gann* 72:798–801, 1981.
61. Shinozuka H, Lombardi B, Abanobi SE. A comparative study of the efficacy of four barbiturates as promoters of the development of γ-glutamyltranspeptidase-positive foci in the liver of carcinogen treated rats. *Carcinogenesis* 3:1017–1020, 1982.
62. Pitot HC, Goldsworthy T, Campbell HA, Poland A. Quantitative evaluation of the promotion by 2,3,7,8-tetrachlorodibenzo-p-dioxin of hepatocarcinogenesis from diethylnitrosamine. *Cancer Res* 40:3616–3620, 1980.
63. Schulte-Hermann R, Ohde G, Schuppler J, Timmermann-Trosiener I. Enhanced proliferation of putative preneoplastic cells in rat liver following treatment with the tumor promoters phenobarbital, hexachlorocyclohexane, steroid compounds and nafenopin. *Cancer Res* 41:2556–2562, 1981.
64. Pereira MA, Herren SL, Britt AL, Khoury MM. Promotion by polychlorinated biphenyls of enzyme-altered foci in rat liver. *Cancer Lett* 15:185–190, 1982.
65. Maeura Y, Williams GM. Enhancing effect of butylated hydroxytoluene on the development of liver altered foci and neoplasms induced by N-2-fluorenylacetamide in rats. *Food Chem Toxicol* 22:191–198, 1984.
66. Numoto S, Mori H, Furuya K, Levine WG, Williams GM. Absence of a promoting or sequential syncarcinogenic effect in rat liver by the carcinogenic hypolipidemic drug nafenopin given after N-2-fluorenylacetamide. *Toxicol Appl Pharmacol* 77:76–85, 1985.
67. Numoto S, Furukawa K, Furuya K, Williams GM. Effects of the hepatocarcinogenic peroxisome-proliferating hypolipidemic agents clofibrate and nafenopin on the rat liver cell membrane enzymes γ-glutamyltranspeptidase and alkaline phosphatase and on the early stages of liver carcinogenesis. *Carcinogenesis* 5:1603–1611, 1984.
68. Williams GM, Maruyama H, Tanaka T. Lack of rapid initiating, promoting or sequential syncarcinogenic effects of di(2-ethylhexyl)phthalate in rat liver carcinogenesis. *Carcinogenesis* 8:875–880, 1987.
69. Remandet B, Gouy D, Berthe J, Mazue G, Williams GM. Lack of initiating or promoting activity of six benzodiazepine tranquilizers in rat liver limited bioassays monitored by histopathology and assay of liver and plasma enzymes. *Fundam Appl Toxicol* 4:152–163, 1984.
70. Williams GM, Telang S, Tong C. Inhibition of intercellular communication between liver cells by the liver tumor promoter 1,1,1-trichloro-2,2-bis(p-chlorophenyl)ethane. *Cancer Lett* 11:339–344, 1981.
71. Telang S, Tong C, Williams GM. Epigenetic membrane effects of a possible tumor promoting type on cultured liver cells by the nongenotoxic organochlorine pesticides chlordane and heptachlor. *Carcinogenesis* 3:1175–1178, 1982.
72. Williams GM, Tong C, Telang S. Polybrominated biphenyls are nongenotoxic and produce an epigenetic membrane effect in cultured liver cells. *Environ Res* 34:310–320, 1984.
73. Williams GM. Epigenetic promoting effects of butylated hydroxyanisole. *Food Chem Toxicol* 24:1163–1166, 1986.

Mouse Liver Carcinogenesis: Mechanisms
and Species Comparisons, pages 147–148

Discussion

This chapter relates to the preceding chapters.

Dr. Robert LeBoeuf: Dr. Williams, were you suggesting based on your foci analysis that all foci have stable genetic change, and therefore all foci represent an initiated population of cells?

Dr. Gary Williams: I do not think we have really done enough to say that all foci are genetically abnormal. For other reasons I think that all foci are indicative of an induced change by the carcinogen; if that is what you mean by initiation, a term that is becoming increasingly nebulous, then I would say yes.

Dr. LeBoeuf: What I am wondering is whether there are phenotypic responses to a toxic environment, if you will, that in some cases are reversible.

Dr. Williams: I would think so, and I would like to make a general comment. In Dr. Goldfarb's talk, he emphasized certain differences that he saw between rats and mice. Later on a gentleman from Monsanto also picked up on that. One thing that we must constantly keep in mind is that we are all working with somewhat different protocols. The protocols that Dr. Goldfarb was alluding to in his rat studies are quite different from the ones we use, and I do not come to the same conclusions from our studies that he did from his. If we had a representative of the Solt-Farber school here, they would have many different things to say about these lesions too. To assess the propensity of a population of carcinogen-affected cells to progress to a tumor, I think you have to consider this in the context of the end result, i.e., what percentage of animals get tumors and what the multiplicity is. Then, assessments of the nature of earlier populations should be in relationship to the biological efficacy of the dose or regimen that was used.

Dr. Stanley Goldfarb: There is one rat model that is close to the infant mouse model. That is Carl Peraino's one-day rat model in which he injects diethylnitrosamine. We have done some experiments with that model. My impression from studying that material is that the GGT-positive foci that develop are much more labile than the lesions we see in the mouse. They look much more irregular. They look like they are remodelling; they do not look anything like what we see in the neonatal mouse model.

Dr. James Popp: I think this whole topic is very important. We must be very careful that the conclusions we draw are based on data that is really showing differences between rats and mice vs. simply differences due to different procedures or protocols. In that regard, additional studies on a rat model that would be comparable to Stan Goldfarb's mouse studies would certainly be warranted.

Dr. Frank Iverson: I've heard a general consensus, at least from the speakers at the conference, that we may be in a position to look at threshold for nongenotoxics. Are we anywhere near a consensus on this?

Dr. Williams: I am firmly convinced that there are thresholds in carcinogenesis.

Dr. Michael Gallo: I think we have to define what we mean by a threshold. Thresholds are wonderful things. We have used them in physiology for a long time. I think we have thresholds in human medicine and in human carcinogenicity, for example, in hormonal tumors. When we talk about thresholds, almost automatically we connect promotion, threshold, and safety factors. I propose that we consider several types of promoters, for example, those that perturb hormonal mechanisms, and those that bind to receptors such as the estrogen, epidermal growth factor, and glucocorticoid receptors. The question about thresholds is one of operation. The partial occupancy theory of Clark in the 1920s and Stevenson in the 1930s says that we have a threshold. It says that the slope of the early reaction is much shallower than the slope in the rest of the reaction once a critical mass is hit. We have exactly the same thing in nerve function. I think many people will agree that there are operational thresholds, but whether there are biophysical thresholds is another question.

Dr. Iverson: I agree with your point, but I was asking specifically about the liver, and I was talking specifically about cellular proliferation.

Dr. Popp: So your question really is, Is there a threshold for the enhancement of cell replication that is chemically induced?

Dr. Gallo: I think it is going to be compound specific.

Dr. Williams: If you examine every carcinogenicity study that has ever been done over a broad dose range, there has been what Dr. Gallo refers to as an operational threshold. Moreover, for nongenotoxic, or epigenetic, agents there is abundant evidence that, if the agent is administered below the biological threshold, there are no tumors no matter what the underlying biological mechanism is—cytotoxicity, enhanced replication, promotion, hormonal effects, etc. I do not think that is just a function of the size of the study. For example, phenacetin is a kidney carcinogen in humans at abuse-level doses, where it causes kidney damage, but is not associated with kidney cancer when used at nonabuse levels, even by millions of people.

Dr. Popp: We have learned here that there is a lot of interest in enhancement of cell replication and its potential importance in hepatocarcinogenesis and the development of lesions. I think additional studies should be done in this area. However, I think we frequently lose sight of what we are really talking about in the general discussions about cell replication and its relationship to hepatocarcinogenesis. Are we talking about cell replication as it may involve the initiation process through interaction with DNA injury? Are we talking about broad organ effects, meaning enhancement of cell replication throughout the entire organ? Or are we talking about those two effects in contrast to enhanced cell replication specifically within the developing lesion? Cell replication may be important at all of these points. It is extremely important for us to make such distinctions when we discuss enhanced cell replication in the carcinogenic process.

Mouse Liver Carcinogenesis: Mechanisms
and Species Comparisons, pages 149–161

The Role of Hyperplasia in Liver Carcinogenesis

Jon C. Mirsalis and Karen L. Steinmetz
Cellular and Genetic Toxicology Department, SRI International, Menlo Park, CA 94025

A wide variety of industrial chemicals and pharmaceuticals have been shown to induce liver tumors in rodents in lifetime studies. Whereas many hepatocarcinogens are clearly genotoxic (e.g., aflatoxin B_1, dimethylnitrosamine, benzidine, 2,6-dinitrotoluene), an even larger group of rodent hepatocarcinogens are apparently nonmutagenic (1). These include chlorinated solvents (carbon tetrachloride, trichloroethylene, 1,1,2-trichloroethane), halogenated biphenyls (polybrominated biphenyls, polychlorinated biphenyls), phthalate esters (di[2-ethylhexyl]phthalate), antihistamines (methapyrilene), sedatives (phenobarbital), and hypolipidemics (WY-14,643). These compounds share a common effect in that all induce significant elevations in S-phase synthesis (SPS) in rodent hepatocytes following in vivo treatment (2–6).

Rodent carcinogenicity data are important components of human risk assessment; therefore, elucidation of the mechanism of hepatocarcinogenesis in rodents is essential for the proper assessment of risk to humans. Genotoxic hepatocarcinogens (e.g., aflatoxin B_1 and dimethylnitrosamine) show similar carcinogenic effects in rodents and humans (7). In contrast, however, fundamental differences appear to exist between rodents and humans in the risk of cancer from nongenotoxic carcinogens. For example, the antiepileptic drug phenobarbital produces liver tumors in both rats and mice, yet chronic use of this drug by epileptics does not result in an increased risk of cancer attributable to chemical exposure (8). Likewise, the antihistamine methapyrilene is a potent hepatocarcinogen in rats (9), but no data have appeared in the literature indicating any risk of cancer in the very large human population that took this drug for many years.

Early studies in our laboratory indicated that genotoxic chemicals can interact with hepatic DNA in vivo, resulting in induction of DNA repair, but many genotoxic compounds induce significant elevations in hepatic SPS as well (10,11). Subsequent to these reports, we observed that many nongenotoxic hepatocarcinogens also induce increased SPS in the liver, increases that correlate well with hepatocarcinogenesis (4,5). We have developed different methodologies for the measurement of hyperplasia in the liver and other tissues and have used these approaches to study the effects of chemicals on increased cell turnover and to relate these effects to carcinogenicity.

In this chapter, we review the basic mechanisms of induction of cell turnover, provide data supporting the assertion that cell turnover alone may lead to tumors, and discuss ways hyperplasia may lead to formation of tumors.

Cell Proliferation as a Tumor-Promoting Effect

The classic two-step model of carcinogenesis requires both initiation and promotion prior to formation of tumors. Initiators are generally genotoxic chemicals that lead to an irreversible alteration in DNA, whereas liver tumor promoters often produce an increase in cell turnover. For example, partial hepatectomy (PH) is an effective tumor promoter in rats that have been initiated with a genotoxic hepatocarcinogen such as diethylnitrosamine (12). Many potent hepatocarcinogens (aflatoxin B_1, 2-acetylaminofluorene, dimethylnitrosamine, dinitrotoluene) induce both genotoxicity and increased cell turnover (10,11).

The importance of cell proliferation in the hepatocarcinogenicity of chemicals is illustrated in the comparison of the effects of two methylating agents, dimethylnitrosamine (DMN) and methylmethane sulfonate (MMS). These compounds induce comparable levels of unscheduled DNA synthesis (UDS) in CD-1 mouse hepatocytes, but only DMN induces SPS in the same hepatocytes (Table 1). DMN produces liver tumors in rodents; MMS does not. These results illustrate the necessity of cell proliferation for the induction of tumors in the liver and further support the theory that DNA adducts alone do not lead to tumors.

Although data suggest that formation of DNA adducts must be followed by cell proliferation for induction of tumors, evidence is growing that cell turnover alone may be hepatocarcinogenic, even in the absence of exogenous genotoxic chemicals. This phenomenon is particularly evident in certain strains of mice (e.g., B6C3F_1) that demonstrate a high spontaneous incidence of liver

Table 1. Induction of UDS and SPS in Male CD-1 Mice following Multiple Treatments with Dimethylnitrosamine and Methylmethane Sulfonate*

Treatment	Dose (mg/kg)	n	NG	% IR	% S Phase
Control/corn oil	0	12	−8.1 ± 0.7	1 ± 0	0.17 ± 0.04
Control/water	0	3	−9.5 ± 1.4	1 ± 0	0.19 ± 0.13
DMN	4	3	11.0 ± 2.0	56 ± 5	1.73 ± 0.61
	8	15	12.3 ± 1.7	55 ± 4	2.40 ± 0.34
MMS	50	3	5.4 ± 1.7	48 ± 10	0.51 ± 0.07
	100	3	18.2 ± 6.9	78 ± 15	0.12 ± 0.03

*Dimethylnitrosamine (DMN) and methylmethane sulfonate (MMS) were administered to male CD-1 mice on three consecutive days. Mice were killed 2 h after the third dose, and hepatocytes were isolated and cultured with [^{3}H]thymidine. Unscheduled DNA synthesis and S-phase synthesis were evaluated on the same slides. Abbreviations: NG, net grains per nucleus; % IR, percent with > 5 NG.

tumors in untreated animals. Newberne and Clark (13) demonstrated that PH alone produced an increase in liver tumors in male B6C3F$_1$ mice: 9 months after PH, 20% of animals had liver tumors compared with 0% in sham-operated controls.

S-Phase Synthesis as a Measure of Hyperplasia

The background level of cell proliferation in adult livers is very low, usually less than 0.2%. A compound may induce a fivefold elevation in cell proliferation, but this would still leave 99% of the liver unaffected. Measurement of hyperplasia by conventional histopathologic methods is difficult because such small changes in cell turnover are not easily discernible. In contrast, measurement of S-phase synthesis as nuclear incorporation of [^{3}H]thymidine during DNA replication provides a very sensitive means for measuring small increases in cell turnover by simple autoradiographic techniques.

We have developed two methods for assessing SPS in tissues following in vivo treatment of rodents with chemicals. One simply employs the in vivo–in vitro hepatocyte DNA repair assay, which is used as a screen for chemicals that induce UDS (5,10). This autoradiographic technique involves in vivo treatment of animals followed by isolation of hepatocytes and in vitro incubation with [^{3}H]thymidine. The in vitro incubation allows maximum incorporation of thymidine as cells are incubated in a uniform monolayer. The disadvantages of this technique are that it is fairly labor intensive and usually requires that each cell type be grown under separate culture conditions. The advantage is that it is

Table 2. Effect of Time and Quantity of [^{3}H]Thymidine Injection on Measurement of S-Phase Synthesis in CD-1 Mouse Liver*

Treatment	[3H]Thymidine dose	Last injection (h)†	(n)‡	% Cells in S-phase (mean ± SE§)
Corn oil	4 × 20 μCi	24	(1)	0.24
Corn oil	4 × 60 μCi	24	(1)	0.15
Corn oil	1 × 60 μCi	2	(1)	0.55
CCl_4	4 × 20 μCi	24	(3)	2.18 ± 0.74
CCl_4	4 × 60 μCi	24	(3)	5.26 ± 1.28
CCl_4	1 × 60 μCi	2	(3)	15.04 ± 5.06

*Male CD-1 mice were treated for 4 consecutive days with corn oil or 35 mg/kg carbon tetrachloride administered by gavage. Animals received intraperitoneal injections of [^{3}H]thymidine on either 4 consecutive days (24 h before they were killed) or a single injection 2 h before they were killed.
†Interval between the last injection and when the animals were killed.
‡Number of treated animals.
§SE represents variation among animals.

the only technique available for simultaneously evaluating DNA repair and DNA replication in the same cell population.

The second method involves in vivo administration of [^{3}H]thymidine during or after treatment with test chemicals and evaluation of SPS in autoradiographic preparations of tissue sections. This approach allows the rapid evaluation of SPS in multiple tissues: histology specimens can be prepared for any tissue of interest and autoradiography performed on the sections. The single disadvantage is that the tissues do not take up sufficient thymidine to accurately measure DNA repair; therefore, genotoxicity cannot be simultaneously evaluated using this approach.

The second technique is particularly useful for animals that are part of a chronic or subchronic toxicology study. At various intervals, animals can be injected with [^{3}H]thymidine and then killed a few hours later. This approach has surprisingly high sensitivity for measurement of SPS. We conducted a small experiment in which animals were administered carbon tetrachloride for 4 days and received either daily [^{3}H]thymidine injections following the dose or a single injection 2 h before they were killed. As shown in Table 2, animals administered injections shortly before they were killed had a much higher percentage of cells in SPS, even though the daily injections result in a greater total thymidine dose over the course of the study. The cumulative dosing approach, using repeated injections or minipumps loaded with [^{3}H]thymidine would of course be preferred when multiple killing times are not possible.

When the genotoxicity of a compound is unknown or equivocal, we recommend the in vivo–in vitro approach, whereas the in vivo assay is more useful for compounds already established as nongenotoxic.

Mechanism of S-Phase Induction

Increased SPS is simply an indication that DNA synthesis has been stimulated, but different mechanisms can bring about this effect. Three basic classes of compounds induce an increase in SPS in the liver.

- Hepatotoxicants. These compounds produce hepatocellular toxicity, leading to cell death, necrosis, and ultimately, regeneration of necrotic tissue. Compounds inducing this regenerative hyperplasia include genotoxic agents such as dinitrotoluene (10), 2-acetylaminofluorene (14), and DMN (15) and nongenotoxic compounds such as carbon tetrachloride (16) and methapyrilene (6).
- Mitogens and inducers. These compounds stimulate hepatic cell turnover in the absence of toxicity, leading to an increase in the liver-to-body weight ratio and an increased number of cells in the liver. This group includes cytochrome P-450 inducers such as phenobarbital (17) and the plasticizer di(2-ethylhexyl)-phthalate (DEHP) (2).
- Inducers of increased ploidy. Some chemicals may act as hepatic mitogens, stimulating hyperplasia; however, through an alternative mechanism, the same chemical may also interfere with cell division, possibly by interference with

spindle-fiber formation. Such interference would result in an increase in the DNA content of cells (i.e., diploid cells becoming tetraploid) without hyperplasia. No clear examples exist of compounds that exert such an effect in liver, although this phenomenon has only recently been investigated (18) and such compounds theoretically may exist.

Regardless of how hyperplasia is induced, evidence suggests that increased cell turnover in the liver results in an increase in liver tumors (19). Hence, the hepatotoxicants methapyrilene and carbon tetrachloride and the mitogens phenobarbital and DEHP are all liver carcinogens in rodents.

Increased cell proliferation has also been observed in tissues other than the liver, and we have studied induction of SPS in kidney tubule epithelial cells (20) and stomach epithelium (21). All these studies indicate correlations between increased hyperplasia and increased neoplasia.

Induction of SPS following Hepatotoxicity

An example of a hepatotoxic agent that induces significant hepatic regeneration is the antihistamine methapyrilene (MP). This compound is a potent rat hepatocarcinogen (9) yet fails to induce positive responses in most assays that measure genotoxicity (4). MP produces significant hepatocellular toxicity, as indicated by the rapid increase in the levels of liver-specific enzymes in the serum (Fig. 1). This increase in serum levels is followed by the onset of DNA synthesis approximately 24 h after treatment. DNA synthesis continues for approximately 48 h and is accompanied by significant histopathologic changes in the liver, including hepatocyte swelling and neutrophil infiltration into the periportal region. A significant elevation in mitotic figures is observed throughout the liver approximately 96 h after treatment.

Excellent correlations are also observed for carcinogenicity and SPS induction by halogenated compounds. Examination of SPS following single-exposure dosing regimens is a useful indicator of the difference between carcinogenic potential of chemicals in mice, even between sexes (Table 3). For example, *bis*(2-chloro-1-methylethyl)ether is a potent hepatocarcinogen in male B6C3F_1 mice but not in female mice. This compound induces an 87-fold elevation in SPS over controls in male mice but no SPS in female mice (4).

Measurement of SPS following Repeat-Dose Exposure

Measurement of SPS following single doses provides information on the potential of a chemical to induce hyperplasia; however, single-dose exposure is not always relevant to the conditions in a chronic study. For example, the plasticizer DEHP induces a burst of SPS following initiation of a chronic feeding study; however, SPS returns to control levels after the first week of dosing (2,3,22). In contrast to the weak hepatocarcinogen DEHP, the hypolipidemic agent WY-

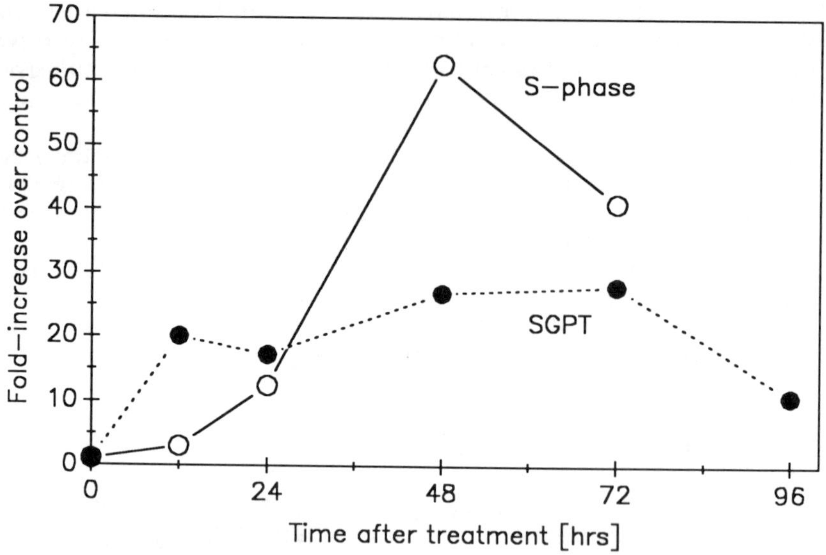

Figure 1. Induction of S-phase synthesis and SGPT in Fischer-344 rats following administration of a single oral dose of 225 mg/kg methapyrilene. Rats were killed at the times indicated after dosing, serum was collected from the orbital sinus for evaluation of SGPT, and S-phase synthesis was measured by the in vivo–in vitro technique.

Table 3. Carcinogenicity, Unscheduled DNA Synthesis (UDS), and S-phase Synthesis (SPS) in B6C3F$_1$ Mice following Treatment with Halogenated Compounds

	Male mouse			Female mouse		
Compound	Liver tumors	UDS	SPS	Liver tumors	UDS	SPS
Carbon tetrachloride	MB	–	+	MB	–	+
bis(2-Chloro-1-methylethyl)ether	MB	–	+	–	–	–
Chloroform	MB	–	+	MB	NT	NT
2,6-Dichloro-*p*-phenylenediamine	MB	–	+	MB	–	+
Polybrominated biphenyl mixture	M	–	+	M	–	+
1,1,2,2-Tetrachloroethane	M	–	–	M	–	E
1,1,2-Trichloroethane	M	–	+	M	–	+
Trichloroethylene	B	–	+	MB	–	E

Abbreviations: B, benign liver tumors; M, malignant liver tumors; NT, not tested; –, no response observed; +, positive response observed; E, equivocal response.

14,643, a potent hepatocarcinogen, induces a sustained level of hepatocellular proliferation for many months following initiation of treatment (3). This sustained proliferation has been postulated as the principal mechanism of tumor induction.

The difference in SPS induction between acute and chronic dosing regimens may explain the discordant SPS and carcinogenicity data observed for some compounds (Table 3). The hepatocarcinogen 1,1,2,2-tetrachloroethane fails to induce SPS in male mice following a single oral dose. Bromoform is a potent single-dose inducer of SPS in male mouse liver but does not produce liver tumors in mice (4). SPS induction under chronic exposure at lower doses may be quite different than these acute, high-dose responses, and studies are under way in our laboratory to determine the effects of these compounds on cell turnover after repeated low doses.

Chronic exposure to chemicals can induce metabolism of chemicals, saturate detoxification mechanisms (e.g., glutathione), or produce toxicity that results in secondary effects (e.g., changes in feeding habits that result in altered metabolism). An example of how chronic exposure can alter the SPS response of a chemical is seen for the pyrrolizidine alkaloid riddelline (23). $B6C3F_1$ mice and F-344 rats were administered riddelline by gavage for either 5 or 30 days, in conjunction with 14-day and 13-week prechronic toxicology studies. Mice treated with riddelline for 5 days did not show an increase in SPS, but treatment for 30 days produced a very significant elevation in SPS in female mice (Table 4). UDS levels also were higher after 30 days of treatment, suggesting that repeat dosing may either induce the metabolism of riddelline or saturate detoxification mechanisms. In either case, these results indicate that single-dose protocols may not always be appropriate for the accurate assessment of S-phase induction.

Table 4. Induction of S-phase Synthesis by Riddelline following 5 or 30 Days of Exposure*

Dose (mg/kg)	Male 5 Days	Male 30 Days	Female 5 Days	Female 30 Days
0.0	0.26 ± 0.04	0.87 ± 0.29	2.54 ± 1.14	5.15 ± 2.05
3.3	0.49 ± 0.18	1.75 ± 0.19	4.30 ± 1.36	20.99 ± 5.31
10.0	0.39 ± 0.17	1.26 ± 0.26	4.46 ± 2.23	0.98 ± 0.45
25.0	0.32 ± 0.07	1.07 ± 0.16	1.35 ± 0.52	0.66 ± 0.13

*Animals were administered riddelline by gavage at the doses indicated for 5 days or approximately 30 days. Animals were killed within 24 h of the final dose, livers were removed, and S-phase synthesis was measured by the in vivo–in vitro technique. Data represent percent of cells in repair ± the standard error (variation between animals); each data point based on 4 or 5 animals at each kill time.

Mechanism of Tumor Induction by Increased Cell Proliferation

Several possible mechanisms exist by which increased cell proliferation may lead to increased neoplasia, and the role of chemically induced cell turnover is probably related to all of the following factors, as well as other factors not yet considered.

Promotion of Spontaneously Initiated Cells

Some strains of mice already have a high spontaneous rate of tumor formation in the liver. The presence of activated oncogenes in mouse liver (as presented by Marshall Anderson at this conference) may be related to this high spontaneous initiation rate. In such mice, addition of a tumor-promoting chemical would be sufficient to produce an increase in tumor induction or a decrease in the time it takes for the first tumors to appear.

Acceleration of the Aging Process

If a cell is programmed for a finite number of cell divisions, increasing cell turnover will expedite the process. At the terminal division, gene function may be radically altered, possibly leading to phenotypic transformation. Alternatively, "terminal" cells may stop dividing, allowing populations of transformed cells to expand rapidly.

Increased Error Rates

Cell division may increase the susceptibility of DNA to mutagenesis by endogenous/exogenous mutagens, increase the error rate of DNA replication or repair, or alter gene regulation or function.

Altered Gene Function

Nonprogrammed cell divisions may result in altered gene function; i.e., chemically induced proliferation may trigger events that do not occur in normal cell division.

Altered DNA Methylation

Cell division that exceeds the normal rate of cell turnover may result in hypomethylation of oncogenes (or other genes), and this could lead to altered gene expression or activation of oncogenes. Some investigators have reported decreased methylation of oncogenes by nongenotoxic chemicals (see chapter by Vorce and Goodman, this volume).

Future Directions

We have only begun to understand the role of cell proliferation in chemical carcinogenicity. There is a clear correlation between cell proliferation and carcinogenicity in the liver and other tissues, but these correlations are imperfect. In addition, we have little information on the true mechanism of tumor induction by stimulation of cell turnover.

Several major areas must be addressed to improve our understanding of this phenomenon.

Cell Proliferation and DNA Methylation

Studies on DNA hypomethylation (Vorce and Goodman, this volume) offer an attractive explanation of how gene regulation may be altered by increased cell turnover. Studies of the link between increased cell turnover, changes in DNA methylation, gene expression, and carcinogenesis are essential.

Burst vs. Sustained Proliferation

The majority of studies have utilized only short dosing periods, usually less than 1 month and often only a few days. The work of Marsman et al. (3) has underscored the importance of sustained proliferation in hepatocarcinogenicity. Further studies are needed to determine the exact role of sustained proliferation in induction of tumors.

Human Models

Evidence now clearly suggests that increased cell turnover in mouse liver results in an increased incidence of tumors. Data on rats are not as clear, and similar data on humans are nonexistent. Establishment of the role, if any, of cell turnover in human cancer will require the development of models to evaluate cell turnover in humans by noninvasive techniques. Some of the newer imaging technologies, such as positron emission tomography, may be suitable for the assessment of DNA synthesis in human tissues and may also be useful for the management of diseases such as hepatitis.

Cell division is a normal process in most tissues and therefore has been overlooked in the search for mechanisms of cancer induction. If one views cancer as simply a normal process that is out of control, the stimulation of normal DNA synthesis to accelerated rates seems to be a plausible component of chemical carcinogenesis. Understanding chemically induced cell proliferation, and the steps that follow it, should provide a tremendous advance in our knowledge of how chemicals cause cancer.

Acknowledgments

The authors gratefully acknowledge the assistance of James Bakke, Carol Hamilton, Kim Tyson, Erica Loh, Marilyn Ramsey, and Dana Spak in the conduct of these studies. This work was supported in part by contracts from the National Toxicology Program/National Institute of Environmental Health Sciences.

Open Forum

Dr. Chris Kemp: You compared control DNA synthesis rates in males and females, and the rates appeared to be similar or higher in females.

Dr. Jon Mirsalis: They are almost always higher in females. That brings up a very good point. We are working with relatively young animals, rarely more than 12–14 weeks old. A typical National Toxicology Program or National Cancer Institute bioassay, on the other hand, uses female mice at an age when their cell proliferation rate is much higher than in males. Earlier today, we saw a slide indicating that female mice have a greater sensitivity to liver carcinogens, even though the males have a higher spontaneous rate. I think this might have something to do with that. This is almost like a newborn mouse assay. We are giving chemicals when there is still a tremendous amount of cell turnover.

Dr. Kemp: How do you explain this in light of the greater susceptibility of male than female mice to spontaneous liver tumor induction?

Dr. Mirsalis: I think those are two completely different mechanisms. I cannot explain why males have a higher rate of spontaneous liver tumor formation than females. I believe more of the compounds in the NTP data base are potent female mouse liver carcinogens than male liver carcinogens. I think this has to do with the susceptibility of the animals at the age when we normally start bioassays. Maybe we should start all bioassays when they are 6 months old.

Dr. Kemp: Do you feel that the DNA synthesis rate would slow down in females at a later time?

Dr. Mirsalis: We have not taken it out any further than about a month, so I do not know.

Dr. Debra Kaden: Have you done any work to see if thresholds exist?

Dr. Mirsalis: Not really, but this is obviously very important. Clearly there is a threshold for hepatotoxic carcinogens. If you give these compounds at subhepatotoxic doses, they are not carcinogenic and they do not induce S-phase synthesis. We have done a little bit of work on thresholds in rats with 4,4′-methylenedianiline, which is a good hepatotoxic agent.

Dr. Kaden: You said that, for in vivo–in vitro assays, 2 h before animals were killed was the best time to give the tritium dose. How does that go along with previous work showing that the peak of DNA synthesis seems to occur from 24–48 h?

Dr. Mirsalis: The 24–48 h period is the time after treatment of chemical.

I am talking about animals that have been on treatment for days or weeks. On a single dose it takes 24–48 h before you bring cells up to DNA synthesis. When we pull an animal off a study, there are cells at that instant in time that are undergoing DNA synthesis. So when we label, we are just looking at a window of time.

Dr. Gabriel Altmann: Did you use both methods—autoradiography of sections and the culture method—for each of these compounds?

Dr. Mirsalis: We did not use both methods with every compound. We have done a comparison with some compounds. For example, with carbon tetrachloride, both methods give the same result. In other words, both methods seem to have about the same sensitivity. I think the histologic method is better because you can identify cell types, and it is easier and less expensive. The disadvantage is that you just cannot get enough label in to measure DNA repair. So if you want to look at both, you have to culture the cells. In our more recent work with histology for the compounds that I did not identify, we have not used the in vivo–in vitro approach.

Dr. Altmann: How does increased turnover lead to an eventual increase in tumor incidence, in your opinion?

Dr. Mirsalis: There clearly is a correlation between increased cell turnover and increased tumors, but I do not know why. I do not think it is cell turnover alone. Cell turnover is probably leading to another event such as altered oncogene activation.

Dr. Altmann: How many of these compounds lead to increased incidence of tumors?

Dr. Mirsalis: Virtually 100%.

Dr. Altmann: Without previous initiation?

Dr. Mirsalis: Yes. I am not aware of a compound that has turned out to be a good S-phase inducer on a repeat treatment dose that does not also produce liver tumors.

Dr. Jerry Styles: When you did your methapyrilene work, did you discriminate between your labeling in mononucleated and binucleated cells?

Dr. Mirsalis: The methapyrilene was all done in vivo–in vitro. We have looked at that in general and it does not seem to be preferentially in binucleated or mononucleated cells. Actually, most of our methapyrilene work was in the rat, which has a much lower percentage of binucleated cells. But we have not done that histologically.

References

1. Tennant RW, Spalding J, Stasiewicz S, Caspary W, Mason JM, Resnick MA. Comparative evaluation of genetic toxicity patterns of carcinogens and noncarcinogens: Strategies for predictive use of short-term assays. *Environ Health Perspect* 75: 87–95, 1987.
2. Butterworth BE, Bermudez E, Smith-Oliver T, Earle L, Cattley R, Martin J, Popp JA, Strom S, Jirtle R, Michalopoulos G. Lack of genotoxic activity of di(2-ethylhexyl)-phthalate in rat and human hepatocytes. *Carcinogenesis* 5:1329–1335, 1984.

3. Marsman DS, Cattley RC, Conway JG, Popp JA. Relationship of hepatic peroxisome proliferation and replicative DNA synthesis to the hepatocarcinogenicity of the peroxisome proliferators di(2-ethylhexyl)-phthalate and [4-chloro-6-(2,3-xylidino)-2-pyrimidinylthio]acetic acid (WY-14,643) in rats. *Cancer Res* 48:6739–6744, 1988.
4. Mirsalis JC. In vivo measurement of unscheduled DNA synthesis and hepatic cell proliferation as an indicator of hepatocarcinogenesis in rodents. *Cell Biol Genet Toxicol* 3:165–173, 1987.
5. Mirsalis JC, Tyson CK, Loh EN, Steinmetz KL, Bakke JP, Hamilton CM, Spak DK, Spalding JW. Induction of hepatic cell proliferation and unscheduled DNA synthesis in mouse hepatocytes following in vivo treatment. *Carcinogenesis* 6:1521–1524, 1985.
6. Steinmetz KL, Tyson CK, Meierhenry EF, Spalding JW, Mirsalis JC. Examination of genotoxicity and morphologic alteration in hepatocytes following in vivo or in vitro exposure to methapyrilene. *Carcinogenesis* 9:959–963, 1988a.
7. International Agency for Research on Cancer. *IARC Monogr Eval Carcinog Risk Chem Man* Vol. 1, 1972.
8. International Agency for Research on Cancer. *IARC Monogr Eval Carcinog Risk Chem Man* 13:159–181, 1977.
9. Lijinsky W, Reuber MD, Blackwell BN. Liver tumors induced in rats by oral administration of the antihistaminic methapyrilene hydrochloride. *Science* 209: 817–819, 1980.
10. Mirsalis JC, Butterworth BE. Induction of unscheduled DNA synthesis in rat hepatocytes following in vivo treatment with dinitrotoluene. *Carcinogenesis* 3:241–245, 1982.
11. Mirsalis JC, Tyson CK, Butterworth BE. Detection of genotoxic carcinogens in the in vivo-in vitro hepatocyte DNA repair assay. *Environ Mutagen* 4:553–562, 1982.
12. Ogawa K, Solt DB, Farber E. Phenotypic diversity as an early property of putative preneoplastic hepatocyte populations in liver carcinogenesis. *Cancer Res* 40:725–733, 1980.
13. Newberne PM, Clark AJ. Promotion of liver tumors in B6C3F1 mice by partial hepatectomy or dietary choline deficiency. *The Toxicologist* 2:63–64, 1982.
14. Mirsalis JC, Tyson CK, Loh EKN, Bakke JP, Hamilton CM, Steinmetz KL. An evaluation of the ability of benzo(a)pyrene, pyrene, 2- and 4-acetylaminofluorene to induce unscheduled DNA synthesis and cell proliferation in the livers of male rats and mice treated in vivo. In Ashby J, de Serres FJ, Shelby MD, Margolin BH, Ishidate M, Becking GC (eds): *Report of the International Program on Chemical Safety's Collaborative Study on In Vivo Assays*, Vol 1. Cambridge: Cambridge University Press, 1988, pp 361–366.
15. Doolittle DJ, Muller G, Scribner HE. A comparative study of hepatic DNA repair, DNA replication and hepatotoxicity in the CD-1 mouse following multiple administrations of dimethylnitrosamine. *Mutat Res* 188:141–147, 1987a.
16. Doolittle DJ, Muller G, Scribner HE. The relationship between hepatotoxicity and induction of replicative DNA synthesis following single or multiple doses of carbon tetrachloride. *J Toxicol Environ Health* 22:63–78, 1987.
17. Doolittle DJ, Muller G, Scribner HE. The in vivo-in vitro hepatocyte assay for assessing DNA repair and DNA replication: Studies in the CD-1 mouse. *Food Chem Toxicol* 25:399–405, 1987.
18. Styles JA, Pritchard N, Kilcullen J. Chemically induced changes in mouse and rat liver cell ploidy ratios. In Ashby J, de Serres FJ, Shelby MD, Margolin BH, Ishidate

M, Becking GC (eds): *Report of the International Program on Chemical Safety's Collaborative Study on In Vivo Assays*, Vol. 2. Cambridge: Cambridge University Press, 1988, pp 52–53.

19. Spalding JW, Mirsalis JC. The high correspondence of S-phase induction in rodent hepatocytes with chemical carcinogenicity in vivo. *In Vitro* 22:15, 1986.
20. Steinmetz KL, Stack CR, Bakke JP, Hamilton CM, Pardo K, Ramsey M, Mirsalis JC. Evaluation of unscheduled DNA synthesis and S-phase synthesis following treatment with p-dichlorobenzene. *The Toxicologist* 8:160, 1988.
21. Bakke JP, Mirsalis JC. Measurement of unscheduled DNA synthesis (UDS) in rat stomach following in vivo exposure. *Environ Mutagen* 11(Suppl):8, 1988.
22. Cattley RC, Conway JG, Popp JA. Association of persistent peroxisome proliferation and oxidative injury with hepatocarcinogenicity in female F-344 rats fed di(2-ethylhexyl)phthalate for 2 years. *Cancer Lett* 38:15–22, 1987.
23. Mirsalis JC, Steinmetz KL, Blazak WF, Bakke JP, Hamilton CM, Deahl JT, Spalding JW. Evaluation of the potential of riddelline to induce unscheduled DNA synthesis, S-phase synthesis or micronuclei following in vivo treatment with multiple doses. *Environ Mol Mutagen* 11(Suppl):71, 1988.

Mouse Liver Carcinogenesis: Mechanisms
and Species Comparisons, pages 163–176

Genetic and Epigenetic Promotion of Murine Hepatocarcinogenesis

Norman R. Drinkwater, Marie H. Hanigan, and Christopher J. Kemp
McArdle Laboratory for Cancer Research, University of Wisconsin, Madison, WI 53706

Four decades of research on skin tumor induction in mice has shown that carcinogenesis occurs in at least two experimentally distinguishable stages, initiation and promotion (1). These results have had extraordinary heuristic value for understanding carcinogenesis in other tissues, including the liver. Studies from several laboratories have provided insight into the molecular details of the initiation step of hepatocarcinogenesis (2–4). Wiseman et al. (2) demonstrated that liver tumors induced in $B6C3F_1$ mice by a variety of chemical carcinogens carry mutationally activated alleles of the c-Ha-*ras* gene. The critical finding that implicated *ras* activation as an initiating event was that the mutational change depended on the carcinogen used to induce the tumors. Further support for this hypothesis was provided by Buchmann et al. (4), who showed that early preneoplastic lesions in the liver also carried mutant c-Ha-*ras* genes. However, it is likely that alternative pathways also exist for initiation of liver tumorigenesis in rodents. For example, approximately 50% of hepatomas induced in mice by *N,N*-diethylnitrosamine (DEN) did not contain mutant *ras* genes (3). Furthermore, *ras* activation has been reported only sporadically for liver tumors induced in rats (3,5).

While initiation of carcinogenesis is a discrete, essentially instantaneous event, the promotion stage is presumably a composite of events resulting in the altered expression of specific genes, proliferation of initiated and preneoplastic cells, and the acquisition of additional genetic defects (6). Studies of the interaction of particular tumor promoters with cellular components have yielded some understanding of the mechanisms underlying this complex series of events. For example, the identification of protein kinase C as the receptor for phorbol ester tumor promoters has provided a basis for the diverse effects of this class of promoters (7). To complement the biochemical approach to studying tumor promotion, we have pursued a genetic approach based on characterizing those host genes that act as physiologic modulators or targets for postinitiation events in carcinogenesis. These genes may be identified from their effects on the susceptibility of the animal to tumor induction.

Inbred mouse strains vary widely in their susceptibilities to spontaneous and chemically induced hepatocarcinogenesis (8). The lifetime incidence of spontaneous liver tumors in untreated male C3H mice exceeds 50%, while fewer than 4% of male C57BL/6 mice develop liver tumors by 2 years of age (9,10). The relative susceptibilities of these two strains to chemically induced hepatocarcinogenesis parallels this difference in spontaneous tumor incidence. The yield of liver tumors in carcinogen-treated C3H mice is from 15- to more than 50-fold greater than that for C57BL/6 mice for a variety of agents, including simple alkylating agents, *N,N*-dialkylnitrosamines, polycyclic hydrocarbons, and aromatic amines (8,11,12). For example, treatment of preweanling male mice with DEN (0.1 μmol/g) resulted in mean tumor multiplicities of 30 and 0.5 tumors/animal for C3H/HeJ and C57BL/6J mice, respectively (12).

We have used two methods to determine the genetic basis for the difference between C3H/HeJ and C57BL/6J mice in susceptibility to hepatocarcinogenesis. In the first approach, the segregation of susceptibility to liver tumor induction by DEN was analyzed for animals obtained in backcrosses between F_1 hybrid and parental inbred mice and in intercrosses between F_1 mice (12). These studies demonstrated that, while multiple loci contributed to the difference between these inbred strains, a single locus, which we have designated *Hcs* (hepatocarcinogen sensitivity), accounted for approximately 85% of the greater susceptibility of C3H/HeJ male mice relative to C57BL/6J mice. This locus is autosomal, and the C3H and C57BL/6 alleles are semidominant, resulting in an intermediate sensitivity in B6C3F_1 hybrid mice. We also compared a series of BXH recombinant inbred mouse strains derived from C57BL/6J and C3H/HeJ mice for their susceptibilities to liver tumor induction by *N*-ethyl-*N*-nitrosourea (ENU) (12). As in the above genetic model, approximately one-half of the recombinant inbred strains were similar in sensitivity to the C3H/HeJ parent, while the remaining strains were relatively resistant to hepatocarcinogenesis.

Major goals of our laboratory are to identify the *Hcs* gene at the molecular level and to understand the biologic basis for the effects of this gene on hepatocarcinogenesis. The studies described here demonstrate that the greater susceptibility of C3H/HeJ male mice relative to C57BL/6J male mice results from the more rapid growth of preneoplastic lesions in the former strain. We also provide evidence for a broader role of the *Hcs* locus in regulating hepatocyte proliferation, obtained in studies of the influence of partial hepatectomy on the development of preneoplastic and neoplastic lesions in C3H/HeJ and C57BL/6J mice.

Materials and Methods

Mice and Carcinogen Treatment

The C57BL/6J and C3H/HeJ mice used in these studies were bred in the laboratory from animals purchased from the Jackson Laboratory (Bar Harbor, ME). The animals were housed in plastic cages on corn cob bedding (Bed-O'Cobs, Anderson Cob Division, Maumee, OH) and provided with Wayne Breeder Blox (Continental Grain, Chicago, IL) and acidified tap water ad libitum.

Growth Kinetics of Preneoplastic Lesions

In the first experiment, 12-day-old male C3H/HeJ and C57BL/6J mice were injected intraperitoneally with ENU (0.5 μmol/g) dissolved in sterile trioctanoin (Pfaltz and Bauer, Stamford, CT) (0.01 ml/g). At 12, 16, or 20 weeks of age, ENU-treated mice (6 to 9 animals/group) were killed by CO_2 asphyxiation, and liver sections were prepared as described below for the analysis of glucose-6-phosphatase (G6P-ase)–deficient foci. Control animals that received trioctanoin at 12 days of age were killed at 20 weeks (20 per group). In a second experiment, similarly treated male C3H/HeJ and C57BL/6J mice (12 to 14 per group) were killed at 12, 18, 24, or 32 weeks of age for analysis of hepatic foci.

[^{3}H]Thymidine Labeling Studies

Male C3H/HeJ and C57BL/6J mice were treated with DEN (0.2 μmol/g) at 12 days of age. At 20 or 28 weeks of age each mouse received a total of six intraperitoneal injections of [methyl-^{3}H]thymidine (0.6 μCi/g) (6.7 Ci/mmol, New England Nuclear, Boston, MA). The [^{3}H]thymidine was injected every 6 h beginning at midnight. Three h after the last injection, the animals were killed, and two serial sections were cut from frozen tissue blocks. The first section was stained for G6P-ase as described below. The second was coated with Kodak NTB-2 emulsion (Eastman Kodak, Rochester, NY), stored at 4°C for 4 weeks, developed, and stained with hematoxylin and eosin. Each G6P-ase–deficient focus was identified, and the corresponding basophilic focus was located on the serial section. We counted the labeled and unlabeled hepatocytes within each focus using a criterion of 5 or more autoradiographic grains over the cell nucleus to classify a hepatocyte as labeled. The labeling index for each focus was determined as the ratio of labeled to total nuclei in the lesion.

Partial Hepatectomy during the Promotion Phase of Carcinogenesis

In each of two independent experiments, 12-day-old male C3H/HeJ and C57BL/6J mice were injected with ENU (0.5 μmol/g) dissolved in sterile trioctanoin. At 6 weeks of age, the animals were anesthetized by intramuscular injection of ketamine (40 μg/g) (Ketaset, Bristol Myers, Syracuse, NY) and xylazine (5 μg/g) (Rompun, Miles Laboratories, Shawnee, KS) and subjected to a two-thirds hepatectomy (13). In the first experiment, groups of 6 to 9 mice were killed at 18, 24, or 32 weeks of age for analysis of G6P-ase–deficient foci. In the second experiment, animals (9–21 per group) were killed at 14, 20, 26, or 32 weeks of age for analysis of hepatic foci. This experiment included similar groups that were subjected to a sham operation in which each animal was anesthetized and the liver exposed and returned to the body cavity. Tumors larger than 2 mm in diameter and visible on the surface of the liver were counted in 32-week-old animals for the second experiment.

Analysis of Hepatic Foci

After the animals were killed and their livers removed, four sections (4 mm thick) from the three largest lobes of the liver were arranged on filter paper and immediately frozen on dry ice. Cryostat sections (10 µm thick) were stained for G6P-ase activity. Enzyme-deficient foci were quantified by microscopic analysis, and the number and size distribution of G6P-ase–deficient foci were determined by stereologic analysis as previously described (14). Only those hepatic foci larger than 0.135 mm in diameter were considered in this analysis (14).

Results and Discussion

Strain-Dependent Growth of Preneoplastic Hepatic Lesions

The observations that C3H/HeJ male mice are more susceptible than C57BL/6J mice to hepatocarcinogenesis by a broad variety of agents (8) and that these two strains do not differ in the levels of hepatic DNA modification after carcinogen treatment (11,12) are consistent with a model in which the *Hcs* gene modulates the promotion phase of carcinogenesis. We have tested this hypothesis by analyzing the development of preneoplastic hepatic lesions in carcinogen-treated C57BL/6J and C3H/HeJ male mice.

Weeks to months prior to the development of hepatic tumors, the livers of carcinogen-treated rodents contain focal hepatocellular lesions that may be identified on the basis of altered histochemical staining (14–17). In the mouse, the most consistent alteration is a decrease in G6P-ase activity; increased glycogen storage or RNA levels and decreased adenosine triphosphatase or iron accumulation may also be observed (16). Several lines of evidence indicate that these hepatic foci represent preneoplastic lesions (14–17): (i) The hepatic foci grow progressively and demonstrate increased levels of DNA synthesis in comparison with the surrounding normal liver. (ii) There is a close correlation between the development of the enzyme-altered foci and the induction of hepatocellular adenomas and carcinomas. (iii) The histochemical phenotypes of the tumors are similar to those observed for the hepatic foci. Analysis of the temporal development of hepatic foci by stereologic methods (18) allows the discrimination of the effects of specific treatments on the initiation and promotion phases of hepatocarcinogenesis. The effective dose of an initiating agent determines the number of hepatic foci but has little effect on the size distribution or growth rate of the lesions (15). In contrast, treatment with promoting agents results in an increase in the size and rate of growth of the hepatic foci (19).

We have compared the kinetics for the development of G6P-ase–deficient hepatic foci in male C3H/HeJ and C57BL/6J mice treated at 12 days of age with ENU (14). The most striking difference between these two strains was in the apparent rate of growth of the preneoplastic lesions (Fig. 1). At each time point, the average size of the foci in C3H/HeJ mice was significantly greater ($P < 0.001$)

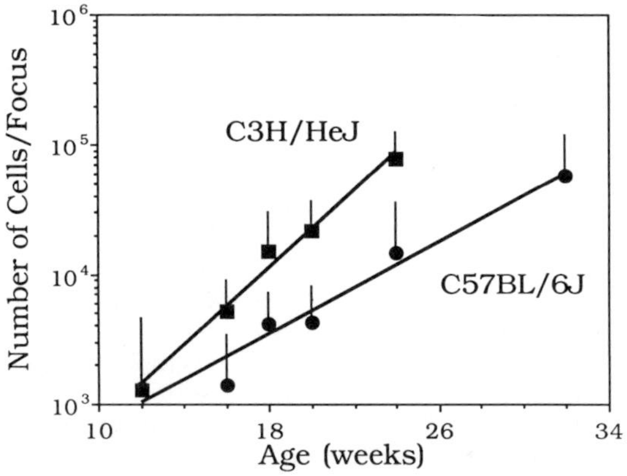

Figure 1. Growth kinetics of preneoplastic hepatic foci in male C3H/HeJ and C57BL/6J mice. Male mice were treated with *N*-ethyl-*N*-nitrosourea (0.5 μmol/g) at 12 days of age and killed at the indicated times for analysis of glucose-6-phosphatase–deficient foci. The mean number of cells/focus is shown for C57BL/6J (circles) and C3H/HeJ (squares) mice. The SD for each point is indicated by the thin vertical bar. The solid lines associated with each set of values were estimated by linear regression.

than for C57BL/6J mice. For example, at 24 weeks of age, the mean number of cells per focus in C3H/HeJ male mice was approximately 78,000, more than fivefold higher than that observed in C57BL/6J mice. From 12 to 32 weeks of age, the G6P-ase–deficient foci in male C3H/HeJ mice exhibited a volume doubling time of 2.0 ± 0.1 weeks, while that for C57BL/6J mice was 3.4 ± 0.4 weeks. Consistent with the stereologic measurements of the growth rates of the preneoplastic lesions in the two strains, the proportion of cells undergoing DNA synthesis in the foci of C3H/HeJ male mice was significantly greater than that for C57BL/6J mice (Table 1, Fig. 2). Thus, the mean labeling index for G6P-ase–deficient foci in C3H/HeJ mice at 20 weeks of age was 12%, and approximately 25% of the foci had labeling indices greater than 16%. In C57BL/6J male mice, the mean labeling index was 8.3%, and fewer than 10% of the foci had a labeling index greater than 16%. Similar results were obtained for hepatic foci in animals at 28 weeks of age (Table 1).

We concluded from these results that the C3H allele of the *Hcs* locus increases susceptibility to hepatocarcinogenesis through effects on the promotion stage of tumor induction (14). This locus apparently modulates a growth regulatory pathway in preneoplastic, and possibly in normal (14), hepatocytes. The more rapid growth of preneoplastic lesions in C3H/HeJ male mice relative to C57BL/6J mice would have an exponential effect on the appearance of the large lesions that are presumably the immediate precursors of liver tumors. In addition, the more rapid proliferation of preneoplastic hepatocytes in the susceptible strain would increase the likelihood of occurrence of any additional events required for conversion to the neoplastic state (15). One prediction of

Table 1. [^{3}H]Thymidine Labeling of Preneoplastic Hepatic Foci in C57BL/6J and C3H/HeJ Male Mice*

Strain	No. of mice	Age (weeks)	Labeling index (%)
C57BL/6J	8	20	8.3 ± 2.0†
	12	28	9.5 ± 2.5
C3H/HeJ	11	20	12 ± 2
	11	28	11 ± 2

*Male mice were treated at 12 days of age with DEN (0.2 μmol/g). At the indicated age, each animal received six intraperitoneal injections of [^{3}H]thymidine beginning 33 h before it was killed. The labeling index is the average proportion of cells within glucose-6-phosphatase–deficient hepatic foci demonstrating significant nuclear incorporation of [^{3}H]thymidine as determined by autoradiography.
†Values are mean ± SEM.

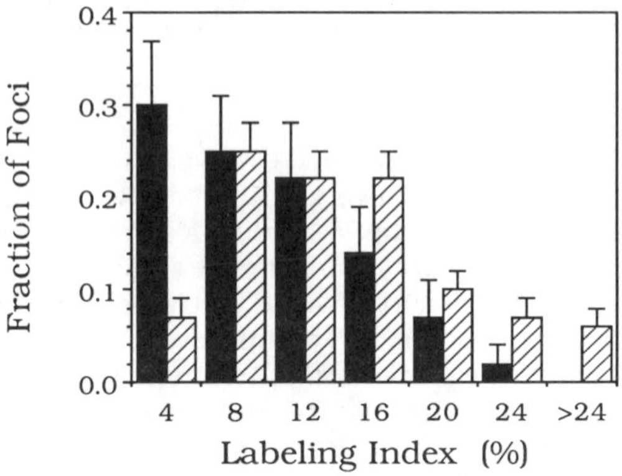

Figure 2. Distributions of [^{3}H]thymidine labeling indices for preneoplastic hepatic foci in C3H/HeJ and C57BL/6J male mice. Data are for glucose-6-phosphatase–deficient hepatic foci in 20-week-old animals from the experiment described in Table 1. The proportion of hepatic foci in each class is represented by a solid (C57BL/6J) or hatched (C3H/HeJ) bar. The SD for each value is indicated above the bar.

this model is that the development of hepatic tumors in C57BL/6J mice would be delayed rather than inhibited relative to that for C3H/HeJ mice. Treatment of preweanling male mice with DEN (0.2 μmol/g) resulted in mean liver tumor multiplicities of 39 and 2.3 for 32-week-old C3H/HeJ and C57BL/6J mice, respectively (12). However, the yield of liver tumors in similarly treated C57BL/6J mice increased with age, such that the mean tumor multiplicity was 20 by 50 weeks of age (C. J. Kemp, C. N. Leary, N. R. Drinkwater, submitted).

Studies from other laboratories have also demonstrated strain-dependent differences in the growth rate or progression of preneoplastic or neoplastic hepatic lesions. Kakizoe and coworkers (20) compared the growth rates of hyperbasophilic foci in C57BL/6 and $B6C3F_1$ mice treated with DEN and obtained results qualitatively similar to those obtained in our studies. Becker (21) previously reported that the rate of malignant progression in tumors induced in $B6C3F_1$ mice was greater than for liver tumors induced in C57BL/6N mice. Dragani et al. (22) observed that hepatic nodules induced in $B6C3F_1$ mice were significantly larger and more numerous than those observed in $(C57BL/6 \times BALB/c)F_1$ mice. In each of these cases, it is likely that the C3H allele of the *Hcs* locus contributed to the observed modulation of postinitiation events in hepatocarcinogenesis.

Influence of Partial Hepatectomy on the Development of Hepatic Foci and Liver Tumors

Partial hepatectomy approximately 24 h prior to carcinogen treatment in adult rats or mice greatly increases the yield of liver tumors relative to those animals receiving the carcinogen alone (23,24). It is likely that this enhancement results from more efficient initiation of carcinogenesis in the operated animals, since timing carcinogen exposure at the peak of regenerative DNA synthesis would increase the likelihood that initiating mutations would be fixed by replication in the target cells. However, some studies indicate that partial hepatectomy may influence hepatocarcinogenesis by mechanisms unrelated to initiation. Pound and McGuire (25) suggested that repeated partial hepatectomy between 5 and 15 weeks after DEN treatment of adult rats resulted in the earlier occurrence of hepatic lesions and, when the reduced number of target cells was taken into account, a greater yield of tumors. More recently, Bartsch et al. (26) demonstrated that partial hepatectomy 8 to 10 weeks prior to chronic exposure of rats to DEN greatly accelerated the appearance of liver tumors. Our studies demonstrate that partial hepatectomy can act as a promoter of hepatocarcinogenesis in C57BL/6J male mice but not C3H/HeJ male mice.

Administration of a partial hepatectomy to C3H/HeJ male mice at 6 weeks of age, 4 weeks after initiation of carcinogenesis by treatment with ENU, resulted in an increase in the average size of G6P-ase–deficient hepatic foci at ages between 14 and 28 weeks, relative to sham-operated animals (Fig. 3). At each time point, the mean number of cells per focus was approximately fourfold higher in hepatectomized than in control animals. However, the growth rates of the preneoplastic lesions were similar for these two groups. The apparent volume doubling times for G6P-ase–deficient foci in hepatectomized and sham-operated animals were 1.9 ± 0.4 and 2.0 ± 0.6 weeks, respectively, and were indistinguishable from that observed in previous studies of unoperated animals. These results indicate that the preneoplastic hepatocytes in C3H/HeJ mice responded to the proliferative stimulus provided by partial hepatectomy during liver regeneration but that there was no long-term effect on the growth rate of the hepatic foci.

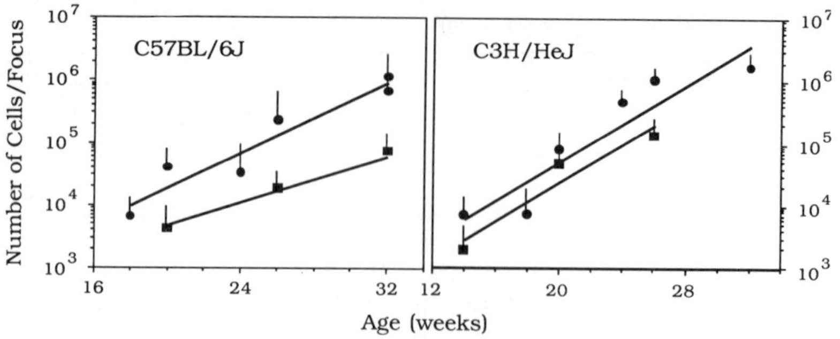

Figure 3. Influence of partial hepatectomy on the growth of preneoplastic hepatic foci in male C3H/HeJ and C57BL/6J mice. The mice were treated with ENU at 12 days of age and subjected to a two-thirds hepatectomy (circles) or sham operation (squares) at 6 weeks of age. Animals were killed at the indicated times for analysis of glucose-6-phosphatase–deficient hepatic foci. The mean number of cells per focus was plotted for each group, along with the SD (thin vertical bar) for each value. The solid line through each set of points was determined by linear regression. The mouse strain is indicated at the upper left of each panel.

Partial hepatectomy had a persistent effect on the growth of G6P-ase–deficient foci in male C57BL/6J mice, in contrast to results with C3H/HeJ mice (Fig. 3). At 18 to 20 weeks of age, the mean number of cells per focus in hepatectomized mice was approximately threefold higher than for sham-operated or unoperated animals; this ratio increased to 12-fold by 32 weeks of age. The apparent volume doubling time for hepatic foci in sham-operated control animals was 3.0 ± 0.1 weeks, similar to that observed for unoperated C57BL/6J male mice (3.4 ± 0.4). The growth of G6P-ase–deficient foci was accelerated by partial hepatectomy, such that the volume doubling time was 2.2 ± 0.3 weeks.

The strain-dependent promotion of hepatocarcinogenesis by partial hepatectomy was clearly demonstrated by analysis of liver tumor development at 32 weeks of age (Table 2). The mean number of liver tumors observed in C3H/HeJ male mice that received a sham operation was 109 ± 50, while that for hepatectomized animals was 48 ± 20. This reduction to 40% of the number of tumors observed in control animals is consistent with the surgical removal of approximately 2/3 of the initiated hepatocytes in the hepatectomized animals. In contrast, partial hepatectomy increased the number of liver tumors observed in ENU-treated C57BL/6J male mice. The mean tumor multiplicities for sham-operated and hepatectomized animals were 1.7 ± 3.6 and 9.7 ± 7.5, respectively. When the number of target cells remaining in the livers of the operated animals is taken into account, this difference reflects an approximately 17-fold increase in the risk for liver tumor development as a consequence of partial hepatectomy.

How partial hepatectomy increases the rate of growth of preneoplastic lesions and promotes liver tumor development in male C57BL/6J mice is unknown. Thompson and coworkers (27) have demonstrated that several cellu-

Table 2. Effect of Partial Hepatectomy on Liver Tumor Induction in ENU-treated C3H/HeJ and C57BL/6J Male Mice[*]

Strain	Partial hepatectomy	Number of mice	Liver tumor multiplicity
C3H/HeJ	–	21	109 (50)[†]
	+	24	48 (20)
C57BL/6J	–	19	1.7 (3.6)
	+	21	9.7 (7.5)

[*]Male mice were injected intraperitoneally with *N*-ethyl-*N*-nitrosourea (0.5 μmol/g) at 12 days of age. At 6 weeks of age, the animals were subjected to a two-thirds hepatectomy (+) or sham operation (–). Liver tumors were enumerated at 32 weeks of age.
[†]Values are mean (SD).

lar proto-oncogenes, including c-*fos*, c-*myc*, and p53, are sequentially expressed in regenerating rat liver. Mead and Fausto (28) also found that TGF-α mRNA levels increase approximately ninefold during the first 24 h after partial hepatectomy and suggested that TGF-α may stimulate hepatocytes to synthesize DNA by an autocrine mechanism. Although alterations in transcription have not been studied in detail in regenerating mouse liver, it is possible that increased transcription of one or more of the above genes may act as a switch to yield a persistent increase in the rate of proliferation of preneoplastic hepatocytes in C57BL/6J mice.

The *Hcs* Gene and Growth Regulation in the Liver

Our studies demonstrate that the C3H allele of the *Hcs* gene increases the rate of growth of preneoplastic hepatic lesions and is, in effect, a genetic promoter of hepatocarcinogenesis (14). Although the data indicate that this gene is involved in the regulation of the proliferation of both normal and preneoplastic hepatoctyes (14), the identity of the *Hcs* gene and its mechanism of action at the cellular level are currently unknown. Results of studies by Condamine and coworkers (29) on liver development and spontaneous liver tumors obtained in allophenic mice derived from mixtures of C3H and C57BL/6 or BALB/c embryos are consistent with the phenotypic expression of this gene in hepatocytes rather than in nonparenchymal cells or in tissues other than the liver. However, studies of regenerating liver have shown that nonparenchymal cells in the liver can provide negative regulatory signals for hepatocytes (30); it is possible that the *Hcs* gene may participate in these paracrine regulatory pathways.

The susceptible allele of the *Hcs* locus acts synergistically with at least one chemical tumor promoter. We have compared C3H/HeJ and C57BL/6J mice for their sensitivities to promotion of ENU-induced hepatocarcinogenesis by the halogenated aromatic hydrocarbon 3,4,5,3′,4′,5′-hexabromobiphenyl (HBB) (31). Although C3H/HeJ mice treated with ENU or ENU in combination with HBB developed substantially more numerous liver tumors than the corresponding groups of C57BL/6J mice, the ratio of tumor multiplicities for animals receiving the promoter to those not treated with HBB was similar for both strains of mice. These observations indicate that promotion of liver tumor induction by the *Hcs* gene and by HBB occur by independent pathways.

In contrast, the efficacy of partial hepatectomy as a promoter depended on the genetic background of the mouse. Thus, partial hepatectomy effectively resulted in a 20-fold increase in liver tumor induction in C57BL/6J male mice but had little influence on the process of hepatocarcinogenesis in C3H/HeJ mice. One interpretation of this result is that the *Hcs* gene and partial hepatectomy may promote hepatocarcinogenesis through the same, or overlapping, pathways for growth regulation. We are currently comparing the time courses for hepatocyte replication and the transcription of specific genes in regenerating livers of C3H/HeJ and C57BL/6J inbred and BXH recombinant inbred mice in an effort to test this hypothesis.

Conclusions

C3H/HeJ male mice are significantly more susceptible to hepatocarcinogenesis than are C57BL/6J mice. A single locus, *Hcs*, is largely responsible for this difference between the two strains. The C3H allele of this locus acts as a promoter of hepatocarcinogenesis by increasing the growth rate of preneoplastic lesions. We have demonstrated that partial hepatectomy administered weeks after carcinogen treatment promotes liver tumor induction in C57BL/6J but not C3H/HeJ mice. The functional interplay between genetic promotion by the *Hcs* locus and epigenetic promotion by partial hepatectomy may indicate that both perturb the same growth regulatory pathway in preneoplastic hepatocytes.

Open Forum

Dr. Daniel Byrd: What chromosome is *Hcs* on? Do you have a location on the chromosome, and are you making congenic strains?

Dr. Norman Drinkwater: I wish I could answer the first two questions. One reason we used recombinant inbred mice to look at tumor induction is that the strain distribution patterns in those animals can be used to map genes. Thus far we do not have a map location for the *Hcs* gene. We have begun to use a series of DNA markers related to endogenous viruses as probes to look for consegration with the *Hcs* gene. We hope to be able to map that gene sometime in the next year and a half. It is difficult to breed a congenic animal because this phenotype is late—32 weeks of age—so approaches based on using tumor multiplicity as the phenotype for selecting animals to breed are somewhat difficult. We attempted to breed congenic animals, but unfortunately, after three or four generations we lost the susceptibility. We will try again as soon as we have a location for the gene because then we can use a linked marker more efficiently.

Dr. Gary Williams: What evidence do you have that the gene for susceptibility is specific for liver tumors? The females of the C3H mouse strain are highly susceptible to mammary tumor induction.

Dr. Drinkwater: In MTV mice, C3H mice are somewhat susceptible to mammary tumor development, but so are several other inbred strains. The basis for the idea that this gene is acting in a tissue-specific way is that there is no regular pattern for increased susceptibility to tumor induction at other sites. One approach to exploring your question would be to look at the distribution of patterns of sensitivity among a variety of inbred strains. You can, in fact, identify strains that are as susceptible as the C3H mouse is for the liver for virtually any tissue site: the Aj mouse for the lung, the C57BL/6 mouse for thymic lymphoma, and so on. By and large, the patterns that you see for susceptibility at other sites do not match that for liver tumor induction, suggesting that those genes that make animals susceptible to tumor induction sites operate in a fairly tissue-specific way. We could ask your question directly by looking at the recombinant inbred strains again for susceptibility to tumor induction at other sites. There are slight differences between C3H and C57BL/6 in lung tumor induction, but they are not very large (no more than a twofold effect), and they do not appear to be correlated to liver tumor susceptibility in RI mice.

Dr. Stanley Goldfarb: I have no ready explanation for why we have different kinetics, but if we looked closely we could probably come up with several. A few potential factors should be mentioned. One is that you are initiating with ethylnitrosourea and we are initiating with diethylnitrosamine.

Dr. Drinkwater: We get similar results with either DEN or ENU.

Dr. Goldfarb: I was going to say that I think that is an unlikely explanation. More likely, it is the difference in diet, which may be important. You use breeder blocks, which have a higher fat content, and we use lab blocks, which have about half the fat. So the energy content was much higher in your system. Also, the tumor doubling time in both your C3H and B6C3F$_1$ strains is much shorter than in our model. Your doubling time is 2 weeks, and ours is 3 weeks. This slower doubling time may have opened up the window a little bit more. Maybe we are able to see this early difference in the first 10–20 weeks. That is a likely explanation, but I do not know why we see the globules and you do not.

Dr. Drinkwater: I agree that diet is probably the most relevant factor. Since the animals are probably going to maintain themselves isocalorically, I do not think there is any substantial difference in animal body weight between your studies and mine. The relative contribution of protein to their caloric intake is higher in the diet you are using than in ours, and that higher protein content may contribute to the accumulation of protein in the B6 mice. The failure to see the drop-off in the growth rate of those lesions is probably related to the fact that we do not see those accumulations of protein droplets.

Dr. Jerry Styles: The mouse does polyploidize fairly spectacularly. As the animal gets older, the number of chromosome copies are going to increase, and hence the number of gene copies. Would you expect therefore that susceptibility would increase as the animal ages?

Dr. Drinkwater: No, I would not. We treat the animals at an age when the liver is essentially diploid and dividing at a rapid rate, so we get efficient initiation of carcinogenesis by treating at 12 days of age. One thing that is not clear to me is the degree to which the early lesions remain diploid. In other words, cells that are initiated at 12 days of age may remain diploid throughout the development of the tumors. So it may be that the gene copy number does not increase with age in the 12-day initiation protocol we are using. It is a different story with adult animals, since these animals have a low overall liver replication rate; some level of cell killing and cell proliferation is therefore probably needed to get initiation.

Acknowledgments

We are grateful to Julie Ginsler, Cynthia Leary, Mary Winkler, and Jane Weeks for their excellent technical assistance. This work was supported by grants CA-07175, CA-22484, and CA-09020 from the National Cancer Institute. M. H. H. is supported by grant PF2895 from the American Cancer Society.

References

1. Boutwell RK. Some biological aspects of skin carcinogenesis. *Prog Exp Tumor Res* 4:207–250, 1964.
2. Wiseman RW, Stowers SJ, Miller EC, Anderson MW, Miller JA. Activating mutations of the c-Ha-*ras* protooncogene in chemically induced hepatomas of the male $B6C3F_1$ mouse. *Proc Natl Acad Sci USA* 83:5825–5829, 1986.
3. Stowers SJ, Wiseman RW, Ward JM, Miller EC, Miller JA, Anderson MW, Eva A. Detection of activated proto-oncogenes in N-nitrosodiethylamine-induced liver tumors: A comparison between $B6C3F_1$ mice and Fischer 344 rats. *Carcinogenesis* 9:271–276, 1988.
4. Buchmann A, Mahr J, Bauer-Hofmann R, Schwarz M. Mutations at codon 61 of the H-ras proto-oncogene in precancerous liver lesions of the $B6C3F_1$ mouse. *Molecular Carcinogenesis*, in press.
5. McMahon G, Hanson L, Lee J, Wogan GN. Identification of an activated c-Ki-ras oncogene in rat liver tumors induced by aflatoxin B_1. *Proc Natl Acad Sci USA* 83: 9418–9422, 1986.
6. Slaga TJ, Klein-Szanto AJP, Boutwell RK, Stevenson DE, Spitzer HL, D'Amato B (eds). *Skin Carcinogenesis: Mechanisms and Human Relevance.* New York: Alan R. Liss, 1989.
7. Nishizuka Y. The role of protein kinase C in cell surface signal transduction and tumor promotion. *Nature* 308:693–698, 1984.
8. Drinkwater NR. Genetic control of hepatocarcinogenesis in inbred mice. In Colburn NH (ed): *Genes and Signal Transduction in Multistage Carcinogenesis*. New York: Marcel Dekker, 1989, pp 3–17.

9. Andervont HB. Studies on the occurrence of spontaneous hepatomas in mice of strains C3H and CBA. *J Natl Cancer Inst* 11:581–592, 1950.
10. Frith CH, Wiley L. Spontaneous hepatocellular neoplasms and hepatic hemangiosarcomas in several strains of mice. *Lab Anim Sci* 32:157–162, 1982.
11. Delclos KB, Tarpley WG, Miller EC, Miller JA. 4-Aminoazobenzene and N,N-dimethyl-4-aminoazobenzene as equipotent hepatic carcinogens in male C57BL/6 × C3H/He F_1 mice and characterization of N-(deoxyguanosin-8-yl)-4-aminoazobenzene as the major persistent hepatic DNA-bound dye in these mice. *Cancer Res* 44:2540–2550, 1984.
12. Drinkwater NR, Ginsler JJ. Genetic control of hepatocarcinogenesis in C57BL/6J and C3H/HeJ inbred mice. *Carcinogenesis* 7:1701–1707, 1986.
13. Higgens GM, Anderson RM. Experimental pathology of the liver. I. Restoration of the liver of the white rat following partial surgical removal. *Arch Pathol* 12:186–202, 1931.
14. Hanigan MH, Kemp CJ, Ginsler JJ, Drinkwater NR. Rapid growth of preneoplastic lesions in hepatocarcinogen-sensitive C3H/HeJ male mice relative to C57BL/6J male mice. *Carcinogenesis* 9:885–890, 1988.
15. Vesselinovitch SD, Mihailovich N. Kinetics of diethylnitrosamine hepatocarcinogenesis in the infant mouse. *Cancer Res* 43:4253–4259, 1983.
16. Vesselinovitch SD, Hacker HJ, Bannasch P. Histochemical characterization of focal hepatic lesions induced by single diethylnitrosamine treatment in infant mice. *Cancer Res* 45:2774–2780, 1985.
17. Moore MR, Drinkwater NR, Miller EC, Miller JA, Pitot HC. Quantitative analysis of the time-dependent development of glucose-6-phosphatase–deficient foci in the livers of mice treated neonatally with diethylnitrosamine. *Cancer Res* 41:1585–1593, 1981.
18. Campbell HA, Pitot HC, Potter VR, Laishes BA. Application of quantitative stereology to the evaluation of enzyme-altered foci in rat liver. *Cancer Res* 42:465– 472, 1982.
19. Pitot HC, Goldsworthy T, Campbell HA, Poland A. Quantitative evaluation of the promotion by 2,3,7,8-tetrachlorodibenzo-p-dioxin of hepatocarcinogenesis from diethylnitrosamine. *Cancer Res* 40:3616–3620, 1980.
20. Kakizoe S, Pugh TD, Goldfarb S. Reduced cell replication rate in hepatocellular carcinomas of C57BL/6 mice: An explanation for the strain's "resistance" to hepatocarcinogenesis (Abstract). *Proc Am Assoc Cancer Res* 29:81, 1988.
21. Becker FF. Morphological classification of mouse liver tumors based on biological characteristics. *Cancer Res* 42:3918–3923, 1982.
22. Dragani TA, Manenti G, Della Porta G. Genetic susceptibility to murine hepatocarcinogenesis is associated with high growth rate of NDEA-initiated hepatocytes. *J Cancer Res Clin Oncol* 113:223–229, 1987.
23. Craddock VM. Liver carcinomas induced in rats by single administration of dimethylnitrosamine after partial hepatectomy. *J Natl Cancer Inst* 47:899–907, 1971.
24. Chernozemski IN, Warwick GP. Liver regeneration and induction of hepatomas in B6AF_1 mice by urethan. *Cancer Res* 30:2685–2690, 1970.
25. Pound AW, McGuire LJ. Repeated partial hepatectomy as a promoting stimulus for carcinogenic response of liver to nitrosamines in rats. *Br J Cancer* 37:585–594, 1978.
26. Bartsch H, Preat V, Aitio A, Cabral JRP, Roberfroid M. Partial hepatectomy of rats ten weeks before carcinogen administration can enhance liver carcinogenesis:

Preliminary observations. *Carcinogenesis* 9:2315–2317, 1988.
27. Thompson NL, Mead JE, Braun L, Goyette M, Shank PR, Fausto N. Sequential protooncogene expression during rat liver regeneration. *Cancer Res* 46:3111–3117, 1988.
28. Mead JE, Fausto N. Transforming growth factor α may be a physiological regulator of liver regeneration by means of an autocrine mechanism. *Proc Natl Acad Sci USA* 86:1558–1562, 1989.
29. Condamine H, Custer RP, Mintz B. Pure-strain and genetically mosaic liver tumors histochemically identified with the (beta)-glucuronidase marker in allophenic mice. *Proc Natl Acad Sci USA* 68:2032–2036, 1971.
30. Braun L, Mead JE, Panzica M, Mikumo R, Bell GI, Fausto N. Transforming growth factor β mRNA increases during liver regeneration: A possible paracrine mechanism of growth regulation. *Proc Natl Acad Sci USA* 85:1539–1543, 1988.
31. Drinkwater NR, Hanigan MH, Kemp CJ. Genetic determinants of hepatocarcinogenesis in the $B6C3F_1$ mouse. *Toxicol Lett,* in press.

Mouse Liver Carcinogenesis: Mechanisms
and Species Comparisons, pages 177–186

Comparative Genetic Mechanisms of 2,3,7,8-Tetrachlorodibenzo-*p*-Dioxin (TCDD)-Induced Tumors

William F. Greenlee, Thomas R. Skopek, Kevin Gaido, and Cheryl Walker
Department of Cellular and Molecular Toxicology, Chemical Industry Institute of Toxicology, Research Triangle Park, NC 27709

2,3,7,8-Tetrachlorodibenzo-*p*-dioxin (TCDD or dioxin), an unwanted contaminant in certain commercial products, has been the focus of intensive investigation for over two decades and serves as the prototype for the halogenated aromatic compounds (HACs), a large group of environmental toxicants that includes the dibenzo-*p*-dioxins, dibenzofurans, biphenyls, and azo- and azoxybenzenes (1–3). Studies in a variety of animal models and in human and animal cells in culture indicate that the pathologic lesions produced by TCDD are separable into two major categories: those that are widely expressed across species such as thymic atrophy, teratogenesis, and wasting; and species-specific responses including hepatotoxicity, edema of the pericardium, hyperkeratosis and chloracne, and the induction of certain tumors (4,5).

Many of the actions of TCDD on target cells are mediated by an intracellular binding protein, the Ah receptor (2,6). This protein was first detected in hepatic cytosol fractions from inbred mice (7) and subsequently has been found in almost all tissues examined (2,8). In most cells the Ah receptor mediates the TCDD-dependent expression of a battery of genes that includes the structural gene for cytochrome P_1-450 (2,9). In certain target cells, including murine and human epidermal keratinocytes, the Ah receptor is postulated to regulate a second gene battery encoding proteins functioning in the control of cell growth and differentiation (6,10).

Regulation of Cytochrome P_1-450 Gene Expression by the Ah Receptor

The regulatory elements for the cytochrome P_1-450 gene from a mouse hepatoma variant cell line have been identified, cloned, and mapped (11–13). Located upstream (5′-ward) of the promoter region are an inhibitory domain that blocks promoter function and at least three dioxin-responsive elements (DREs) shown to have the properties of transcriptional enhancers (11–14). Evidence from exonuclease protection (15) and gel retardation (14) assays shows that the TCDD-Ah receptor complex interacts directly with the DREs to stimulate transcription of the cytochrome P_1-450 gene. Recently, a 7-bp nucleotide sequence, 5′-T(A/T)GCGTG-3′, within the DREs has been identified as the DNA binding site for the TCDD-Ah receptor complex (16).

Qualitatively similar observations have been reported for the upstream regulatory regions for the human cytochrome P_1-450 gene (17,18) and the rat cytochrome P-450c gene (19). Both the mouse and human cytochrome P_1-450 regulatory regions are responsive to TCDD when transfected into cells from either species (12,18). These observations suggest a similar mechanism for the TCDD-dependent regulation of cytochrome P_1-450 gene expression in rodent and human cells.

Carcinogenicity in Rodent Models

Several 2-year bioassays of TCDD have been carried out in rodents. The results of three of these studies are summarized in Table 1. A daily intake of 0.1 μg/kg of TCDD in the diet results in the appearance of tumors in both rats and mice. In the study carried out in Sprague-Dawley rats, females were found to be more sensitive than males. At a dose of 0.1 μg/kg/day, nearly one-half of the female rats developed either hepatocellular carcinomas (11/49), stratified squamous cell carcinomas of the hard palate and nasal turbinates (4/49), or keratinizing squamous cell carcinomas of the lung (7/49) (20). Although limited dose-response and structure-activity data are available, it is assumed that the induction of these tumors by TCDD is mediated at least in part by the Ah receptor.

Based on the lifetime doses producing a comparable total tumor incidence, TCDD has been estimated to be a more potent rodent carcinogen than aflatoxin B_1 (21). However, TCDD does not display significant covalent binding

Table 1. Carcinogenicity of TCDD in Rodent Models

Species (strain)	Calculated daily dose of TCDD (μg/kg/day)	Target organ
Rat* (Sprague-Dawley)	0.1	Liver, oropharynx, lung
	0.01	Lung
	0.001	No increase in tumors
Mouse† ($B6C3F_1$)	0.07	Liver
	0.007	No increase in tumors
Mouse‡ (Swiss)	0.1	Liver
	0.001	No increase in tumors

*Kociba et al. (20).
†National Toxicology Program (35).
‡Toth et al. (36).

activity to protein, RNA, and DNA (observed binding was at least four to six orders of magnitude lower than observed for most chemical carcinogens) (21–23) and is inactive in various bacterial mutagenicity test systems (24,25). These observations suggest that TCDD is not genotoxic.

Activity of TCDD in Tumor Promotion Models

Tumor-promoting activity for TCDD has been demonstrated in several initiation-promotion model systems. TCDD promotes the formation of hepatocellular carcinomas in rats treated with an initiating dose of diethylnitrosamine (26), papillomas in *N*-methyl-*N'*-nitro-*N*-nitrosoguanidine (MNNG)-initiated skin of HRS/J hairless mice (27), and transformed foci in MNNG-treated C3H-10T1/2 cells (28). Comparison of the promoting activities of TCDD and the prototype promoting agent 12-*O*-tetradecanoylphorbol-13-acetate (TPA) in mouse skin and C3H10T1/2 cells indicates that TCDD is approximately two (in vivo) to four (in vitro) orders of magnitude more potent than its phorboid counterpart (Table 2).

The involvement of the Ah receptor in the promoting activity of TCDD in both in vivo and in vitro model systems is inferred by the reported dose dependence and stereospecificity (26–28). In rat liver, the Ah receptor appears to be the major genetic determinant conferring susceptibility to TCDD. In contrast, genetic analysis of the promoting activity of TCDD in the skin of hairless mice (HRS/J) indicates a requirement for at least two regulatory genes, *Ah* and *hr* (27). The *Ah* gene, the putative structural gene for the TCDD receptor protein, determines sensitivity to TCDD, whereas the *hr* gene determines the extent of cellular responses mediated by the Ah receptor (2). TCDD is only active as a tumor promoter in the skin of mice bearing the recessive mutation *hr/hr*.

Table 2. Comparison of the Relative Promoting Activities of TCDD and TPA

Model	Initiator	Promoter (dose)	Incidence of tumors or transformed foci	TCDD/TPA
Mouse skin*	MNNG	TCDD (7.5 ng)	13/17†	0.0025
	MNNG	TPA (3 μg)	13/18†	
C3H10T1/2 cells‡	MNNG	TCDD (40 πM)	25/84§	0.0001
		TPA (400 nM)	24/47§	

*Poland et al. (27).
†Number of surviving mice with tumors/total number of surviving mice.
‡Abernethy et al. (28).
§Total number of dishes with transformed foci/total number of dishes scored.

TCDD-Induced Growth Dysfunction in Human Epidermal Cells

The phorbol ester tumor promoters have been shown to induce specific changes in cell growth and differentiation. In mouse skin, the dual actions of phorboids on the differentiation of normal cells and clonal expansion of initiated cells appear to play a key role in the promotion of tumors (29,30). Studies carried out in cultures of human epidermal cells and squamous cell carcinoma (SCC) lines indicate that TCDD, like the phorbol ester tumor promoters, can alter normal patterns of cell growth and differentiation (6). The treatment of confluent cultures of normal human epidermal cells with TCDD results in an increase in the number of cells undergoing terminal differentiation (31). This response observed in vitro appears to be analogous to the epidermal hyperkeratinization reported in animal models (10) and in humans exposed to TCDD-contaminated 2,4,5-trichlorophenol (32).

The biochemical and molecular mechanisms for the TCDD-induced hyperkeratinization observed in normal human epidermal cells have been studied in human SCC lines that display predominantly either enhanced differentiation (line SCC-12F) or altered proliferation (line SCC-9) when treated with TCDD (6). Based on evidence obtained from these model systems and analysis of the actions of TCDD in murine skin (10), it has been postulated that TCDD-induced alterations in cell growth and differentiation result from the Ah receptor-dependent regulation of gene(s) encoding proteins that, in turn, control the activity of cellular growth regulatory receptor systems (2,6) (Fig. 1).

The Search for TCDD-Responsive Growth Control Genes

Two approaches are being used to isolate and clone TCDD-responsive growth regulatory genes. One cloning strategy assumes that all genes regulated by the Ah receptor will contain upstream (5′-ward) regulatory regions highly homologous to the DREs of the cytochrome P_1-450 gene (Fig. 1). Oligonucleotide (30-mer) probes to the published human cytochrome P_1-450 DRE sequences have been synthesized and used to screen a genomic library prepared from the human keratinocyte line SCC-12F. From this initial oligonucleotide screening, 50 recombinant plaques that hybridized with the DRE oligonucleotide probes under high-stringency conditions were isolated and purified. Secondary screening of these plaques with radiolabeled RNA from control and TCDD-treated cells is in progress to identify potential TCDD-responsive growth regulatory clones.

A second cloning strategy is modeled after the approach used to identify serum-responsive growth regulatory genes (33). A cDNA library is prepared from TCDD-treated cells and then screened by subtractive or differential hybridization using radiolabeled single-stranded cDNA from control and TCDD-treated cells. cDNA clones hybridizing with radiolabeled single-stranded cDNA prepared from TCDD-treated, but not control, cells should represent a TCDD-inducible gene. An analogous approach, screening a cDNA library prepared from control cells, would be used to identify genes suppressed by TCDD. In this case, TCDD-suppressible genes would hybridize with single-stranded cDNA prepared from control, but not TCDD-treated, cells.

Validation of in vitro cell models used to isolate TCDD-responsive genes by differential hybridization screening of cDNA libraries can be obtained by monitoring the expression of a gene known to be induced by TCDD, such as the structural gene for cytochrome P_1-450. Analysis of cytochrome P_1-450 gene expression in normal human epidermal cells and in two SCC-12F subclones (designated c12c2 and c6c1) are shown in Figure 2. TCDD treatment results in a time-dependent increase in cytochrome P_1-450 mRNA in both the normal cells and SCC-12F subclones, indicating that these cells possess a functional Ah receptor-dependent gene regulatory system.

Concluding Comments

TCDD is a potent carcinogen in rodents; however, the evidence from published epidemiologic studies indicates that TCDD does not appear to be a human carcinogen (see ref. 5 for a review of these data). Currently, the EPA regulates TCDD as a potential human carcinogen and has established the acceptable daily intake (ADI) at 6 fg/kg/day (34). This standard assumes that TCDD is a complete carcinogen and was determined using a linear multistage (no-threshold) model based upon the rat liver carcinogenicity data of Kociba et al. (20). The assumption that TCDD is a complete carcinogen is not consistent with the available animal data demonstrating tumor-promoting activity. The Canadian Ministry of Environment (Environment Canada) assumes that TCDD is a reversible tumor promoter with a no observable effect level of 0.001 μg/kg/day (see Table 1). Using a safety factor of 100, the ADI calculated for TCDD by

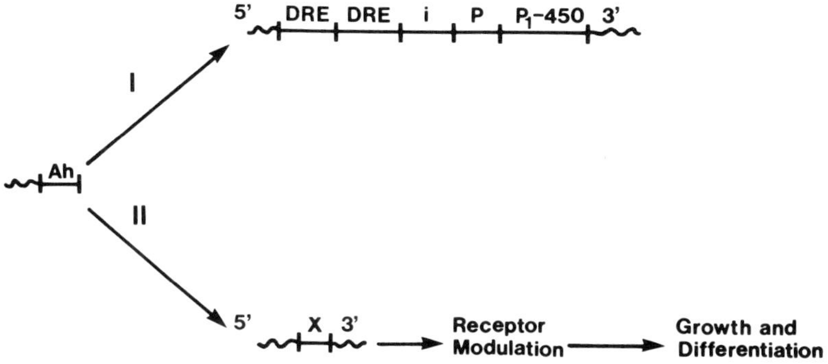

Figure 1. Proposed model for the regulation of multiple responses by the Ah receptor. Abbreviations: DRE, dioxin-responsive regulatory elements (DNA binding sites for the Ah receptor–TCDD complex are located within the DRE); i, inhibitory domain; P, RNA polymerase binding site; X, putative growth regulatory gene(s) controlled by the Ah receptor–TCDD complex.

Environment Canada is 10 pg/kg/day, a value 1,670 times higher than the ADI adopted by the EPA.

Regulation of TCDD as a reversible tumor promoter is consistent with the available animal data (26,27). However, the issue of both tissue and species extrapolation needs to be addressed. In the skin of hairless (HRS/J) mice, the promotion of papillomas by TCDD only occurs in animals homozygous at the *hr* locus, whereas in rat liver, multiple gene loci do not appear to be involved in determining responsiveness to TCDD. The inability of epidemiologic investiga-

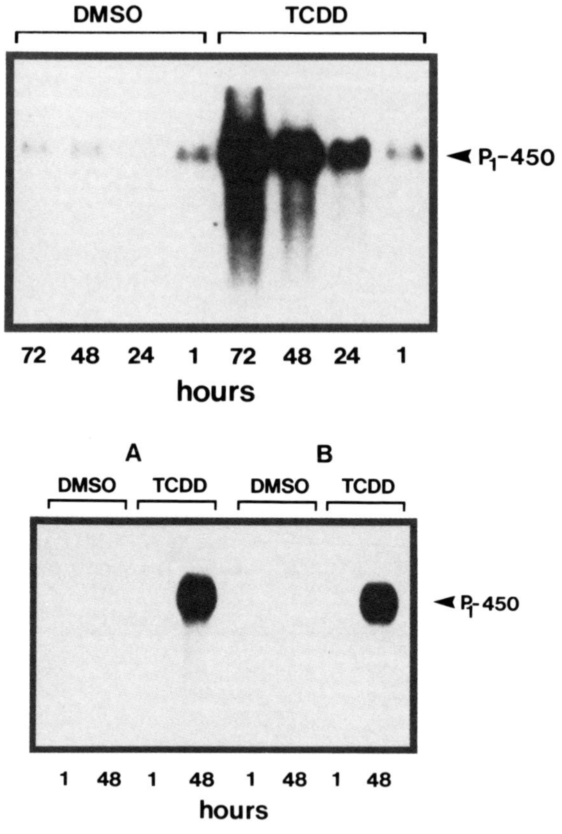

Figure 2. Northern analysis of cytochrome P_1-450 expression in normal (top panel) and transformed (bottom panel) human keratinocytes. Confluent cultures were treated with TCDD (10 nM) or solvent vehicle (0.1% dimethyl sulfoxide [DMSO]) for the time periods indicated. Poly A^+ RNA was isolated (37,38), size fractionated on 1% formaldehyde agarose gels, and transferred to nitrocellulose (39). The nitrocellulose filters were baked and probed for expression of cytochrome P_1-450 ^{32}P-labeled (40) 860-bp *Pst*I fragment of hpP-450 6 (41). Hybridization and washing conditions were as described (42). In the bottom panel, A represents the SCC-12F subclone designated c6c1 and B, the subclone designated c12c2.

tions to definitively link TCDD, an extraordinarily potent animal carcinogen, to the formation of tumors in humans suggests that genetic mechanisms may exist in human cells analogous to those operative in the skin of hairless mice. The resolution of these issues requires further knowledge of the molecular mechanisms of tumor promotion and elucidation of the genetic determinants of TCDD promoter activity in both rodent and human cell models.

Acknowledgments

The cytochrome P_1-450 gene probe was kindly provided by Dr. R. H. Tukey. Kevin Gaido is supported by a CIIT postdoctoral fellowship award. We also acknowledge the excellent technical support provided by Karen Dold, Lisa Ross, and Wendy Stewart.

Open Forum

Dr. Debra Kaden: TCDD clearly has a lot of species-specificity. I notice that you are looking mainly at human cDNA libraries. Might there be some type of subtractive hybridization or expression experiment that you could do to see why there is such a species specificity, and then possibly pinpoint what some of the key genes involved might be?

Dr. William Greenlee: We wanted to take a system that we could control well and that would allow us to isolate the important growth-response genes. At that point, it would be appropriate to ask whether similar genes are regulated in rodent models. More important, if you can isolate and clone a given gene or family of genes using techniques such as in situ hybridization, you can go back to whole animal models and ask questions about changes in the expression of these genes in vivo.

Dr. Kaden: My concern is that you are clearly seeing a tumor response in animals, yet we do not really know what is going on in humans. So if you are looking for a tumor-promoting effect, then maybe the human system is not the best system to start with.

Dr. Greenlee: The most characteristic response in humans to TCDD is the skin disorder, chloracne, which basically results from altered proliferation and differentiation. Metaplasias occur in the sebaceous glands, as well as hyperkeratinization in the interfollicular epidermis. So we took the approach of trying to establish a human epithelial cell culture model that permitted characterization of trichloroacetic acid (TCA)-induced changes in growth and differentiation, which is what tumor promoters, as best we know, are all about anyway, whether in the rodent or the human. We would like to understand how proliferation and differentiation are regulated by TCA in this human model; then we can tackle the problem of tissue (and species) specificity relative to what we know at the cellular and molecular level in a well-studied model system.

Dr. Michael Gallo: A lot of the TCDD specificity is a function of dose. We are looking at effects in epithelial cells in the dermal cell lines.

References

1. Kimbrough RD. The toxicity of polychlorinated polycyclic compounds and related chemicals. *CRC Crit Rev Toxicol* 2:445–498, 1974.
2. Poland A, Knutson JC. 2,3,7,8-Tetrachlorodibenzo-p-dioxin and related halogenated aromatic hydrocarbons: Examination of the mechanism of toxicity. *Annu Rev Pharmacol Toxicol* 22:517–554, 1982.
3. Safe SH. Comparative toxicology and mechanism of action of polychlorinated dibenzo-p-dioxins and dibenzofurans. *Annu Rev Pharmacol Toxicol* 26:371–399, 1986.
4. Poland A, Greenlee WF, Kende AS. Studies on the mechanism of action of the chlorinated dibenzo-p-dioxins and related compounds. *Ann NY Acad Sci* 320:214–230, 1979.
5. U.S. Environmental Protection Agency. Polyhalogenated dibenzo-p-dioxins/dibenzofurans: Testing and reporting requirements. *Federal Register* 52:21412–21442, 1987.
6. Greenlee WF, Osborne R, Dold KM, Hudson LG, Young MJ, Toscano WA Jr. Altered regulation of epidermal cell proliferation and differentiation by 2,3,7,8-tetrachlorodibenzo-p-dioxin (TCDD). In Hodgson E, Bend JR, Philpot RM (eds): *Reviews in Biochemical Toxicology*, Vol. 8. New York: Elsevier, 1987, pp 1–35.
7. Poland A, Glover E, Kende AS. Stereospecific, high affinity binding of 2,3,7,8-tetrachlorodibenzo-p-dioxin by hepatic cytosol: Evidence that the binding species is receptor for induction of aryl hydrocarbon hydroxylase. *J Biol Chem* 251:4936–4946, 1976.
8. Gasiewicz TA. Receptors for 2,3,7,8-tetrachlorodibenzo-p-dioxin: Their inter- and intra-species distribution and relationship to the toxicity of this compound. *Proc 13th Annu Conf Environ Toxicol*, Dayton, OH, pp 259–269, 1983.
9. Whitlock JP Jr. The regulation of cytochrome P-450 gene expression. *Annu Rev Pharmacol Toxicol* 26:333–369, 1986.
10. Knutson JC, Poland A. Response of murine epidermis to 2,3,7,8-tetrachlorodibenzo-p-dioxin: Interaction of the ah and hr loci. *Cell* 30:225–234, 1982.
11. Jones PBC, Galeazzi DR, Fisher JM, Whitlock JP Jr. Control of cytochrome P_1-450 gene expression by dioxin. *Science* 227:1499–1502, 1985.
12. Jones PBC, Durrin LK, Galeazzi DR, Whitlock JP Jr. Control of cytochrome P_1-450 gene expression: Analysis of a dioxin-responsive enhancer system. *Proc Natl Acad Sci USA* 83:2802–2806, 1986.
13. Jones PBC, Durrin LK, Fisher JM, Whitlock JP Jr. Control of gene expression by 2,3,7,8-tetrachlorodibenzo-p-dioxin: Multiple dioxin-responsive domains 5′-ward of the cytochrome P_1-450 gene. *J Biol Chem* 261:6647–6650, 1986.
14. Denison MS, Fisher JM, Whitlock JP Jr. Inducible, receptor-dependent protein-DNA interactions at a dioxin-responsive transcriptional enhancer. *Proc Natl Acad Sci USA* 85:2528–2532, 1988.
15. Durrin LK, Whitlock JP Jr. In situ protein-DNA interactions at a dioxin-responsive enhancer associated with the cytochrome P_1-450 gene. *Mol Cell Biol* 7:3008–3011, 1987.

16. Denison MS, Fisher JM, Whitlock JP Jr. The DNA recognition site for the dioxin-*Ah* receptor complex: Nucleotide sequence and functional analysis. *J Biol Chem* 263:17221–17224, 1988.
17. Neuhold LA, Gonzalez FJ, Jaiswal AK, Nebert DW. Dioxin-inducible enhancer region upstream from the mouse P_1-450 gene and interaction with a heterologous SV40 promoter. *DNA* 5:403–411, 1986.
18. Jaiswal AK, Gonzalez FJ, Nebert DW. Comparison of human and mouse P_1-450 upstream regulatory sequences in liver- and nonliver-derived cell lines. *Mol Endocrinol* 1:312–320, 1987.
19. Sogawa K, Fujisawa-Sehara A, Yamane M, Fujii-Kuriyama Y. Location of regulatory elements responsible for drug induction in the rat cytochrome P-450c gene. *Proc Natl Acad Sci USA* 83:8044–8048, 1986.
20. Kociba RJ, Keyes DG, Beyer JE, Carreon RM, Wade CE, Dittenber DA, Kalnins RP, Frauson LE, Park CN, Barnard SD, Hummel RA, Humiston CG. Results of a two-year chronic toxicity and oncogenicity study of 2,3,7,8-tetrachlorodibenzo-p-dioxin in rats. *Toxicol Appl Pharmacol* 46:279–303, 1978.
21. Poland A, Glover E. An estimate of the maximum *in vivo* covalent binding of 2,3,7,8-tetrachlorodibenzo-p-dioxin to rat liver protein, ribosomal RNA, and DNA. *Cancer Res* 39:3341–3344, 1979.
22. Rose JQ, Ramsey JC, Wentzler TH, Hummel RA, Gehring PJ. The fate of 2,3,7,8-tetrachlorodibenzo-p-dioxin following single and repeated oral doses to the rat. *Toxicol Appl Pharmacol* 36:209–226, 1976.
23. Vinopal JH, Casida JE. Metabolic stability of 2,3,7,8-tetrachlorodibenzo-p-dioxin in mammalian liver microsomal systems and in living mice. *Arch Environ Contam Toxicol* 1:122–132, 1973.
24. Wassom JS, Huff JE, Loprieno N. A review of the genetic toxicology of chlorinated dibenzo-p-dioxins. *Mutat Res* 47:141–160, 1977–1978.
25. Geiger LE, Neal RA. Mutagenicity testing of 2,3,7,8-tetrachlorodibenzo-p-dioxin in histidine auxotrophs of *Salmonella typhimurium*. *Toxicol Appl Pharmacol* 59:125–129, 1981.
26. Pitot HC, Goldsworthy T, Campbell HA, Poland A. Quantitative evaluation of the promotion by 2,3,7,8-tetrachlorodibenzo-p-dioxin of hepatocarcinogenesis from diethylnitrosamine. *Cancer Res* 40:3616–3620, 1980.
27. Poland A, Palen D, Glover E. Tumour promotion by TCDD in skin of HRS/J hairless mice. *Nature* 300:271–273, 1982.
28. Abernethy DJ, Greenlee WF, Huband JC, Boreiko CJ. 2,3,7,8-Tetrachlorodibenzo-p-dioxin (TCDD) promotes the transformation of C3H10T1/2 cells. *Carcinogenesis* 6:651–653, 1985.
29. Yuspa SH, Morgan DL. Mouse skin cells resistant to terminal differentiation associated with initiation of carcinogenesis. *Nature* 293:72–74, 1981.
30. Akhurst RJ, Fee F, Balmain A. Localized production of TGF-β mRNA in tumour promoter-stimulated mouse epidermis. *Nature* 331:363–365, 1988.
31. Osborne R, Greenlee WF. 2,3,7,8-Tetrachlorodibenzo-p-dioxin (TCDD) enhances terminal differentiation of cultured human epidermal cells. *Toxicol Appl Pharmacol* 77:434–443, 1985.
32. Suskind RR. Chloracne, "the hallmark of dioxin intoxication." *Scand J Work Environ Health* 11:165–171, 1985.
33. Almendral JM, Sommer D, MacDonald-Bravo H, Burckhardt J, Perera J, Bravo R. Complexity of the early genetic response to growth factors in mouse fibroblasts. *Mol Cell Biol* 8:2140–2148, 1988.

34. U.S. Environmental Protection Agency. *Health Assessment Document for Polychlorinated Dibenzo-p-dioxins (Final Report).* (EPA-600/8-84-014F). Cincinnati, OH: U.S. Environmentalal Protection Agency, Office of Environmental Assessment, 1987.
35. National Toxicology Program. *Carcinogenesis Bioassay of 2,3,7,8-Tetrachlorodibenzo-p-dioxin in Osborne-Mendel Rats and B6C3F1 Mice (Gavage Study).* (NTP Technical Report 209). Research Triangle Park, NC: National Toxicology Program, 1982.
36. Toth K, Somfai-Relle S, Sugar J, Bence J. Carcinogenicity testing of herbicide 2,4,5-trichlorophenoxyethanol containing dioxin and of pure dioxin in Swiss mice. *Nature* 278:548–549, 1979.
37. Aviv H, Leder P. Purification of biologically active globin messenger RNA by chromatography on oligothymidylic acid-cellulose. *Proc Natl Acad Sci USA* 69:1408–1412, 1972.
38. Chirgwin JM, Przybyla AE, MacDonald RJ, Rutter WJ. Isolation of biologically active ribonucleic acid from sources enriched in ribonuclease. *Biochemistry* 18:5294–5299, 1979.
39. Lehrach H, Diamond D, Wozney JM, Boedtker H. RNA molecular weight determinations by gel electrophoresis under denaturing conditions: A critical reexamination. *Biochemistry* 16:4743–4751, 1977.
40. Rigby P, Dieckmann M, Rhodes C, Berg P. Labeling deoxyribonucleic acid to high specific activity *in vitro* by nick translocation with DNA polymerase I. *J Mol Biol* 113:237–251, 1977.
41. Quattrochi LC, Okino ST, Pendurthi UR, Tukey RH. Cloning and isolation of human cytochrome P-450 cDNAs homologous to dioxin-inducible rabbit mRNAs encoding P-450 4 and P-450 6. *DNA* 4:395–400, 1985.
42. Walker C, Nettesheim P, Barrett JC, Jirik FR, Sorge J, Joyce M, Gilmer T. Mouse retroviral sequences acquired by cell lines after passaging through nude mice detected by hybridization of the *fms* probe pSM3. *Cancer Res* 49:625–628, 1989.

Mouse Liver Carcinogenesis: Mechanisms
and Species Comparisons, pages 187–202

The Effects of TCDD on Receptors for Epidermal Growth Factor, Glucocorticoid, and Estrogen in Ah-Responsive and -Nonresponsive Congenic Mice and the Effects of TCDD on Estradiol Metabolism in a Liver Tumor Promotion Model in Female Rats

Joyce A. Goldstein, F. H. Lin, Sidney J. Stohs, Mark Graham, George Clarke, Linda Birnbaum, and George Lucier
National Institute of Environmental Health Sciences, Research Triangle Park, NC 27709

2,3,7,8-Tetrachlorodibenzo-*p*-dioxin (TCDD) is one of the most toxic chemicals known to man (1). It is one of a family of isosteric chemicals that produce similar types of toxicity. The toxic effects are species specific, but include thymic atrophy, lymphoid involution, immunotoxicity, teratogenicity, hepatic porphyria, liver lesions, chloracne in species such as hairless mice, rabbits, and primates, and a wasting syndrome that is followed by death. TCDD is also a potent carcinogen in both the mouse and the rat (2,3) and a promoter of hepatocarcinogenesis in the rat (4).

Extensive structure-activity relationships have shown that the most toxic members of the class of chemicals are planar, coplanar, or nearly planar compounds with halogen atoms occupying at least three of the four lateral positions in the molecule (Fig. 1) (1). All of these compounds are known to interact with a high-affinity cytosolic receptor known as the Ah receptor (5). The ligand receptor complex is thought to mediate all or most of the known biologic effects of TCDD via induction of a battery of structural genes. One of the structural genes shown to be transcriptionally activated by TCDD is the gene for the enzyme cytochrome P_1-450 (6–8) and its associated monooxygenase activities, such as aryl hydrocarbon hydroxylase (AHH) and ethoxyresorufin *O*-deethylase (EROD) (1). The Ah receptor is very stereospecific, and there is a good correlation between the affinity of various ligands for the Ah receptor, the potency of these chemicals as inducers of AHH activity, and their ability to produce various toxic responses (5,6,9).

Inducibility of cytochrome P_1-450 by TCDD is controlled by a single gene locus or a group of closely linked loci in mice, known as the Ah locus (10). P_1-

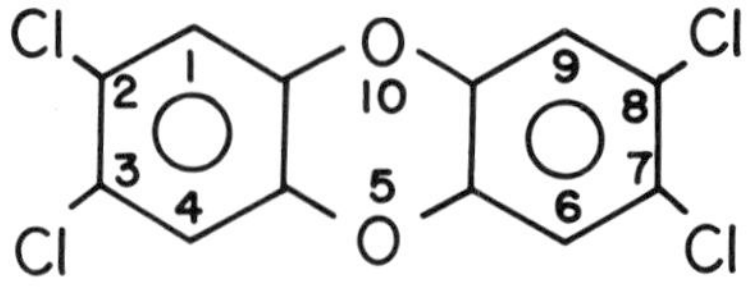

2,3,7,8 - tetrachlorodibenzo-p-dioxin

Figure 1. Structure of 2,3,7,8-tetrachlorodibenzo-*p*-dioxin (TCDD).

450 is inducible by TCDD and 3-methylcholanthrene (3-MC) in certain inbred strains of mice, such as the C57BL/6, while this enzyme is not induced by 3-MC in other inbred, nonresponsive strains (prototype DBA/2). Nonresponsive strains are also 10-fold less sensitive to induction of P_1-450 by TCDD. Inducibility has been shown to be inherited as a simple autosomal dominant trait. The evidence suggests that the defect in nonresponsive mice is an altered receptor with a lower affinity for TCDD. For example, classic receptor binding assays demonstrate that livers of Ah-responsive mice contain a high-affinity Ah receptor, while those of nonresponsive mice contain fewer receptors or receptors with less affinity for TCDD (5).

Recently, TCDD has been shown to decrease the binding of a number of endogenous ligands to their receptors. For example, TCDD has been shown to decrease the binding of epidermal growth factor (EGF) to its receptor in both rat liver plasma membranes (11) and in cultured human epidermal cells (12). This decrease appeared to reflect a decrease in the number of EGF receptors. However, surprisingly, TCDD treatment did not alter the levels of EGF mRNA in human epidermal cells (13). TCDD also decreased glucocorticoid receptor levels in mouse and rat liver (14) and estrogen receptor levels in the uterus and liver of female rats (G. I. Sunahara, G. W. Lucier, Z. McCoy, E. H. Bresnick, E. R. Sanchez, K. G. Nelson, submitted). The decrease in glucocorticoid receptor binding appeared to reflect a decrease in the total number of receptor sites, but western blot analysis of hepatic cytosol from TCDD-treated rats did not show a decrease in the amount of immunoreactive glucocorticoid receptor protein (15). Therefore, the effects of TCDD on these receptor systems, unlike many of the other biologic responses to TCDD, may not reflect effects on transcription of the structural genes for these proteins. Thus, it is of particular interest to know whether the decrease in the binding capacity of these receptors is mediated via the Ah receptor. This question is especially important since these receptors and their ligands are all involved in aspects of cell differentiation and could therefore potentially be involved in the toxicity and/or carcinogenicity of TCDD.

Two studies are described in this review. In the first, we examined whether the effects of TCDD on EGF, glucocorticoid, and estrogen receptor binding required the presence of the Ah receptor. In these studies, we compared the dose-response for the effects of TCDD on EGF, glucocorticoid, and estrogen receptor binding, as well as EROD activity, in congenic C57BL/6J mice that were either

Ah responsive ($Ah^{b/b}$) or Ah nonresponsive ($Ah^{d/d}$). These congenic strains of mice differ only at the Ah locus and a limited number of closely linked loci. For a review of the genetics of congenic mice, the reader is referred to Flaherty (16). Activity of EROD (a P_1-450 mediated enzyme) is known to be mediated by the Ah locus (10) and served as a positive control. Since TCDD could also potentially affect estrogen action by increasing hepatic 17β-estradiol metabolism, a second set of studies examined the effects of TCDD on estrogen metabolism in livers of rats treated with TCDD either acutely or subchronically in a liver tumor promotion model.

Effects of TCDD on Various Endogenous Receptors in Ah-Responsive and -Nonresponsive Mice

Methods

Female wild-type C57BL/6J mice ($Ah^{b/b}$) were obtained from Jackson Laboratories, Bar Harbor, ME. Female congenic B6-$Ah^{d/d}$ mice were originally obtained from Dr. Dan Nebert of the National Institutes of Health (NIH) and were NIH sublines but were then bred by Dr. Alan Poland at McArdle Laboratory, University of Wisconsin, into C57BL/6J (Jackson Laboratories, Bar Harbor, ME) by backcross-intercross. These B6.D2-$Ah^{d/d}$ (NE5) mice were obtained from Dr. Poland and maintained in a National Institute of Environmental Health Sciences colony by backcross-intercross cycles at Research Triangle Institute (17). Mice were treated with a single oral dose of TCDD and killed at 7 days or the times indicated.

Estrogen receptor binding and glucocorticoid receptor binding were measured in hepatic cytosol as previously described (15,18). EGF receptor binding in hepatic plasma membranes was measured as described by Sunahara et al. (19). Liver microsomal EROD activity was measured as described previously (20).

Results and Discussion

The effects of TCDD on EGF binding to plasma membranes from livers of Ah-responsive wild-type C57BL/6J mice is shown in a typical Scatchard plot in Figure 2a. EGF binding can be separated into two components, high- and low-affinity binding. TCDD decreased the maximum binding capacity for both the high- and low-affinity binding sites by approximately 80%–90%, indicating a loss of both binding sites following TCDD treatment. It did not change the dissociation constant (Kd) for EGF binding. TCDD also decreased the maximum binding capacity of the cytosolic glucocorticoid receptor, although the decrease (~30%) was much less than the decrease in EGF binding (Fig. 2b).

Figure 3 shows the time course for the effects of TCDD on hepatic EROD activity and the β_{max} for hepatic glucocorticoid receptor binding and EGF bind-

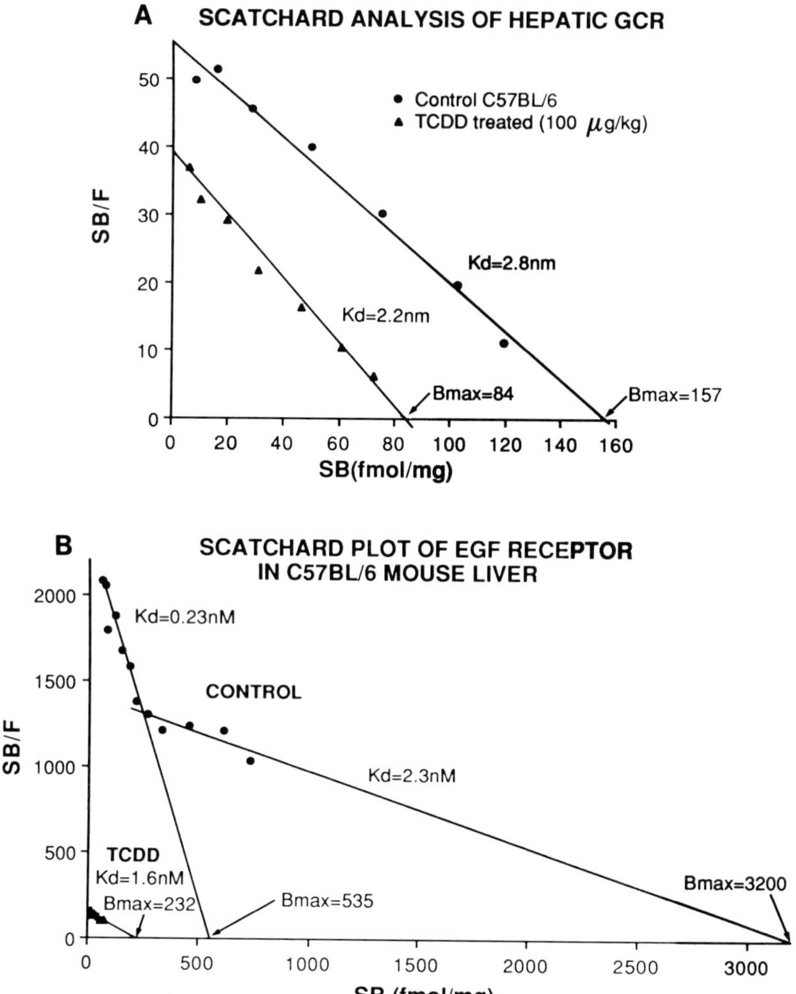

Figure 2. Scatchard analysis showing effects of TCDD (100 μg/kg) on binding of [^{3}H]dexamethasone to the glucocorticoid receptor (A) and [^{125}I]epidermal growth factor (EGF) to the EGF receptor (B) in wild-type C57BL/6J mice. Mice were given 100 μg/kg TCDD and killed 7 days later.

ing in wild-type $Ah^{b/b}$ C57BL/6J mice. The effects of TCDD on hepatic glucocorticoid receptor binding occurred much earlier (maximum at 6 h) than the effects on the EGF receptor (7 days) or EROD activity (~12 h). However, maximal effects on all three parameters were seen at 7 days; therefore dose-response studies in Ah-responsive and -nonresponsive mice were carried out at this time point.

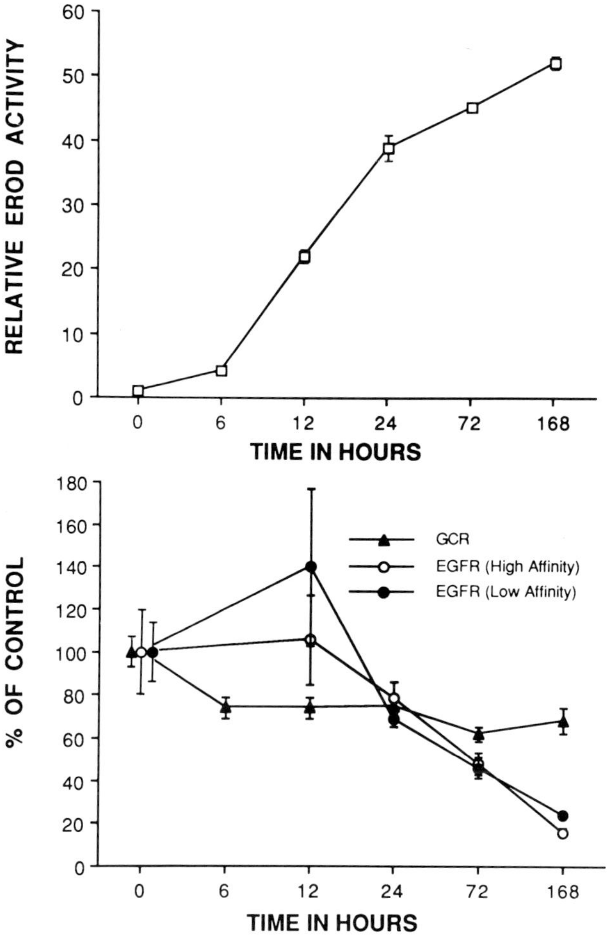

Figure 3. Time course for the effects of TCDD (30 μg/kg) on hepatic ethoxyresorufin *O*-deethylase activity (EROD) (top), glucocorticoid receptor (GCR) binding, and EGF binding (bottom) in liver plasma membranes of wild-type ($Ah^{b/b}$) C57BL/6J mice. Values represent the fold-increases over control for EROD (means ± SE). The β_{max} for GCR and EGF receptor binding were calculated by Scatchard analysis. Values for each treatment group represent the β_{max} as percentages of control values (means ± SE).

Table 1 shows that the ED_{50} for the induction of hepatic EROD was approximately 10-fold lower in responsive $Ah^{b/b}$ than nonresponsive $Ah^{d/d}$ mice, as would be expected from previous studies. The ED_{50} for the maximum binding capacity (β_{max}) for both the low- and high-affinity components of the EGF receptor was also 10-fold lower in livers of Ah-responsive mice than Ah-nonresponsive mice, indicating that both these effects are mediated via the Ah receptor.

Table 1. The ED_{50} for the Effects of TCDD on the Induction of EROD Activity and the Decrease in EGF Receptor Binding in Livers of Responsive ($Ah^{b/b}$) and Nonresponsive ($Ah^{d/d}$) Mice*

		EGF binding (β_{max})	
Treatment group	EROD activity	High affinity	Low affinity
C57BL/6J $Ah^{b/b}$	1.6	0.7	0.6
Congenic $Ah^{d/d}$	15.0	7.6	7.0

*Wild-type $Ah^{b/b}$ and congenic nonresponsive ($Ah^{d/d}$) mice were dosed orally with TCDD in corn oil (0.3–300 μg/kg) and killed 7 days later. The ED_{50} was estimated from the dose-response curve.

Abbreviations: EROD, ethoxyresorufin *O*-deethylase; EGF, epidermal growth factor.

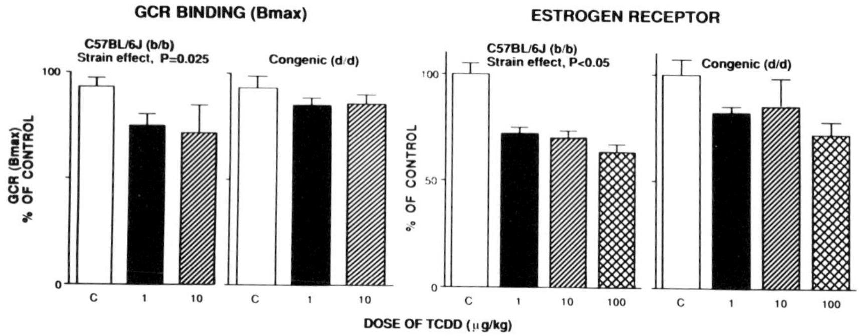

Figure 4. Effects of TCDD on hepatic estrogen receptor binding and glucocorticoid (β_{max}) in wild-type Ah-responsive (b/b) and -nonresponsive congenic (d/d) mice. Values represent percentages of control values (means ± SE).

We then examined the effects of the Ah locus on the maximum binding capacity of the hepatic estrogen and glucocorticoid receptors. The dose-response curves were quite shallow; therefore, it was difficult to estimate an ED_{50}. Figure 4 shows that TCDD decreased [^{3}H]estradiol binding to this receptor by ~30% in wild-type responsive mice and 15% in nonresponsive congenic mice. However, the effects of TCDD on the estrogen receptor were significantly greater in Ah-responsive mice than in Ah-nonresponsive mice ($P = 0.01$).

The effects of TCDD on the glucocorticoid receptor were also relatively modest (a maximum decrease of ~30%). There was a significant strain difference between the effects of TCDD in Ah-responsive and -nonresponsive mice in one of two experiments (Fig. 4). The results of sucrose gradient analysis of glucocorticoid receptor binding for this experiment also suggested that the effect was greater in Ah-responsive mice than in Ah-nonresponsive mice. However, in an earlier experiment the dose-response curves were very similar, and there was no statistical difference between strains. The difficulties in verifying whether the two strains differed resulted from the fact that the dose-response curve for the effects of TCDD on the glucocorticoid receptor was extremely shallow. We are repeating these experiments with larger numbers of animals to verify whether this effect is indeed associated with the Ah locus.

In conclusion, these experiments show that TCDD decreases the binding of EGF to its hepatic plasma membrane receptor by ~90% in C57BL/6J mice and also decreases binding to the estrogen and glucocorticoid receptor by ~30%. Moreover, studies in congenic strains of mice differing only at the Ah locus and closely linked loci indicate that the effects of TCDD on the EGF receptor and the estrogen receptor appear to be mediated by the Ah receptor. The strain differences for the effects on the glucocorticoid receptor were less pronounced, but it appears likely that this effect is also mediated by the Ah receptor.

Estrogen Metabolism in a Liver Tumor Promotion Model in the Rat

TCDD could conceivably affect the action of estrogens in two ways. First, a number of laboratories have confirmed that TCDD decreases both hepatic and uterine estrogen receptor levels (15,21, and this chapter). Second, TCDD could conceivably increase metabolism of endogenous estrogens such as estradiol by increasing cytochrome P-450 mediated metabolism of estrogens.

Many estrogens, including 17β-estradiol, are hydroxylated at the 2- or 4-positions of the A-ring, leading to the production of catechol estrogens (22–24). 2-Hydroxylation is the major pathway of estradiol metabolism (23). The formation of catechol estrogen metabolites has been proposed to play a role in the carcinogenic action of estrogens (25–27), by either the covalent binding of reactive metabolites (28–30) or the generation of active oxygen species by catechol-quinone oxidation reactions (31).

2-Hydroxylation is catalyzed primarily by the hepatic P-450 system (22,32). A number of forms of cytochrome P-450 have been shown to 2-hydroxylate estrogen in rat liver, including two male-specific forms, P-450h and P-450 PCN-E (32–34). In addition, P-450d, a TCDD-inducible isozyme of P-450, hydroxylates 17β-estradiol in the 2-position in a reconstituted enzyme system (33). Therefore, it is conceivable that formation of catechol estrogens could be one factor involved in the carcinogenic actions of TCDD in the female rat. However, the role of cytochrome P-450d in the 2-hydroxylation of estradiol (E_2OHase) in rat liver has been questioned (32). Therefore, we recently examined the effects of TCDD on E_2O-Hase activity in rat liver microsomes and the contribution of induction of cytochrome P-450d to this process (35). In this study, TCDD was administered both acutely and chronically in a liver tumor promotion study in female rats. Since the role of P-450d in 17β-estradiol hydroxylation is controversial, and since P-450 reductase (the flavoprotein involved in the reduction of P-450) has been reported to be rate-limiting for some steps in liver microsomal oxidation reactions (36), we also examined whether P-450 reductase limited the hydroxylation of E_2OHase in TCDD-induced microsomes.

Methods

TCDD was administered acutely to 8-week-old Sprague-Dawley (CD) rats of both sexes at a maximal inducing dose of 10 μg/kg, animals were killed 4 days later, and microsomes were prepared. TCDD was also administered in a tumor promotion model to female Sprague-Dawley (CD) rats that were first dosed with saline or 200 mg/kg diethylnitrosamine (DEN) as the initiator. TCDD was then administered as the promoter biweekly at doses of 0.14, 0.42, and 1.4 μg/kg (equivalent to 10, 30, or 100 ng/kg/day) beginning 15 days after initiation, and the mice were killed after 30 or 60 weeks of promotion. Samples of liver were fixed for routine staining with hematoxylin and eosin or for γ-glutamyltranspeptidase (GGT) assays (37). The number and size of GGT-positive foci were determined stereologically.

E_2OHase was measured as previously described (35). P-450d was determined by radioimmunoassay (38).

Results and Discussion

A single dose of 10 μg/kg TCDD produced maximal induction of cytochrome P-450c and P-450d in young adult female rats and increased liver microsomal E_2OHase approximately twofold (Fig. 5). This induction of E_2OHase effect was more pronounced when exogenous purified cytochrome c reductase was added to microsomes. In contrast, basal levels of E_2OHase were higher in livers of male rats, and TCDD increased E_2OHase only when exogenous reductase was added to liver microsomes before the assay. Antibody to P-450d completely inhibited E_2OHase in hepatic microsomes from TCDD-treated male and female rats to levels seen in liver microsomes from control rats (35). These results indicate that TCDD increases E_2OHase activity in liver microsomes of female rats and that this increase can be totally attributed to an increase in cytochrome P-450d in the liver. Male rats have higher constitutive E_2OHase levels because of the presence of other constitutive male enzymes, presumably P-450h and P-450 PCN-E, as reported by Dannan et al. (32). Therefore, cytochrome c reductase appears to be rate limiting for the metabolism of 17β-estradiol in livers of TCDD-treated male rats.

It is not yet clear whether the effects of TCDD on E_2OHase are substantial enough to modulate the biologic effects of estrogens. In fact, Shiverick and Muther (39) reported that 10 μg/kg TCDD increased rather than decreased plasma concentrations of exogenous estrone in sexually immature female rats. However, studies by Gierghy and coworkers (40,41) show that TCDD does in-

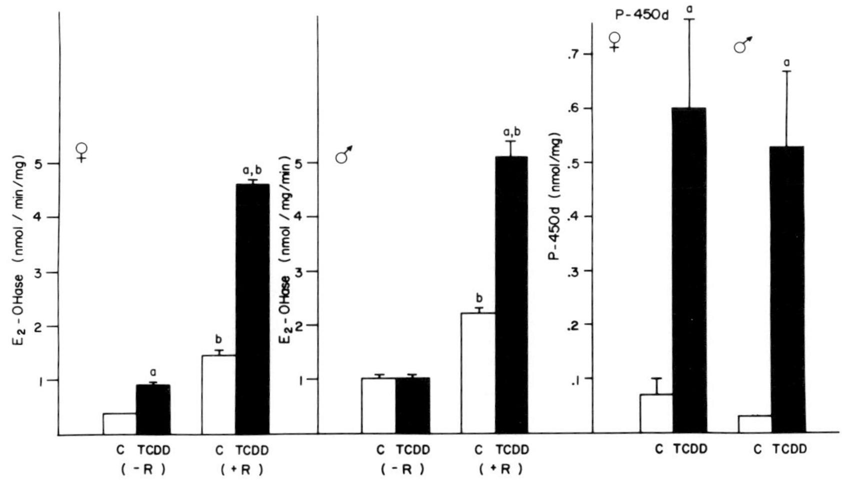

Figure 5. Effects of TCDD on liver microsomal P-450d and on E_2OHase activity in adult male and female rats. Adult male and female rats were given a single oral dose of TCDD (10 µg/kg) and killed 4 days later. Liver microsomes were assayed for E_2OHase activity with (+) or without (–R) preincubation with exogenous purified reductase (10,000 U/ml). Liver microsomal content of P-450d was assayed by radioimmunoassay. Results represent the means ± SE. [a]Significantly different from controls ($P < 0.05$), Student's t-test. [b]Significantly different from levels of E_2OHase measured in the same treatment group assayed in the absence of reductase ($P < 0.05$). Reprinted from Graham et al. *Carcinogenesis* 9:1935–1941, 1988, with permission.

crease 2-hydroxylation of 17β-estradiol and decreases estradiol concentrations in MCF-7 cells, a human breast cancer cell line. This increase in estradiol 2-hydroxylation is concomitant with a decrease in the effectiveness of estradiol in increasing the secretion of tissue plasminogen activator (42). TCDD also supresses postconfluent cell accumulation in cultures of MCF-7 cells, and the authors suggested that this effect might be due to estrogen depletion (43). It seems likely, however, that the diminution of the response of the rat and mouse uterus to estrogens (15,20) is primarily due to a decrease in the estrogen receptor, although increases in estradiol metabolism could contribute to this effect. Moreover, increases in estradiol metabolism could increase the hepatic concentration of catechol estrogens, which are known to be more reactive than their parent estrogens and to bind to macromolecules (27–30). Therefore, we also examined the ability of the liver of female rats to 2-hydroxylate 17β-estradiol in a liver tumor promotion study, using TCDD as the tumor promoter.

Table 2 shows the effects of TCDD on GGT-positive foci in a liver tumor promotion model in the female Sprague-Dawley rat. A biweekly dose of 1.4 µg/kg TCDD increased the proportion of the liver occupied by GGT-positive foci at 30 weeks. A slight increase that was not statistically significant was also seen at a dose of 0.42 µg/kg. No tumors were noted at this time, but tumors were seen in five of eight DEN-initiated rats receiving the 1.4 µg/kg dose of TCDD at 60

Table 2. Effect of TCDD (1.4 μg/kg Biweekly) on Hepatocellular GGT-Positive Foci and Tumors in Female Sprague-Dawley Rats after Initiation with a Necrogenic Dose of DEN*

	30 weeks		60 weeks		
			No. rats with		
Treatment group	Liver volume with foci (%)	No. rats with tumors	Adenomas	Carcinomas	Combined
Saline controls	0 ± 0	0/4	0/5	0/5	0/5
DEN controls	0.24 ± 0.08†	0/4	0/5	1/5‡	1/5
Saline/0.14 μg/kg TCDD	0.007 ± 0.005	0/4	—	—	—
Saline/0.42 μg/kg TCDD	0.05 ± 0.05	0/4	—	—	—
Saline/1.4 μg/kg TCDD	0.15 ± 0.15	0/4	0/7	0/7	0/7
DEN/0.14 μg/kg TCDD	0.07 ± 0.03†	0/4	—	—	—
DEN/0.42 μg/kg TCDD	0.54 ± 0.44†	0/4	—	—	—
DEN/1.4 μg/kg TCDD	1.51 ± 0.4†§	0/4	5/8¶	1/8	5/8¶

*Female Sprague-Dawley rats (71 days old) were treated with a necrogenic dose of diethylnitrosamine (DEN) (200 mg/kg) and then given the indicated biweekly oral doses of TCDD for 30 weeks or 60 weeks.
†Significantly greater than saline controls ($P < 0.05$), Mann-Whitney U-test.
§Significantly greater than DEN controls ($P < 0.05$), Mann-Whitney U-test.
‡Hepatocholangiocarcinoma.
¶P value = 0.044 vs. saline controls and 0.18 vs. DEN/controls by Fisher's exact test.
Reprinted from Graham et al. *Carcinogenesis* 9:1935–1941, 1988, with permission.

weeks compared with one of five treated with DEN alone. Moreover, the foci in the TCDD/DEN–treated animals were large and essentially confluent at 60 weeks, but not in the animals treated with DEN alone. These results indicate that TCDD acted as a tumor promoter in this model.

In the tumor promotion study, TCDD induced E_2OHase maximally in livers of female rats at doses of 0.14 μg/kg for 30 weeks (Fig. 6). These increases were significant only when liver microsomes were assayed in the presence of optimal amounts of exogenous reductase. However, reductase levels in liver microsomes of the saline control and DEN control animals were 23% and 50% lower, respectively, than reductase levels in young adult female rats. Therefore, reductase was more limiting in liver microsomes of young adult rats, and this influenced the results.

Initiation with DEN appeared to potentiate the effects of the low dose of TCDD on both E_2OHase activity and P-450d. Although it is possible that the increase in P-450d activity in TCDD/DEN–treated animals occurs specifically in the preneoplastic nodules, this seems unlikely since another laboratory (44) has reported a decrease in both P-450c and P-450d in neoplastic nodules. Moreover, preliminary immunostaining results in our laboratory using an antibody to P-

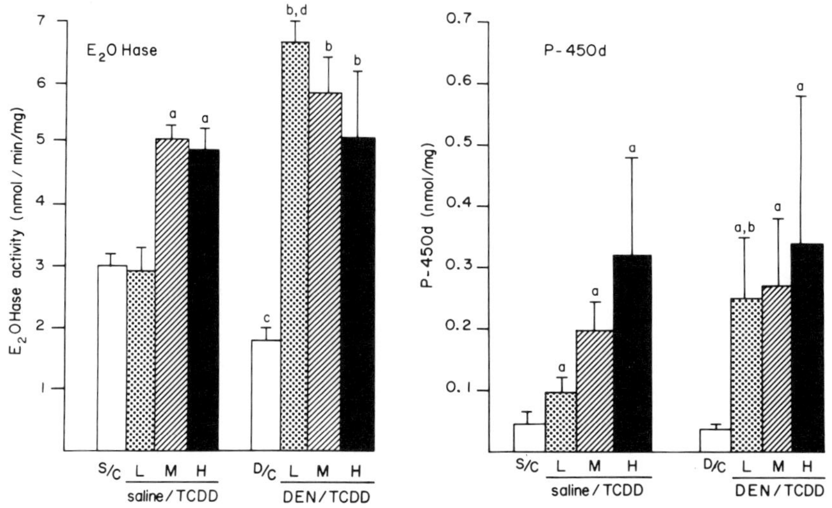

Figure 6. E_2OHase activity and P-450d content of liver microsomes of female rats given saline (S) or a single initiating dose of diethylnitrosamine (DEN) and biweekly doses of TCDD as a promoter for 30 weeks. Female rats (71 days old) were dosed intraperitoneally with 200 mg/kg DEN. TCDD was then administered (po) at doses of 0.14 (L), 0.42 (M), or 1.4 (H) μg/kg at biweekly intervals starting 15 days after initiation with DEN, and rats were killed after 30 weeks of promotion. Liver microsomal E_2OHase activity was assayed in the presence of reductase. P-450d content was assayed by radioimmunoassay. Values represent the means ± SE of values for four animals per group. [a]Significantly different from the corresponding control group ($P < 0.05$), Student's t-test. [b]DEN/TCDD treatment group values significantly higher than those for the group receiving a corresponding dose of saline/TCDD ($P < 0.05$). [c]DEN control values significantly lower than saline control values ($P < 0.05$). [d]DEN/TCDD treatment group values significantly higher than for animals receiving a corresponding dose of saline/TCDD ($P < 0.05$). Reprinted from Graham et al. *Carcinogenesis* 9:1935–1941, 1988, with permission.

450d indicated no increase in P-450d in GGT-positive foci compared with the surrounding hepatocytes in TCDD-promoted animals (unpublished data).

TCDD is a well-known promoter of DEN-initiated hepatocarcinogenesis (4). Moreover, TCDD has been reported to be a complete hepatocarcinogen in bioassays in female rats but not in male rats (2,3). Therefore, it is conceivable that increases in reactive catechol or quinone metabolites of estrogens may play a role in the carcinogenic effects of TCDD. The current studies show that TCDD does increase the 2-hydroxylation of estradiol in liver microsomes of female rats. This process would result in increased production of catechol estrogens. In addition, our results suggest that both the decrease in estrogen receptors and increase in hepatic metabolism of estrogens might play a role in the TCDD-induced diminution of the responsiveness of both rats and mice to estrogens (15,21).

Open Forum

Dr. Dennis Parke: From all these fascinating receptor studies, do you have any idea what the natural ligand for the Ah receptor is? The Ah is a very powerful receptor.

Dr. Joyce Goldstein: Not really. Several laboratories have been working on that. It is very difficult to purify the Ah receptor. Alan Poland's laboratory is attempting to do this. If someone ever purifies the receptor enough to clone it, then they might be able to tell what it is.

Dr. Parke: Some 10 years ago or more, we thought the Ah receptor would be a general steroid receptor. But then binding experiments showed that steroids did not bind as well as TCDD.

Dr. Goldstein: Right. No endogenous substances have ever been found that could compete with TCDD.

Dr. Parke: I do not see why they should compete. This is a misconception. TCDD is planar. It is resistant to metabolism because of those four chlorines; and this enables it to take an unfair advantage over the endogenous steroids and polypeptides that might well bind to the Ah receptor.

Dr. Goldstein: Yes, but you can compete it off with nonchlorinated compounds like 3-methylcholanthrene.

Dr. Parke: Yes, well maybe they are not fair, either. Anyway, your paper was fascinating.

Dr. Frederick Becker: Dr. Goldstein, in the data that you presented, the effects on the C57BL/6J or wild-type mouse seemed quite dramatic. This mouse showed all the receptivity, etc., of other mice. Has anyone ever done a long-term study to demonstrate whether TCDD is tumorigenic in the C57BL/6 mouse?

Dr. Goldstein: No, I don't think so.

Dr. Michael Gallo: I don't think so, either.

Dr. Becker: So you have a mouse in which you have all the receptivity and the sequelae or cascade that follow TCDD exposure, but that is a low-incidence mouse. By the way, I wish this conference would eliminate the term "resistant" mouse, because it is not a resistant mouse—it is just a low-incidence mouse. Resistance connotes some sort of supermouse that does not get tumors. This mouse gets tumors if it is exposed to a carcinogen, but it does not seem to get tumors if it is exposed to a promoter. It would be very interesting to see if TCDD induces tumors in the liver of the C57BL/6J mouse.

Dr. Goldstein: Yes, it would be.

Dr. Gallo: Just a note on that. In the less susceptible strains—the Hr– and Hr+—we have seen that if you crank the dose up enough in the hr– locus, you get hyperkeratinization. So I think you are absolutely right. Recently, we have shown that if you give copper with diethylnitrosamine, you decrease the acute toxicity of DEN, because the copper decreases that reductas— so I think that you are on to something as far as the reductase stuff goes.

Dr. Norman Drinkwater: We have compared tumor promotion in congenic C57BL/6, Ah^b, and Ah^d mice as well as C3H or $B6C3F_1$ mice that carry the B or D alleles for the Ah receptor. The result is that for both the less susceptible and more susceptible mice, hexabromobiphenyl, which is a long-lived congener of TCDD, was able to promote tumor induction after initiation with diethylnitrosamine at 12 days of age. The ability to promote tumor induction in fact was dependent on having the allele for the Ah locus.

Dr. Goldstein: So it did associate with the Ah locus?

Dr. Drinkwater: Yes. The Ah^b mice on either a C57BL/6 or a $B6C3F_1$ background were susceptible to promotion by hexabromobiphenyl, but the Ah^d mice were not.

Dr. Becker: Was that after initiation?

Dr. Drinkwater: Yes. We ran the obvious controls not using the initiating agent, and we did see some increase in tumors in the mice with the $B6C3F_1$ background relative to untreated animals, but the numbers are small enough that it is not easy to say what is going on.

Dr. Becker: Was the congenic strain derived from C57BL/6 and C3H?

Dr. Drinkwater: Yes.

References

1. Goldstein JA, Safe S. Mechanism of action and structure-activity relationships for the chlorinated dibenzo-*p*-dioxins and related compounds. In Kimbrough RD, Jensen S (eds): *Halogenated Biphenyls, Terphenyls, Naphthalenes, Dibenzodioxins and Related Products*. Amsterdam, New York: Elsevier, 1989, pp 239–293.
2. Kociba RJ, Keyes DG, Beyer JE, Carreon RM, Wade CE, Dittinber DA, Kalnins RP, Frauson LE, Park CN, Barnard SD, Hummel RA, Huminston CG. Results of a two-year chronic toxicity and oncogenicity study of 2,3,7,8-tetrachlorodibenzo-p-dioxin in rats. *Toxicol Appl Pharmacol* 46:279–303, 1978.
3. National Toxicology Program. Carcinogenesis bioassay of 2,3,7,8-tetrachlorodibenzo-*p*-dioxin in Osborne-Mendel rats and B6C3F1 mice (gavage study). TR 209. Research Triangle Park, NC: Carcinogenesis Testing Program, National Institute of Environmental Health Sciences, 1982.
4. Pitot HC, Goldsworthy T, Campbell HA, Poland A. Quantitative evaluation of the promotion by 2,3,7,8-tetrachlorodibenzo-p-dioxin of hepatocarcinogenesis from diethylnitrosamine. *Cancer Res* 40:3616–3620, 1980.
5. Poland A, Glover E, Kende AS. Stereospecific, high affinity binding of 2,3,7,8-tetrachlorodibenzo-*p*-dioxin by hepatic cytosol. *J Biol Chem* 251:4936–4946, 1976.
6. Poland A, Glover E. Chlorinated dibenzo-*p*-dioxins: Potent inducers of δ-aminolevulinic acid synthetase and aryl hydrocarbon hydroxylase. II. A study of the structure activity relationship. *Mol Pharmacol* 9:736–747, 1973.
7. Jones PBC, Galeazzi DR, Fisher JM, Whitlock JP Jr. Control of cytochrome P_1-450 gene expression by dioxin. *Science* 227:1499–1502, 1985.
8. Gonzalez FJ, Jukey RH, Nebert DW. Structural gene products of the Ah locus: Transcriptional regulation of cytochrome P_1-450 and P_3-450 mRNA levels by 3-methylcholanthrene. *Mol Pharmacol* 26:117–121, 1984.
9. Poland A, Glover E. 2,3,7,8-Tetrachlorodibenzo-*p*-dioxin: Segregation of toxicity with the Ah locus. *Mol Pharmacol* 17:68–94, 1980.

10. Poland A, Glover E, Robinson JR, Nebert DW. Genetic expression of aryl hydrocarbon hydroxylase activity. *J Biol Chem* 249:5599–5606, 1974.
11. Madhukar BV, Brewster DW, Matsumra F. Effects of in vivo administered 2,3,7,8-tetrachlorodibenzo-*p*-dioxin on receptor binding of epidermal growth factor in the hepatic plasma membrane of rat, guinea pig, mouse, and hamster. *Proc Natl Acad Sci USA* 81:7407–7411, 1984.
12. Osborne R, Greenlee WF. 2,3,7,8-Tetrachlorodibenzo-*p*-dioxin (TCDD) enhanced terminal differentiation of cultured human epidermal cells. *Toxicol Appl Pharmacol* 77:434–443, 1985.
13. Osborne R, Cook JC, Dold KM, Ross L, Gaido K, Greenlee WF. TCDD receptor: Mechanisms of altered growth regulation in normal and transformed human keratinocytes. *Progress in Cancer Research and Therapy* 34:407–416, 1988.
14. Ryan RP, Nelson KG, Lucier GW, Birnbaum LS, Sunahara GI. 2,2,4,7,8-Pentachlorodibenzofuran and 1,2,3,4,7,8-hexachlorodibenzofuran decrease glucocorticoid receptor binding in mouse liver and placental cytosol (Abstract). *The Toxicologist* 7:125, 1987.
15. Romkes M, Safe S. Comparative activities of 2,3,7,8-tetrachlorodibenzo-*p*-dioxin and progesterone as antiestrogens in the female rat uterus. *Toxicol Appl Pharmacol* 92:368–380, 1988.
16. Flaherty L. Congenic Strains. In Foster HL, Small JD, Fox JG (eds): *The Mouse in Biomedical Research, Vol 1. History, Genetics, and Wild Mice.* New York: Academic Press, 1981, pp 215– 222.
17. Birnbaum L. Distribution and excretion of 2,3,7,8-tetrachlorodibenzo-*p*-dioxin in congenic strains of mice which differ at the Ah locus. *Drug Metab Dispos* 14:34–40, 1986.
18. Sunahara GI, Nelson KG, Wong TK, Lucier GW. Decreased human birth weights after in utero exposure to PCBs and PCDFs are associated with decreased placental EGF-stimulated receptor autophosphorylation capacity. *Mol Pharmacol* 32:572–578, 1987.
19. Powell-Jones W, Thompson T, Nayfeh SN, Lucier GW. Sex differences in estrogen binding by cytosolic and nuclear components of rat liver. *J Steroid Biochem* 13:219–229, 1980.
20. Pohl RA, Fouts JR. A rapid method for assaying the metabolism of 7-ethoxyresorufin by microsomal subcellular fractions. *Anal Biochem* 107:150–155, 1980.
21. Gallo MA, Hesse EJ, MacDonald GJ, Umbreit TH. Interactive effects of estradiol and 2,3,7,8-tetrachlorodibenzo-*p*-dioxin on hepatic cytochrome P-450 and mouse uterus. *Toxicol Lett* 32:123–132, 1986.
22. Fishman J, Guzik H, Hellman L. Aromatic ring hydroxylation of estradiol in man. *Biochemistry* 9:1593–1598, 1970.
23. Fishman J. Aromatic hydroxylation of estrogens. *Annu Rev Physiol* 45:61–72, 1983.
24. Barbieri RL, Canick JA, Ryan KJ. Estrogen 2-hydroxylase: Activity in rat tissues. *Steroids* 32:529–538, 1978.
25. Li SA, Klicka JK, Li JJ. Estrogen 2- and 4-hydroxylase activity, catechol estrogen formation, and implications for estrogen carcinogenesis in the hamster kidney. *Cancer Res* 45:181–185, 1985.
26. Metzler M. Metabolism of stilbene estrogens and steroidal estrogens in relation to carcinogenicity. *Arch Toxicol* 55:104–109, 1984.
27. Liehr JG. 2-Fluoroestradiol: Separation of estrogenicity from carcinogenicity. *Mol Pharmacol* 23:278–281, 1983.
28. Tsibris JCM, McGuire P. Microsomal activation of estrogens and binding to rat

liver DNA in vivo by selective induction of microsomal and nuclear aryl hydrocarbon hydroxylase activity. *Cancer Res* 38:4640–4644, 1977.
29. Liehr JG, Randerath K, Randerath E. Target organ-specific covalent DNA damage preceding diethylstilbestrol-induced carcinogenesis. *Carcinogenesis* 6:1067–1069, 1985.
30. Jaggi W, Lutz WK, Schlatter C. Covalent binding of ethinyl estradiol and estrone to rat liver DNA in vivo. *Chem Biol Interact* 23:13–18, 1978.
31. Cerutti P. Repairable damage in DNA. In Hanawalt P, Friedberg E, Foc C (eds): *DNA Repair Mechanisms*. New York: Academic Press, 1978, pp 1–14.
32. Dannan GA, Porubeck DJ, Nelson SD, Waxman DJ, Guengerich FP. 17β-Estradiol 2- and 4-hydroxylation catalyzed by rat hepatic cytochrome P-450: Roles of individual forms, inductive effects, developmental patterns, and alterations by gonadectomy. *Endocrinology* 188:1952–1960, 1986.
33. Ryan DE, Thomas PE, Levin W. Purification and characterization of a minor form of hepatic microsomal cytochrome P-450 from rats treated with polychlorinated biphenyls. *Arch Biochem Biophys* 216:272–288, 1982.
34. Cheng KC, Schenkman JB. Metabolism of progesterone and estradiol by microsomes and purified cytochrome P-450 RLM3 and RLM5. *Drug Metab Dispos* 12: 222–234, 1984.
35. Graham M, Lucier GW, Linko P, Maronpot RR, Goldstein JA. Increases in cytochrome P-450 mediated 17β-estradiol 2-hydroxylase activity in rat liver microsomes after both acute administration and subchronic administration of 2,3,7,8-tetrachlorodibenzo-p-dioxin in a two-stage hepatocarcinogenesis model. *Carcinogenesis* 9:1935–1941, 1988.
36. Domin BA, Philpot RM. The effect of substrate on the expression of activity catalyzed by cytochrome P-450: Metabolism mediated by rabbit isozyme 6 in pulmonary microsomal and reconstituted monooxygenase systems. *Arch Biochem Biophys* 246:1128–1142, 1986.
37. Campen DB, Sloop TC, Maronpot RR, Lucier GW. Continued development of hepatic γ-glutamyltranspeptidase-positive foci upon withdrawal of 17α-ethinylestradiol in diethylnitrosamine-initiated rats. *Cancer Res* 47:2328–2333, 1987.
38. Luster MJ, Lawson LD, Linko P, Goldstein JA. Immunochemical evidence for two 3-methylcholanthrene-inducible forms of cytochrome P-448 in rat liver microsomes using a double-antibody radioimmunoassay procedure. *Mol Pharmacol* 23:252–257, 1983.
39. Shiverick KT, Thomas FM. Effects of 2,3,7,8-tetrachlorodibenzo-*p*-dioxin on serum concentrations and the uterotrophic action of exogenous estrone in rats. *Toxicol Appl Pharmacol* 65:170–176, 1982.
40. Gierthy JF, Lincoln DW II, Kampcik SJ, Dickerman HW, Bradlow HL, Niwa T, Swaneck GE. Enhancement of 2- and 16α-estradiol hydroxylation in MCF-7 human breast cancer cells by 2,3,7,8-tetrachlorodibenzo-*p*-dioxin. *Biochem Biophys Res Commun* 157:515–520, 1988.
41. Gierthy JF, Lincoln DW, Kampcik SJ, Dickerman HW, McKenna L, Bradlow HL, Niwa T, Swaneck GE. 2,3,7,8-Tetrachlorodibenzo-*p*-dioxin enhances estradiol hydroxylation and suppresses estrogen-dependent growth of MCF-7 human breast cancer cells in culture (Abstract). *The Toxicologist* 9:136, 1989.
42. Gierthy JF, Lincoln DW, Gillespie MB, Seeger JI, Martinez HL, Dickerman HW, Kumar SA. Suppression of estrogen-regulated extracellular tissue plasminogen activator activity of MCF-7 cells by 2,3,7,8-tetrachlorodibenzo-*p*-dioxin. *Cancer Res* 47:6198–6203, 1987.

43. Gierthy JF, Lincoln DW II. Inhibition of postconfluent focus production in cultures of MCF-7 human breast cancer cells by 2,3,7,8-tetrachlorodibenzo-*p*-dioxin. *Breast Cancer Res Treat* 12:227–233, 1988.
44. Kunz HW, Buchmann A, Schwarz M, Schmitt R, Kuhlmann WD, Wolf CR, Oesch F. Expression and inducibility of drug-metabolizing enzymes in preneoplastic and neoplastic lesions of rat liver during nitrosamine-induced hepatocarcinogenesis. *Arch Toxicol* 60:198–203, 1987.

Mouse Liver Carcinogenesis: Mechanisms
and Species Comparisons, pages 203–214

The Androgen Receptor and Liver Tumor Development in Mice

Christopher J. Kemp and Norman R. Drinkwater
McArdle Laboratory for Cancer Research, University of Wisconsin, Madison, WI 53706

It has long been known that male mice are more susceptible than female mice to the development of spontaneous and some chemically induced liver tumors (1–6). Studies of liver tumor induction after perinatal carcinogen treatment have demonstrated that this sexual dimorphism results from the opposing effects of estrogens and androgens acting during the promotion stage of hepatocarcinogenesis (7,8). In these studies, mice received a single dose of a carcinogen at an early age, and the subsequent development of preneoplastic foci and liver tumors was analyzed throughout the adult life of the animals, independent of further carcinogen treatment (7–10). Vesselinovitch et al. (7) demonstrated that castration of male mice following perinatal treatment with *N,N*-diethylnitrosamine (DEN) decreased both the rate of growth of preneoplastic foci and the multiplicity of liver tumors from those of intact males, and conversely, ovariectomy of female mice increased tumor development over that observed for normal females. Administration of testosterone to ovariectomized female mice resulted in a further increase in tumor multiplicity (8, and C. J. Kemp, C. N. Leary, N. R. Drinkwater, submitted). These experiments demonstrated that testosterone is a potent tumor promoter in mice, and this explains, at least in part, the greater susceptibility of male mice.

We performed a series of experiments designed to elucidate the mechanism of action of testosterone in its role as a tumor promoter. Specifically, we wanted to determine whether testosterone acted through the androgen receptor and, if so, whether this effect was direct, at the level of the hepatocyte, or indirect. To address these questions, we studied liver tumor development in Tfm (testicular feminization) mutant mice, which lack functional androgen receptors (11,12). Although the precise molecular defect of this mutation is unknown, cytosol prepared from tissues of these mutant mice has little or no high-affinity androgen receptor binding (11,12, and C. J. Kemp, C. N. Leary, N. R. Drinkwater, submitted), and it was recently shown that these mice do not have detectable levels of androgen receptor mRNA in either the liver or the kidney (13). The Tfm mutation and the androgen receptor gene are located on the X chromosome (13,14); thus, Tfm/Y males are hemizygous and express no wild-type receptor. These mice are androgen insensitive and do not develop those phenotypes that require a functional androgen receptor, such as male secondary sex characteristics (15). On the other hand, Tfm/+ heterozygous females are mosaic as a result of random X-inactivation, and individual cells or clones of cells in the adult express either the normal wild-type receptor or the nonfunctional mutant receptor (16,17).

The androgen receptor belongs to the superfamily of steroid receptor genes and shares extensive structural homology with estrogen, progesterone, glucocorticoid, and mineralocorticoid receptors (13,18,19). The classic model for the mechanism of action of the steroid hormones postulates that they bind to their respective receptors within the target cell. This hormone-receptor complex then induces or represses the expression of particular genes, which leads to differentiation, proliferation, or physiologic responses in the target tissue.

There is a wealth of evidence implicating several classes of steroids and steroid receptors in the carcinogenic process. For example, it is clear that estrogens, acting through estrogen receptors, play a major role in the growth of breast cancers (20). However, how the hormone-receptor complex triggers cell proliferation is not fully understood. It has recently been suggested that one or more growth factors, including platelet-derived growth factor, epidermal growth factor, transforming growth factors-α and -β, or insulin-like growth factor-1, may mediate these proliferative effects of estrogens by an autocrine or paracrine mechanism (20,21).

Cancer of the prostate is the best studied androgen-dependent neoplasm. It is well established that the growth of the prostate gland and some prostate carcinomas in vivo is androgen dependent (22). However, testosterone has no direct proliferative effect on pure populations of prostate epithelial cells in vitro (23). Tissue reconstruction studies suggested that the effect of testosterone on the proliferation of these cells may be indirect, mediated by growth factors secreted from the supporting mesenchymal tissue (24). Although the molecular pathways by which steroids influence tumorigenesis are still far from understood, there is increasing evidence that the proliferative effects of steroids may be mediated by secreted growth factors.

In this chapter, we present evidence indicating that, in the intact animal, the liver tumor–promoting effects of testosterone are mediated through the androgen receptor. However, despite the presence of functional androgen receptors within the target hepatocytes, testosterone appears to act indirectly as a tumor promoter.

Materials and Methods

Androgen Receptor Assay

The apparent affinity and abundance of the hepatic androgen receptors were determined as previously described (6, and C. J. Kemp, C. N. Leary, N. R. Drinkwater, submitted). Livers from untreated adult mice were homogenized in buffer and centrifuged to obtain cytosol free of lipids, nuclei, and microsomes.

Radiolabeled [^{3}H]mibolerone (from 0.1 to 8 nM) was added to aliquots of cytosol with or without a 300-fold excess of unlabeled mibolerone and allowed to bind for 20 h at 4°C. Bound [^{3}H]mibolerone was separated from free ligand by hydroxylapatite filtration, and radioactivity was measured by scintillation counting. Binding parameters were estimated by the method of Scatchard (25).

Mice and Tumor Induction

The mice used in these experiments were bred from stocks of C57BL/6J +/Tfm female mice and +/Y male mice and C3H/HeJ mice obtained from the Jackson Laboratory (Bar Harbor, ME). In the first experiment, C57BL/6J mice were treated at 12 days of age with DEN (0.2 μmol/g intraperitoneally) and killed at 16, 24, or 32 weeks of age for analysis of preneoplastic foci or at 50 weeks of age for enumeration of liver tumors. Preneoplastic foci were identified by staining frozen liver sections for glucose-6-phosphatase activity, and the volume doubling time of these foci was estimated as previously described (26, and C. J. Kemp, C. N. Leary, N. R. Drinkwater, submitted). Tumors were identified as white nodules greater than 2 mm in diameter visible on the surface of the liver.

In the second experiment, a large number of tumors was required for androgen receptor analysis, and the Tfm mutation was studied on the more susceptible (C57BL/6J × C3H/HeJ) F_1 background (27). B6C3F_1 wild-type and Tfm/+ female mice were treated at 12 days of age with DEN (0.1 μmol/g) and ovariectomized at 8 weeks of age. Half the females subsequently received biweekly injections of testosterone (0.15 mg/mouse, subcutaneously). Testosterone injections were discontinued 2 weeks before the mice were killed, to avoid competition with the radiolabeled ligand in the androgen receptor–binding assays. Mice that received testosterone were killed at 38 weeks of age; untreated mice were killed at 50 weeks. Liver tumors were counted, and the androgen receptor levels of all tumors heavier than 0.1 g were determined as described (C. J. Kemp, C. N. Leary, N. R. Drinkwater, submitted). Briefly, individual tumors were dissected free, homogenized in buffer, and centrifuged to obtain cytosol. Specific binding was determined by incubation of triplicate samples of cytosol with a saturating concentration of radiolabeled mibolerone (2 nM), with or without a 300-fold excess of unlabeled mibolerone. After a 20-h incubation at 4°C, bound mibolerone was separated from free ligand by hydroxylapatite filtration, and radioactivity was measured by scintillation counting. Specific binding was expressed as fmol receptor/mg cytosolic protein.

Results and Discussion

Analysis of Hepatic Androgen Receptors

The binding properties of androgen receptors in hepatic cytosol from wild-type and Tfm mutant mice are shown in Table 1. All groups except Tfm/

Table 1. Abundance and Affinity (K_d) of Androgen Receptors from Liver Cytosol*

Genotype	Androgen receptor fmol/mg protein	Androgen receptor K_d (nM)	Liver weight
+/+	32.0 (2.0)†	0.23 (0.03)	1.3 (0.1)
Tfm/+	13.0 (2.5)	0.39 (0.01)	1.5 (0.2)
Tfm/Y	2.7 (1.6)	3.30 (1.10)	2.2 (0.2)
+/Y	13.0 (6.0)	0.41 (0.01)	2.1 (0.2)
+/Y (castrate)‡	26.9 (1.1)	0.37 (0.04)	—§

*Apparent abundance and affinity were determined by Scatchard analysis using 0.1–8.0 nM [^{3}H]mibolerone.
†Values are mean (SD).
‡Adult male mice were castrated from 1 to 2 weeks prior to receptor assay.
§Not determined.

Y males had abundant, high-affinity androgen receptors. The low level of specific binding seen in Tfm/Y mice was of a much lower affinity (apparent K_d = 3.3 nM) than that from wild-type mice (K_d = 0.2–0.4 nM) and demonstrated that hepatocytes from Tfm/Y male mice lack functional androgen receptors. The lower level of receptor binding in Tfm/+ females (13 fmol/mg) compared with normal females (32 fmol/mg) is presumably a reflection of the mosaicism of the Tfm/+ liver; roughly 50% of the hepatocytes expressed functional receptors, and 50% had no functional receptors. The level of androgen receptors observed in male mice (13 fmol/mg) may be an underestimation of total cellular receptor because our assay only detects cytosolic receptors. Castration of male mice resulted in a twofold increase in cytosolic receptor content, to a level similar to that observed in wild-type females (Table 1). A similar increase in cytosolic androgen receptor concentration following castration was observed in kidney tissue (28) and was attributed to a shift in location of receptors from the activated or bound state in the nucleus to the unoccupied state in the cytosol.

As mentioned in the introduction, many sexually dimorphic phenotypes are feminized in Tfm/Y male mice because of the lack of functional androgen receptors. It was thus surprising to observe that liver weight in Tfm/Y males (2.2 ± 0.2 g) was similar to that for wild-type males (2.1 ± 0.2 g) and 60% larger than that for females (1.4 ± 0.2 g) (Table 1). Analysis of the total lipid content of the livers revealed that Tfm/Y male mice averaged 13 ± 3% dry weight lipid/wet weight liver, which was twice as high as for females (6.6 ± 0.3%). Histologic staining with oil red O showed much larger and more numerous intracellular deposits of lipids in hepatocytes from Tfm/Y male mice than in hepatocytes from normal male or female mice. This accumulation of lipids in hepatocytes from Tfm/Y male mice produced extensive cell hypertrophy that could account for the greater liver weights of Tfm males relative to females.

Table 2. Growth Rate of Preneoplastic Foci and Liver Tumor Induction by DEN in C57BL/6J-Tfm Mice*

Genotype	Number of mice	Focus volume doubling time (weeks)†	Liver tumor multiplicity
+/+	20	12.0 (3.0)	0.6 (0.9)
Tfm/+	26	8.3 (1.9)	1.5 (2.0)
Tfm/Y	23	6.6 (1.2)‡	0.7 (1.4)
+/Y	23	3.5 (0.4)§	20.0 (10.0)¶

*Mice were injected at 12 days of age with DEN (0.2 μmol/g) and killed at 16, 24, or 32 weeks of age for analysis of preneoplastic foci or at 50 weeks of age for liver tumor enumeration.
†The foci growth rates were calculated by linear regression of the data presented in Figure 1.
‡$P < 0.06$ vs. +/+ females.
§$P < 0.01$ vs. females and Tfm males.
¶$P < 0.05$ vs. females and Tfm males.

Tumor Promotion by Testosterone Is Androgen Receptor Dependent

Table 2 shows the liver tumor multiplicities in 50-week-old wild-type and Tfm mice treated as preweanlings with DEN. As in many previous studies (1–6), +/Y male mice were much more susceptible (averaging 20 tumors per animal) than female mice (averaging 0.6 to 1.6 tumors per animal). Tfm/Y male mice were completely feminized with respect to liver tumor induction, averaging only 0.7 tumors per animal. To examine in more detail the basis for this difference in susceptibility, we analyzed the development of preneoplastic foci in the same four groups of mice. Consistent with the tumor data, the foci grew more rapidly in +/Y male mice than in either the Tfm/Y males or females (Fig. 1 and Table 2). The estimated volume doubling time (T_d) of the foci in +/Y male mice ($T_d = 3.5$ weeks) was 1.9-fold faster than in Tfm/Y male mice ($T_d = 6.6$ weeks) and 2.4- to 3.5-fold faster than in female mice ($T_d = 8.3$ to 12 weeks).

These effects of the Tfm mutation, i.e., a decrease in the growth rate of preneoplastic foci and a decrease in liver tumor multiplicity, are similar to those seen following castration of +/Y male mice (7) and indicate that the tumor-promoting effects of testosterone are mediated through the androgen receptor.

Testosterone Does Not Act Directly as a Tumor Promoter

The second series of experiments was designed to determine whether the tumor-promoting effects of testosterone occurred directly, at the level of the hepatocyte, or indirectly, through a paracrine or endocrine pathway. Because

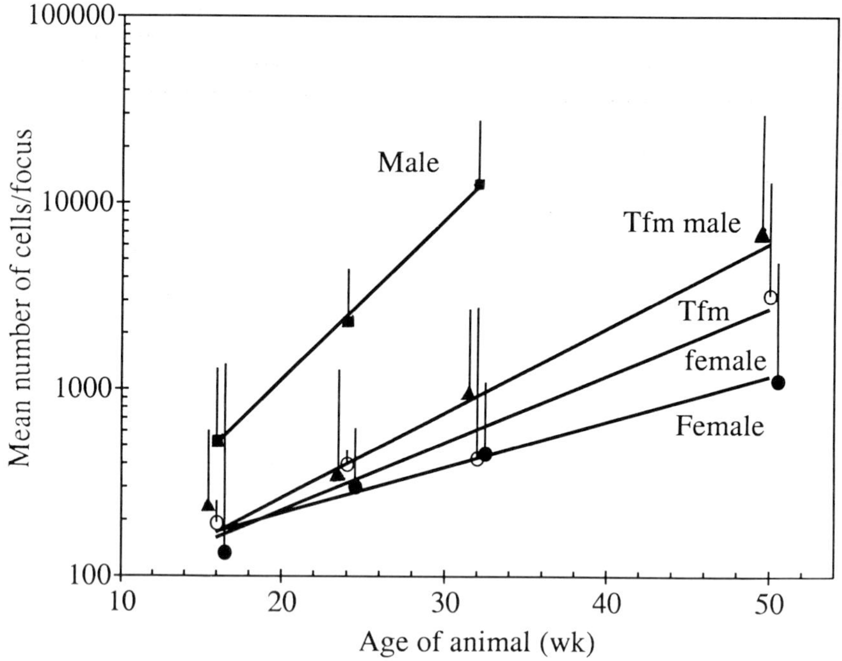

Figure 1. Growth rate of preneoplastic foci in C57BL/6J mice treated with DEN. Frozen liver sections were stained for glucose-6-phosphatase. Each point represents the mean number of cells per focus for 12–15 animals. Some points were moved slightly along the time axis to accomodate the standard error bars. Closed circles: +/+ females; open circles: Tfm/+ females; triangles: Tfm/Y males; boxes: +/Y males.

there are two distinct populations of hepatocytes in the livers of Tfm/+ female mice, those expressing normal androgen receptors and those expressing nonfunctional receptors, we could distinguish between these two models by treating Tfm/+ mosaic mice with testosterone and measuring the androgen receptor levels in the resulting tumors. If testosterone acted directly, only hepatocytes with functional receptors would respond, and the majority of the testosterone-promoted tumors would contain wild-type receptors. This model also predicted that fewer tumors would arise in Tfm/+ females than in wild-type females, since there would be approximately half as many cells at risk in the former group. The indirect model predicts that testosterone treatment would stimulate both cell types to develop into tumors and that roughly half the tumors should be receptor positive and half should be receptor negative. The results of these experiments (Table 3) demonstrate that testosterone treatment increased tumor multiplicity equally in wild-type and Tfm/+ female mice (31 and 32 tumors per animal, respectively, at 38 weeks of age) relative to females not treated with testosterone (17 tumors per animal at 50 weeks of age). The latter group was killed

Table 3. Liver Tumor Induction and Androgen Receptors in B6C3F$_1$ female mice treated with DEN and testosterone*

Genotype	Number of mice	Tumor multiplicity	Age (weeks)	Androgen receptor abundance (fmol/mg)†	
				Normal	Tumor
Tfm/+	12	17 (5)‡	50	14 (6)	2.4 (5.4)
Tfm/+ plus T*	11	32 (12)	38	15 (6)	3.8 (5.1)
+/+ plus T*	22	31 (15)	38	26 (4)	8.7 (10)

*Twelve-day-old female mice were injected with DEN (0.1 μmol/g intraperitoneally) and ovariectomized at 8 weeks of age. Groups 2 and 3 also received biweekly injections of testosterone (0.15 mg/mouse subcutaneously) from 10 weeks of age until 2 weeks before they were killed.

†Specific binding was determined by incubating cytosol prepared from tumors or normal liver with 2 nM [^{3}H]mibolerone with or without 300 nM radioinert mibolerone.

‡Values are mean (SD).

3 months later than the testosterone-treated groups because at 38 weeks of age, they had an insufficient number of liver tumors for receptor analysis.

Androgen receptor levels in normal liver tissue from these three groups of mice (Table 3) were similar to those for untreated mice (Table 1): Tfm/+ females with or without testosterone treatment averaged 14–15 fmol receptor/mg protein, which was 55%–60% of the level seen in normal females treated with testosterone (26 fmol/mg). Surprisingly, the androgen receptor level in the majority of tumors was much lower than in the surrounding normal liver tissue (Table 3). The mean receptor levels in tumors were 2.4, 3.8, and 8.7 fmol/mg, respectively, for Tfm/+ females, Tfm/+ females treated with testosterone, and for wild-type females treated with testosterone. Specific binding was also determined in tumors and normal liver tissue from Tfm/Y male mice and was always less than 1 fmol/mg (Table 4). Using this value to differentiate receptor-positive from receptor-negative tumors, we found that 93% (of the expected 100%) of the tumors from normal females treated with testosterone were receptor positive, 58% were receptor positive in Tfm/+ females treated with testosterone, and 47% (of the expected 50%) were receptor positive in Tfm/+ females not treated with testosterone. Thus, testosterone treatment significantly increased the number of liver tumors in Tfm/+ females (Table 3) but did not increase the percentage of tumors that expressed functional androgen receptors (Table 4) over the level for untreated Tfm/+ females. These results indicate that receptor-positive and receptor-negative hepatocytes are equally likely to develop into tumors in response to testosterone and that testosterone acts indirectly.

In a separate experiment, we measured the androgen receptor levels in six liver tumors from B6C3F$_1$ male mice that had been treated as preweanlings with DEN and castrated at least 2 weeks before they were killed. The mean receptor

Table 4. Distribution of Androgen Receptor Levels in Liver Tumors*

Genotype	No. of tumors	% Tumors with receptor levels (fmol/mg) of <1	1–4	4–10	10–20	20–30
Tfm/Y	9	100	0	0	0	0
Tfm/+	25	52	36	8	0	4
Tfm/+ plus T	33	42	36	3	18	0
+/+ plus T	38	8	31	29	13	18

*These data are from the experiment described in Table 3. Twelve-day-old mice were injected with DEN (0.1 μmol/g intraperitoneally), and females were ovariectomized at 8 weeks of age. Groups 3 and 4 also received biweekly injections of testosterone.

concentration in the tumors was 11 ± 8 fmol/mg and was 17 ± 5 fmol/mg for the surrounding normal liver tissue. All six tumors had greater than 1 fmol/mg receptor. Thus, liver tumors from male mice and female mice have reduced levels of androgen receptors.

In summary, three lines of evidence support the indirect model for the tumor-promoting effects of testosterone. Testosterone treatment increased the number of liver tumors equally in wild-type and Tfm/+ female mice; the proportion of tumors that expressed androgen receptors was not increased by testosterone in Tfm/+ female mice; and the majority of tumors, the development of which was enhanced by testosterone, had much lower levels of androgen receptors than the surrounding normal liver tissue.

The loss of androgen receptors in mouse liver tumors was surprising, as studies of rats (29) and humans (30–32) have demonstrated androgen receptor levels in liver tumors are equal to or greater than those in normal liver. It is not known whether this represents significant species differences in the regulation of receptor levels or differences in tumor etiology. The receptor loss in mouse liver tumors could be a nonspecific consequence of tumor progression and/or hepatocyte dedifferentiation and indicates that at this stage of tumor development, many of these cells have probably lost their ability to respond directly to androgens. It would be of interest to determine the mechanism and timing of this receptor loss.

Conclusions

Testosterone acts as a liver tumor promoter by stimulating the growth of preneoplastic lesions (7,8), thus increasing the rate of appearance of tumors in male mice. By studying liver tumor development in Tfm/Y mutant mice, we have shown that this effect is mediated through the androgen receptor. Despite the presence of functional androgen receptors in normal hepatocytes, testosterone does not act directly; it stimulated both receptor-positive and -negative hepatocytes to develop into tumors. This implies that testosterone acts indirectly as a tumor promoter. In the postulated indirect models, testosterone would stimulate the release of hepatotrophic growth factors from an androgen receptor–containing tissue, either within the liver (e.g., nonparenchymal liver cells) or distant from the liver (e.g., the hypothalamus or the submaxillary gland). These factors, in turn, would act in a paracrine or endocrine fashion to mediate the proliferative effects of testosterone. Epidermal growth factor, derived from the submaxillary gland, is one such candidate. This factor is present in male mice at much higher levels than in females (33–35), its production is both testosterone and androgen receptor dependent (34,35), and it is mitogenic to hepatocytes in culture (36). Preliminary results of an experiment in our laboratory indicate that removal of the submaxillary gland in male mice may significantly decrease the yield of liver tumors.

Acknowledgments

We thank Cindy Leary, Mary Winkler, and Jane Weeks for technical assistance. This work was supported by grants CA-07175, CA-22484, and CA-09020 from the National Cancer Institute.

Open Forum

Dr. Ernest McConnell: What is your definition of a tumor?

Dr. Christopher Kemp: A white nodule that is larger than 2 mm in size.

Dr. McConnell: Was the morphology of the tumors different in receptor-positive and receptor-negative animals?

Dr. Kemp: No. But I did find that the level of lipids in the livers of these animals was around 12%, while in normal males it is only about 5%. This may be worth following up.

Dr. Gary Williams: Did you do any enzyme histochemical reactions other than the glucose-6-phosphate dehydrogenase reaction to determine if there were any phenotypic differences in the foci or the tumors, apart from just hematoxylin and eosin?

Dr. Kemp: No.

Dr. Williams: You pointed out that male mice were more susceptible to hepatocarcinogenesis than female mice, which is also true for rats and, to some extent, hamsters. However, this cannot be ascribed entirely to events in the promotional phase. There are major differences between males and females in the activation of carcinogens. In many cases these differences are the main determinant of the sex differences in hepatocarcinogenesis.

Dr. Frederick Becker: Could you review the distribution of androgen-negative and androgen-positive tumors in the normal male?

Dr. Kemp: As in the females, androgen receptor levels in males are lower in tumors than in normal liver tissue.

Dr. Becker: But aren't there androgen receptor–negative tumors in the wild-type male?

Dr. Kemp: I analyzed six tumors and all had greater than 1.0 fmol of receptor, indicating they were receptor positive. But five of six had lower levels of receptor than normal liver tissue.

Dr. Becker: Obviously, that came from an androgen-positive receptor premalignant cell, because they are all positive. So there is a third possibility to explain the androgen-receptor negative tumors: downregulation of the androgen-receptor gene in these tumors as a result of cancerization in the malignant phenotype. We know that these tumors often downregulate to the point of disappearance. Van Potter and Henry Pitot showed the diversification of phenotype; one common means of diversification was downregulation. I think you have to consider this as a possible explanation.

Dr. Kemp: I absolutely agree. We would like to obtain an antibody to the receptor and stain preneoplastic glucose-6-phosphatase–deficient lesions and determine when during their development the receptor is downregulated or decreased.

Dr. Michael Gallo: When you look at what we have done with binding assays vs. an antibody for the estrogen receptor you get a different picture. Do you know anything about the aromatase activity in these cells?

Dr. Kemp: If aromatase requires an androgen receptor to be expressed, then it is probably not expressed.

Dr. Gallo: Aromatase is the P-450–mediated enzyme that takes testosterone to estradiol. What I wanted to ask was, Have you eliminated the possibility that what you may be looking at is catechol estrogen activation?

Dr. Kemp: I cannot answer that right now.

References

1. Burns EL, Schenken JR. Spontaneous primary hepatomas in mice of strain C3H: A study of incidence, sex distribution, and morbid anatomy. *Am J Cancer* 39:25–35, 1940.
2. Andervont HB. Studies on the occurrence of spontaneous hepatomas in mice of strains C3H and CBA. *J Natl Cancer Inst* 11:581–592, 1950.
3. Agnew LRC, Gardner WU. The incidence of spontaneous hepatomas in C3H, C3H (low milk factor), and CBA mice and the effect of estrogen and androgen on the occurrence of these tumors in C3H mice. *Cancer Res* 12:757–761, 1952.
4. Hancock RL, Dickie MM. Biochemical, pathological, and genetic aspects of a spontaneous mouse hepatoma. *J Natl Cancer Inst* 43:407–415, 1969.
5. Toh YC. Sex differences and carcinogenesis. *Adv Cancer Res* 18:184–195, 1973.
6. Kemp CJ, Drinkwater NR. Genetic variation in liver tumor susceptibility, plasma testosterone levels, and androgen receptor binding in six inbred strains of mice. *Cancer Res*, in press.

7. Vesselinovitch SD, Mihailovich N, Rao KVN, Goldfarb S. Relevance of basophilic foci to promoting effect of sex hormones on hepatocarcinogenesis. *Carcinogenesis* 7:127–131, 1982.
8. Moore MR, Drinkwater NR, Miller EC, Miller JA, Pitot HC. Quantitative analysis of the time-dependent development of glucose-6-phosphatase-deficient foci in the livers of mice treated neonatally with diethylnitrosamine. *Cancer Res* 41:1585–1593, 1981.
9. Vesselinovitch SD, Hacker HJ, Bannasch P. Histochemical characterization of focal hepatic lesions induced by single diethylnitrosamine treatment in infant mice. *Cancer Res* 45:2774–2780, 1985.
10. Bannasch P. Preneoplastic lesions as end-points in carcinogenicity testing. I. Hepatic preneoplasia. *Carcinogenesis* 7:689–695, 1986.
11. Attardi B, Ohno S. Cytosol androgen receptor from kidney of normal and testicular feminized (Tfm) mice. *Cell* 2:205–212, 1974.
12. Gehring U, Tomkins GM. Characterization of a hormone receptor defect in the androgen-insensitivity mutant. *Cell* 3:59–64, 1974.
13. Lubahn DB, Joseph DR, Sullivan PM, Willard HF, French FS, Wilson EM. Cloning of the human androgen receptor complementary DNA and localization to the X chromosome. *Science* 240:327–330, 1988.
14. Lyon MF, Hawkes SG. X-linked gene for testicular feminization in the mouse. *Nature* 227:1217–1219, 1970.
15. Goldstein JL, Wilson JD. Studies on the pathogenesis of the pseudohermaphroditism in the mouse with testicular feminization. *J Clin Invest* 51:1647–1658, 1972.
16. Tettenborn U, Dofuku R, Ohno S. Noninducible phenotype exhibited by a proportion of female mice heterozygous for the X-linked testicular feminization mutation. *Nature New Biol* 234:37–40, 1971.
17. Drews U. Direct and mediated effects of testosterone: The development of intersexes in sex reversed mosaic mice, heterozygous for testicular feminization. *Anat Embryol* 146:325–340, 1975.
18. Chang C, Kokontis J, Liao S. Molecular cloning of human and rat complementary DNA encoding androgen receptors. *Science* 240:324–326, 1988.
19. Evans RM. The steroid and thyroid hormone receptor superfamily. *Science* 240:889–896, 1988.
20. Dickson RB, Lippman ME. Estrogenic regulation of growth and polypeptide growth factor secretion in human breast carcinoma. *Endocrine Rev* 8:29–43, 1987.
21. Lippman ME, Dickson RB, Gelmann EP, Rosen N, Knabbe C, Bates S, Valverius E, Bronzert D, Huff K, Kasid A. Human breast carcinoma cells produce growth regulatory peptides. In Bresciani F, King RJB, Lippman ME, Raynaud JP (eds): *Hormones and Cancer*, Vol. 3. New York: Raven Press, 1988, pp 203–213.
22. Bosland MC. *Prostatic Carcinogenesis*. Netherlands: Krips Repro Meppel, 1989.
23. McKeehan WL, Adams PS, Rosser MP. Direct mitogenic effects of insulin, epidermal growth factor, glucocorticoid, cholera toxin, unknown pituitary factors and possibly prolactin, but not androgen, on normal rat prostate epithelial cells in serum-free, primary cell culture. *Cancer Res* 44:1998–2010, 1984.
24. Cunha GR, Cooke PS, Mee S, Higgins SJ, Donjacour AA, Bigsby RM, Young P. Mesenchymal-epithelial interactions in androgen-induced prostate growth and development. In Bresciani F, King RJB, Lippman ME, Raynaud JP (eds): *Hormones and Cancer*, Vol. 3. New York: Raven Press, 1988, pp 460–473.
25. Scatchard G. The attractions of proteins for small molecules and ions. *Ann NY Acad Sci* 51:660–666, 1949.

26. Hanigan MH, Kemp CJ, Ginsler JJ, Drinkwater NR. Rapid growth of preneoplastic lesions in hepatocarcinogen-sensitive C3H/HeJ male mice relative to C57BL/6J male mice. *Carcinogenesis* 9:885–890, 1988.
27. Drinkwater NR, Ginsler JJ. Genetic control of hepatocarcinogenesis in C57BL/6J and C3H/HeJ inbred mice. *Carcinogenesis* 7:1701–1707, 1986.
28. Isomaa V, Pajunen AE, Bardin CW, Janne OA. Nuclear androgen receptors in the mouse kidney: Validation of a new assay. *Endocrinology* 111:833–843, 1982.
29. Ostrowski JL, Ingleton PM, Underwood JC, Parsons MA. Increased hepatic androgen receptor expression in female rats during diethylnitrosamine liver carcinogenesis. *Gastroenterology* 94:1193–1200, 1988.
30. Nagasue N, Ito A, Yukaya H, Ogawa Y. Androgen receptors in hepatocellular carcinoma and surrounding parenchyma. *Gastroenterology* 89:643–647, 1985.
31. Wilkinson ML, Iqbal MJ, Williams R. Characterisation of high affinity binding sites of androgens in primary hepatocellular carcinoma. *Clin Chim Acta* 152:105–113, 1985.
32. Ohnishi S, Murakami T, Moriyama T, Mitamura K, Imawari M. Androgen and estrogen receptors in hepatocellular carcinoma and in the surrounding noncancerous liver tissue. *Hepatology* 6:440–443, 1986.
33. Perheentupa J, Lakshmanan J, Hoath SB, Beri U, Kim H, Macaso T, Fisher DA. Epidermal growth factor measurements in mouse plasma: Method, ontogeny, and sex difference. *Am J Physiol* 248:E391–E396, 1985.
34. Byyny RL, Orth DN, Cohen S, Doyne ES. Epidermal growth factor: Effects of androgens and adrenergic agents. *Endocrinology* 95:776–782, 1974.
35. Barthe PL, Bullock LP, Mowszowicz I, Bardin CW, Orth DN. Submaxillary gland epidermal growth factor: A sensitive index of biologic androgen activity. *Endocrinology* 95:1019–1024, 1974.
36. Richman RA, Claus TH, Pilkis SJ, Friedman DL. Hormonal stimulation of DNA synthesis in primary cultures of adult rat hepatocytes. *Proc Natl Acad Sci USA* 73:3589–3593, 1976.

Mouse Liver Carcinogenesis: Mechanisms
and Species Comparisons, pages 215–230

Role of Cytochromes P-450 in Mouse Liver Tumor Production

Dennis V. Parke and Costas Ioannides
Department of Biochemistry, University of Surrey, Guildford, Surrey, UK GU2 5XH

In the safety evaluation of pesticides, drugs, and other chemicals, life-span studies in experimental animals for potential carcinogenicity are mandated by most national and international regulatory bodies. Although one rodent and one other species were originally required for these life-span carcinogenicity studies, the two-species rodent assay using rats and mice has been increasingly favored, largely because of the short life spans of these species and, hence, the lower cost of the studies. However, this choice of species has little scientific validity, for neither species has been shown to be an appropriate surrogate model for evaluating the potential tumorigenicity of chemicals in man. Moreover, recent statistical evaluations of the predictive performance of these two rodent species have shown poor correlation between the incidence of chemically induced carcinogenicity in rats and that in mice, especially when individual sexes are considered (1). It has thus been concluded that, as there is little correlation between the predictive values for rats and mice, there is little likelihood of either rodent species' being a reliable indicator for the prediction of chemically induced malignancy in humans (1,2).

In particular, a number of chemicals of different classes, including pesticides, industrial chemicals, drugs, and food additives (but especially halogenated compounds), that give rise to liver tumors in mice are considered to be of doubtful significance to man. Mouse liver tumors are often unique in that they may be induced by chemicals, e.g., phenobarbital, diazepam, and butylated hydroxyanisole (BHA), that do not induce tumors in other murine tissues, do not induce tumors in the liver, or any other tissue, of rats or other animal species, and from epidemiologic considerations, are generally regarded as unlikely to be human carcinogens. The chlorinated hydrocarbon pesticides, such as dichlorodiphenyltrichloroethane (DDT) and dieldrin, present a typical case of the rodent two-species assay yielding conflicting results, as they give rise to a dose-related increase in liver tumors in the mouse, but not in the rat or any other species studied. The Joint Meeting of the WHO and FAO on Pesticide Residues (3) has addressed this problem at several annual meetings and concluded that in many cases when a chemical provokes an increase in liver tumors in mice but not in rats, the positive results in mice are likely to be spurious, caused by physiologic and biochemical characteristics unique to mice; they recommend further testing of the chemical in other species, such as the hamster and/or guinea pig.

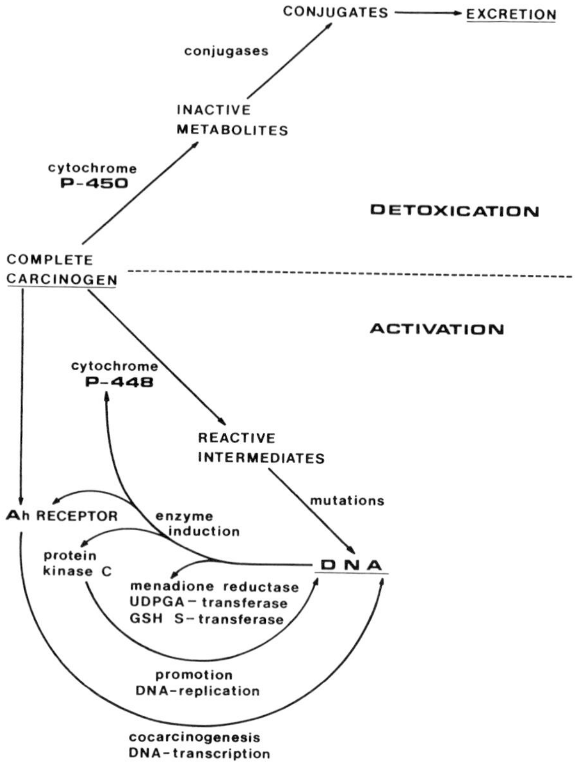

Figure 1. Mechanisms of chemical carcinogenesis and promotion: metabolic detoxication and activation.

Mechanisms of Chemical Carcinogenesis

Current theories of chemical carcinogenesis recognize a sequence of events, namely, initiation, promotion, and progression. These are associated with a series of interrelated mechanisms, including: metabolic oxygenation of the chemical, particularly in conformationally hindered positions, to form reactive intermediates, such as epoxides, quinones, nitroso compounds, and other ultimate carcinogens (activation); covalent binding of the activated metabolite to nuclear DNA, resulting in base changes and DNA rearrangements (mutagenesis); interaction of the chemical or metabolite with the Ah receptor, initiating DNA transcription of mRNA for synthesis of cytochrome P450 I (P-448), the Ah receptor protein, and other xenobiotic-metabolizing enzymes (cocarcinogenesis); and interaction of the chemical or a metabolite with the Ah or other receptors, resulting in genomal derepression of protein kinase C, leading to oncogene activation, DNA replication, mitosis, and cell proliferation (promotion) (Fig. 1).

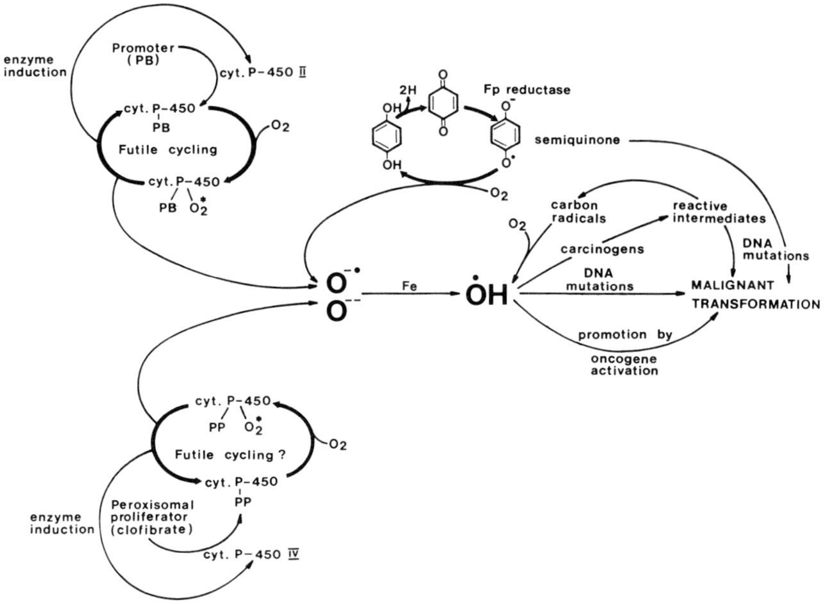

Figure 2. Mechanisms of oxygen radical–mediated chemical carcinogenesis: redox cycling and futile cycling. Abbreviations: PB, phenobarbitone; PP, peroxisomal proliferator; Fp, flavoprotein.

In addition to these events (in which the carcinogenic chemical is activated by oxidative metabolism and then interacts directly or through a metabolite to evoke DNA damage, genomal derepression, oncogene activation, and DNA replication), a further series of indirect events may be involved. These all concern radical mechanisms, especially the actions of the reactive oxygen radicals, which are known to be responsible for the carcinogenic effects of ionizing radiation. These indirect events are similarly associated with a series of interrelated mechanisms that include: (i) the formation of reactive oxygen radicals (e.g., superoxy anion, O_2–·) from tissue oxygen, in the redox cycling of free radical–reactive intermediates (semiquinones, nitroso compounds, oxy radicals), or the formation of superoxide or peroxide anions from futile cycling of the cytochromes P-450, and the conversion of superoxy anion and peroxide to the highly reactive hydroxyl radical (·OH) in the presence of iron; (ii) DNA damage by ·OH and other radicals, involving single- or double-stranded breaks, base modifications, and rearrangements, resulting in mutagenesis (4); (iii) epigenetic modulation of gene expression by oxidative stress, involving changes in structural or enzymic proteins, leading to modification of regulatory molecules, oncogene activation, and DNA replication (5,6); and (iv) activation of endogenous and xenobiotic chemicals by ·OH-mediated oxygenations and polymerizations to form novel oxidants, mutagens, and ultimate carcinogens (7,8) (see Fig. 2).

Table 1. Inducible Proteins of the Mammalian Cytochrome P-450 Superfamily

Family		Inducing agents	Former names
P450 I	A1	PAH	Rat c, mouse P_1, human P_1
	A2	Aromatic amides, isosafrole	Rat d, mouse P_3, human P_3
P450 II	A	PAH	
	B1	Phenobarbital	
	B2		
	E	Ethanol	Rat j, human j
P450 III	A1	Steroids	Rat pcn 1
	A2		Rat pcn 2
P450 IV	A1	Peroxisomal proliferators	Rat LA_{ω}

Abbreviation: PAH, polycyclic aromatic hydrocarbons.

Cytochromes P-450 and the Activation of Carcinogens

However, the most significant events in the determination of tumorigenicity for many chemicals are the activation and/or detoxification processes the body's enzymes use to determine the chemicals' fate before they interact with DNA or cytosolic receptors or are eliminated from the cell. The alternatives of metabolic activation and detoxification present the biologic paradox of cancer, namely, that the enzymes that activate carcinogens are members of the same family as the enzymes that result in their detoxification, that is, the cytochromes P-450. These enzymes, the mixed-function oxidases, ubiquitous to all tissues but found mostly in the liver, are members of a large superfamily of cytochromes responsible for oxygenations necessary for the biosynthesis of cholesterol, bile acids, steroid hormones, and eicosanoids and for the oxygenation of xenobiotic chemicals in phase 1 of their metabolic detoxification and activation (9,10) (see Table 1). Recent studies have shown that certain families of these cytochromes effect the oxygenation of chemicals that generally results in activation and the formation of reactive intermediates, consequent toxicity, and malignancy; the other families, in contrast, effect oxygenations that are followed by conjugation of the new oxygen function in phase 2 metabolism, with consequent detoxification and elimination (11).

The foremost of the families of the cytochromes P-450 known to activate chemicals and to cause chemical carcinogenicity is that of the cytochromes P450 I (P-448) (11). These cytochromes (P-450c and d of the rat, and P_1 450 and P_3 450 of mice) accept only planar, rigid molecules, which they oxygenate in conformationally hindered positions that do not readily allow subsequent phase 2 metabolism and detoxification (12,13). These oxygenated reactive intermediates consequently interact with vital intracellular macromolecules, namely, proteins, RNA, and DNA, with resulting toxicity and tumorigenicity. In contrast, many of the other families of cytochrome P-450 accept globular, flexible molecules as substrates, oxygenating them in nonhindered positions and yielding products suitable for phase 2 conjugations, thus resulting in the detoxification of the chemical (13,14).

Chemicals that are substrates for the cytochromes P450 I also interact with the Ah receptor, which, like the cytochromes P450 I, accepts only planar substrates, e.g., polycyclic aromatic hydrocarbons and 2,3,7,8-tetrachlorodibenzo-*p*-dioxin (TCDD) (14). Interaction with this receptor results in genomal derepression, with transcription of RNAs specific for the biosynthesis of the cytochromes P450 I, the Ah receptor, and other proteins but not for other cytochrome P-450s or the cytochrome P-450 reductase (9). Other receptors, such as the epidermal growth factor receptor and steroid receptors, which probably have affinities for planar chemicals, may modulate DNA replication and mitosis by protein kinase C activation, although this may also be modulated directly by the Ah receptor (9,15) (see Fig. 1). Interaction of the P450 I substrates with these receptors results in progressively increased activation of chemicals by the induced P450 I (cocarcinogenesis) and, consequently, progressively increased DNA replication and mitosis (promotion) (16).

The cytochromes P450 I form epoxides from planar molecules, such as benzo[*a*]pyrene, in which the oxygen bridge is conformationally hindered; hence these epoxides are poor substrates for epoxide hydrolase (17), the enzyme that metabolizes nonhindered epoxides to dihydrodiols that are subsequently conjugated, detoxified, and excreted. Other isozymes of cytochromes P-450 can also form hindered epoxides (e.g., dieldrin) by the metabolism of globular molecules. And although these epoxides are also poor substrates for epoxide hydrolase (18), they do not react readily with DNA or other cellular macromolecules, including the Ah receptor but, being resistant to conjugation, detoxication, and excretion, are highly persistent.

Oxygen Toxicity and Body Weight

It has long been known that the oxidative metabolism of drugs and other xenobiotics by the cytochromes P-450, the flavoprotein oxidases, hemoglobin, and other enzymes, varies inversely with the body size of the animal species under study; for many xenobiotics the rate of oxidative metabolism in the mouse is some 50 times greater than that in man (19,20). This is because the smaller the animal, the greater the rate of metabolic oxidation and the greater the rate of tissue oxygen uptake (21). A further consequence of this phenomenon is that small animals have high rates of activation of oxygen to free radicals and, since oxygen radicals are highly toxic, shorter life spans, than larger species.

This phenomenon, often ignored by toxicologists, makes the mouse a particularly vulnerable species to oxygen toxicity, as evidenced by its brief life span, the high incidence of spontaneous tumorigenicity, and the protective effects of antioxidants and radical scavengers (5,6). Furthermore, the high rates of oxygenation of xenobiotic chemicals in mouse, whether resulting in detoxification or activation, result in more extensive metabolism of xenobiotics; smaller amounts of chemicals—especially volatile ones—are excreted unchanged. For example, chloroform, designated as a carcinogen in the mouse, is oxidatively metabolized to about 80% of the dose in this species, while in the rat and squirrel monkey, oxidative metabolism accounts for only 66% and 18%, respectively, the remainder being exhaled unchanged (22). A consideration of the species differences in the toxicokinetics of chloroform clearly indicates that for this chemical, the mouse is a most inappropriate animal model for testing potential carcinogenicity in man.

The high tissue oxygen uptake of small rodents also results in the greater production of oxygen radicals, which exposes the mouse to a greater risk of radical-induced tissue damage, mutations, and malignancy than occurs with larger species. Furthermore, oxygen radicals are capable of directly activating carcinogens without the intervention of cytochrome P450 I, e.g., the activation of benzidine in the presence of hydrogen peroxide (23), so that high oxygen radical production (as occurs in the mouse) can increase the induction of liver tumors by direct damage to DNA, the direct activation of chemical carcinogens, and the oxidation of glutathione, removing the major protection of the mouse against ultimate carcinogens, mutagens, and other reactive intermediates.

Species Differences in Cytochrome P-450 Oxygenations

Rates of mixed-function oxidation of xenobiotic chemicals in the mouse are higher than those in rats or man and, with the possible exception of the hamster, are higher than in all other mammalian species (24) (see Table 2). Furthermore, an inverse correlation has been demonstrated between species life span, or body weight, and the ability of cultured fibroblasts to oxidatively activate the carcinogen, 7,12-dimethylbenz[*a*]anthracene to mutagens that covalently bind to DNA (25). This is due not only to the higher rates of tissue oxygen uptake but also to higher tissue activities of the cytochromes P-450.

Studies of species differences in the rates of activation of chemicals have indicated that some strains of mice also have relatively high levels of hepatic cytochrome P450 I activity, at least compared with the rat and guinea pig (see Table 3). Cytochrome P450 I A1, together with P450 II E, is responsible for the activation of the drug paracetamol to the hepatotoxic quinoneimine reactive intermediate (26). The rates of metabolic activation of [^{3}H]paracetamol and covalent binding of the radiolabeled active metabolite in the livers of the hamster, mouse, and rat were 2.1, 1.6, and 0.8 ng/mg protein/h, indicating a higher level of activation of paracetamol in mouse liver than in the rat (26). Similarly, the activation of the hepatocarcinogen 2-acetylaminofluorene by cytochrome P450 I A2 (P-450 d of rat), which effects the *N*-hydroxylation of planar aromatic amides, is high in the mouse. Hepatic 2-acetamidofluorene *N*-hydroxylase activities in vitro in various species are: hamster, 127; mouse, 108; rabbit, 97; rat, 60; and guinea pig, 2 pmol/min/nmol cytochrome P-450 (27). In the case of another activating cytochrome P-450, namely, P450 II E, which metabolically

Table 2. Hepatic Drug-Metabolizing Enzyme Activities of Rodents and Humans In Vitro*

Species	Enzyme activities (nmol product/min/mg protein)				
	Microsomal mono-oxygenase	Microsomal epoxide hydrolase	Cytosolic GSH S-transferase	Epoxide hydrolase/ monooxy-genase	GSH-S-transferase/ epoxide hydrolase
Man	0.42 ± 0.16	8.2 ± 2.4	1,650 ± 480	20	200
Rat (Sprague-Dawley)	0.81 ± 0.12	8.5 ± 0.3	1,380 ± 110	10	150
Hamster (Syrian golden)	2.57 ± 0.58	9.7 ± 1.0	4,200 ± 110	4	450
Mouse (NMRI)	1.76 ± 0.12	2.0 ± 0.2	5,290 ± 430	1	2,600

*The substrate for monooxygenase was 7-ethoxycoumarin, for epoxide hydrolase was benzo[*a*]pyrene 4,5-oxide, and for GSH S-transferase was 2,4-dinitro-chlorobenzene. Values are means ± SD for 5 animals and 15 human subjects.
(Data are derived from ref. 24.)

Table 3. Basal Levels of Cytochrome P-450 Activities in Liver*

Species	Enzymic activities (units/mg microsomal protein)		
	Aryl hydrocarbon hydroxylase	Biphenyl 2-hydroxylase	Biphenyl 4-hydroxylase
Mouse (C57 BL/6N)	240	1,050	2,580
Mouse (DBA/2N)	220	1,060	2,330
Rat (Sprague-Dawley)	61	<50	900
Hamster (Syrian)	370	922	2,200
Guinea pig	103	127	930

*Aryl hydrocarbon hydroxylase (AHH) and biphenyl 2-hydroxylase are good indicators of cytochrome P450 I (MC-inducible P-488) activity, and biphenyl 4-hydroxylase is an indicator of cytochrome P450 II B plus cytochromes P450 I (inducible by PB).
(Data are derived from ref. 27.)

activates dialkyl nitrosamines and benzene, the mouse again exhibits higher specific activity than the rat or man, although even higher activity is seen in the hamster (see Table 4) (28).

There are, however, wide genetic variations in the mouse with respect to cytochrome P450 I and Ah receptor activity, and both correlate well with the susceptibility of the strain to chemical carcinogenicity.

Table 4. Activation of Nitrosamines to Mutagens by Hepatic S9 Fractions from Various Animal Species

	Histidine revertants/plate*			
Carcinogen	Mouse	Hamster	Rat	Human
Dimethylnitrosamine	70	123	19	16
Methylethylnitrosamine	69	151	31	10
Dipropylnitrosamine	53	101	14	36

*Spontaneous reversion rates have been subtracted (TA 100).
(Data are derived from ref. 28.)

Table 5. Species Differences in Rates of Methylene Dichloride Metabolism

Species	P-450 pathway (detoxication) V_{max} (mg/h/kg)	GSH S-transferase pathway (activation) V_{max} (mg/h/kg)	Ratio of GSH/P-450 pathways (activation/) detoxication)
Mouse	12.4	1,208	98
Rat	2.7	91	34
Hamster	6.8	26	4
Man	1.4	3.4	2.5

*V_{max} is the maximum specific velocity of the enzymic reaction.
(Data are derived from ref. 34.)

Species Differences in Conjugase Activities

In the detoxification of epoxides, many of which are reactive intermediates, mutagens, and ultimate carcinogens formed by oxidative metabolism involving the cytochromes P450 I, two major families of conjugases are involved. These are the epoxide hydrolases, which use water as the conjugating material, and the glutathione S-transferases, which use glutathione (GSH), the valuable intracellular redox buffer and the ultimate defense against the toxicity of chemical oxidants and oxygen radicals. Glutathione and the GSH-dependent enzymes, such as the GSH S-transferases and GSH peroxidase, provide a primary defense against toxic chemicals and carcinogens by GSH conjugation of epoxides and quinones and protect cells against toxic oxidants and oxygen radicals by reducing these (29). Substances that elevate GSH protect against tumor promotion (30), while depletion of cellular GSH by administration of the GSH-depleting agent diethyl maleate (31) or by high doses of toxic chemicals (32) enhances the risk of tissue injury and tumor promotion.

Table 6. Effects of Repeated Dosing with Lindane (50–360 ppm in diet) for 3 to 90 Days

Liver function change	Relative effect
Increased liver weight	CF1 > B6 = rat
7-Ethoxycoumarin *O*-deethylase	CF1 = B6 > rat
GSH S-transferase	CF1 > rat > B6
Epoxide hydrolase	CF1 = B6 < rat
UDPGA-transferase	CF1 < B6 < rat

(Data are derived from ref. 35.)

As Pacifici et al. (33) have shown, the activities of epoxide hydrolase in livers of different species increase in the order mouse < rat < dog < monkey < man, while the converse is observed with hepatic glutathione S-transferase activities. Consequently, the ratio of liver glutathione S-transferase to epoxide hydrolase activities increases from 2 in man, 0.5 in monkeys, and 2 in dogs to 17 in rats, and 80 in mice. Thus, while man uses ubiquitous, abundant water to detoxify reactive epoxides, the mouse utilizes its limited reserves of glutathione for this role, making the mouse even more vulnerable to the cytotoxic and genotoxic effects of reactive intermediates and oxygen radicals.

The studies of Lorenz et al. (24) (see Table 2) show that mice have high microsomal monooxygenase activity, low epoxide hydrolase, and high glutathione S-transferase activities. The ratios of epoxide hydrolase/monooxygenase and GSH S-transferase/epoxide hydrolase, which are indicators of efficient epoxide detoxification and lack of glutathione conservation, respectively, are relatively similar in rats and man but are more than 10-fold different in the mouse.

Similarly, a recent study has indicated that the anomalous and unique induction of mouse liver tumors after exposure to methylene dichloride is due to the high rate of metabolism of the solvent by a GSH S-transferase pathway (see Table 5) (34). For while the P-450 monooxygenase pathway appears to lead to the detoxification of methylene dichloride, the glutathione S-transferase pathway results in its activation. The ratio of these two opposing metabolic pathways (GSH/P-450) is three times greater in the mouse than in the rat and 40 times greater in the mouse than in man. Moreover, in man a greater proportion of methylene dichloride is exhaled unchanged than in the rat or mouse.

The glutathione transferases are also involved in the metabolism of the pesticide lindane, which causes hepatocellular carcinoma in CF1 mice at 400 ppm, benign hepatomas in $B6C3F_1$ mice at 160 ppm, but no liver tumors in rats even at 470 ppm (35). Again, although the mechanism of lindane-induced tumorigenesis is not known, the greater GSH S-transferase and P-450 monooxygenase (7-ethoxycoumarin *O*-deethylase) activities, resulting in GSH depletion, and the smaller epoxide hydrolase and UDPG-glucuronyl transferase activities in mouse liver and in the CF1 mouse in particular could explain the unique susceptibility of this species (see Table 6).

Genetic Differences in Mouse Tumorigenesis

Genetic differences in the risk of cancer have been demonstrated extensively in mice, and the elegant studies of Nebert and his colleagues (9) have shown that a major part of these differences in the susceptibility of different strains of mice to chemically induced malignancy is attributable to differences in the Ah locus. The Ah locus, which mediates P450 I monooxygenase activity, is highly inducible in the responsive C57 BL/6N (B6) strain of mouse but not in the nonresponsive DBA/2N strain, and the decreased induction of P450 I in DBA/2 mice has been shown to be due to an Ah cytosolic receptor defect (36). The Ah receptor regulates transcription of at least one of the P450 I enzymes, a UDP-glucuronyltransferase, a glutathione S-transferase, an NAD(P)H-menadione oxidoreductase, and the Ah receptor protein. Furthermore, cytochrome P450 I induction is associated with increased activity of protein kinase C and several other enzymes, which may similarly be under Ah receptor control (9). Aryl hydrocarbon hydroxylase activity (benzo[*a*]pyrene hydroxylase; AHH) is induced five- to sixfold by 3-methylcholanthrene (MC) in B6 mice but is not increased at all in the DBA mouse (27). And the activation of the protein pyrolysis product, the mutagen Trp-P-2 (3-amino-1-methyl-5H-pyrido[4,3b]indole), and the P450 I-catalyzed formation of the proximate mutagen *N*-hydroxy-Trp-P-2 by liver microsomes was markedly increased by MC pretreatment in the MC-responsive B6 mice but not in the nonresponsive DBA mice (37,38). In contrast, DBA mice are susceptible to two-stage liver carcinogenesis (diethylnitrosamine initiation treatment followed by phenobarbital (PB) promotion), while B6 mice are resistant to the PB promotion.

The explanation of this apparent anachronism is that different cytochromes P-450 are involved. The activation of Trp-P-2 is catalyzed by cytochrome P450 I (P-448), which is highly inducible in B6 mice but not in DBA mice. The activation of diethylnitrosamine to the proximate initiating agent is catalyzed by P450 II E (P-450j), and the promotional effect of PB probably depends on the redox cycling of P450 II B. Hepatic concentrations of the various cytochromes P-450 are genetically determined, and an increase in one particular isozyme generally means a decrease in the others.

Studies in humans have shown a highly significant correlation between high AHH induction in polycyclic aromatic hydrocarbon–stimulated transformed lymphocytes and an increased risk of cigarette smoke–induced bronchiogenic carcinoma (39).

Unique Mechanism of Mouse Tumorigenesis

Liver tumors in various strains of mice are known to be associated with prolonged exposure to chemicals such as phenobarbitone, DDT, dieldrin, butylated hydroxytoluene (BHT), and other chemicals that induce hepatic microsomal drug-metabolizing enzymes (40–42). Only mice appear to be thus affected, although these same compounds can promote the formation of liver tumors in rats from lesions previously initiated by hepatocarcinogens, such as 2-acetylaminofluorene or diethylnitrosamine (43). The common feature of the diverse chemicals that induce mouse liver tumors or promote previously initiated rat liver tumors is that they are microsomal enzyme inducers of the phenobarbital type; that is, they induce cytochrome P450 II B. In contrast, complete carcinogens are of the polycyclic aromatic hydrocarbon type and induce cytochrome P450 I by interaction with the Ah receptor, which simultaneously promotes the protein kinase C cascade, oncogene activation, and DNA replication. The PB-type enzyme inducers have an affinity for P450 II B but are metabolically oxygenated with difficulty, so that the activated oxygen of the P-450 is released as superoxy anion instead of being inserted into the drug substrate, and futile cycling occurs (see Fig. 2) (44). The superoxy anion subsequently yields hydroxyl radicals known to be promoters of the carcinogenic process (5,6). Similarly, the peroxisomal proliferating chemicals, such as clofibrate and diethylhexyl phthalate, which give rise to rodent liver tumors, are microsomal enzyme inducers of cytochrome P450 IV (45) and result in the production of peroxide (46) and thus reactive oxygen radicals, which can function as tumor promoters.

The reasons only mice appear susceptible to the induction of liver tumors by these P450 II B–inducing chemicals may be that (i) being a very small species, the mouse has a very high tissue oxygen uptake and, consequently, a high rate of spontaneous oxygen radical production, especially in the liver; (ii) the relatively high hepatic activity level of the activating cytochromes P450 I causes mice (compared with rat and man) to experience high levels of tumor initiation due to the high rate of formation of reactive intermediates/ultimate carcinogens from environmental pollutants; and (iii) by virtue of the high ratio of GSH S-transferase/epoxide hydrolase activities in mouse liver, hepatic GSH is readily depleted in the detoxification of reactive intermediates, and the liver is consequently highly vulnerable to the tumor-promoting effects of oxygen radicals. Consequently, although the mouse may be an excellent species for studying molecular, cellular, and genetic sequences in the development of chemically induced malignancy, it is, by virtue of the unique attributes previously described, quite unsuited as a surrogate model for the evaluation of the tumorigenic potential of chemicals in man.

The primary event in most instances of chemical carcinogenesis—whether it be detoxification, activation to a mutagen or ultimate carcinogen, enzyme induction and futile cycling, peroxisomal proliferation and peroxide formation, or combinations of these—is the interaction of the chemical with one or more specific cytochromes P-450. This determines detoxication or activation, safety or carcinogenic hazard. Such valuable information can now be obtained from determination of the molecular orbitals and electronic structures of chemicals, and from these data the cytochrome P-450 with which they interact may be ascertained by comparison with known specific substrates of different forms of the cytochromes P-450 (2,12,13). Determination of the form of cytochrome P-450 that metabolizes a given chemical enables prediction of its probable toxicity. For example, P450 I accepts planar molecules and activates them to com-

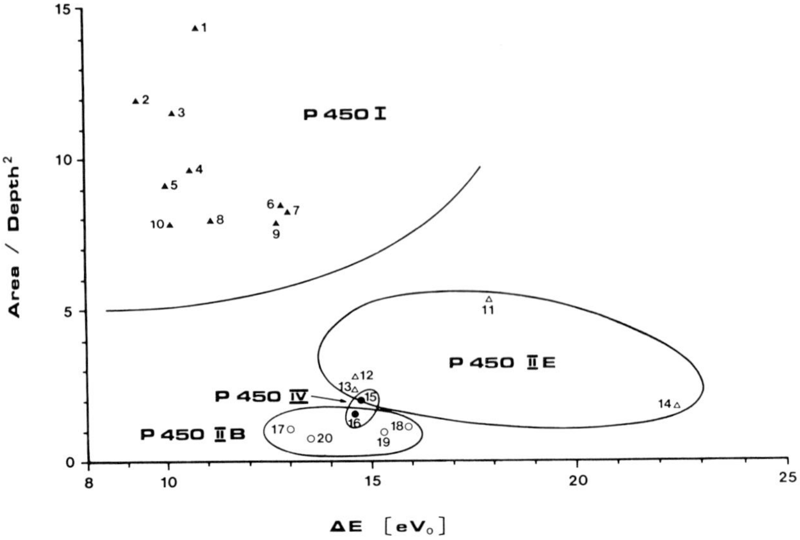

Figure 3. Molecular conformation and electronic structure parameters of chemicals and their relationship to specific forms of the cytochromes P-450. Values for the molecular shape parameters (area/depth^2) are plotted against the electronic frontier orbital energies (ΔE = E(HOMO)–E(LEMO)) for the following chemicals: (1) dibenz[*a,h*]anthracene; (2) benzo[*a*]pyrene; (3) 1-nitropyrene; (4) *N,N*-dimethylaminoazobenzene; (5) 2-aminoanthracene; (6) Trp-P-1 (3-amino-1,4-dimethyl-5H-pyrido[4,3b]indole); (7) 4-aminobiphenyl; (8) 5,12-dimethylchrysene; (9) β-naphthylamine; (10) 3-methylcholanthrene; (11) benzene; (12) dimethylnitrosamine; (13) nitrosopiperidine; (14) ethanol; (15) clofibrate; (16) monethylhexyl phthalate; (17) aldrin; (18) phenobarbitone; (19) phenytoin; and (20) DDT. The boundaries drawn for the different cytochromes P-450 are arbitrary but specify the form of cytochrome P-450 that preferentially metabolizes the different chemicals enclosed. Overlap of boundaries occurs but has been minimized for clarity by excluding plots for those chemicals known to be substrates for two or more different forms of cytochrome P-450, e.g., paracetamol (a/d^2 = 4.8; ΔE = 14.4). P450 I substrates (closed triangles), P450 II B substrates (open circles), P450 II E substrates (open triangles), and P450 IV substrates (closed circles). (Data are derived from refs. 2,12,13).

plete carcinogens; P450 II B generally detoxifies chemicals but can cause futile cycling, which may result in tumor promotion in rodent species; P450 II E also results in oxygen radical generation, possibly by futile cycling; and P450 IV metabolizes peroxisomal proliferators, which form peroxides that may also result in rodent liver tumors (see Fig. 3). In this way, frank chemical carcinogenicity, mouse liver tumorigenesis, postinitiation liver tumor promotion, and peroxisomal proliferation can be predicted from molecular conformation and electronic structure rapidly and at low cost, with limited biologic material, and as only a knowledge of the molecular structure is necessary, even before the chemical itself has been synthesized.

Open Forum

Dr. Philip Leder: Dr. Parke, given that the peroxisomal inducers elevate the levels of hydrogen peroxide intracellularly and given the evidence for free-radical induction of mutagenicity, why do you think it is that none of the peroxisomal inducers have been shown to be mutagenic?

Dr. Dennis Parke: We still do not understand the mechanism of hydrogen peroxide formation. In the short-term mutagenesis tests, we have the microsomal fraction. Maybe mitochondria or even the peroxisomes are needed to get mutagenesis. I think you have got to look at the organelles and the enzymes in the S9 mix. Does anyone know what the peroxisomal content of the S9 mix is?

Comment: Some experiments with hepatocytes have shown the presence of peroxisomes.

Dr. Parke: And it still is not mutagenic? I suppose the catalase is working very effectively in that case. I really don't know. But there are very good protective mechansims against the toxicity of peroxide. Of course, the peroxide must be converted into hydroxyl radicals or singlet oxygen to be mutagenic, and that requires iron.

Dr. John Garst: I want to thank you for such a tremendous advertisement for the use of primates. The mouse certainly is not a good animal, and I wanted to alert you to the possibility that the chimpanzee is an animal worth using. We have about 2,000 monkeys and chimpanzees available at our facility. In particular, of these, the chimpanzee is certainly an underutilized resource for metabolism studies. In several cases, pharmaceutical companies have found metabolites in man that they have found in no other species except the chimp.

Dr. Parke: Dick Adams of the National Cancer Institute has promoted primates over the past 10–15 years. He spoke to a private meeting of Congress about this 10 years ago. He made the most eloquent and elegant plea for the use of primates in safety evaluation studies.

Dr. George Wolff: In a lindane chronic bioassay at the National Center for Toxicological Research, we found that the most susceptible genotype actually had a lower glutathione transferase activity than the less susceptible, and also that lindane decreased the activity of glutathione transferase. Do you have any comments on how that fits with your data?

Dr. Parke: Other work in Germany shows that lindane is activated by glutathione transferase. Certainly lindane does go through a dehalogenation pathway involving glutathione. I don't know whether it involves glutathione transferase. I imagine that there is probably an activation pathway giving rise to a radical mechanism and a deactivation pathway giving rise to a glutathione derivative that then forms a mercapturate that is detoxified. I demonstrated the converse of what you suggest, so there are two points of view to this.

Dr. Randall Ruch: Would you speculate on how the increased oxygen radicals that you have proposed to occur in the mouse with these chlorinated and halogenated compounds might cause cancer, despite the fact that these compounds are not genotoxic?

Dr. Parke: The oxygen radicals themselves in the in vivo system could cause genetic damage or could oxygenate endogenous chemicals like amino acids. It has been shown that oxygenation creates oxygenated polycyclics. They look very much like steroid hormones. Oxygen radicals per se could convert androgens to estrogens in vivo. So there are many different mechanisms. Generators of oxygen radicals are much more active than the cytochromes in affecting activating oxygenations.

References

1. DiCarlo FJ. Carcinogenesis bioassay data: Correlation by species and sex. *Drug Metab Rev* 15:409–413, 1984.
2. Parke DV, Lewis DFV, Ioannides C. Current procedures for the evaluation of chemical safety. In Richardson ML (ed): *Risk Assessment of Chemicals in the Environment.* London: Royal Society of Chemistry, 1988, pp 45–72.
3. Joint Meeting on Pesticide Residues. Etiology and pathogenesis of liver tumours. Pesticide residues in food, 1977. FAO Plant Production and Protection Paper 10 Rev., Rome: FAO, 1978, pp 7,8.
4. Hsie AW, Recio L, Katz DS, Lee CQ, Wagner M, Schenley RL. Evidence for reactive oxygen species inducing mutations in mammalian cells. *Proc Natl Acad Sci USA* 83:9616–9620, 1986.
5. Troll W, Wiesner R. The role of oxygen radicals as a possible mechanism of tumor promotion. *Annu Rev Pharmacol Toxicol* 25:509–528, 1985.
6. Kensler TW, Taffe BG. Free radicals in tumour promotion. *Adv Free Radical Biol Med* 2:347–387, 1986.
7. Gower JD, Wills ED. The generation of oxidation products of benzo[a]pyrene by lipid peroxidation: A study using γ-irradiation. *Carcinogenesis* 5:1183–1189, 1984.
8. Kensler TW, Trush MA. Role of oxygen radicals in tumor promotion. *Environ Mutagen* 6:593–616, 1984.
9. Nebert DW, Gonzalez FJ. P-450 genes: Structure, evolution, and regulation. *Annu Rev Biochem* 56:945–993, 1987.
10. Nebert DW, Adesnik M, Coon MJ, Estabrook RW, Gonzalez FJ, Guengerich FP, Gunsalus IC, Johnson EF, Kemper B, Levin W, Phillips IR, Sato R, Waterman MR. The P-450 gene superfamily: Recommended nomenclature. *DNA* 6:1–11, 1987.
11. Ioannides C, Lum PY, Parke DV. Cytochrome P-448 and the activation of toxic chemicals and carcinogens. *Xenobiotica* 14:119–137, 1984.
12. Lewis DFV, Ioannides C, Parke DV. Molecular dimensions of the substrate binding site of cytochrome P-448. *Biochem Pharmacol* 35:2179–2185, 1986.
13. Lewis DFV, Ioannides C, Parke DV. Structural requirements for substrates of cytochromes P-450 and P-448. *Chem Biol Interact* 64:39–60, 1987.
14. Parke DV, Ioannides C, Lewis DFV. Metabolic activation of carcinogens and toxic chemicals. *Hum Toxicol* 7:397–404, 1988.
15. King CS, Cooper JA. Effects of protein kinase C activation after epidermal growth factor binding on epidermal growth factor receptor phosphorylation. *J Biol Chem* 261:10073–10078, 1986.

16. Parke DV, Ioannides C. The cytochromes P-448: A unique family of enzymes involved in chemical toxicity and carcinogenesis. *Biochem Pharmacol* 36:4197–4207, 1987.
17. Levin W, Wood AW, Lu AYH. Role of purified cytochrome P-448 and epoxide hydrolase in the activation and detoxification of benzo(a)pyrene. In Jerina DM (ed): *Drug Metabolism Concepts*. ACS Symposium Series No. 44. American Chemical Society, 1977, pp 99–125.
18. Walker CH, Timms CW, Wolf CR, Oesch F. The hydration of sterically hindered epoxides by epoxide hydrolase of the rat and rabbit. *Biochem Pharmacol* 35:499–503, 1986.
19. Walker CH. Species differences in microsomal monooxygenase activity and the relationship of biological half-lives. *Drug Metab Rev* 7:295–323, 1978.
20. Davidson IWF, Parker JC, Beliles RP. Biological basis for extrapolation across mammalian species. *Regul Toxicol Pharmacol* 6:211–237, 1986.
21. Booth J, Boyland E, Cooling C. The respiration of human liver tissue. *Biochem Pharmacol* 16:721–724, 1967.
22. Brown DM, Langley PF, Smith D, Taylor DC. Metabolism of chloroform. I. The metabolism of [^{14}C]chloroform by different species. *Xenobiotica* 4:151–163, 1974.
23. Josephy PD, Subden RE. Hydrogen peroxide-dependent activation of benzidine to mutagenic species. *Mutat Res* 141:23–28, 1984.
24. Lorenz J, Glatt HR, Fleischmann R, Ferlinz R, Oesch F. Drug metabolism in man and its relationship to that in three rodent species: Monooxygenase, epoxide hydrolase, and glutathione S-transferase activities in subcellular fractions of lung and liver. *Biochem Med* 32:43–56, 1984.
25. Schwartz AG, Moore CJ. Inverse correlation between species life-span and capacity of cultured fibroblasts to metabolize polycyclic hydrocarbon carcinogens. *Fed Proc* 38:1989–1992, 1979.
26. Ioannides C, Steele CM, Parke DV. Species variation in the metabolic activation of paracetamol to toxic intermediates: Role of cytochromes P-450 and P-448. *Toxicol Lett* 16:55–61, 1983.
27. Thorgeirsson SS, Atlas SA, Boobis AR, Felton JS. Species differences in the substrate specificity of hepatic cytochrome P-448 from polycyclic-treated animals. *Biochem Pharmacol* 28:217–226, 1979.
28. Phillipson CE, Ioannides C. Metabolic activation of nitrosamines to mutagens by various animal species including man. *Biochem Pharmacol* 34:441–442, 1985.
29. Meister A, Anderson ME. Glutathione. *Annu Rev Biochem* 52:711–760, 1983.
30. Perchellet JP, Owen MD, Posey TD, Orten DK, Schneider BA. Inhibitory effects of glutathione level-raising agents and D-alpha-tocopherol on ornithine decarboxylase induction and mouse skin tumor promotion by 12-O-tetradecanoylphorbol-13-acetate. *Carcinogenesis* 6:567–573, 1985.
31. Kluwe WM, Hook JB. Potentiation of acute chloroform nephrotoxicity by the glutathione depletor diethyl maleate and protection by the microsomal enzyme inhibitor piperonyl butoxide. *Toxicol Appl Pharmacol* 59:457–466, 1981.
32. Tong S, Masson HA, Ioannides C, Bechtel WD, Parke DV. Effects of brotizolam on mixed-function oxidases and glutathione metabolism in the rat. *Xenobiotica* 16: 595–604, 1986.
33. Pacifici GM, Boobis AR, Brodie MJ, McManus ME, Davies DS. Tissue and species differences in enzymes of epoxide metabolism. *Xenobiotica* 11:73–79, 1981.
34. European Chemical Industry Ecology and Toxicology Centre. Methylene dichloride (dichloromethane): Human risk assessment using experimental animal data.

Technical Report No. 32. Brussels: European Chemical Industry Ecology and Toxicology Centre, 1988.
35. Oesch F, Friedberg T, Herbst M, Paul W, Wilhelm N, Bentley P. Effects of lindane treatment on drug metabolizing enzymes and liver weight of CF1 mice in which it evoked hepatomas and in nonsusceptible rodents. *Chem Biol Interact* 40:1–14, 1982.
36. Nebert DW, Robinson JR, Niwa A, Kumaki K, Poland AP. Genetic expression of aryl hydrocarbon hydroxylase activity in the mouse. *J Cell Physiol* 85:393–414, 1975.
37. Yamazoe Y, Yamaguchi N, Kamataki T, Kato R. Metabolic activation of Trp-P-2, a mutagenic amine from tryptophan-pyrolysate, by liver microsomes from 3-methylcholanthrene-responsive and non-responsive mice. *Xenobiotica* 10:483–494, 1980.
38. Kato R. Metabolic activation of mutagenic heterocyclic aromatic amines from protein pyrolysates. *CRC Crit Rev Toxicol* 16:307–348, 1986.
39. Kouri RE, McKinney CE, Slomiany DJ, Snodgrass DR, Wray NP, McLemore TL. Positive correlation between high aryl hydrocarbon hydroxylase activity and primary lung cancer as analyzed in cryopreserved lymphocytes. *Cancer Res* 42:5030–5036, 1982.
40. Peraino C, Fry RJM, Staffeldt E. Enhancement of spontaneous hepatic tumorigenesis in C3H mice by dietary phenobarbital. *JNCI* 51:1349–1350, 1973.
41. Thorpe E, Walker AIT. The toxicology of dieldrin (HEOD). II. Comparative long-term oral toxicity studies in mice with dieldrin, DDT, phenobarbitone, β-BHC and γ-BHC. *Food Cosmet Toxicol* 11:433–442, 1973.
42. Turusov VS, Day NE, Tomatis L, Gati E, Charles RT. Tumors in CF-1 mice exposed for six consecutive generations to DDT. *JNCI* 51:983–995, 1973.
43. Peraino C, Fry RJM, Staffeld E, Christopher JP. Comparative enhancing effects of phenobarbital, amobarbital, diphenylhydantoin, and dichlorodiphenyltrichloroethane on 2-acetylaminofluorene-induced hepatic tumorigenesis in the rat. *Cancer Res* 35:2884–2890, 1975.
44. Ingelman-Sundberg M, Hagbjork AL. On the significance of the cytochrome P-450-dependent hydroxyl radical-mediated oxygenation mechanism. *Xenobiotica* 12:673–686, 1982.
45. Lake BG, Gray TJB, Pels Rijcken WR, Beamand JA, Gangolli SD. The effect of hypolipidaemic agents on peroxisomal β-oxidation and mixed-function oxidase activities in primary cultures of rat hepatocytes: Relationship between induction of palmitoyl-CoA oxidation and lauric acid hydroxylation. *Xenobiotica* 14:269–276, 1984.
46. Reddy JK, Lalwani ND, Reddy MK, Qureshi SA. Excessive accumulation of autofluorescent lipofuscin in the liver during hepatocarcinogenesis by methyl clofenapate and other hypolipidaemic peroxisome proliferators. *Cancer Res* 42:259–266, 1982.

Mouse Liver Carcinogenesis: Mechanisms
and Species Comparisons, pages 231–248

Mallory Bodies in Hepatomas and Hyperplastic Nodules: In Vitro and In Vivo Studies

M. Cadrin, H. Kawahara,* M. Ohta,† Y. Katsuma,†*
*N. Marceau,‡ and S. W. French**
**Department of Pathology, University of Ottawa, Ontario, Canada K1H 8M5,*
†Department of Internal Medicine, Kyoto Prefectural University, Kyoto 602, Japan,
and ‡Laval University Center for Cancer Research, Quebec, Canada G1R 2J6

Mallory bodies (MBs) are encountered in 19% of hepatomas in man (1). They are more frequent in mouse hepatomas (71%), hyperplastic nodules (86%), and preneoplastic foci in mice fed griseofulvin and are never found in spontaneous liver tumors in control mice (2). Mice fed dieldrin also form MBs in their liver tumors (3). Thus it can be said that MB formation is a predictable feature of some drug-induced hepatic tumors in mice.

We have made several attempts to characterize the MB-forming cells derived from the livers of griseofulvin-fed mice in tissue culture to determine the phenotypic changes associated with MB formation in liver tumors. We found that tissue cultures of hepatocytes teased out from liver tumors in mice fed griseofulvin retained MBs up to 14 days but eliminated the MBs by autophagocytosis (4,5) so that the formation of MBs by the tumor cells maintained in serial passages was not observed. Primary tissue cultures of liver tumor cells from griseofulvin-fed mice and liver slices that contained MBs showed similar intermediate-filament (IF) cytoskeletal arrangements when the cells were extracted with Triton X-100 and viewed unembedded by transmission electron microscopy. Further, MBs were seen to be suspended by a rich network of IFs, similar to that seen in normal hepatocytes when viewed in three dimensions using stereopairs. Tumor nodules from livers of griseofulvin-fed mice were split into several parts so that 3-day primary cultures could be compared with liver slices by light microscopy, electron microscopy, and immunofluorescence after detergent extraction. Mouse hepatoma cells showed four different fluorescent staining patterns when stained with anticytokeratin antibodies, i.e., type I cytoplasm +, MBs –; type II cytoplasm +, MBs +; type III cytoplasm –, MBs +; and type IV cytoplasm –, MBs –. Human liver samples showed similar staining patterns. The extracted cytoskeletons from the tissue slices and tissue cultures of the liver tumors showed a rich network of IFs despite the fact that these filaments failed to stain with antibodies to cytokeratins in type III and IV cells. This indicated that the IFs had lost their antigenicity (epitopes) to the antibodies, possibly because of conformational changes or a switch in gene expression of cytokeratins. Most recently, we have observed ubiquitination of MBs and a loss of ubiquitin from the cytoplasm in liver tumor cells from mice fed griseofulvin, both in their liver slices (6) and cells in tissue culture (unpublished data). These results indicate that when the MB-containing hepatocytes lost the antigenicity of IFs to antibodies to cytokeratins, they also lost their positive staining property with antibodies to ubiquitin. The problem remains of determining whether the loss of staining by immunofluorescence of the IFs in MB-containing cells is due to a change in gene expression so that the IFs were formed by different IF proteins or staining was masked by a conformational change. To investigate this, we studied tumor cells in primary culture and slices of liver tumors using immunogold labeling of the IFs with antibodies to cytokeratins and correlated the results with those from gel electrophoresis and immunoblotting of proteins derived from the cytoskeleton residue of the extracted cells. The reversibility of the loss of the staining for cytokeratin antibodies was tested by incubating tumor cells in tissue culture in the presence or absence of griseofulvin in the medium. Tumor cell structural polarity and function were also tested in vitro.

Methods

Mice and Tissues

Twelve adult male C3H mice were fed 2.5% (wt/wt) griseofulvin in a semisynthetic complete diet (Taklad Test Diet, Madison, WI) for 12 months. Griseofulvin was obtained as a gift from Schering (Kenilworth, NJ). Ten control mice were fed the diet without griseofulvin. Their livers were removed under anesthesia with pentobarbital. Both tumor and non-tumor tissues were cut into small pieces, snap-frozen in isopentane precooled with liquid nitrogen at –155°C, and then stored in a freezer maintained at –80°C before use.

Frozen Sections

Frozen sections of mouse liver were cut 8 μm thick in a cryostat at –25°C and mounted on grids covered by formvar on coverslips. Thinner sections, 4 μm thick, were prepared for toluidine blue staining and hematoxylin and eosin staining for light microscopy and immunofluorescent observations. Twelve-micrometer sections were used for biochemical studies. Light microscopic observation showed that the liver tumors of griseofulvin-fed mice consisted of hyperplastic nodules and hepatocellular carcinomas (HCC). HCC were diagnosed when a definitive trabecular pattern was observed.

Liver Cell Cultures

Hepatocytes were isolated by 0.1% collagenase perfusion method. Cells were cultured in serum-free William's E medium supplemented with dexamethasone (10 μm) and insulin (100 ng/ml) and under a humid atmosphere containing 5% CO_2 in air at 37°C (7). Primary cultures were studied at 3 h and 24 h incubation with or without griseofulvin added to the medium (2×10^{-4} M in 0.1% DMSO).

Extraction

Cytoskeletons of liver frozen sections were extracted with detergent and then digested with nucleases following the method of Fey et al. (8), with the following modification (9–13): sections were extracted in cytoskeletal (CSK) buffer (100 mM NaCl, 300 mM sucrose, 10 mM PIPES (pH 6.8), 3 mM $MgCl_2$, 0.5% Triton X-100, 1.2 mM phenylmethylsulfonyl fluoride, and 0.1 mM iodoacetamide) for 5 min at 0°C, eluted with CSK-AS buffer (CSK buffer with 250 mM $(NH_4)_2SO_4$ substituted for 100 mM NaCl) for 5 min at 0°C, and then treated with CSK buffer containing 100 μg/ml bovine pancreatic DNAse-1 and 100 mg/ml bovine pancreatic RNAse-A for 10 min and rinsed with the same buffer containing an additional 250 mM $(NH_4)_2SO_4$ for 5 min at room temperature.

Immunofluorescent Staining

Cryostat sections were extracted as mentioned above and rinsed with phosphate-buffered saline (PBS). Then, the samples were incubated with a first antibody for 60 min at room temperature, rinsed with PBS, and incubated with a second antibody conjugated with fluorescent isothiocyanate (FITC). The labeled samples were rinsed with PBS, mounted, and photographed using Zeiss epifluorescence optics.

No difference in the staining pattern of cytokeratins was observed between the unextracted, CSK-extracted, and fully extracted sections, indicating no artifact was introduced by the procedure. Control sections, for which the first antibody was omitted, were always run in parallel to assess nonspecific binding of the antibody. No background nonspecific binding was observed.

Tissue-cultured hepatocytes were fixed in cold ethanol, and double-label fluorescence staining was performed: the samples were incubated with TROMA 1 or TROMA 2 followed by the FITC-conjugated second antibody and then incubated with rhodamine-phalloidin for localization of actin.

Electron Microscopy

After extraction, samples were fixed with 2.5% glutaraldehyde for 30 min at room temperature. After fixation, they were dehydrated through a graded series of ethanol, dried through the CO_2 critical point, and coated with carbon. Samples were examined with a Phillips EM 420 transmission electron microscope at 60–80 kV. Stereopairs were prepared using a goniometer stage. Tissue cultures were also cut en face after embedding in plastic, the coverslips were removed by the pop-off technique, and then the tissue was viewed by transmission electron microscopy.

Immunoelectron Microscopy

After extraction, but without fixation, the samples were incubated with the primary antibody for 60 min at room temperature, rinsed with PBS, and incubated with the second antibody labeled with colloidal gold for 60 min at room temperature. The labeled samples were rinsed with PBS and fixed with 2.5% glutaraldehyde for 30 min at room temperature. Controls with the first antibody omitted were always run to evaluate nonspecific binding of immunogold to the cytoskeleton. A few immunogold particles bound to the formvar over the grid independent of the cytoskeleton. Cultured cells grown on formvar-covered gold grids were extracted serially in CSK buffers and 0.5% Triton X-100. The cells were incubated with TROMA 1 or TROMA 2 followed by 10 nm immunogold reagent, washed, and fixed. The samples were dehydrated by ethanol series, dried at the critical point, and coated with carbon.

Secretion of Fluorescein Diacetate

Fluorescein diacetate (FD) (25 μg/ml) dissolved in 0.1% DMSO was added to the culture medium. Hepatocytes were incubated with FD, and secretion of FD into the bile canaliculi was visualized by fluorescence microscopy at 5, 15, 30, and 60 min.

Secretion of Horseradish Peroxidase

Cultured hepatocytes were incubated with horseradish peroxidase (HRP) (10 mg/ml) for 3 min. At 0, 30, and 60 min after rinsing HRP from the cells, the cells were fixed with 2.5% glutaraldehyde and 0.8% paraformaldehyde, and thin sections were made by the pop-off en face method.

Primary Antibodies

Rat IgG anti-mouse cytokeratin TROMA 1, TROMA 2, and TROMA 3 isolated from mouse trophectoderm (a generous gift from Dr. R. Kempler to Dr. N. Marceau) were used to stain CKs 55, 49, and 44, respectively. These antibodies have previously been characterized by immunoblots (14,15). Mouse IgG anti-human epidermal keratin antibody AE3, which is directed against 58 kDa and 65–67 kDa polypeptides, was used (Hybritech, La Jolla, CA). Mouse IgG anti-human epidermal keratin antibody EAB903, which is directed against 57 and 66 kDa polypeptides, also was used (Enzo, New York, NY).

Secondary Antibodies

FITC-conjugated rabbit anti-rat IgG and FITC-conjugated rabbit anti-mouse IgG (Zymed Laboratories, South San Francisco, CA) were used at a dilution of 1:20 in PBS. Rabbit anti-rat IgG–coated colloidal gold (5 nm) (Janssen Pharmaceutica, Brussels, Belgium) was used at a dilution of 1:10 in PBS. Control studies were performed by substituting PBS. Biotin-conjugated goat anti-mouse IgG and biotin-conjugated goat anti-rat IgG (Jackson ImmunoResearch Laboratories, West Grove, PA) were the third set of secondary antibodies used.

Gel Electrophoresis and Immunoblotting

Whole liver cells and extracted cytokeratins solubilized in 0.05 M Tris HCl, pH 6.8, 2% SDS, and 5% β-mercaptoethanol were run on a 10% polyacrylamide gel. Gels were stained with Coomassie brilliant blue or with silver or prepared for immunoblotting. For immunoblotting, gels were incubated for 2 h in a buffer containing 4 M urea, 10 mM Tris-HCl, pH 7.5, 50 mM NaCl, 2 mM EDTA, and 0.1 mM dithiothreitol in order to partially reconstitute the original conformation of the proteins or were immediately equilibrated for 30 min in transfer buffer containing 190 mM glycine and 23 mM Tris, pH 8.8, and transferred electrophoretically to nitrocellulose paper. The quality of the transfer was monitored by staining the paper with 1.2% Ponceau S in 3% trichloroacetate. The saturation of nonspecific binding sites was carried out by incubating the nitrocellulose paper in PBS containing 0.5% skim milk (Difco) for 2 h at room temperature. The paper was then washed in PBS 0.1% Tween 20 incubated with the different primary antibodies diluted in PBS 0.5% skim milk for 2 h at 37°C, washed with PBS 0.1% Tween 20, incubated with their respective biotinylated secondary antibodies diluted in PBS 0.5% skim milk, washed with PBS 0.1% Tween 20, incubated with the streptavidin peroxidase (Jackson ImmunoResearch Laboratories) diluted in PBS 0.5% skim milk, washed with PBS Tween 20, and washed with PBS, and the peroxidase was located by incubating with 4-chloronaphthol 0.4 mg/ml in PBS with 0.03% H_2O_2.

Results

Immunofluorescence Staining

In the livers of control mice, TROMA 1 (CK 55) stained IFs located at the border and in the cytoplasm of hepatocytes and IFs in the bile duct epithelium. The staining pattern of the cytoplasm of hepatocytes appeared filamentous. The bile canalicular sheath was easily identified in cross and longitudinal sections as a distinctive branching tubular structure in control livers (Fig. 1a). The staining pattern with TROMA 2 (CK 49) was similar to that with TROMA 1 (Fig. 1b), except the filamentous pattern was less dense. TROMA 3 (CK 44) stained only IFs of bile duct epithelium (Fig. 1c). Virtually no specific fluorescence was observed on the fully extracted sections with AE3 and EAB903 (Fig. 1d). Controls without primary antibodies showed no staining by immunofluorescence.

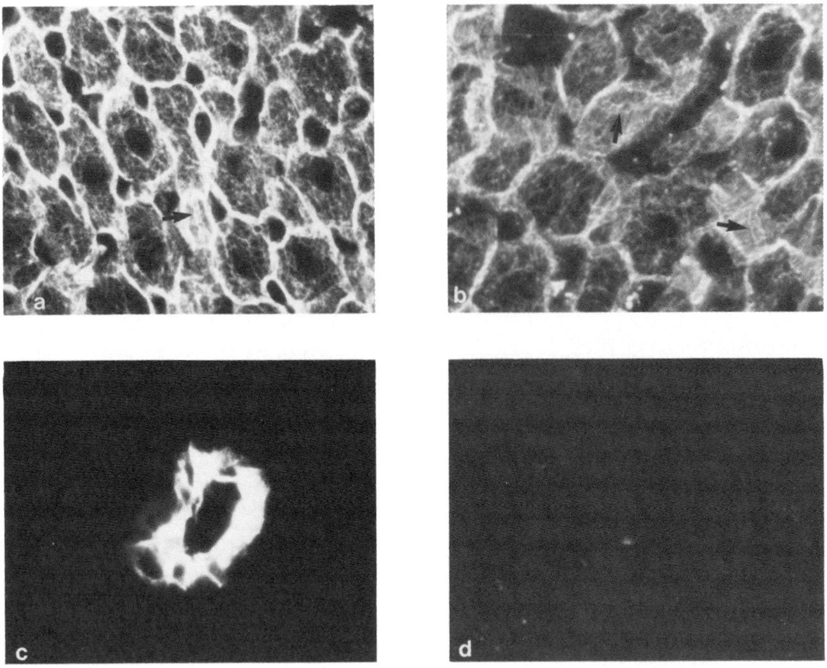

Figure 1. Immunofluorescence of control mouse liver stained with TROMA 1 (a), TROMA 2 (b), TROMA 3 (c), and EAB903 (d). (a) TROMA 1 stains the cell border, the cytoplasm, and the pericanalicular sheath. The cytoplasmic staining is filamentous. Note that the pericanalicular sheath branches when cut longitudinally (arrow); ×620. (b) Like TROMA 1, TROMA 2 stains the cytoplasm, the cell border, and the pericanalicular sheath (arrows). However, the staining pattern with TROMA 2 in the cytoplasm was somewhat less densely packed with stained filaments when compared with that stained with TROMA 1; ×990. (c) Only the bile duct epithelium is heavily stained with TROMA 3; ×620. (d) No cytokeratin labeling was observed with EAB903; ×620.

Hyperplastic nodules from the griseofulvin-treated mouse livers were identified by the absence of bile duct epithelium and protoporphyrin pigment and their circumscribed periphery. Three different staining patterns of nodules were observed with TROMA 1 and TROMA 2. Type A hepatocytes formed a mosaic pattern of stained and unstained cells (Figs. 2a, b). Cells stained with variable intensity with or without the presence of stained MBs. In type B hepatocytes, the liver cells failed to show cytoplasmic staining, but the MBs were stained intensively. Numerous, very small MBs were located at the periphery of the cell (Fig. 2c) in some cases. All the type C hepatocytes contained stained IFs in the cytoplasm with or without stained MBs. In an HCC, the hepatocytes forming a trabecular pattern contained stained MBs and cytoplasmic IFs (Fig. 2d). TROMA 3, which stained bile duct epithelium (CK 44) in the control liver, stained the cytoplasmic cytokeratin and MBs in some hepatocytes in hyperplastic nodules

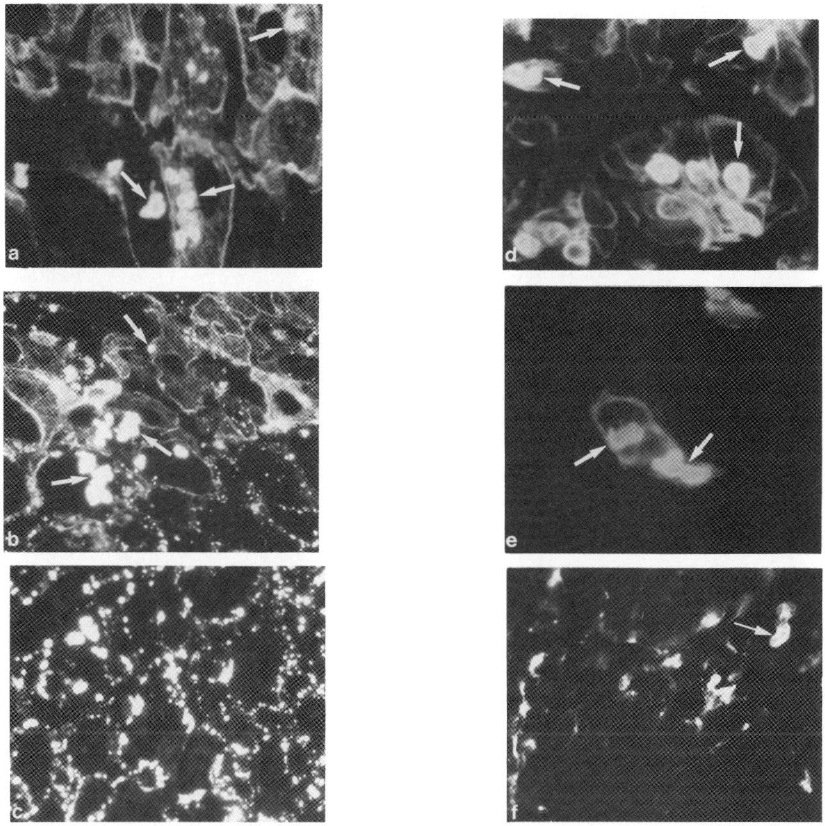

Figure 2. Immunofluorescence staining of tumor hepatocytes from a griseofulvin-fed mouse whose liver tissue was stained with TROMA 1 (a,c,d), TROMA 2 (b), TROMA 3 (e), and EAB903 (f). (a) The hepatocytes in a hyperplastic nodule stain with variable intensity in the cytoplasm with or without the presence of stained Mallory bodies (arrows); ×620. (b) TROMA 2 staining of the same tissue as in (a). The hepatocytes also stain with variable intensity in the cytoplasm, with or without the presence of stained Mallory bodies (arrows); ×380. (c) Tumor cells in a hyperplastic nodule are not stained. Mallory bodies are stained. Some Mallory bodies are very small and are localized at the cell periphery; ×240. (d) All hepatocytes in an HCC stain strongly in the cytoplasm in the presence or absence of stained Mallory bodies (arrows). The tumor cells form a trabecular pattern; ×620. (e) Some tumor cells in an HCC and some Mallory bodies (arrows) are stained when the same tissue as shown in (d) is stained with TROMA 3; ×620. (f) EAB903 showed staining of both the cytoplasm and the MBs (arrow); ×620.

and in an HCC (Fig. 2e). AE3 and EAB903 stained large MBs located in the center of hepatocytes, as well as smaller MBs located at the periphery (Fig. 2f).

The non-tumor-bearing livers of the griseofulvin-fed mice were characterized by an increase in bile ducts that were stained by TROMA 1, TROMA 2, and

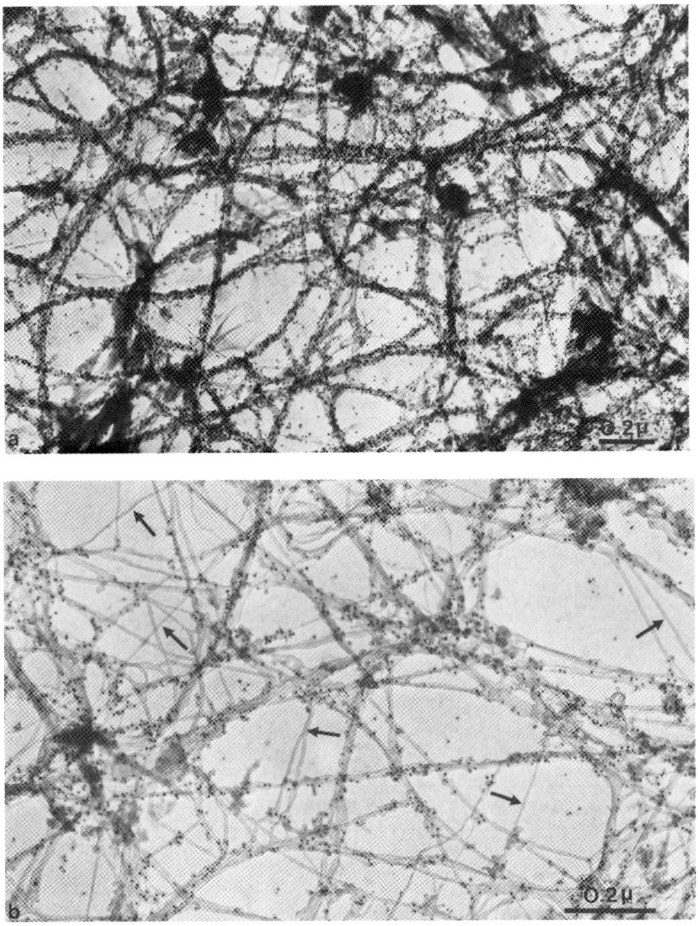

Figure 3. Immunoelectron microscopy of the control mouse liver labeled with TROMA 1 (5 nm gold) (a) and TROMA 2 (5 nm gold) (b). (a) Most of the IFs of the cytoplasm are stained; ×65,000. (b) A larger percentage of IFs (arrows) were unstained with TROMA 2 compared with TROMA 1; ×105,000.

TROMA 3. Hepatocytes stained with TROMA 1 and 2 showed a heterogeneous labeling pattern. The three different patterns observed in the hyperplastic nodules were observed in different areas within the non-tumor liver areas (not shown). Labeling of the cells with TROMA 3 showed CK 44 in a few hepatocytes that contained stained MBs. AE3 and EAB903 stained the same way as in hyperplastic nodules, where MBs were positive with some diffuse cytoplasmic labeling (Fig. 2f).

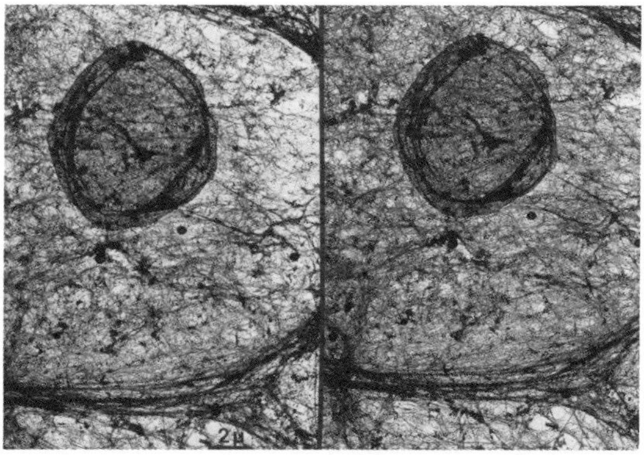

Figure 4. Stereopair of the non-tumor liver tissue of griseofulvin-fed mouse after extraction. Note the dense meshwork of IFs in the cytoplasm as compared with the control; ×4,900. Reprinted with permission from *Electron Microscopy Reviews*, French et al., Copyright 1989, Pergamon Press PLC, in press.

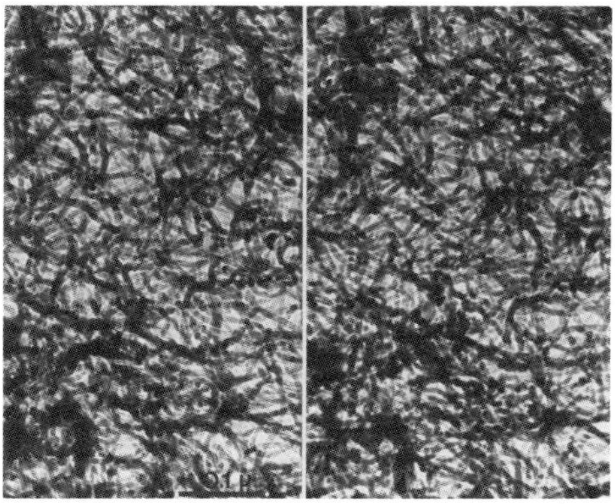

Figure 5. Stereopair of a Mallory body in the tumor tissue from a griseofulvin-fed mouse liver after extraction. Filaments seem to be long. Branching of these filaments is not seen; ×166,000.

Immunoelectron Microscopy

As described previously (9), extracted frozen sections of normal mouse liver cells retained their characteristic polygonal shape, with a thick IF network at the cell border and fewer IFs in the cytoplasm. The pericanalicular sheaths around the bile canaliculi (BC) were identified as branching tubes composed of IFs in continuity with the filament network located at the cell border of the hepatocytes, as viewed in stereopairs. Immunogold labeling with TROMA 1 was detected on almost all the IFs in the cell. IFs at the cell cortex, at the pericanalicu-

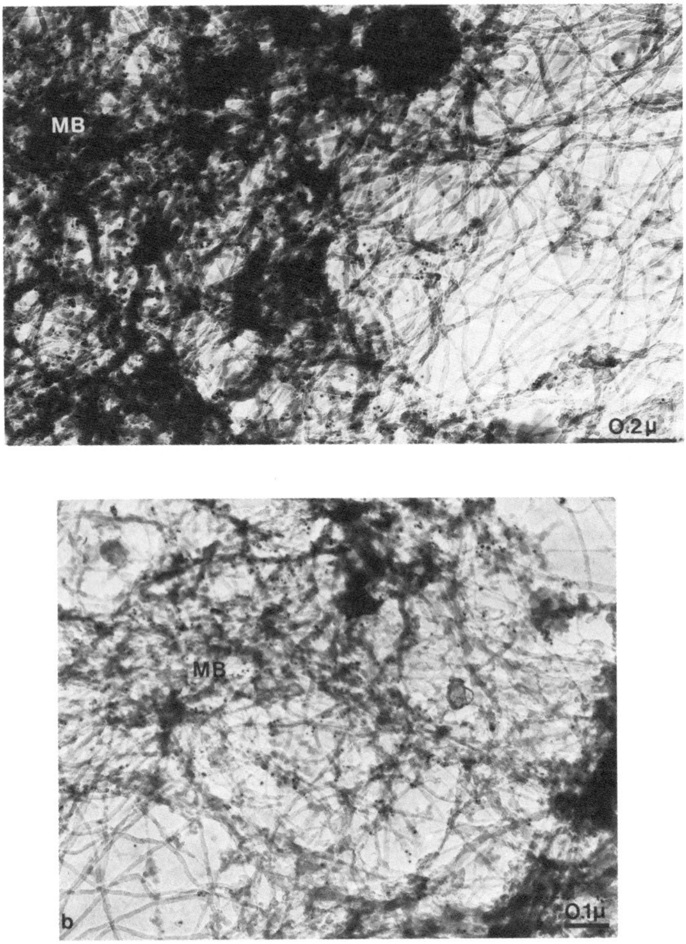

Figure 6. Immunoelectron microscopy of Mallory body and IFs in tumor hepatocytes in a hyperplastic nodule from a griseofulvin-fed mouse stained with TROMA 1 (5 nm gold) (a) and TROMA 2 (5 nm gold) (b). (a) The MB is stained, but the IFs are not stained; ×112,000. (b) The MB is stained, but the IFs are not stained with TROMA 2; ×98,000.

lar sheath, in the cytoplasm, and at the perinuclear region were coated with gold (Fig. 3a). TROMA 2 gold-labeled filaments had the same distribution as the ones labeled with TROMA 1. However, TROMA 2 did not stain as high a percentage of filaments as TROMA 1 (Fig. 3b). This could mean that some filaments were missing TROMA 2 epitopes or that the TROMA 2 epitopes were masked by conformational changes (16).

In the frozen liver sections of the griseofulvin-fed mice, the ultrastructural features of IFs and MBs were similar to those described previously (9). The density of the IF network varied within the cell and from cell to cell, but in general it appeared to be more dense in these cells than in controls (Fig. 4). MBs were identified as irregularly shaped electron-dense masses; their characteristic filamentous nature was discernible at the edge. Filaments of some MBs were randomly arranged and nonbranching when viewed as stereopairs (Fig. 5). Numerous IFs, which often formed bundles, radiated from the MBs and formed a complex network within the rest of the cell. All the liver cells contained a cytoplasmic IF network. Cells devoid of IFs (empty cells) were not observed.

Griseofulvin liver sections labeled with immunogold TROMA 1 showed the same pattern of staining seen by immunofluorescence, i.e., a heterogeneous pattern of staining. In the type A liver cell pattern, the MBs were labeled with variable intensity. The cytoplasmic IFs appeared as a mixture of labeled and unlabeled IFs in some hepatocytes. In other hepatocytes, there was a complete absence of labeling (Fig. 6a). TROMA 2 also stained MBs with variable intensity when we used immunogold. The cytoplasmic IFs were labeled or unlabeled (Fig. 6b) with the antibody, as seen with TROMA 1 labeling. With the type B staining pattern, MBs were also stained with variable intensity; however, none of the cytoplasmic IFs were labeled with TROMA 1 or TROMA 2. With the type C pattern, TROMA 1 stained MBs and IFs. Thus, we have shown that, depending on the staining pattern in control and griseofulvin-treated hepatocytes, the IFs that were otherwise structurally identical either did or did not label with anticytokeratin antibodies. The number that did not label markedly increased in the griseofulvin-treated hepatocytes associated with MB formation. This established the fact the hepatocytes that failed to stain, the so-called empty cells, were filled with unstained IFs.

Immunofluorescence of the cultured hepatocytes gave the same results seen in the tissue sections. At 3 h after seeding, the cytokeratin IFs in control cultured hepatocytes were stained throughout the cytoplasm with TROMA 1. Cytokeratin IFs in the hepatocytes containing MBs were not stained at 3 h of incubation, but the cytokeratin IF network was stained normally at 24 h. This change was inhibited by griseofulvin-containing medium (2×10^{-4} in 0.1% DMSO) at 24 h incubation (16, and H. Kawahara, N. Marceau, S. W. French, submitted).

Double-label fluorescence staining of the cells in tissue culture showed intercellular BC were present in the control hepatocytes, and the intracytoplasmic BC were present in the hepatocytes from both the griseofulvin-fed and control mice at 24 h incubation. Immunogold labeling of the control hepatocytes showed that the cytokeratin IFs formed pericanalicular sheaths between hepatocytes and spherical sheaths in the cytoplasm. In the hepatocytes from griseofulvin-fed mice, MBs appeared to connect with the cytokeratin IFs in the cytoplasm and were labeled with TROMA 1. Both the intracytoplasmic and intercellular BC stained with rhodamine-phalloidin, indicating the presence of actin (16, and H. Kawahara, N. Marceau, S. W. French, submitted).

Functional Studies

The studies on the secretion of FD in control hepatocytes showed that uptake of FD occurred by 5 min of incubation, and the secretion of FD into the intercellular and intracellular BC was seen at 30 min. In the hepatocytes from the griseofulvin-fed mice, the uptake of FD was normal, and the secretion of FD into the intracytoplasmic BC was seen at 30 min. The studies on the secretion of HRP in control hepatocytes showed that many vesicles containing HRP were seen around the intracytoplasmic BC at 60 min after incubation with HRP. In the hepatocytes with MBs from griseofulvin-fed mice, the uptake of HRP was markedly reduced in the cytoplasm, and the tubulovesicular system was empty.

Immunoblots

Because of the heterogeneity in the composition of the liver cells in the griseofulvin-fed mice, we performed our biochemical studies on extracted cytokeratin cytoskeleton from liver cryosections. In all cases, some sections of the liver were stained for immunofluorescence, the others were extracted, and the enriched cytokeratin fraction was run on a gel and used for immunoblotting. In control sections of liver, the antibody TROMA 1 strongly reacted with a 55-kDa cytokeratin and, in some cases, weakly reacted with a 49-kDa cytokeratin (Fig. 7, lanes 1–3). The absence of the 49-kDa band in lane 3 could have occurred because insufficient protein was entered into the gel, although it was also missing from the unextracted liver homogenate (Fig. 8, lane 2). TROMA 2 reacted only with 49-kDa cytokeratin (Fig. 7, lane 4). TROMA 3 detected a weak band of 44 kDa or no protein in control liver (Fig. 7, lane 5). AE3 detected CK 55 and higher molecular weight (M_r) cytokeratins (Fig. 7, lane 6). Since the control strip stained with only the second antibody, these high M_r bands were regarded as nonspecific staining.

In griseofulvin-treated mouse livers, we looked at sections showing differing immunofluorescence patterns: a liver of type A (i.e., mosaic pattern) and a portion of a liver of type B (i.e., cells with no cytoplasmic IF labeling). In the case of cytokeratin extracted from type A liver and processed for immunoblotting, we obtained with TROMA 1 the same patten as in the control (Fig. 7, lane 1′) or the presence of an additional band at an M_r of 60 kDa (Fig. 7, lane 3′). This polypeptide reacted intensely with TROMA 1 in comparison with CKs 55 and 49 (Fig. 7, lane 3′). The 60-kDa band in lane 3′ is quite broad and was intensely labeled. It appears less broad and intense when stained with Coomassie blue in the gel (Fig. 8). This band could be ubiquitinated CK 55, since this would add 5 kDa to the molecular weight of CK 55. We have noted in a separate study of immunoblots from griseofulvin-treated livers that there are many bands above 40 kDa that are heavily ubiquitinated (6). TROMA 2 reacted only with CK 49 (Fig. 7, lane 4′). TROMA 3 reacted only with CK 44, which is normally only pre-

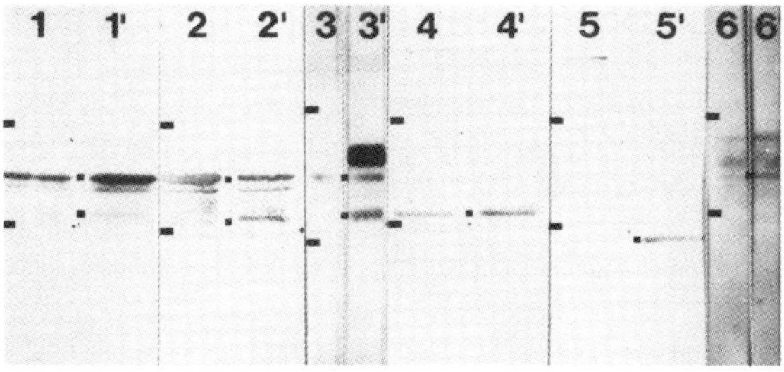

Figure 7. Immunoblot of cytokeratin-enriched material from extracted control liver (lanes 1 to 7) and from liver of a griseofulvin-treated mouse (lanes 1′ to 7′). Lanes 1, 1′, 2, 2′, 3, and 3′, TROMA 1 labeling; lanes 4 and 4′, TROMA 2; lanes 5 and 5′, TROMA 3; lanes 6 and 6′, AE3. Standard M_r of 66 and 45 kDa are indicated by straight bar. The dots denote cytokeratin polypeptides Endo A (55 kDa), Endo B (49 kDa), and Endo C (44 kDa). The arrows show the modified polypeptides in the liver of a griseofulvin-treated mouse.

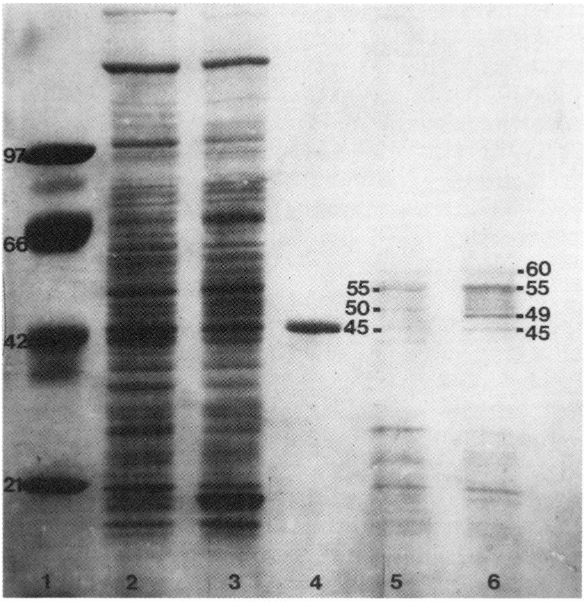

Figure 8. Coomassie blue stain of control liver and griseofulvin-treated liver after detergent extraction. Lane 1 is standards, lane 2 is unextracted liver of control mouse, lane 3 is unextracted liver of griseofulvin-treated mouse, lane 4 is the actin standard, lane 5 is the extracted liver from the control mouse that is shown in lane 3 in Figure 7, and lane 6 is the extracted liver from the griseofulvin-treated mouse that is shown in lane 3′ in Figure 7.

sent in bile duct epithelial cells (Fig. 7, lane 5′). AE3 gave similar reactions as in control liver samples (Fig. 7, lane 6′). Biochemical analysis of type B liver sections showed a different reaction than the control to TROMA 1 (Fig. 7, lane 2′). TROMA 2 and 3 showed no differences from the control liver (not shown). TROMA 1 reacted with one more polypeptide than was found in the control sample. TROMA 1 reacted with the CK 55 band and stained slightly the CK 49 band but also stained a 50-kDa polypeptide in the griseofulvin sample.

Discussion

Changes in the organization of cytokeratin IFs in the livers of griseofulvin-fed mice were studied extensively by immunofluorescence and electron microscopy. Immunofluorescence investigations using various antibodies have shown positive staining with cytokeratin antibodies in control hepatocytes. However, MB-containing cells failed to show staining of cytoplasmic cytokeratin "empty cells" (17–22). It has been proposed that in the course of griseofulvin intoxication, small MB granules enlarge and coalesce, IF bundles disappear progressively, and large MBs appear as irregular inclusions in cells that are otherwise devoid of IF staining with antibodies to the cytokeratin component D (23). We also observed by immunofluorescence that some MB-containing cells lose their cytoplasmic staining with TROMA 1 and TROMA 2. Our IEM results clearly show that the liver cells containing labeled MBs have a complex cytoplasmic IF network that failed to bind the antibodies TROMA 1 and TROMA 2. This result indicates that negative staining by immunofluorescence of the IFs within the cytoplasm of the MB-containing cells is not due to the lack of cytoplasmic IFs, as suggested by others (17,20,22).

To explain why the IFs failed to stain with antibody to cytokeratin, we postulate two mechanisms that could occur in these cells during transformation. The first is a selective masking of IF epitopes through conformational changes in IF polypeptide organization. This type of change has already been proposed (17) to explain the staining of MBs with cytokeratin antibodies that stain cytokeratins on immunoblots but not in the liver cell cytoplasm. The second is a synthesis of a new IF protein, which replaces the normal cytokeratin protein and is not normally expressed in hepatocytes. In this study, evidence supports both hypotheses. Our results obtained with the anti-human cytokeratin antibodies AE3 and EAB903 show that after griseofulvin feeding, these antibodies stained the MBs but they also stained the cytoplasm of the hepatocytes weakly. This could be due to unmasking of epitopes normally hidden. By immunoblot, we showed that these cytokeratin antibodies detected exactly the same cytokeratin antigens in the control and the treated animals and that CK 55 was recognized by these antibodies. This indicates that no new cytokeratin in these griseofulvin liver sections was detected by these antibodies, so that the staining of MBs and IFs in the griseofulvin-treated mouse liver cells is the result of structural modification of the cytokeratin (unmasking) similar to that reported previously (17). Our tissue culture studies also suggested unmasking had occurred, since the nonstaining empty cells recovered their normal staining by 24 h incubation and this return to normal was prevented by griseofulvin in vitro.

On the other hand, evidence for the modification of cytokeratins in the MB-containing cells was obtained on immunoblots with TROMA 1 and TROMA 2 on extracts of tumors showing the type A and type B liver-staining pattern. Because of the heterogeneity of the liver and to correlate biochemical data with changes in structure, we performed our analysis on small pieces of the liver that were cryosectioned; a portion of the section was used for microscopy and immunofluorescence, and the other part was used for immunoblots. In the case of the type A liver pattern, we observed by immunoblot a normal cytokeratin composition for cells that showed a variable cytokeratin staining pattern by immunofluorescence. On the other hand, TROMA 1 showed the presence of a 60-kDa protein. This protein's intense reaction with the antibody suggested that a new cytokeratin had been expressed that shared an epitope with CK 55. Or this polypeptide could represent a posttranslational modification of CK 55 that increased its molecular weight, i.e., ubiquitination (6,22). In the case of the type B hepatocyte staining pattern, we observed on the immunoblot that the two normal cytokeratins were present in the liver tissue. However, TROMA 1 stained a band at 50 kDa that was not present in the control liver. The presence of the 50-kDa band may indicate that intracellular proteolysis occurred and contributed to the structural changes in cytokeratins during MB formation.

Evidence that liver cells express new cytokeratins in response to griseofulvin was demonstrated in this study, as well as in a previous study (6). We observed that some hepatocytes with MBs stained with TROMA 3, an antibody specific for cytokeratin 19, a 44-kDa cytokeratin. This antibody reacts only with bile duct epithelial cells in normal liver. Therefore, the staining of hepatocytes with TROMA 3 indicates that a change in cytokeratin expression has occurred in these cells. Analogous results have been observed with an antibody against human epidermal cytokeratin (9). This antibody stained only bile duct epithelium in controls and stained some MB-containing cells after griseofulvin treatment. It has been suggested by Butron-Vila et al. (25) and Desmet (26) that some hepatocytes change their phenotypic expression and acquire characteristics of bile duct epithelial cells due to the stimulus of persistent cholestasis. Butron-Vila et al. (25) described the appearance of cytokeratin IFs of the ductular type in hepatocytes in cholestasis. Ray (27) reported similar changes in alcoholic hepatitis, in which MBs are formed in centrilobular hepatocytes, where bile duct epithelia are not seen and therefore cannot be mistaken for hepatocytes. Desmet (26) coined the phrase "bile duct metaplasia" to describe this phenomenon.

It seems likely from the results obtained in our study that MB formation in griseofulvin liver hepatocytes is a complex event, involving multiple cellular modifications. Our study failed to show the presence of the insoluble 65-kDa protein that was described by Denk et al. (28) and Franke et al. (29). None of our antibodies reacted with such a polypeptide, despite the fact that the sections were very rich in MBs according to immunofluorescence. We report here that some of the changes in cytokeratin organization and conformation may be due to posttranslation modification of cytokeratins. Moreover, we report evidence that altered gene expression during MB formation may lead to synthesis of new cytokeratins not normally expressed by hepatocytes.

Open Forum

Dr. Dennis Parke: Are there any chemicals other than griseofulvin and methanol that induce Mallory body formation?

Dr. Samuel French: Dieldrin in the mouse. Yokon described another chemical that produces Mallory bodies but no tumors in mice. But dieldrin and griseofulvin are faithful in producing tumors with Mallory bodies.

Dr. Don Stevenson: If I understand what these relate to, in a sense Mallory bodies are aggregations of the cytoskeleton. So this is simply affecting the spatial distribution of the internal organelles of the cell?

Dr. French: Yes.

Dr. Stevenson: Is that likely to affect the transmissional diffusion of metabolites and enzymes and so forth between the various organelles?

Dr. French: I think so. Certainly we show evidence that the cells fail to undergo pinocytosis, which is the main avenue for cells taking up proteins and other elements into the cytoplasm. So there is a block in uptake in these hepatocytes that is presumably related to the change in intermediate filaments. I did not emphasize it, but the distribution of actin also changes, so that actin is present all the way around the cell instead of being localized around the canaliculi.

Dr. Gabriel Altmann: When the Mallory bodies disappear after you discontinue feeding, do the nodules disappear at the same time?

Dr. French: Everything disappears—the GGT-positive foci, the nodules, and the Mallory bodies. The total disappearance takes about 7 months. It is a slow process.

Dr. Stanley Goldfarb: I was intrigued with your early slide on the strain effect. As I understood it, the hepatomas in the black 6 mice tended to have more Mallory bodies after dieldrin than in the B6C3 and the C3H.

Dr. French: That's right.

Dr. Goldfarb: That's interesting in view of data I presented showing that the globules accumulate earlier in C57BL/6 mice induced in a neonatal model. Do you think it is possible that this reflects some predisposition to an injury phenomenon in the neoplasms in this strain vs. the F_1C3H mouse?

Dr. French: Your globules were secretory globules.

Dr. Goldfarb: I am just saying that there was some general predisposition toward injury—some accelerated phenomenon—in those tumors in the black 6 mouse.

Dr. French: I must admit I don't know the answer, but obviously there is a parallel there.

Dr. Parke: I have a last question. Alcohol is known to induce P-450 IIE.

Griseofulvin is also a very potent inducer of the cytochromes P-450, but I don't know which isomer. Does anyone know? It would be very interesting if both of these induced the same cytochrome.

References

1. Nakanuma Y, Ohta G. Is Mallory body formation a preneoplastic change? A study of 181 cases of liver bearing hepatocellular carcinoma and 82 cases of cirrhosis. *Cancer* 55:2400–2404, 1985.
2. Tazawa J, Irie T, French SW. Mallory body formation runs parallel to γ-glutamyl transferase induction in hepatocytes of griseofulvin-fed mice. *Hepatology* 6:989–1001, 1983.
3. Meierhenry EF, Ruebner BH, Gershwin ME, Hsieh LS, French SW. Mallory body formation in hepatic nodules of mice ingesting dieldrin. *Lab Invest* 44:392–396, 1981.
4. French SW, Swierenga SHH, Okanoue T, Marceau N. Cytoskeleton of the liver cell in health and disease. In Farber E, Phillips MJ (eds): *Pathogenesis of Liver Disease.* IAP Monograph No. 28. New York: Williams & Wilkins, 1987, pp 95–112.
5. French SW, Katsuma Y, Ray MB, Swierenga SHH. Cytoskeletal pathology induced by ethanol. *Ann NY Acad Sci* 492:262–276, 1987.
6. Ohta M, Marceau N, Perry G, Manetto V, Gambetti P, Autilio-Gambetti L, Metuzals J, Kawahara H, Cadrin M, French SW. Ubiquitin is present on the cytokeratin intermediate filaments and Mallory bodies of hepatocytes. *Lab Invest* 59:848–856, 1988.
7. Marceau N, Goyette R, Pelletier G, Antakly T. Hormonally-induced changes in the cytoskeleton organization of adult and newborn rat hepatocytes cultured on fibronectin precoated substratum: Effect of dexamethasone and insulin. *Cell Mol Biol* 29:421–435, 1983.
8. Fey EG, Wan KM, Penman S. Epithelial cytoskeletal framework and nuclear matrix-intermediate filament scaffold: Three dimensional organization and protein composition. *J Cell Biol* 98:1973–1984, 1984.
9. Katsuma Y, Swierenga SHH, Khettry U, Marceau N, French SW. Changes in the cytokeratin intermediate filament cytoskeleton associated with Mallory body formation in mouse and human liver. *Hepatology* 7:1215–1223, 1987.
10. Katsuma Y, Marceau N, Ohta M, French SW. Cytokeratin intermediate filaments of rat hepatocytes: Different cytoskeletal domains and their three-dimensional structure. *Hepatology* 8:559–568, 1988.
11. Katsuma Y, Swierenga SHH, Marceau N, French SW. Connections of intermediate filaments with the nuclear lamina and the cell periphery. *Biol Cell* 59:193–204, 1987.
12. Katsuma Y, Swierenga SHH, Marceau N, French SW. Selective rearrangement of cytokeratin filaments in cultured liver epithelial cells induced by nickel. *J Hepatol* 5:344–354, 1987.
13. Ohta M, Marceau N, French SW. Experimental cholestasis: Effect on the pericanalicular sheath of intermediate filaments (Abstract). *Hepatology* 7:1076, 1987.
14. Boller K, Kemler R, Baribault H, Doetschman T. Differential distribution of cytokeratins after microinjection of anti-cytokeratin monoclonal antibodies. *Eur J Cell Biol* 43:459–468, 1987.

15. Kemler R, Brulet P, Schnebelen M-T, Gaillard J, Jacob F. Reactivity of monoclonal antibodies against intermediate filament proteins during embryonic development. *J Embryol Exp Morphol* 64:45–60, 1981.
16. Kawahara H, Marceau N, French SW. The secretory functions in cultured hepatocytes from griseofulvin-fed mice. *FASEB J* 3:A329, 1989.
17. Hazan R, Denk H, Franke WW, Lackinger E, Schiller DL. Change of cytokeratin organization during development of Mallory bodies as revealed by a monoclonal antibody. *Lab Invest* 54: 543–553, 1986.
18. Denk H, Franke WW, Dragosics B, Zeiler I. Pathology of cytoskeleton of liver cells: Demonstration of Mallory bodies (alcoholic hyalin) in murine and human hepatocytes by immunofluorescence microscopy using antibodies to cytokeratin polypeptides from hepatocytes. *Hepatology* 1:9–20, 1981.
19. Kimoff RJ, Huang S. Immunocytochemical and immunoelectron microscopic studies on Mallory bodies. *Lab Invest* 45:491–503, 1981.
20. Denk H, Lackinger E, Vennigerholz F. Pathology of the cytoskeleton of hepatocytes. *Prog Liver Dis* 8:237–251, 1986.
21. Tsunoo C, Harwood TR, Arak S, Yokoo H. Cytoskeletal alterations leading to Mallory body formation in livers of mice fed 3,5-diethoxycarbonyl-1,4-dihydrocollidine. *J Hepatol* 5:85–97, 1987.
22. Denk H, Krepler R, Lackinger E, Artlieb U, Franke WW. Biochemical and immunocytochemical analysis of the intermediate filament cytoskeleton in human hepatocellular carcinomas and in hepatic neoplastic nodules of mice. *Lab Invest* 46:584–596, 1982.
23. Denk H, Lackinger E, Cowin P, Franke WW. Maintenance of desmosomes in mouse hepatocytes after drug-induced rearrangement of cytokeratin filament material: Demonstration of independence of desmosomes and intermediate-sized filaments. *Exp Cell Res* 161:161–171, 1985.
24. Lowe J, Blanchard A, Morrell K, Lennox G, Reynolds L, Billett M, Landon M, Mayer RJ. Ubiquitin is a common factor in intermediate filament inclusion bodies of diverse type in man, including those of Parkinson's disease, Pick's disease, and Alzheimer's disease, as well as Rosenthal fibers in cerebellar astrocytomas, cytoplasmic bodies in muscle, and Mallory bodies in alcoholic liver disease. *J Pathol* 155:9–15, 1988.
25. Butron-Vila MM, Haot J, Desmet VJ. Cholestatic features in focal nodular hyperplasia of the liver. Liver 4:387–395, 1984.
26. Desmet V. Intrahepatic bile ducts under the lens. *J Hepatol* 1:545–559, 1985.
27. Ray MB. Distribution patterns of cytokeratin antigen determinants in alcoholic and nonalcoholic liver diseases. *Hum Pathol* 18:61–66, 1987.
28. Denk H, Krepler R, Lackinger E, Artlieb U, Franke WW. Immunological and biochemical characterization of the keratin-related component of Mallory bodies: A pathologic pattern of hepatocytic cytokeratins. *Liver* 2:165–175, 1982.
29. Franke WW, Denk H, Schmid E, Osborn M, Weber K. Ultrastructural, biochemical, and immunologic characterization of Mallory bodies in livers of griseofulvin-treated mice: Fimbriated rods of filaments containing prekeratin-like polypeptides. *Lab Invest* 40:207–220, 1979.

Mouse Liver Carcinogenesis: Mechanisms
and Species Comparisons, pages 249–257

Heme Metabolism in Hepatocellular Carcinoma

Daniel L. Stout and Frederick F. Becker
Section of Experimental Pathology, The University of Texas
M. D. Anderson Cancer Center, Houston, TX 77030

In the latter part of the 1960s, a series of papers describing cytologic and biochemical changes in rat liver during feeding of the hepatotoxins dieldrin and 3′-methyl-4-dimethylaminoazobenzene were published (1–3). They reported that during the early stages of feeding, proliferation of endoplasmic reticulum occurred. Concomitant with hypertrophy, cytochrome P-450, aniline hydroxylase, and *p*-nitroreductase were increased. As the feeding period was extended, P-450 and its associated catalytic activities declined. Since that time, there have been numerous reports of similar findings associated with chemical hepatocellular carcinogenesis in the rat (4,5). Chemically induced rat liver cancers (PHC) and their putatively premalignant progenitors, hyperplastic nodules (HN), have been subjected to extensive biochemical characterization during the last several years. With few exceptions, decreased phase I (6–8) and increased phase II (9–15) drug-metabolizing activity were found to be characteristic of emergent HN and PHC. The widely used liver carcinogen 2-acetylaminofluorene inhibits the liver's mitogenic response to partial hepatectomy (16). This phenomenon has been linked to the production of cytotoxic metabolites by phase I enzymes. Phenotypically altered hepatocytes sustain less DNA modification when challenged with carcinogens requiring metabolic activation than do normal hepatocytes (17). Because cells demonstrating the resistant phenotype can proliferate in the presence of cytotoxic chemicals, it has been suggested that these cells are the product of adaptation to a toxic milieu (18).

In recent studies utilizing spontaneous and chemically induced mouse liver tumors (MLT), we measured the activities of 12 enzymes that have been reported to be altered in rat HN and PHC (19,20). The enzyme patterns were identical in mouse tumors of both etiologies. In general, the enzyme alterations in MLT also mirrored those seen in HN and PHC; the phase I monooxygenase functions were depressed, while phase II enzymes, such as DT diaphorase and glutathione S-transferase, were elevated. However, in contrast to HN and PHC, in which epoxide hydrase and UDP-glucuronosyltransferase are strikingly elevated (13,15), epoxide hydrase was unchanged in MLT while UDP-glucuronosyltransferase was decreased. Thus, it would seem that some phenotypic alterations perceived to have a significant role in carcinogenic progression may be of questionable significance. We have concluded from these findings that the phenotype of the resistant hepatocyte is inherent in the genetic program for the hepatocarcinogenesis process and that it is expressed in the absence of chemical exposure.

Because the hemoproteins P-450, b5, and catalase were demonstrated to be reduced in rat and mouse liver lesions (20,21), we questioned whether reduced heme synthesis or increased heme catabolism was the basis for the hemoprotein decline. This idea was supported by our more recent findings: aminolevulinic acid synthase, the first and rate-limiting enzyme in the heme synthetic pathway, was reduced in PHC and MLT, while the catabolic enzyme heme oxygenase (HO) was increased (20,21). In the study we describe here, heme synthesis by MLT and normal liver was measured in vivo.

Materials and Methods

Chemicals

Coproporphyrin I was from Sigma (St. Louis, MO). [^{55}Fe]ferric chloride, 2 mCi/μg, was from ICN Radiochemicals (Irvine, CA). [2-^{14}C]glycine, 41 mCi/mmol, was from Amersham (Arlington Heights, IL).

Animals

Male C3H/HeN mice were obtained from the Animal Program of the Division of Cancer Treatment, National Cancer Institute, maintained in a temperature-, humidity-, and light-controlled room, and allowed free access to water and Purina Rodent Laboratory Chow (Pearland Feed Co., Pearland, TX). Starting at 1 year of age, each mouse was bled monthly, and its circulating level of alpha-fetoprotein was determined by radioimmunoassay (22). Values of 1 μg/ml or higher were invariably associated with sufficient liver tumor mass for analysis.

Tissue Preparation

After the mice were killed by cervical dislocation, their livers were perfused in situ via the portal vein with cold 0.25 M sucrose, 0.02 M Tris, pH 7.4, containing 10 units of heparin/ml to remove the blood. Tumors with extensively vascularized capsules often had high residual blood levels and were discarded. After being freed from surrounding liver, the tumors were sectioned and examined to verify absence of necrosis or pools of blood. The tissues were homogenized in 4 vol 0.15 NaCl.

Heme Extraction

Heme was extracted from 200 μl aliquots of tissue homogenates by vigorous mixing with 5.0 ml of ethyl acetate–glacial acetic acid (4:1 vol/vol). Two additional extracts were done with 2.5 ml of the same reagent, and the extracts pooled. The extracts were washed twice with water and once with 1.5 N HCl (23).

Heme Determinations

The heme and porphyrin contents of the tissue homogenates and extracts were quantitated fluorometrically by the method of Morrison (24). Aliquots of material were added to saturated oxalic acid and allowed to stand at room temperature or heated at 90°C for 40 min before measuring the fluorescence at 662 nm while exciting the aliquots at 407 nm. Coproporphyrin was used to generate standard curves.

Iron Determinations

The iron content of the tissue homogenates was determined colorimetrically with the iron reagent bathophenanthroline disulfonic acid, after hydrolyzing the samples in 6 N HCl overnight, evaporating to dryness, and dissolving in water (25).

Results

To determine an appropriate time to examine heme synthesis, we injected tumor-free mice with mixtures of $^{55}FeCl_3$ and [2-^{14}C]glycine. At 1, 6, and 24 h, the animals were killed, and heme was extracted from the liver homogenates. The results are shown in Table 1. The amount of ^{55}Fe in heme increased with time up to 24 h. However, the amount of ^{14}C in heme was greatest at 6 h; at 24 h it was evident that the label was being lost due to heme turnover (26). Therefore, tumor-bearing mice were killed 6 h following injection of the radioisotope mixture. The following data were derived from three mice that were treated on the same day and presented MLT that met the experimental criteria; i.e., they had not yet developed a vascularized capsule and were free of cysts and necrosis. They were representative of several other experimental trials. One MLT and one sample of normal background tissue (BKG) was taken from each of the three mice. Hence, the statistical significance was determined by using the paired t-statistic rather than the t-statistic for two means. Incorporation of ^{55}Fe and [2-^{14}C]glycine into heme extracted from BKG and MLT is shown in Table 2. Heme from the two tissues showed similar levels of glycine incorporation after 6 h, but the ^{55}Fe content of heme from MLT was only about one-half that in heme from BKG ($P < 0.02$). Aliquots of each homogenate were solubilized and counted to determine whether the tissues differed in uptake of radiolabeled glycine and iron. The amount of label/mg protein is shown in Table 3. The amount of [2-^{14}C]glycine in homogenates from BKG and MLT was identical, but homogenates from BKG contained about 50% more ^{55}Fe than those from MLT. These findings are consistent with the reports that siderosis cannot be induced in MLT and PHC

Table 1. Radiolabel in Heme Extracted from Normal Mouse Liver 1, 6, and 20 H after Intraperitoneal Injection of Label*

Time	dpm/nmol Heme	
	^{55}Fe	^{14}C
1 h	123 ± 8	474 ± 29
6 h	584 ± 63	1,687 ± 197
20 h	1,276 ± 100	991 ± 69

*Six mice were injected intraperitoneally with 50 μCi of ^{55}Fe and [2-^{14}C]glycine. Heme was extracted from liver homogenates prepared from two mice at each time point.

Table 2. Incorporation of ^{55}Fe and ^{14}C into Heme from Mouse Liver Tumors and Background Liver*

Tissue	dpm/nmol Heme		
	^{55}Fe	^{14}C	^{55}Fe/^{14}C
BKG (3)	661 ± 37	1,777 ± 442	0.39 ± 0.12
MLT (3)	366 ± 207	1,844 ± 371	0.19 ± 0.07
$P <$	.01	N.S.	.002

*Heme was extracted from BKG and MLT 6 h after intraperitoneal injection of ^{55}Fe and [^{14}C]glycine. P values were calculated from the paired t-statistic. N.S., $P > .05$.

Table 3. Isotope Content of Mouse Liver Tumors and Background Liver 6 H after Administration of ^{55}Fe and [^{14}C]glycine*

Tissue	dpm ^{55}Fe (mg protein)	dpm ^{14}C (mg protein)
BKG	48,706 ± 13,223	32,161 ± 1,427
MLT	32,561 ± 5,565	31,992 ± 808
$P <$	N.S.	N.S.

*Homogenates of BKG and MLT prepared 6 h after administration of isotope mixtures were hydrolyzed overnight in HCl. The clarified hydrolysates were counted to determine dpm/mg protein. P values were calculated from the paired t-statistic. N.S., $P > .05$.

and suggest that such lesions are aberrant in iron uptake or storage (27,28). Because the rate of heme synthesis in MLT appeared comparable to that in BKG, it could be inferred that MLT maintain iron stores adequate for heme synthesis. Table 4 shows the results of determinations of total iron and heme in these

Table 4. Iron and Heme Content of Mouse Liver Tumors and Background Liver*

Tissue	μg Iron/mg protein	nmol Heme/mg protein
BKG (3)	1.62 ± 0.22	3.03 ± 0.50
MLT (3)	2.59 ± 0.68	4.42 ± 1.97
P <	.04	N.S.

*Total iron and heme were determined in the liver homogenates prepared from treated mice. *P* values were calculated from the paired t-statistic. N.S., $P > .05$.

samples. The total iron and heme content per milligram protein was about 50% greater in MLT than in BKG. These differences were not statistically significant, however.

Discussion

The family of cytochrome P-450s are the most abundant hemoproteins in the liver. It has been estimated that 70% of the heme synthesized by the liver is incorporated into P-450 (26). Acute exposure to chemical carcinogens can induce the liver to synthesize even greater quantities of P-450. Chronic exposure, however, can lead to diminution of hepatic cytochromes and their associated monooxygenase activities. It has been postulated that this phenotypic alteration of the hepatocyte represents an adaptive response that benefits the cell by reducing its potential to form cytotoxic metabolites from the chemical milieu (17). In previous studies of spontaneous and chemically induced MLT, we demonstrated that the resistant hepatocyte phenotype arose spontaneously. Therefore, we concluded that expression of this phenotype was linked to the genetic program for hepatocarcinogenesis.

Although the patterns of altered enzyme activity demonstrated by MLT and PHC were very similar, they were not identical (19). The striking feature common to MLT and PHC was the diminution of all nonmitochondrial hemoproteins (20,21). These findings suggested that hemoprotein decline during hepatocarcinogenesis might result from some alteration of heme metabolism that limited the availability of heme. This prospect was supported by one of our earlier findings: the P-450 content of MLT harvested from mice fed phenobarbital increased above basal levels but never exceeded the basal levels seen in livers of untreated mice (29). A similar response to phenobarbital treatment has also been demonstrated for HN (30). We subsequently reported that 5′-aminolevulinic acid synthase, the first enzyme in the heme synthetic pathway, was significantly reduced in MLT and PHC, while the heme catabolizing enzyme HO was significantly increased (20,21).

The effect of limiting amounts of heme on hemoprotein status in MLT would be twofold. The more obvious consequence is that of stoichiometry;

there may not be enough heme to saturate all the heme-binding sites in the existing apoproteins. The second effect of diminished heme would be a down-regulation of the synthesis of the apoproteins. As mentioned earlier, the total amount of microsomal protein in hepatic lesions is reduced (31). It has been shown by use of antibodies against various species of P-450 that in HN, the apocytochromes themselves are diminished (32). Transcription of P-450 and catalase mRNA has been demonstrated to be positively regulated by heme (33–36). Therefore, a decline in intracellular heme levels could be expected to result in decreased synthesis of apocytochromes and apocatalase.

Our data indicated that the amount of [2-^{14}C]glycine incorporated into heme by MLT and BKG was identical. However, heme extracted from MLT contained less ^{55}Fe than did heme from BKG. The latter result can be explained in part by the finding that MLT took up less ^{55}Fe than BKG. This result is consistent with those of other investigators. Eriksson et al. (37) reported decreased uptake of transferrin by HN, associated with reduced numbers of cell surface receptors for this iron-binding protein. Williams et al. (27,28) have reported that liver lesions in the mouse and rat were resistant to induced siderosis. A second factor contributing to reduced ^{55}Fe incorporation into heme by MLT is that of its greater total cellular iron content. Because MLT can synthesize heme despite reduced iron uptake, it follows that a portion of its measurable total iron resides in a pool utilized for heme synthesis. Thus, the net effect is a significant reduction of specific activity of the ^{55}Fe in this iron pool.

Contrary to our hypothesis, our results indicate that the rates of heme synthesis in MLT and BKG are comparable. The notion that decreased expression of hemoprotein in MLT results from a paucity of heme seems untenable at present. The mechanism for decreased apoprotein synthesis, therefore, may represent an inherent component of the cancer phenotype.

Acknowledgments

We thank Carol Rosenblum, Sylvia Capetillo, and Lisa Thurman for laboratory assistance. This work was supported by the W. C. Hill Fund.

Open Forum

Dr. Dennis Parke: Have you measured ferrochelatase activity?

Dr. Daniel Stout: We have attempted it. We have had more success using zinc than iron. The enzyme can more readily insert zinc into porphyrins than it can iron. Presumably this also occurs in nature: Transferrin binds zinc as well as iron and transports it to the cell. At least a certain percentage of it finds its way into porphyrin, but nobody knows the fate of such molecules.

Dr. Parke: One of the great bonuses of coming to meetings like this is that many of us who come with problems meet someone else working in a completely different field who supplies the answer. Dr. Stout has provided the answer to one of my problems. When we inject iron nitrilotriacetate into mice we can produce damage to the liver and kidneys, acute lethal injury, and hepatic necrosis. We have lowered the dose, but we have not yet produced tumors—maybe we haven't gone on long enough. We find exactly the same pattern that you described: a loss of cytochrome P-450, an increase in heme oxygenase, an increase or a normal incorporation of ^{14}C into the porphyrins, but no heme synthesis. I just wonder whether what you are looking at is what we are doing: producing oxygen radicals.

Dr. Stout: Many of these enzymes are quite sensitive to other drugs. A number of halogenated chemicals induce porphyria in individuals exposed to them, and there are several agents that induce heme oxygenase—phenobarbital for instance. It behaves quite a bit like cytochromes, but it is apparently more exotic in its regulation.

References

1. Arcasoy M, Smuckler EA, Benditt EP. Acute effects of 3′-methyl-4-dimethylaminoazobenzene intoxication on rat liver: Structural and functional changes in the endoplasmic reticulum and NADPH-related electron transport. *Am J Pathol* 62:841–867, 1968.
2. Hutterer F, Klion FM, Wengraf A, Schaffner F, Popper H. Hepatocellular adaptation and injury: Structural and biochemical changes following dieldrin and methyl butter yellow. *Lab Invest* 20:455–464, 1969.
3. Hutterer F, Schaffner F, Klion FM, Popper H. Hypertrophic, hypoactive smooth endoplasmic reticulum: A sensitive indication of hepatotoxicity exemplified by dieldrin. *Science* 161:1017–1019, 1968.
4. Meyer DI, Barber AA. Changes in microsomal enzyme activities during DAB carcinogenesis. *Chem Biol Interact* 7:231–240, 1973.
5. Gravela E, Feo F, Canuto RA, Garcea R, Gabriel L. Functional and structural alterations of liver ergastoplasmic membranes during DL-ethionine hepatocarcinogenesis. *Cancer Res* 35:3041–3047, 1975.
6. Oyanagui Y, Sato N, Hagihara B. Spectrophotometric analysis of cytochromes in rat liver during carcinogenesis. *Cancer Res* 34:458–462, 1974.
7. Astrom A, DePierre JW, Eriksson L. Characterization of drug-metabolizing systems in hyperplastic nodules from the livers of rats receiving 2-acetylaminofluorene in their diet. *Carcinogenesis* 4:577–581, 1983.
8. Farber E. The biochemistry of preneoplastic liver: A common metabolic pattern in hepatocyte nodules. *Can J Biochem Cell Biol* 62:486–494, 1984.
9. Levin W, Lu AYH, Thomas PE, Ryan D, Kizer DE, Griffin MJ. Identification of epoxide hydrase as the preneoplastic antigen in rat liver hyperplastic nodules. *Proc Natl Acad Sci USA* 75:3240–3243, 1978.
10. Schor NA. The use of D-T diaphorase for the detection of foci of early neoplastic transformation in rat liver. *Cancer Lett* 5:167–171, 1978.
11. Yin Z, Sato K, Tsuda H, Ito N. Changes in activities of uridine diphosphate-glucuronyltransferases during chemical hepatocarcinogenesis. *Gann* 73:239–248, 1982.

12. Kitahara A, Yamazaki T, Ishikawa T, Camba EA, Sato K. Changes in activities of glutathione peroxidase and glutathione reductase during chemical hepatocarcinogensis in the rat. *Gann* 74:649–655, 1983.
13. Sato K, Kitahara A, Yin Z, Ebina T, Satoh K, Tsuda H, Ito N, Dempo K. Molecular forms of glutathione S-transferase and UDP-glucuronyltransferase as hepatic preneoplastic marker enzymes. *Ann NY Acad Sci* 417:213–223, 1983.
14. Sato K, Kitahara A, Satoh K, Ishikawa T, Tatematsu M, Ito N. The placental form of glutathione S-transferase as a new marker protein for preneoplasia in rat chemical hepatocarcinogenesis. *Gann* 75:199–202, 1984.
15. Kizer DE, Clouse JA, Ringer DP, Hanson-Painton O, Vaz AD, Palakodety RB, Griffin MJ. Assessment of rat liver microsomal epoxide hydrolase as a marker of hepatocarcinogenesis. *Biochem Pharmacol* 34:1795–1800, 1985.
16. Jackson CD, Irving CC. Effects of 2-acetylaminofluorene on liver cell proliferation after partial hepatectomy of female rats. *Cancer Res* 33:397–401, 1973.
17. Farber E, Parker S, Gruenstein M. The resistance of putative premalignant liver cell populations, hyperplastic nodules, to the acute cytotoxic effects of some hepatocarcinogens. *Cancer Res* 36:3879–3887, 1976.
18. Eriksson L, Ahluwalia M, Spiewak J, Lee G, Sarma DSR, Roomi MJ, Farber E. Distinctive biochemical pattern associated with resistance of hepatocytes in hepatocyte nodules during liver carcinogenesis. *Environ Health Perspect* 49:171–174, 1983.
19. Stout DL, Becker FF. Xenobiotic metabolizing enzymes in genetically and chemically initiated mouse liver tumors. *Cancer Res* 46:2693–2696, 1986.
20. Stout DL, Becker FF. Heme enzyme patterns in genetically and chemically initiated mouse liver tumors. *Cancer Res* 46:2756–2759, 1986.
21. Stout DL, Becker FF. Heme enzyme patterns in rat liver nodules and tumors. *Cancer Res* 47:963–966, 1987.
22. Sell S. Radioimmunoassay of rat alpha-fetoprotein. *Cancer Res* 33:1010–1015, 1973.
23. Bloomer DR, Morton KOA. A radiochemical assay for heme synthase activity. *Enzyme* 28:220–232, 1982.
24. Morrison GR. Fluorometric microdetermination of heme protein. *Anal Chem* 37: 1124–1126, 1965.
25. Landers JW, Zak B. Determination of serum iron by bathophenanthroline disulfonic acid. *Am J Clin Pathol* 29:590–592, 1958.
26. Shiraki H, Guengerich FP. Turnover of membrane proteins: Kinetics of induction and degradation of seven forms of rat liver microsomal cytochrome P-450, NADPH-cytochrome P-450 reductase, and epoxide hydrolase. *Arch Biochem Biophys* 235:86–96, 1984.
27. Williams GM, Yamamoto RS. Absence of stainable iron from preneoplastic and neoplastic lesions in rat liver with 8-hydroxyquinoline-induced siderosis. *JNCI* 49: 685–692, 1972.
28. Williams GM, Hirota N, Rice JM. The resistance of spontaneous mouse heptocellular neoplasms to iron accumulation during rapid iron loading by parenteral administration and their transplantability. *Am J Pathol* 94:65–74, 1979.
29. Becker FF, Stout DL. A constitutive deficiency in the monooxygenase system of spontaneous mouse liver tumors. *Carcinogenesis* 5:785–788, 1984.
30. Feo F, Canuto RA, Garcea R, Brossa O, Caselli GC. Phenobarbital stimulation of cytochrome P-450 and aminopyrine N-demethylase in hyperplastic liver nodules during LD-ethionine carcinogenesis. *Cancer Lett* 5:25–30, 1978.

31. Fox OF, Donaldson SG, Griffin MJ, Carubelli R. Changes in the relative distribution and properties of smooth endoplasmic reticulum subfractions isolated from the livers of rats fed 2-acetylaminofluorene. *Cancer Lett* 24:29–36, 1984.
32. Buchmann A, Kuhlmann W, Schwarz M, Kunz W, Wolf CR, Moll E, Friedberg T, Oesch F. Regulation and expression of four cytochrome P-450 isoenzymes, NADPH-cytochrome P-450 reductase, the glutathione transferases B and C and microsomal epoxide hydrolase in preneoplastic and neoplastic lesions in rat liver. *Carcinogenesis* 6:513–521, 1985.
33. Sakamoto T, Higashi T. Studies on rat liver catalase. X. Effect of hemin and an inhibitor on the translation of catalase messenger RNA1. *J Biochem* 85:389–396, 1979.
34. Dwarki VJ, Francis VNK, Bhat GJ, Padmanaban G. Regulation of cytochrome P-450 messenger RNA and apoprotein levels by heme. *J Biol Chem* 262:16958–16962, 1987.
35. Jayarama-Bhat GJ, Padmanaban G. Heme regulates cytochrome P-450 gene transcription elongation. *Biochem Biophys Res Commun* 151:737–742, 1988.
36. Bhat GJ, Padmanaban G. Heme is a positive regulator of cytochrome P-450 gene transcription. *Arch Biochem Biophys* 264:584–590, 1988.
37. Eriksson LC, Torndal U-B, Andersson GN. The transferrin receptor in hepatocyte nodules: Binding properties, subcellular distribution and endocytosis. *Carcinogenesis* 7:1467–1474, 1986.

Mouse Liver Carcinogenesis: Mechanisms
and Species Comparisons, pages 259–276

Cell-to-Cell Communication: Relationship of Stem Cells to the Carcinogenic Process

J. E. Trosko, C. C. Chang, and B. V. Madhukar
Department of Pediatrics/Human Development, Michigan State University, East Lansing, MI 48824

Discovery consists of seeing what everybody has seen and thinking what nobody has thought.

—Albert Szent-Gyorgyi (1)

That rather self-serving quotation from Albert Szent-Gyorgyi was chosen to highlight this brief analysis of the problem of understanding the elements of a biologically based cancer risk-assessment model. It should go without saying that understanding all the mechanisms related to multistep carcinogenesis is not an easy task. The known complex interactions of genetic, developmental, sex, dietary, and environmental factors in carcinogenesis in both laboratory animals and human beings precludes direct and easy extrapolation from controlled in vitro and experimental animal bioassay systems to the uncontrolled human situation (2,3).

While recent advances in modern molecular biology related to development of the concept of oncogenes (4–7), cancer suppressor genes (8–13), experimental initiation/promotion/progression model systems (14–16), and basic studies on the control of cell growth (i.e., growth regulators and second messengers) (17) have given us valuable information related to the cancer process, no clear picture or consensus view has emerged.

For the sake of taking a different view of the information gathered, we will challenge some of the basic assumptions related to the current paradigm guiding the cancer research field, namely, "carcinogenesis as mutagenesis" (18). Put succinctly, while the bulk of the molecular, biochemical, and cellular studies related to "carcinogens" and carcinogenesis have been done on the predominate cells of target organs (i.e., late progenitor and differentiated cells such as the hepatocytes), the target cell for carcinogenesis (i.e., the early stem or progenitor cell) has not been used.

While the stem cell theory of cancer (19,20), the cancer as "a disease of differentiation" (21,22), and "oncogeny as partially blocked ontogeny" (23) theories have been postulated for years, and while stem cells' role in blood cancers is well accepted (24,25), the idea that they might play a role in solid cancers such as hepatocellular carcinoma seems to have been largely ignored (26,27).

To the extent that the stem cell theory has support, that support comes from several observations, namely the clonal nature of tumors; the general similarity between stem cells and tumor cells in tissue origin, extensive proliferative potential, and tissue-specific differentiation potential; the implication of a small target size for tumor control with radiotherapy; the demonstration that clonogenic potential, self-renewal capacity, and cell differentiation features are restricted to subpopulations of cells in tumors; and the ability to induce terminal differentiation of some neoplastic cells in vitro by various natural differentiation factors or exogenous chemical compounds (see ref. 28).

Clearly, to put this idea in perspective, the counterproposal that the carcinogenesis process can cause the differentiated cell to dedifferentiate has been as difficult to refute as the stem cell hypothesis has been to support (Fig. 1). One of the reasons is isolation of pure populations of epithelial stem cells has not been routinely possible, let alone the ability to control their differentiation or to transform these cells into carcinomas in an in vitro system.

When one examines the early premalignant lesions or frank carcinoma, the observation that immature or undifferentiated cells exist has been interpreted to indicate that they represent the dedifferentiated transformed cells, or

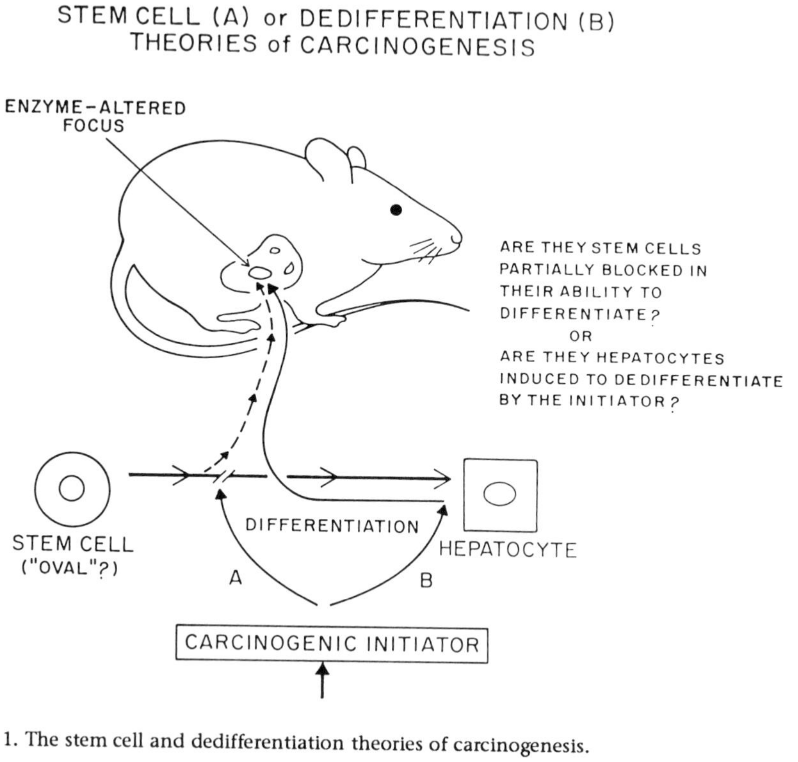

Figure 1. The stem cell and dedifferentiation theories of carcinogenesis.

even if the tumor cells were derived from stem cells, the differentiated cell and the stem cell are equally susceptible to the carcinogen. If the latter interpretation is accepted, studying the predominant and easily obtainable differentiated cell (i.e., hepatocytes) will indicate what might be happening in the stem cell.

While we cannot rule out the dedifferentiation theory, we are taking a fresh view of the idea that molecular, biochemical, and cellular studies of the hepatocytes will not, per se, be indicative of how carcinogens and oncogenes behave in the stem cells. It is our assumption that the correlation of molecular, biochemical, or cellular events in differentiated cells, such as the hepatocyte, may not correlate with, or be relevant to, the carcinogenesis process in vivo ("nonresults") (F. F. Becker, presentation at Mouse Liver Carcinogenesis symposium, Nov. 30–Dec. 3, 1988, Austin, Texas). Rather, not until these studies are done with stem cells will it be possible to test either the stem cell or the dedifferentiation theories of carcinogenesis.

Regulation of Stem Cell Proliferation and Differentiation

By definition, a stem cell is one that, after cell division, has the ability to produce one daughter cell that can eventually terminally differentiate and another daughter cell that can maintain the ability to self-renew the stem property. The fertilized egg is the totipotent stem cell for all eventual differentiated cells of the multicellular organism. During organogenesis, these totipotent stem cells must "commit" to give rise to lineages of tissue-specific progenitor cells, which act as the stem cells for specific tissues.

In an organism such as a human being, the coordinated regulation of growth, differentiation, and adaptive functions of the differentiated cells is mediated by extracellular, intracellular, and intercellular communication mechanisms (Fig. 2). Of the several specialized modes of regulating the delicate balance between controlling proliferation of a stem cell and its differentiated daughter cells, organs such as the skin and liver must be examined. In the skin, as the stem cell divides, one daughter cell is committed to terminally differentiate by going through a transition series of phenotypic changes. Because terminal differentiation in this cell leads to eventual cell death and removal, the other daughter cell remains in the undifferentiated state to start the process over again. With this constant need for cell replacement, the nature of the regulation between the differentiated daughter and the stem cell is probably under a different kind of control than the cells in the liver. The skin could be thought of as an open-ended system.

In the liver, cell proliferation and differentiation must be strictly controlled to maintain the volume of this organ. One has to assume that there are stem cells for the liver or alternatively that the hepatocytes themselves, although differentiated, are not terminally differentiated (without the capacity to dedifferentiate or proliferate). Assuming for the sake of discussion that the oval cell (or some equivalent progenitor) is the stem cell for the various cells in the liver (29–35), one could speculate on the nature of the homeostatic control in closed organ systems such as the liver.

Of the three generic forms of cell-to-cell communication in Figure 3, we know hormones and growth factors (i.e., epidermal growth factor, transforming growth factor-beta) are extracellular communicators that can trigger specific second messengers by activating membrane or cytosolic receptor/enzymes to produce alterations in intracellular cyclic AMP, Ca^{2+}, pH, and phosphorylation of enzymes (i.e., intracellular communication). In turn, in solid tissues such as the liver, the hepatocytes (36), as well as the epithelial cells (37), are coupled by the gap-junction structure, which allows metabolic equilibration or coupling to occur (intercellular communication).

It has been demonstrated that modulation of these gap junctions (up- or downregulation) seems to be correlated or causally related to the changes in intracellular messages induced by extracellular endogenous or exogenous messages (i.e., hormones or tumor-promoting chemicals) (38,39). Indeed, in the liver, cell death and cell removal by partial hepatectomy have been shown to downregulate gap junctions (36,40).

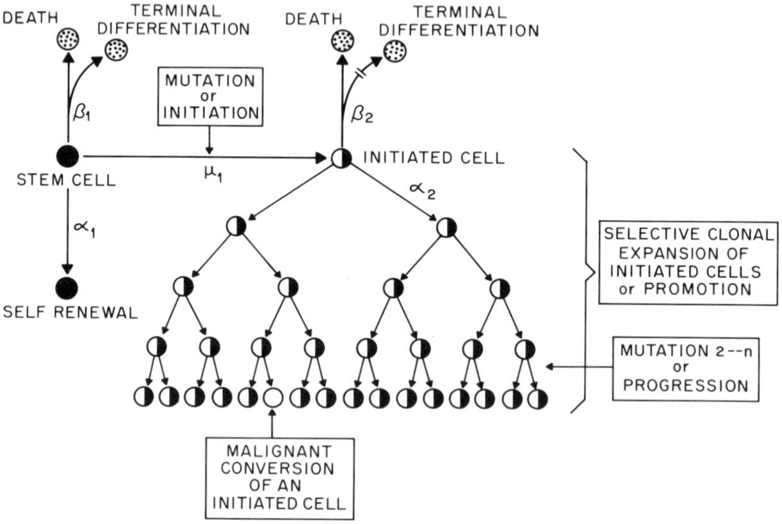

Figure 2. The heuristic schemata characterizes the postulated link between extracellular communication and intercellular communication via various intracellular transmembrane signalling mechanisms. It provides an integrating view of how the neuroendocrine immune system (mind or brain/body connection) and other multisystem coordinations could occur. While not shown here, activation or altered expression of various oncogenes (and anti-oncogenes) could also contribute to the regulation of gap-junction function.

Considering the possible communication relationship between the differentiated cells of a contained organ and their presumptive stem cells, one might speculate that while the two must be communicating to maintain a constant volume, the nature of the communication must be via extracellular messages rather than by direct gap-junctional intercellular communication because if the stem cell directly communicated with its differentiated daughter, the differentiated products could alter stem cell physiology. Physical separation or sequestration of the daughter cell seems to occur with a presumptive human kidney epithelial stem cell (28). Also, if the oval or epithelial cell of the liver is the precursor to the hepatocyte (29–35), there is evidence that while these two cell types communicate via gap junctions with like-type cells, they do not gap junctionally communicate with each other. Physical barriers, such as extracellular matrices (29) and cell adhesion molecules (41), as well as dissimilar gap-junction proteins (42), might prevent such intercellular communication.

Since these differentiated stem and daughter cells probably communicate via extracellular messages, one could speculate on the kinds that might be operative. While most attention to growth regulators has been on "positive" signals, earlier theoretical and recent experimental work have pointed to a cybernetic system of positive and negative growth regulatory signals (27,43,44).

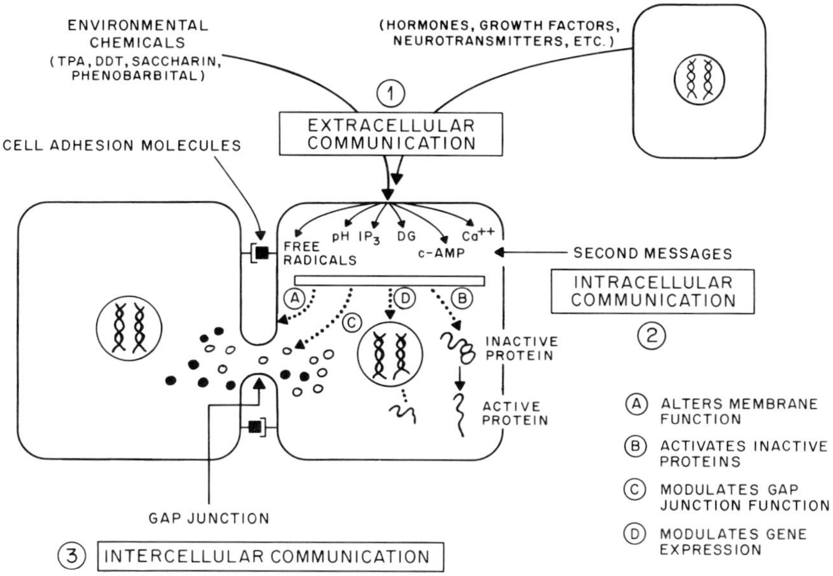

Figure 3. The initiation/promotion/progression model of carcinogenesis. β_1 = rate of terminal differentiation and death of stem cell; β_2 = rate of death, but not of terminal differentiation of the initiated cell (broken arrow); α_1 = rate of cell division of stem cells; α_2 = rate of cell division of initiated cells; μ_1 = rate of the molecular event leading to initiation (i.e., possibly mutation); μ_2 = rate at which second event occurs within an initiated cell. (Reprinted from J.E. Trosko et al. in E.L. Hertzberg, R. Johnson (eds): *Gap Junctions*. New York: Alan R. Liss, 1988; used with permission.)

While the specific in situ signals are not known for any tissue, let alone between stem cells and the differentiated daughters within a tissue, several complex interactions have been delineated, as in the lymphoreticular tissue (45,46). In addition, several negative growth regulators have been associated with a differentiated cell of a tissue and a precursor cell of the same cell or tissue lineage (47–52). Recently, transforming growth factor beta (TGF-β), which has both positive and negative effects on different cell types (53,54), has been shown to inhibit DNA synthesis and cell growth in rat liver epithelial or oval cells (55,56, and unpublished results) and human keratinocytes (57).

The complex communication between stem cells and their differentiated daughter cells via a negative growth regulator would depend on the production of the daughter cell, its reception by the stem cell, and its transmembrane transduction in that cell. To overcome a negative influence on growth and differentiation of the stem cell, positive growth factors (e.g., EGF) or the reduction of the negative growth regulator (i.e., removal or death of the differentiated daughter producing the negative factor), a reduction of receptor for the negative

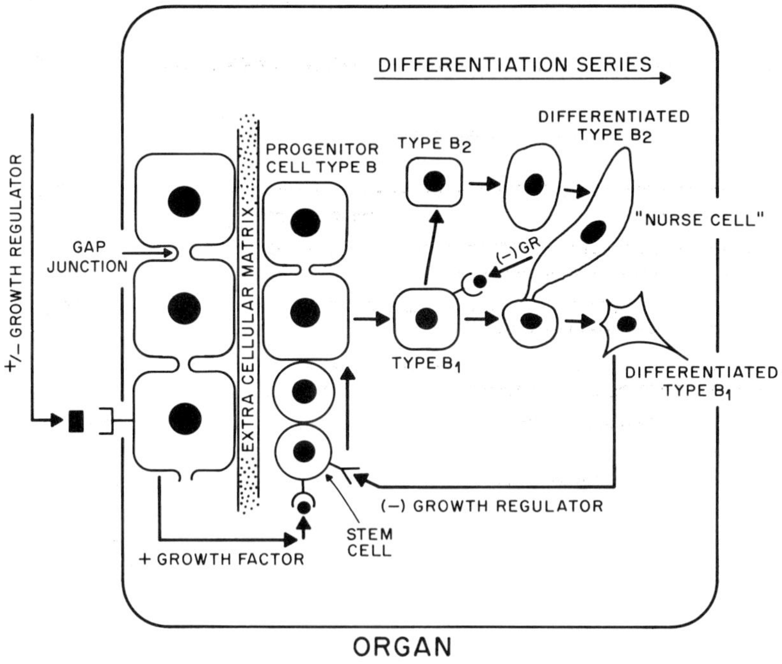

Figure 4. The interrelationship of intercellular communication by positive and negative growth regulators on stem cell growth and differentiation and intercellular communication via gap junctions in a closed organ system. (Reprinted from J.E. Trosko and C.C. Chang in C.J. Gehrs (ed): *Bioindicators: Exposure and Effect*. Chelsea, MI: Lewis Publishers, Inc., in press; used with permission.)

regulator, or a defect in the transduction of the negative transmembrane signal would allow the stem cell to proliferate (Fig. 4). For example TGF-β, while suppressing the growth of blast-type cells, has no effect on the tumor cells of that differentiation lineage (51).

Modulation of Gap-Junctional Intercellular Communication during Carcinogenesis

Since the discovery of gap-junctional communication and the delineation of its presumed roles in regulating cell proliferation, differentiation, and the coordination of physiologic activities of many cell types (58,63), it has been speculated that its dysfunction in cells would lead to many kinds of diseases based on disruption of these fundamental functions (64). Many studies (58,65), but not all (66,68), have shown gap-junction function/structures to be abnormal in many cells derived from tumors or transformed cultures. Cancer is a disease of growth control and of differentiation. If, of course, gap-junctional communication is needed for contact-inhibited growth and for the regulation of differentiation, then its dysfunction in neoplastic cells would be consistent with the hypothesis that this is a crucial phenotype. Some apparent discrepancies have been explained away by the fact that not all cells derived from tumors are tumorigenic, the manner of determining gap junctions (i.e., ultrastructural) might not reflect their functionality, that some tumor cells, while communicating with themselves, could not communicate with surrounding normal cells (selective communication) (64) and the manner of measuring the cells in vitro might influence the communication patterns of the cell in a manner not found in vivo (i.e., endogenous hormones might downregulate gap junctions in vivo, while in vitro, the gap junctions were upregulated because those factors were not present in the culture medium).

A major new insight to the role of gap-junctional communication came when many chemical tumor promoters were shown to modulate gap-junction function in vitro (40,70,71). In addition, fewer gap junctions were seen when mouse skin was treated with phorbol esters, a powerful tumor promoter (72), and in livers subjected to partial hepatectomies (a tumor-promoting stimulus) (73). Moreover, gap-junction protein and its message was shown to decrease in liver during hepatocarcinogenesis (74,76). Finally, the metastatic potential of several tumor cell types has been shown to increase with decreasing gap-junctional communication (65,77,78).

With the introduction of the concept of oncogenes (see ref. 78 for review) and the discovery of various oncogenes coding for different proteins affecting growth factor receptors, growth factors, transmembrane signalling elements or nuclear proteins (80), it seemed reasonable to ask whether any of these could affect gap-junction function, directly or indirectly (81). In the last few years, several oncogenic products (i.e., *src*, H-*ras*, *mos*, *neu*, T antigens of the polyoma virus, but not *myc*) (82–91) have been associated with a downregulation of gap junctions. We have recently shown that downregulation of gap junctions in rat epithelial, oval, cells with the expressed H-*ras* gene can give rise to tumors when injected back into the Fischer-344 rat liver (92).

Finally, with the demonstration that several growth factors (i.e., EGF, TGF-β, bovine pituitary extract) can downregulate gap junctions in normal human keratinocytes (57), a possible linkage between several theories of carcinogenesis can be made, integrating the multistage, initiation/promotion/progression model with the oncogene theory and the stem cell theory with the phenomenon of gap-junctional intercellular communication (93).

A New View of Old Observations Regarding the Initiation/ Promotion Model of Liver Carcinogenesis

There seems to be little doubt that the initiation/promotion/progression concept is applicable to liver carcinogenesis (15). Given the power of the concepts of "carcinogenesis as mutagenesis" (18) and of genotoxicity (94), it has generally been assumed because of the apparent irreversible nature of initiators of liver cancer (i.e., acetylaminofluorene, AAF), mutations have been induced during the initiation phase. With the demonstration by Miller and Miller (95) of chemical metabolism of carcinogens, the differential effect of carcinogens on organs or species seemed to be explained by cells of target tissues metabolizing these chemicals to DNA-damaging, mutagenic electrophilic carcinogens. Early demonstration that specific ultimate electrophils of this metabolism could damage DNA in cells without metabolic capabilities was consistent with this idea (96).

More recently, the presumptive lesions in liver that ultimately are precursors to the hepatocarcinomas (i.e., enzyme-altered foci, nodules) have been noted to be resistant to many carcinogens (97) and not to form DNA lesions, even as a result of the agent inducing the foci (98). The traditional, obvious explanation has been that the hepatocytes, because of their metabolizing capabilities, have been altered by some genotoxic lesion to dedifferentiate and have lost some of their differentiated functions, such as drug metabolism, although they maintain their ability to proliferate to form foci; some go on to form cancers. It is interesting that AAF and some of its metabolites appear to be stronger promoters than initiators (99,100) and that in the bladder of many species, AAF–DNA adducts in non-stem cells do not correlate with relative susceptibilities to tumors (101).

Based on this explanation, attempts to correlate molecular lesions, i.e., specific DNA adducts, or biochemical markers in the hepatocyte with the carcinogenic end point, i.e., malignant tumors, have not always been successful.

We offer an alternative hypothesis—that the oval or similar precursor cell might be the target cell for the carcinogen—although we admit that no evidence precludes the hepatocytes from being a target for carcinogenesis. However, assuming it is the oval cell that is the target cell, many observations related to the stem cell theory, the role of oncogenes in cancer, initiation/promotion of foci, and chemical carcinogenesis might make sense.

It has been shown that furan and furfural can induce liver tumors in mice (102). DNA from these tumors, as well as from spontaneous tumors in these mice, were shown, after transfection, to transform NIH 3T3 cells. The transforming DNA contained an activated H-*ras* in both the spontaneous and furan/furfural-induced tumors. Analysis of the mutation spectrum in the spontaneous and induced tumors interpreted these as being different (albeit, the numbers leading to this conclusion were very small). Assuming the spontaneous and furan-induced mutation spectra in the H-*ras* oncogene of these tumors to be different, the researchers concluded that furan and furfural were mutagenic, in spite of the fact that there is no evidence of these chemicals' mutagenic potential in short-term tests.

How does one reconcile these facts? Given the power of the carcinogenesis as mutagenesis paradigm, it is natural for most to rationalize these findings by speculating that furan and furfural are metabolized in vivo to form mutagenic metabolites. An alternative view is that these chemicals are not mutagenic but rather that they stimulate the growth of the oval cells. During replication of the oval cells, spontaneous mutations occur (both in untreated livers and in furan/furfural-treated livers). If one assumes mutations could occur randomly in any codon of the H-*ras* gene, only those occurring in codon 12 or 61 would confer a selective growth advantage to a cell in a normal liver environment. However, in the furan/furfural-treated rat, the environment of the liver might favor mutations at codons in the H-*ras* gene other than 12 or 61. In effect, these carcinogens, as well as many other nongenotoxic carcinogens, might only be selecting for cells having expressed, overamplified, and activated oncogenes induced during the mitogenic stimulus caused by the chemical. The overexpression, mutation, or amplification of oncogenes might be an indirect consequence of the fact that these chemicals might be blocking contact inhibition or mitogenic suppression of the stem cells by normal cell communication mechanisms.

Brookes et al. (103) have presented evidence that DNA from DMBA-transformed primary mouse bladder cells, which transformed NIH 3T3 cells after transfection, contained mutated Ki-*ras* genes. However, their results suggested that the mutations in the activated Ki-*ras* might not be the direct consequence of the interaction of activated DMBA metabolites with the Ki-*ras* gene.

This interpretation, if correct, points out several important issues, namely; one should be studying the effect of known or potential carcinogens on stem cells, not the differentiated cells of a tissue; the effect of chemicals on cell communication mechanisms, as well as on mutagenic processes; and the mitogenic potential of chemicals in organs (which has threshold-type kinetics).

In many in vitro systems, at least two oncogenes are needed for complete transformation (104), or once a cell is immortalized by chemicals or radiation (105), only a *ras*-type oncogene is needed for carcinogenic transformation. Recently, Braun et al. (106), Huber and Cordingley (107), Seyama (108), Burt et al. (109), and de Feijter et al. (92) have shown that an expressed mutated H-*ras* was sufficient for the conversion of rat liver epithelial or oval cells. In addition, it was shown that c-*myc* was elevated in oval cells of non-initiated liver (110). Since one would predict c-*myc* to be expressed in those proliferating and stem-like cells as part of their normal function because it is related to early oncogenic influence of cell proliferation (111), the expression of a mutated *ras* in these cells, (associated with downregulation of gap junctions) in a manner similar to and synergistic with tumor promoters (112) might be sufficient for tumorigenic transformation.

Another important observation consistent with the stem cell theory of cancer and with the idea that only one oncogene, such as an expressed/mutated H-*ras*, was that transfection of multipotential murine stem cells, followed by induction of differentiation, produced transformed fibroblasts expressing a high level of the p21 protein product (113). Those cells not transformed but differentiated expressed a low level of p21. One assumes the c-*myc* gene was expressed in these multipotential stem cells. Transfection of the same H-*ras* gene into the differentiated fibroblasts produced no transformants. We, of course, assume c-*myc* would not be expressed in these differentiated cells and/or that the transfected H-*ras* in the differentiated cells might be methylated and not expressed.

Nongenotoxic carcinogens (e.g., DDT, dieldrin, phenobarbital, phthalates), which can inhibit intercellular communication, might be expected to interact with oval-like cells in the liver that already express a high level of *myc*. This could lead to foci of enzyme-altered cells. When the nongenotoxic chemical is removed, we would expect gap junctions to upregulate, the phenotype to revert back to normal, and some of the oval cells to differentiate into hepatocytes (i.e., remodel).

If during the nongenotoxic stimulation of oval cell growth or inhibition of apoptosis (114,115), spontaneous mutations occur that keep the *myc* level constitutively high, these might represent the few foci or nodules that do not revert. Only after additional genetic hits (i.e., *ras* expression) would the cell have a constitutive and stable way to block gap-junction function. By blocking intercellular communication, the transfer of growth-suppressing molecules within a colony of cells would be prevented, hence lead to cell growth.

Gap junctions, by being upregulated in cancer cells (116,117), have been correlated with the suppression of cell growth. In fact, it has been speculated that the design of chemotherapeutic drugs ought to consider upregulation of gap-junctional communication in noncommunicating cancer cells (119).

In summary, a new interpretation of the carcinogenic process has been offered for hepatocellular carcinogenesis (or all cancers). Specifically, it assumes that the stem or early progenitor cell in tissue, including the liver, is the target cell for carcinogenesis. The c-*myc* genes in these cells normally would be expressible, since they would be the precursors for needed cell proliferation. The differentiated cells, including the hepatocytes, while capable of metabolizing chemicals to genotoxic molecules would, at best, die, rather than mutate. This cell death would act as a promoting or mitogenic signal for the chemical-resistant stem cell. Nongenotoxic/noncytotoxic chemicals that induce proliferation of stem cells or prevent apoptosis would increase populations of these undifferentiated cells (enzyme-altered foci, nodules, papillomas, polyps). As long as the exogenous or endogenous promoter or inhibitor of intercellular communication is present, these foci of undifferentiated cells proliferate. When these chemicals are removed, the normal cells in these foci regress due to the upregulation of gap junctions. Only during cell proliferation might a spontaneous or induced mutation occur (usually unrelated directly to the chemical that needs metabolism). These are the true initiated cells after removal of the promoter. These cells contain an irreversible lesion that upon restimulation clonally expands to form another focus. If, in these cells, another mutation occurs that stably downregulates the gap junction, the cell becomes promoter-independent (i.e., converted to the progression phase). These cells would not regress until or unless the gap junctions were upregulated, bypassing the genetic defect downregulating the gap junction.

This view of liver (or any other organ) carcinogenesis is clearly hypothetical. However, many observations generated testing the alternative view, namely that the differentiated hepatocyte is the target cell, seem to be consistent with the stem cell (or oval cell in this case) theory of carcinogenesis. Several specific hypotheses can be tested now that oval cell and other stem cell cultures and new technologies of molecular biology have been developed.

Acknowledgments

The authors wish to acknowledge our appreciation to our coworkers Beth Lockwood, Heather Rupp, and William Paradee and our former colleague Dr. Saw Yin Oh for their enthusiastic and skilled contributions to the work on which this manuscript is based. We wish also to thank Darla Conley for her cheerful attitude and excellent typing skills. Research was supported by grants from the National Cancer Institute (CA 21104), the U.S. Air Force Office of Scientific Research [USAFOSR-86-0084], and National Institute of Environmental Health Sciences [ES04911] to J.E.T.

References

1. Foster R. *Innovation: The Attacker's Advantage.* New York: Summit Books, 1986, pp 82–83.
2. Trosko JE, Chang CC. The role of inhibited intercellular communication in carcinogenesis: Implications for risk assessment from exposure to chemicals. In Travis CC (ed): *Biologically Based Methods for Cancer Risk Assessment.* New York: Plenum Press, in press.
3. Trosko JE. A failed paradigm: Carcinogenesis is more than mutagenesis. *Mutagenesis* 3:363–366, 1988.
4. Bishop JM. Viral oncogenes. *Cell* 42:23–38, 1985.
5. Weinberg RA. The action of oncogenes in the cytoplasm and nucleus. *Science* 230: 770–776, 1985.
6. Marshall CJ. Oncogenes. *J Cell Sci* 4:417–430, 1986.

7. Barbacid M. Oncogenes and human cancer: Cause or consequence? *Carcinogenesis* 7:1037–1042, 1986.
8. Klein G. The approaching era of the tumor suppressor genes. *Science* 238:1539–1545, 1987.
9. Koi M, Barrett JC. Loss of tumor-suppressive function during chemically induced neoplastic progression of Syrian hamster embryo cells. *Proc Natl Acad Sci USA* 83: 5992–5996, 1986.
10. Harris H. The analysis of malignancy by cell fusion: The position in 1988. *Cancer Res* 48:3302–3306, 1988.
11. Willman CL, Fenoglio-Preiser CM. Oncogenes, suppressor genes, and carcinogenesis. *Hum Pathol* 18:895–902, 1987.
12. Weinberg RA. Finding the anti-oncogene. *Scientific American* 259:44–57, 1988.
13. Geiser AG, Der CJ, Marshall CJ, Stanbridge EJ. Suppression of tumorigenicity with continued expression of the c-Ha-ras oncogene in EJ bladder carcinoma-human fibroblast hybrid cells. *Proc Natl Acad Sci USA* 83:5209–5213, 1986.
14. Boutwell RK. The function and mechanisms of promoters of carcinogenesis. *CRC Crit Rev Toxicol* 2:419–443, 1974.
15. Pitot HC, Goldsworthy T, Moran S. The natural history of carcinogenesis: Implications of experimental carcinogenesis in the genesis of human cancer. *J Supramol Struct Cell Biochem* 17:133–144, 1981.
16. Weinstein IB, Cattoni-Celli S, Kirschmeier P, Lambert M, Hsiao W, Backer J, Jeffrey A. Multistage carcinogenesis involves multiple genes and multiple mechanisms. *J Cell Physiol* 3:127–137, 1984.
17. Moolenaar WH, Defize LHK, Delaat SW. Ionic signalling by growth factor receptors. *J Exp Biol* 124:359–373, 1986.
18. Ames BN, Durston WE, Yamasaki E, Lee FD. Carcinogens are mutagens: A simple test system combining liver homogenates for activation and bacteria for detection. *Proc Natl Acad Sci USA* 70:2281–2285, 1973.
19. Till JE. Stem cells in differentiation and neoplasia. *J Cell Physiol* 1:3–11, 1982.
20. Kondo S. Carcinogenesis in relation to the stem-cell mutation hypothesis. *Differentiation* 24:1–8, 1983.
21. Markert C. Neoplasia: A disease of cell differentiation. *Cancer Res* 28:1908–1914, 1968.
22. Pierce GB. Neoplasms, differentiation and mutations. *Am J Pathol* 77:103–118, 1980.
23. Potter VR. Phenotypic diversity in experimental hepatomas: The concept of partially blocked ontogeny. *Br J Cancer* 38:1–23, 1978.
24. Nowell PC. Clonal evolution of tumor cell populations. *Science* 194:23–28, 1976.
25. Greaves MF. Differentiation-linked leukemogenesis in lymphocytes. *Science* 234: 697–704, 1986.
26. Buick RN, Pollak MN. Perspectives on clonogenic tumor cells, stem cells and oncogenes. *Cancer Res* 44:4909–4918, 1984.
27. Trosko JE, Chang CC. Stem cell theory of carcinogenesis. *Toxicol Lett*, in press.
28. Chang CC. Contact insensitivity of a subpopulation of normal human fetal kidney epithelial cells and of human carcinoma cell lines. *Cancer Res* 47:1634–1645, 1987.
29. Reid LM, Abreu SL, Montgomery K. Extracellular matrix and hormonal regulation of synthesis and abundance of messenger RNA's in cultured liver cells. In Arias IM, Jakoby WB, Popper H, Schachter D, Shafritz DA (eds): *The Liver: Biology and*

Pathobiology. New York: Raven Press, 1988, pp 717–737.
30. Tsao MS, Smith JD, Nelson KG, Grisham JW. A diploid epithelial cell line from normal adult rat liver with phenotypic properties of oval cells. *Exp Cell Res* 154:38–52, 1984.
31. Braun L, Goyette M, Yaswen P, Thompson NL, Fausto N. Growth in culture and tumorigenicity after transfection with the ras oncogene of liver epithelial cells from carcinogen-treated rats. *Cancer Res* 47:4116–4124, 1987.
32. Germain L, Noel M, Gourdeau H, Marceau N. Promotion of growth and differentiation of rat ductular oval cells in primary culture. *Cancer Res* 48:368–378, 1988.
33. Evarts RP, Nagy P, Marsden E, Thorgeirsson SS. A precursor-product relationship exists between oval cells and hepatocytes in rat liver. *Carcinogenesis* 8:1737–1740, 1987.
34. Evarts RP, Nagy P, Marsden E, Thorgeirsson SS. In situ hybridization studies on expression of albumin and α-fetoprotein during the early stage of neoplastic transformation in rat liver. *Cancer Res* 47:5469–5475, 1987.
35. Germain I, Blouin MJ, Marceau N. Biliary epithelial and hepatocytic cell lineage relationships in embryonic rat liver as determined by the differentiated expression of cytokeratins, α-fetoprotein, albumin and cell surface-exposed components. *Cancer Res* 48:4909–4918, 1988.
36. Yancey SB, Easter D, Revel JP. Cytological changes in gap junctions during liver regeneration. *J Ultrastruct Res* 67:229–242, 1979.
37. Jone C, Trosko JE, Chang CC. Characterization of a rat liver epithelial cell line to detect inhibitors of metabolic cooperation. *In Vitro Cell Dev Biol* 23:214–220, 1987.
38. Spray DC, Bennett MVL. Physiology and pharmacology of gap junctions. *Annu Rev Physiol* 47:281–303, 1985.
39. Trosko JE, Chang CC. Nongenotoxic mechanisms in carcinogenesis: Role of inhibited intercellular communication. In Hart RW (ed): *Banbury Report 31. Carcinogen Risk Assessment: New Directions in the Qualitative and Quantitative Aspects.* Cold Spring Harbor, NY: Cold Spring Harbor Laboratory, 1988, pp 139–170.
40. Traub O, Druge P, Willecke K. Degradation and resynthesis of gap junction protein in plasma membranes of regenerating liver after partial hepatectomy or cholestasis. *Proc Natl Acad Sci USA* 80:755–759, 1983.
41. Kanno Y, Sasaki Y, Shiba Y, Yoshida-Noro C, Takeichi M. Monoclonal antibody ECCD-1 inhibits intercellular communication in teratocarcinoma PCC3 cells. *Exp Cell Res* 152:270–274, 1984.
42. Evans WH. Gap junctions: Towards a molecular structure. *Bioessays* 8:3–10, 1988.
43. Iverson OH. The chalones. In Baserga R (ed): *Handbook of Experimental Pharmacology*, Vol. 57. Berlin: Springer-Verlag, 1981, pp 491–550.
44. Potter VR. Cancer as a problem in intercellular communication: Regulation of growth-inhibiting factors. In Cohn WE (ed): *Nucleic Acid Research and Molecular Biology*, Vol. 29. New York: Academic Press, 1983, pp 161–173.
45. Sachs L. The molecular control of blood cell development. *Science* 238:1374–1379, 1987.
46. Morstyn G, Burgess AW. Hemopoietic growth factors: A review. *Cancer Res* 48: 5624–5637, 1988.
47. deRooij DG, Lok D, Weeuk D. Feedback regulation of the proliferation of the undifferentiated spermatogonia in the Chinese hamster by the differentiation spermatogonia. *Cell Tissue Kinet* 18:71–81, 1985.
48. Kohase M, Henriksen-Destefano D, Sehgal PB, Vilcek J. Dexamethasone inhibits

feedback regulation of the mitogenic activity of tumor necrosis factor, interleukin-1, and epidermal growth factor in human fibroblasts. *J Cell Physiol* 132:271–278, 1987.
49. Keski-Oja J, Moses HL. Growth inhibitory polypeptides in the regulation of cell proliferation. *Med Biol* 65:13–20, 1987.
50. Rennick D, Yang G, Muller-Sieburg C, Smith C, Arai N, Takabe Y, Gemmell L. Interleukin 4 (B-cell stimulatory factor) can enhance or antagonize the factor-dependent growth of hemopoietic progenitor cells. *Proc Natl Acad Sci USA* 84:6889–6893, 1987.
51. Kimchi A, Wang XF, Weinberg RA, Cheifetz S, Massague J. Absence of TGF-β receptors and growth inhibitory responses in retinoblastoma cells. *Science* 240:196–198, 1988.
52. Del Rizzo DF, Eskinazi D, Axelrad AA. Negative regulation of DNA synthesis in early erythropoietic progenitor cells (BFU-E) by a protein purified from medium of C57BL/6 mouse marrow cells. *Proc Natl Acad Sci USA* 85:4320–4324, 1988.
53. Moses HL, Coffey RJ, Leof EB, Lyons RM, Keskroja J. Transforming growth factor-β regulation of cell proliferation. *J Cell Physiol* 5:1–7, 1987.
54. Sporn MB, Roberts AB. Peptide growth factors are multifunctional. *Nature* 332: 217–218, 1988.
55. McMahon JB, Richards WL, del Campo AA, Song MKH, Thorgeirsson SS. Differential effects of transforming growth factor-β on proliferation of normal and malignant rat liver epithelial cells in culture. *Cancer Res* 46:4665–4671, 1986.
56. Liu C, Tsao MS, Grisham JW. Transforming growth factors produced by normal and neoplastically transformed rat liver epithelial cells in culture. *Cancer Res* 48: 850–855, 1988.
57. Madhukar BV, Oh SY, Chang CC, Wade M, Trosko JE. Altered regulation of intercellular communication by epidermal growth factor, transforming growth factor-β and peptide hormones in normal human keratinocytes. *Carcinogenesis* 10:13–20, 1989.
58. Loewenstein WR. Junctional intercellular communication and the control of growth. *Biochim Biophys Acta* 605:1–65, 1979.
59. Schultz RM. Roles of cell to cell communication in development. *Biol Reprod* 32:27–42, 1985.
60. Neyton J, Trautmann A. Physiological modulation of gap junction permeability. *J Exp Biol* 124:93–114, 1986.
61. Pitts JD, Finbow ME. The gap junction. *J Cell Sci* 4:239–266, 1986.
62. Sheridan JD. Cell communication and growth. In DeMello WC (ed): *Cell to Cell Communication*. New York: Plenum Press, 1987, pp 187–222.
63. Hertzberg EL, Johnson RG (eds): *Modern Cell Biology*, Vol. 7. *Gap Junctions*. New York: Alan R. Liss, 1988.
64. Trosko JE, Chang CC. Adaptive and nonadaptive consequences of chemical inhibition of intercellular communication. *Pharmacol Rev* 36(Suppl):137–144, 1984.
65. Kanno Y. Modulation of cell communication and carcinogenesis. *Jpn J Physiol* 35: 693–707, 1985.
66. Larsen WJ, Skowron-Lamaeth C, Carron CP. Gap junction modulation: Possible role in tumor cell behavior. In Milman HA, Elmore E (eds): *Biochemical Mechanisms and Regulation of Intercellular Communication*. New Jersey: Princeton Scientific, 1987, pp 151–180.
67. Finbow ME, Yancey SB. The role of intercellular junctions. In Clemens MJ (ed):

Biochemistry of Cellular Regulation, Vol. 4. Boca Raton, FL: CRC Press, 1981, pp 215–249.
68. Weinstein RS, Pauli BU. Cell junctions and the biological behaviour of cancer. In Bock G, Clark S (eds): *Junctional Complexes of Epithelial Cells*. Chichester: John Wiley & Sons, 1987, pp 240–260.
69. Mesnil M, Yamasaki H. Selective gap-junctional communication capacity of transformed and non-transformed rat liver epithelial cell lines. *Carcinogenesis* 9:1499–1502, 1988.
70. Yotti LP, Chang CC, Trosko JE. Elimination of metabolic cooperation in Chinese hamster cells by a tumor promoter. *Science* 206:1089–1091, 1979.
71. Murray AW, Fitzgerald DJ. Tumor promoters inhibit metabolic cooperation in cocultures of epidermal and 3T3 cells. *Biochem Biophys Res Commun* 91:395–401, 1979.
72. Kalimi GH, Sirsat SM. The relevance of gap junctions to stage 1 tumor promotion in mouse epidermis. *Carcinogenesis* 5:1671–1677, 1984.
73. Pound AW, McGuire LJ. Repeated partial hepatectomy as a promoting stimulus for carcinogenic response of liver to nitrosamines in rats. *Br J Cancer* 37:585–594, 1978.
74. Beer DG, Neveu MJ, Paul DL, Rapp UR, Pitot HC. Expression of the c-raf protooncogene, γ-glutamyltranspeptidase, and gap junction protein in rat liver neoplasms. *Cancer Res* 48:1610–1617, 1988.
75. Janssen-Timmen U, Traub O, Dermietzel R, Rabes HM, Willecke K. Reduced number of gap junctions in rat hepatocarcinomas detected by monoclonal antibody. *Carcinogenesis* 7:1475–1482, 1986.
76. Sugie S, Mori H, Takahashi M. Effect of in vivo exposure to the liver tumor promoters phenobarbital or DDT on the gap junctions of rat hepatocytes: A quantitative freeze fracture analysis. *Carcinogenesis* 8:45–51, 1987.
77. Nicolson CL, Dulksi KM, Trosko JE. Loss of intercellular junctional communication correlates with metastatic potential in mammary adenocarcinoma cells. *Proc Natl Acad Sci USA* 85:473–476, 1988.
78. Hamada J, Takeichi N, Kobayashi H. Metastatic capacity and intercellular communication between normal cells and metastatic cell clones derived from a rat mammary carcinoma. *Cancer Res* 48:5129–5132, 1988.
79. Bishop JM. Viral oncogenes. *Cell* 42:23–38, 1985.
80. Weinberg RA. The action of oncogenes in the cytoplasm and nucleus. *Science* 230: 770–776, 1985.
81. Trosko JE, Jone C, Chang CC. Oncogenes, inhibited intercellular communication and tumor promotion. In Fujiki H, Hecker E, Moore RE, Sugimura T, Weinstein IB (eds): *Cellular Interactions by Environmental Tumor Promoters*. Tokyo: Japan Scientific Societies Press, 1984, pp 101–113.
82 Atkinson MM, Sheridan J. Decreased junctional permeability in cells transfected by 3 different viral oncogenes: A quantitative video analysis. *J Cell Biol* 99:401, 1984.
83. Atkinson MM, Anderson SK, Sheridan JD. Modification of gap junctions in cells transferred by a temperature-sensitive mutant of Rous sarcoma virus. *J Membr Biol* 9:53–64, 1986.
84. Azarnia R, Loewenstein WR. Intercellular communication and the control of growth. X. Alteration of junctional permeability by the src gene. A study with temperature-sensitive mutant Rous sarcoma virus. *J Membr Biol* 82:191–205, 1984.

85. Azarnia R, Loewenstein WR. Intercellular communication and the control of growth. XII. Alteration of junctional permeability by simian virus 40: Roles of the large and small T antigens. *J Membr Biol* 82:213–220, 1984.
86. Azarnia R, Loewenstein WR. Polyomavirus middle T antigen down regulates junctional cell to cell communication. *Mol Cell Biol* 7:946–950, 1987.
87. Azarnia R, Reddy S, Kimiecki TE, Shalloway D, Loewenstein WR. The cellular src gene product regulates junctional cell to cell communication. *Science* 239:398–400, 1988.
88. Chang CC, Trosko JE, Kung HJ, Bombick D, Matsumura F. Potential role of the src gene product in inhibition of gap junctional communication in NIH 3T3 cells. *Proc Natl Acad Sci USA* 82:5360–5384, 1985.
89. El-Fouly MH, Warren ST, Trosko JE, Chang CC. Inhibition of gap junction-mediated intercellular communication in cells transfected with the human H-ras oncogene (Abstract). *Am J Hum Genet* 39:30, 1986.
90. Bignami M, Rosa S, Falcone G, Tuto F, Katoh F, Yamasaki H. Specific viral oncogenes cause differentiated effects on cell to cell communication, relevant to the suppression of the transformed phenotype by normal cells. *Molecular Carcinogenesis* 1:67–75, 1988.
91. El-Fouly MH, Trosko JE, Chang CC. Phenotypic transformation and inhibition of gap junctional intercellular communication in epithelial and mesenchymal cells by the neu oncogene. Conference on Toxicologic Implications of Altered Gap Junctional Intercellular Communication, East Lansing, MI, September 28–30, 1988, p 17.
92. deFeijter E, Weghorst CM, Ruch R, Klaunig J, Chang CC, Madhukar BV, El-Fouly M, Trosko JE. Reduced gap junctional intercellular communication and tumorigenicity of rat liver epithelial cells. Conference on Toxicologic Implications of Altered Gap Junctional Intercellular Communication, East Lansing, MI, September 28–30, 1988, p 25.
93. Trosko JE, Chang CC. Chemical, oncogene and growth factor inhibition of gap junctional intercellular communication: An integrative hypothesis of carcinogenesis. *CRC Crit Rev Oncogenesis*, in press.
94. Ehrenberg L, Brookes P, Druckery H, Lagerlof, Litwin J, Williams G. The relation of cancer induction and genetic damage. *Ambio Special Report* 3:15–16, 1973.
95. Miller JA, Miller EC. Ultimate chemical carcinogens as reactive mutagenic electrophils. In Hiatt HH, Watson JD, Winsten JA (eds): *Origins of Human Cancer*, Vol. A. Cold Spring Harbor, NY: Cold Spring Harbor Laboratory, 1977, pp 605–628.
96. Lieberman NW, Baney RN, Lee RE, Sell S, Farber E. Studies on DNA repair in human lymphocytes treated with ultimate carcinogens and alkylating agents. *Cancer Res* 3:1292–1306, 1971.
97. Farber E, Sarma DSR. Chemical carcinogenesis: The liver as a model. *Pathol Immunopathol Res* 5:1–28, 1986.
98. Huitfeldt HS, Hunt JM, Pitot HC, Poirier MC. Lack of acetylaminofluorene-DNA adduct formation in enzyme-altered foci of rat liver. *Carcinogenesis* 9:647–652, 1988.
99. Kroese ED, vande Poll MLM, Mulder GJ, Meerman JHN. The role of N-sulfation in the N-hydroxy-2-acetylaminofluorene-mediated outgrowth of diethylnitrosamine-initiated hepatocytes to γ-glutamyltranspeptidase-positive foci in male rat liver. *Carcinogenesis* 9:1953–1958, 1988.
100. Kuchlbauer J, Romen W, Neumann HG. Syncarcinogenic effects on the initiation

of rat liver tumors by trans-4-acetylaminostilbene and 2-acetylaminofluorene. *Carcinogenesis* 6:1337–1342, 1985.

101. Schut HAJ, Daniel FB, Schenck KM, Loeb TR, Stoner GD. Metabolism and DNA adduct formation of 2-acetylaminofluorene by bladder explants from human, dog, monkey, hamster and rat. *Carcinogenesis* 5:1287–1292, 1984.
102. Reynolds SH, Stowers SJ, Patterson RM, Maronpot RR, Aaronson SA, Anderson MW. Activated oncogenes in B6C3F1 mouse liver tumors: Implications for risk assessment. *Science* 237:1309–1316, 1987.
103. Brookes P, Cooper CS, Ellis MV, Warren W, Gardner E, Summerhayes IC. Activated Ki-ras genes in bladder epithelial cell lines transformed by treatment of primary mouse bladder explant cultures with 7,12-dimethylbenz[a]anthracene. *Molecular Carcinogenesis* 1:82–88, 1988.
104. Land H, Parada LF, Weinberg RA. Tumorigenic conversion of primary embryo fibroblasts requires at least two cooperating oncogenes. *Nature* 304:596–602, 1983.
105. Newbold RF, Overell RW. Fibroblast immortality is a prerequisite for transformation by EJ c-Ha-ras oncogene. *Nature* 304:648–651, 1983.
106. Braun L, Goyette M, Yaswen P, Thompson NL, Fausto N. Growth in culture and tumorigenicity after transfection with the ras oncogene of liver epithelial cells from carcinogen-treated rats. *Cancer Res* 47:4116–4124, 1987.
107. Huber BE, Cordingley MG. Expression and phenotypic alterations by an inducible transforming ras oncogene introduced into rat liver epithelial cells. *Oncogene* 3:245–256, 1988.
108. Seyama T, Godwin AK, DiPietro M, Winokur TS, Lebovitz RM, Lieberman MW. In vitro and in vivo regulation of liver epithelial cells carrying a metallothionein-ras T24 fusion gene. *Molecular Carcinogenesis* 1:89–95, 1988.
109. Burt RK, Garfield S, Johnson K, Thorgeirsson SS. Transformation of rat liver epithelial cells with v-H-ras or v-raf causes expression of MDR-1, glutathione-S-transferase-P and increased resistance to cytotoxic chemicals. *Carcinogenesis* 9:2329–2332, 1988.
110. Nagy P, Evarts RP, Marsden E, Roach J, Thorgeirsson SS. Cellular distribution of c-myc transcripts during chemical hepatocarcinogenesis in rats. *Cancer Res* 48:5522–5527, 1988.
111. Kelly K, Siebenlist U. The regulation and expression of c-myc in normal and malignant cells. *Annu Rev Immunol* 4:317–338, 1986.
112. Dotto GP, Parada LF, Weinberg RA. Specific growth response of ras-transformed embryo fibroblasts to tumor promoters. *Nature* 318:472–475, 1985.
113. Bell JC, Jardine K, McBurney MW. Lineage-specific transformation after differentiation of multipotential murine stem cells containing a human oncogene. *Mol Cell Biol* 6:617–625, 1986.
114 Wyllie AH. Apoptosis: Cell death in tissue regulation. *J Pathol* 153:313–316, 1987.
115. Bursch W, Lauer B, Timmermann-Trosiener I, Barthel G, Schuppler J, Schulte-Hermann R. Controlled death (apoptosis) of normal and putative preneoplastic cells in rat liver following withdrawal of tumor promoters. *Carcinogenesis* 5:453–458, 1984.
116. Wiener EC, Loewenstein WR. Phosphorylation of lens membranes with a cyclic AMP-dependent protein kinase purified from the bovine lens. *Nature* 305:433–435, 1983.
117. Mehta PP, Bertram JS, Loewenstein WR. Growth inhibition of transformed cells

correlates with their junctional communication with normal cells. *Cell* 44:187–196, 1986.
118. Trosko JE. Mechanisms of tumor promotion: Possible role of inhibited intercellular communication. *Eur J Cancer Clin Oncol* 23:599–601, 1987.
119. Trosko JE, Chang CC, Madhukar BV, Oh SY, Bomick D, El-Fouly MH. Modulation of gap junction intercellular communication by tumor promoting chemicals, oncogenes, and growth factors during carcinogenesis. In Hertzberg EL, Johnson RG (eds): *Gap Junctions,* Vol. 7. New York: Alan R. Liss, 1988, pp 435–448.
120. Trosko JE, Chang CC. The role of inhibited intercellular communication in carcinogenesis: Implications for risk assessment from exposure to chemicals. In Gehrs CJ (ed): *Bioindicators: Exposure and Effect.* Chelsea, MI: Lewis Publishers, in press.

Mouse Liver Carcinogenesis: Mechanisms
and Species Comparisons, pages 277–291

Gap-Junctional Intercellular Communication and Murine Hepatic Carcinogenesis

James E. Klaunig,[*†] Randall J. Ruch,[*] James A. Hampton,[*]
Christopher M. Weghorst,[*] and James A. Hartnett[*]*
*[*]Toxicology Program, Department of Pathology and [†]Department of Pharmacology,
Medical College of Ohio, Toledo, OH 43614*

Chemical carcinogenesis is a multistep process that has been mechanistically subdivided into three major stages: tumor initiation, promotion, and progression. Initiation of a cell results in a mutational change in cellular DNA, usually through direct interaction of a chemical (or its metabolites) with the DNA. Tumor promotion is produced when agents (promoters) increase the tumorigenic response of an initiating carcinogen when administered after the initiator (1). Tumor progression involves the development of malignant neoplasms from benign lesions. In general, tumor promoters appear to facilitate the expansion of the initiated cell population through induction of cell proliferation. Subsequent induced or spontaneous mutations may be required for further succession of preneoplastic, initiated cells to become a neoplasm. The selective induction of cell division in initiated cells by tumor promoters may make the initiated cells more "prone" to these mutations (2). While DNA damage (initiation) by chemical carcinogens has been extensively defined, how tumor promoters work is less clear.

Cells in an organism can communicate in many ways, including neural innervation, release of soluble factors that affect cells at a distance (e.g., hormones), or through direct contact. The latter category includes intercellular communication through gap junctions. Gap junctions consist of aggregates of proteinaceous particles that span the membranes of opposing cells, forming a junction. Each particle or connexon is tightly bound to a similar particle from the opposing cell. Within the paired connexons, an aqueous channel of approximately 1–2 nm connects the cytoplasm of the two cells and can be seen on electron microscopy (3). Each connexon consists of a hexamer of equivalent proteins of 27 kDa–32 kDa (in liver) that form a rosette around the central hydrophilic pore (4). These pores or channels permit the passive diffusion of a variety of low molecular weight materials (less than 1 kDa) between cells. The functions of gap-junction intercellular communication appear to include: the maintenance of cellular homeostasis through exchange of ions, nutrients, and precursors; the coordination and enhancement of tissue response to external stimuli (e.g., hormones) through passage of cyclic nucleotides or other second messengers from stimulated to nonstimulated cells; the control and coordination of embryonic development through the provision of pathways for electrical and chemical gradients; the establishment of electrical conduits for cardiocyte and neuron electrical cell coupling; and the control of cellular proliferation through the provision of pathways for mitogenic inhibitory or stimulatory molecules (e.g., cAMP) to pass between cells (5).

Loewenstein (5,6) has hypothesized that intercellular communication functions in the control of cellular growth and that the loss of intercellular communication aids the development of cancer. While some indirect evidence supports his contention, it has not been directly demonstrated that intercellular communication serves as a primary control of cellular growth. The evidence in favor of Loewenstein's hypothesis includes rapidly dividing cells such as hepatocytes in regenerating, partially hepatectomized livers (7); epidermal cells along a wound border (8); and that most transformed cells have both less capacity for intercellular communication and fewer and smaller gap junctions than normal cells (5). Also, the degree of loss of intercellular communication correlates with the degree of transformation. Normal colonic mucosal cells readily pass microinjected fluorescent dye between them, premalignant mucosal cells have an intermediate capacity, while malignant cells failed to pass any dye (9). Also, when tumor cells in culture or tumors in vivo are treated with agents such as cAMP analogues that can induce their differentiation and/or inhibit their growth, intercellular communication is established, and gap junctions appear (10). In addition, regulatory molecules for cellular replication and differentiation, such as cAMP and cGMP, can pass through gap junctions (5). In normal cells, intercellular communication may serve as a conduit for inhibitors and/or effectors of cellular replication and differentiation, such as ions and cyclic nucleotides. Intercellular communication would permit the dissipation of these agents between cells. If intercellular communication were blocked by a tumor promoter, a cell would be isolated from its neighbors and the inhibitor/effector content of the cell might be altered to stimulate cell replication or alter differentiation. Initiated cells might be more susceptible to the inhibition of intercellular communication by xenobiotics such as cytotoxins or tumor promoters, less capable of establishing intercellular communication with their uninitiated neighbors, and/or might be more or less responsive to the growth and differentiation signals passed by intercellular communication from adjacent cells.

In 1979 two laboratories working independently found that the tumor promoter 12-*O*-tetradecanoylphorbol-13-acetate (TPA) inhibited intercellular communication in cultured cells (11,12). This initial finding has led to a number of investigations of the effects of tumor promoters on intercellular communication. Several cell culture systems have been utilized, including liver cell lines and primary cultured hepatocytes, to define the effects of tumor promoters on hepatic intercellular communication (13–20). A number of hepatic tumor promoters have been shown to inhibit intercellular communication in cultured hepatocytes. There appears to be a good correlation between the ability of a chemical compound to act as a hepatic tumor promoter and its ability to inhibit intercellular communication in cultured hepatocytes. Tumor promoters may prevent the exchange of growth regulatory molecules between hepatocytes by the inhibition of intercellular communication. This inhibition of growth control may therefore allow the hepatocytes to divide. If the cells in which gap-junctional intercellular communication is inhibited are preneoplastic, then this selective inhibition of growth control may allow these cells to proliferate and clonally expand. This increased division would allow preneoplastic cells to acquire additional mutations and eventually free these cells of growth control mechanisms, allowing their progression into neoplasia.

Several workers have also found a modification of gap-junction expression in neoplastic hepatic lesions. Beer et al. (21) have shown a distinct decrease in the immunohistochemical staining of gap-junction protein in rat hepatic tumors coupled with a concomitant decrease in the amount of gap-junction mRNA compared with normal rat liver. Janssen-Timmen et al. have similarly demonstrated a decrease in gap-junction number by immunohistochemical methods in rat liver tumors (22).

Materials and Methods

Animals

Four-week-old male $B6C3F_1$ mice were purchased from Charles River Breeding Laboratory (Wilmington, MA). Mice were kept four to a cage in polycarbonate cages containing corncob bedding and were given food (Purina rodent chow) and water ad libitum. Hepatic tumors were produced in half the mice by exposing them to diethylnitrosamine (DEN) in their drinking water (25 mg/l) for 4 weeks starting at 4 weeks of age, followed by treatment with deionized drinking water for an additional 20 weeks. Control mice (no DEN treatment) were given deionized water only.

Liver Cell Culture

Mouse hepatocytes were isolated using the two-stage portal perfusion technique previously described (23,24). Isolated cells were plated at a density of 1×10^6 viable cells per culture dish in L-15 culture medium supplemented with 10% fetal bovine serum (FBS), gentamicin (25 μg/ml), glucose (1 mg/ml), and dexamethasone (1 μM).

Isolation and Culture of Precision-Cut Liver Slices

A major criticism in the use of hepatocyte culture systems in toxicologic/carcinogenesis studies is that the in vitro cell culture systems do not fully reflect the in vivo liver. In vivo liver has a number of other cell types (Kupffer cells, endothelial cells, Ito cells, bile ductal cells, etc.) that may play a role in the carcinogenesis/toxicity process. Also, the pharmacokinetics of carcinogenic and toxic compounds in vivo may be completely different from that seen in vitro.

These factors may be important both in the cellular mechanisms of action of the toxin and in the dose of compound that will produce an effect. It is conceivable that cultured isolated hepatocytes could respond quite differently to a toxic compound than does the in vivo liver. Recently, Brendal and coworkers (25) developed a tissue slicer for thin (200–300 μm in thickness) "precision-cut liver slices" that maintain their viability in culture. The use of these liver slices provides an important link between in vivo studies and in vitro isolated cultured hepatocytes because liver slices contain all the normal liver cells in their proper in vivo architectural arrangement. Also, since slices can be obtained relatively easily and quickly (a few minutes), animals can be treated with toxic compounds in vivo, the liver can be removed and sliced, and mechanistic studies can be performed in vitro (without the several hours' delay required to produce cultured hepatocytes). In addition, in vitro comparisons between species, including human, can be readily made. Using this slicer, we have developed methods for the maintenance and assessment of the viability of these slices in culture. We have examined the effects of selected hepatotoxins on cell viability in short-term cultures of rodent, monkey, and human precision liver slices.

In the current studies, liver for slicing was isolated from mice immediately after euthanasia. The entire liver was removed and placed in cold L-15 medium supplemented with gentamicin (25 μg/ml), glucose (1 mg/ml), and dexamethasone (1 μM). The liver was then cut into 1-cm cubes with a razor blade placed in the slicer and sliced. Sections 200–300 μm thick were obtained. These slices were carefully collected and placed in a 100 × 20 mm culture dish containing L-15 medium. Slices were washed three times with cold medium to remove blood and dead cells. The slices were then placed, three per dish, into a 60-mm culture dish. Three milliliters of L-15 medium were placed in each dish. Dishes were maintained at 36.5°C in a humidified, 100% air incubator in a rocking organ culture system. Viability and liver cell function were assessed by examining hematoxylin and eosin–stained formalin-fixed paraffin-embedded sections of the slices, lactate dehydrogenease (LDH) leakage (15,16), and protein synthesis (26).

Intercellular Communication in Cultured Mouse Hepatocytes and Liver Slices

Intercellular communication in cultured mouse liver slices and cells was evaluated by dye coupling. Cultured mouse hepatocytes (donor cells) were microinjected with Lucifer yellow CH dye in 0.1 M LiCl at 3 nA for 1 min as previously described (18). Five minutes after the donor cell dye injection, dye coupling from the injected donor cell to adjacent recipient cells was evaluated. Intercellular communication was expressed as the percentage of dye-coupled recipients of the total number of possible recipients (those cells in contact with the donor cell). Dye coupling was determined for each time period and each concentration of compound examined.

Dye coupling was also examined in slices of normal and preneoplastic/neoplastic mouse liver. Three DEN-treated mice and three untreated mice were killed at 28 weeks of age. Livers from these mice were sliced and cultured as described above. Hepatocytes within the slices were then evaluated for dye coupling. Donor cells were injected with Lucifer yellow iontophoretically for 60 sec. Dye was allowed to spread to adjacent recipient cells for an additional 4 min, after which the injected slice was either placed in phosphate-buffered formalin or fresh frozen in liquid nitrogen. Frozen slices were sectioned on a cryostat, and fixed slices were processed and embedded in JB-4 glycol methacrylate. Dye coupling of hepatocytes was evaluated in unstained JB-4 plastic sections by fluorescent microscopy. The donor cell and any dye-coupled recipient cells fluoresced and were readily apparent. Using our protocol, the donor cell fluoresces brightly and is easily identifiable. These cells were photographed. The sections were then stained with hematoxylin, and hepatic morphology was demonstrable. Even after the staining with hematoxylin, the fluorescent cells were apparent. Therefore, dye coupling in either preneoplastic, neoplastic, or normal cells was easily qualitatively assessed.

Experimental Design

Dye Coupling in Cultured Mouse Hepatocytes. Initial studies examined the formation and maintenance of gap junction–mediated intercellular communication (by dye coupling) in freshly isolated and cultured mouse hepatocytes. Freshly isolated B6C3F$_1$ mouse liver cells were plated onto 60-mm culture dishes containing 5 ml of L-15 medium, and dye coupling was assessed in the cultured cells after 2, 4, 6, 8, 12, 24, 48, 72, and 96 h of culture.

Influence of Culture Medium and Medium Supplements on Intercellular Communication in Cultured Mouse Hepatocytes. Previous studies by our group have noted an effect of the type of culture medium and culture medium supplements on the ability of cultured rat hepatocytes to establish and maintain intercellular communication in primary culture. In this study, we examined these parameters in cultured mouse hepatocytes. Freshly isolated, individual mouse hepatocytes were plated on 60-mm culture dishes in either L-15 medium or RPMI-1640 culture medium supplemented with 10% FBS, gentamicin (25 μg/ml), and glucose (1 mg/ml), with or without dexamethasone (1 μM). Dye coupling was evaluated in the cultured cells after 6, 24, and 48 h of culture.

Effect of Hepatic Promoters on Murine Hepatocyte Dye Coupling. The effect of several nongenotoxic hepatic carcinogens on dye coupling in cultured mouse hepatocytes was examined. Freshly isolated mouse hepatocytes were plated at 1×10^6 viable cells per culture dish in 5 ml of supplemented L-15 medium. After 24 h of culture, mouse hepatocytes were treated with one of three sublethal concentrations of each compound for 4 h. Dye coupling was assessed after the 4-h treatment. Untreated control cultures and solvent-treated control cultures were also examined for dye coupling.

Effects of Hepatic Tumor Promoters on Untreated and Tumor-Containing Mouse Liver Slices. Liver slices that contained tumors and those that did not were treated with either hexachlorocyclohexane (lindane) (25 µg/ml in dimethyl sulfoxide), dimethyl sulfoxide only, or nothing. After 4 h of culture, untreated, solvent-treated, and lindane-treated slices were evaluated for dye coupling with Lucifer yellow dye. LDH leakage, protein synthesis, and histologic evaluations were performed on similarly treated liver slices from the same mouse.

Immunohistochemical Demonstration of Gap-Junction Protein in Normal and Tumor-Containing Mouse Liver Slices. In dye-injected liver slices that were fresh frozen, polyclonal antibody to gap-junction protein (32 kDa) (provided by Dr. David Paul) (4), was used to immunohistochemically demonstrate the presence of gap junctions in normal, preneoplastic, and neoplastic livers. Since these slices were also injected with dye, it was possible to correlate the amount of dye coupling with the amount of gap-junction protein present in the cells. Fresh-frozen sections were cut from the slices on a cryostat. Sections were treated with cold acetone for 5 min, air dried, hydrated, and stained for the presence of anti-connexon 32 by immunofluoresence.

Results

The formation and maintenance of intercellular communication in cultured mouse hepatocytes is shown in Figure 1. Freshly isolated, individual hepatocytes displayed measurable dye coupling soon after (within 30 min) placement in culture. Dye coupling progressively increased in the cultured hepatocytes, with a maximum coupling of 93% at 24 h postplating. Dye coupling displayed a slight, gradual decline during the remaining sampling times, reaching 82.6% at 96 h.

The type of medium used to culture the hepatocytes had a pronounced influence on the ability of the murine hepatocytes to establish and maintain dye coupling in primary culture (Fig. 2). Cells plated in L-15 medium rapidly formed and maintained dye coupling in culture. In contrast, cells grown in RPMI-1640 medium showed a decrease in dye coupling after 24–48 h in culture. The ability of the cultured mouse hepatocytes to maintain dye coupling was dependent on the presence of dexamethasone in the culture medium (Fig. 2). Omission of dexamethasone from the culture medium resulted in a significant decrease in dye coupling. The decrease observed with RPMI-1640 after 24 h of culture was prevented by the addition of 5 mM L-alanine and 5 mM sodium pyruvate to the culture medium (Table 1).

The effect of selected hepatic tumor promoters on cell-to-cell communication (as measured by dye coupling) in male B6C3F$_1$ mouse hepatocytes is shown in Table 2. All the compounds examined displayed a dose-dependent inhibition of intercellular communication after 4 h of treatment.

In mouse liver slices, dye coupling was readily apparent in untreated normal hepatocytes (Table 3). Extensive transfer of dye from the donor cell to adjacent primary, secondary, and tertiary cells was observed. Significantly less dye coupling was seen in neoplastic lesions (adenomas) (Table 3) than in the normal

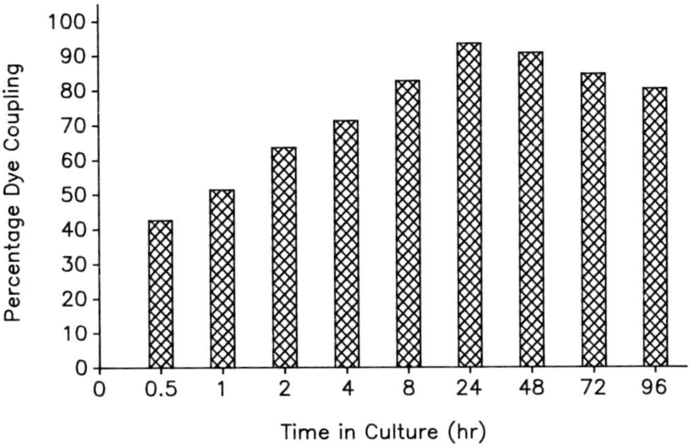

Figure 1. Dye coupling levels in nontreated mouse hepatocytes after various culture durations.

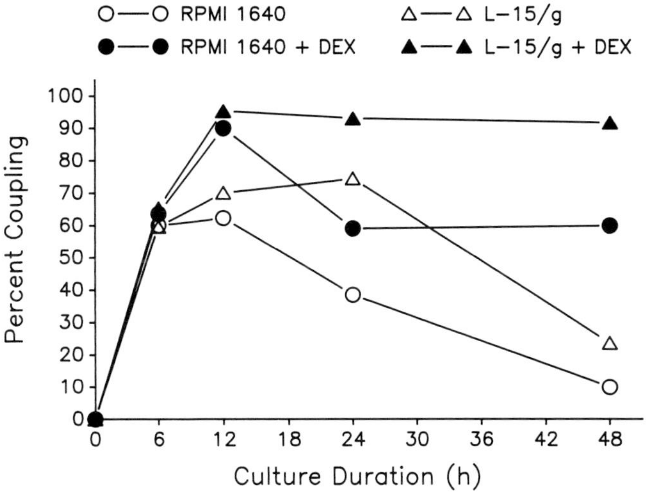

Figure 2. Dependence of mouse hepatocyte dye coupling on culture medium and dexamethasone (DEX; 1 μM) over 48 h culture duration.

liver. The spread of dye in the adenomas was approximately 40%–50% of that seen in the normal liver. In addition to the decrease in the number of cells that appeared dye coupled, the amount of dye that transferred and the rate of transfer in the neoplastic cells appeared to be less than that seen in the normal liver slices. Dye coupling in preneoplastic lesions displayed a heterogeneous pattern that differed between foci. In some foci, dye coupling appeared similar in extent and rate to that seen in the normal liver. In other foci, the pattern of dye coupling more closely resembled that seen in the neoplastic lesions, with

Table 1. Effects of Sodium Pyruvate and L-Alanine on Mouse Hepatocyte Intercellular Communication (Dye-Coupling) in RPMI-1640 Medium after 24 and 48 H in Culture

	Percent coupling	
Medium	24 h	48 h
RPMI only	56.5	20.0
RPMI + Pyr	85.1*	81.3*
RPMI + Ala	81.8*	80.7*
RPMI + Pyr + Ala	90.2*	85.7*
L-15/g + Dex	93.9*	85.7*

*$P < 0.05$ vs. RPMI-only group (Chi-squared test).
Abbreviations: RPMI, RPMI-1640 culture medium; Pyr, sodium pyruvate, 5 mM; Ala, L-alanine, 5 mM; DEX, dexamethasone, 1 μM.

a significant decrease in the number of dye-coupled cells and the apparent rate of dye transfer.

Treatment of normal liver slices for 4 h with lindane (25 μg/ml) resulted in a significant decrease in the amount and extent of dye coupling from that seen in the normal untreated liver (Table 3). Treatment of neoplastic and preneoplastic lesions with lindane resulted in a complete inhibition of dye coupling between the cells within the lesion. No heterogeneity of response to the inhibition of dye coupling induced by lindane was observed between foci; however, the sample of injected foci may not have been large enough to adequately define this feature.

The results of the immunofluoresent staining of the normal, preneoplastic, and neoplastic hepatocyte-containing liver slices are shown in Table 4. Normal hepatocytes displayed extensive immunofluorescent staining of the gap-junction protein antibody. This appeared as numerous punctate immunofluorescent plaques on the membranes of adjacent cells. Neoplastic lesions showed a significant decrease in the amount of immunofluorescent-stained material. Occasional immunofluorescent-stained plaques were apparent on the membranes between neoplastic cells within the neoplastic lesions, but for the most part the adenomas displayed significantly less staining than normal liver. Preneoplastic foci in the mouse liver slices exhibited a similar heterogeneity in immunofluorescent staining with the gap-junction antibody to that seen with the dye coupling (Table 4). Some of the foci showed a pattern of gap-junction antibody staining similar to that of normal liver, while other foci displayed a significant decrease in gap-junction staining, similar to that exhibited by the neoplastic lesions. There appeared to be a correlation between the observed heterogeneity in the preneoplastic foci for the immunofluoresent staining to the antibody to connexon 32 and the heterogeneity in the dye coupling with Lucifer yellow. Those foci with reduced dye coupling between their cells also exhibited decreased immunofluorescent staining.

Table 2. Inhibition of Intercellular Communication in Mouse Hepatocytes by Hepatic Tumor Promoters

Compound	Concentration	Percent communication
Untreated	—	94.7
DMSO	0.1%	95.8
Phenobarbital	20 μg/ml	72.6*
	100 μg/ml	67.4*
	500 μg/ml	59.4*
DDT	10 μg/ml	88.3
	20 μg/ml	63.4*
	30 μg/ml	34.5*
Lindane	15 μg/ml	89.3
	25 μg/ml	54.2*
	35 μg/ml	31.6*
MEHP	5 μM	91.3
	10 μM	76.0*
	50 μM	66.7*
TCE	10 μM	92.6
	50 μM	80.4*
	100 μM	72.3*
TCA	50 μM	94.3
	100 μM	72.4*
	500 μM	63.7*
Nafenopin	10 μM	92.6
	25 μM	68.7*
	50 μM	42.3*
Aroclor 1254	1 μM	83.4*
	10 μM	76.2*
	20 μM	69.1*

*$P < 0.05$ vs. untreated control group.
Abbreviations: DDT 1,1-*bis*(4-chlorophenyl)-2,2,2-trichloroethane; MEHP, monoethylhexylphthalate; TCE, trichloroethylene; TCA, trichloroacetic acid.

Table 3. Dye Coupling in Liver Slices from Mice with Tumors

	Dye coupling	
Liver status	Untreated	Lindane-treated
Normal tissue	++++	+
Foci	+ or +++	−
Adenomas	+	−

Table 4. Dye Coupling and Immunohistochemical Demonstration of Gap Junctions in Normal and Preneoplastic Mouse Liver

Liver tissue	Dye coupling	Gap-junction (32-kDa) protein
Normal	++++	++++
Foci	+ or +++	+/– or ++
Adenomas	+	+/–

Discussion

Intercellular communication was present (as measured by dye coupling) between cultured mouse hepatocytes during the first 4 days of culture. The percentage of hepatocytes that exhibited dye coupling was dependent upon both the type of culture medium used and the presence of culture medium supplements, especially the glucocorticoid dexamethasone. Dye coupling was also enhanced in RPMI-1640 medium by supplementation with sodium pyruvate and L-alanine. These observed differences in the extent of dye coupling exhibited under different culture medium conditions may be related to hepatocyte differentiation, since the addition of glucocorticoid hormones to hepatocyte culture medium has been shown to enhance the expression of differentiated functions in these cells (27). These supplements might help maintain the hepatocyte differentiated state and thus gap-junctional communication, decrease gap-junctional protein degradation, and/or increase gap-junctional protein synthesis.

The inhibition of intercellular communication in mouse hepatocytes was observed with several chemically unrelated tumor promoters. These data and previous studies by our group (15–20) suggest that the inhibition of cultured hepatocyte intercellular communication might be predictive of liver carcinogenicity by tumor promoters and be a way hepatic tumor promoters function in producing their carcinogenic response. This effect of hepatic tumor promoters and nongenotoxic carcinogens is not restricted to cultured cells but has been demonstrated by Sugie et al. (28) and Yamasaki and coworkers (29) in vivo. The greater effects of lindane on dye coupling between preneoplastic and neoplastic hepatocytes compared with normal cells suggest that preneoplastic and neoplastic cells are more sensitive to the hepatic promoter. The gap-junction protein antibody stained punctate regions of the plasma membrane of apposing cells. Normal liver showed extensive immunohistochemical staining, while hepatocytes from preneoplastic cells displayed only faint, infrequent staining. This correlates with the observations of other workers on the rat liver (21,22). Also, despite the decrease in demonstrable gap-junction protein by immunohistochemical staining with the 32-kDa antibody, measurable dye coupling was seen in the preneoplastic and neoplastic cells. This may be due to the fact that dye coupling can be maintained between hepatocytes, even with a significant decrease in the number of gap junctions. Only a small number of connexons may need to be functional for dye coupling to occur. Another possible explanation for this effect is that the gap-junction protein is present in the preneoplastic and neoplastic cells but is not immunologically identifiable as the liver 32-kDa protein. There are other forms of gap-junction connexon protein (43 kDa and 21 kDa) that are produced in adult tissue. The expression of these different forms may allow for the continued maintenance of dye coupling without the immunological demonstration of the 32-kDa connexon protein.

In the liver, not all preneoplastic foci advance to the neoplastic stage. The phenotypic and genotypic factors and characteristics that determine whether preneoplastic foci advance to neoplasia are important in understanding the mechanisms of the hepatic carcinogenesis process. Several workers have noted a heterogeneity among preneoplastic hepatic lesions for a number of histochemical markers (30). Our current studies suggest a heterogeneity among preneoplastic hepatic mouse lesions in their morphologic and functional expression of intercellular communication. Those preneoplastic lesions that contain decreased gap-junctional intercellular communication and/or a decrease in the number of gap junctions may be more susceptible to subsequent growth stimulation. This growth stimulation may occur through exogenous chemicals (i.e., tumor promoters) or endogenous factors (i.e., hormonal environment) that are capable of inhibiting gap-junction intercellular communication. Inhibition of intercellular communication in these preneoplastic cells effectively isolates them from the normal cellular growth control mechanisms and allows for the selective expansion of this preneoplastic cell population and eventual advancement to neoplasia.

There is strong evidence that the regulation of gap-junctional intercellular communication plays a role in the process of hepatocarcinogenesis. Our group and others previously demonstrated that a number of nongenotoxic carcinogens at sublethal concentrations inhibit gap-junctional intercellular communication in cultured rodent hepatocytes. This inhibition is both dose- and time-dependent and rapidly reversible upon removal of the promoting agent. Whether the inhibition of intercellular communication in vivo results in the isolation and preferential proliferation of preneoplastic cells remains to be resolved. Previous studies in the rat (21,22) and our work in the mouse demonstrate a decrease in the expression of gap-junction protein in preneoplastic and neoplastic liver lesions. These findings, again, suggest that decreased intercellular communication plays a role in the hepatocarcinogenesis process. The physiologic result of this decrease in intercellular communication in hepatocarcinogenesis remains to be resolved. Decreased intercellular communication may be essential to the carcinogenic process or may simply be an acquired, nonspecific functional change in preneoplastic and neoplastic lesions. The important role that gap-junctional intercellular communication plays in the normal growth, differentiation, and physiologic functioning of cells points to a similarly important role for the observed decrease in intercellular communication in hepatic neoplastic and preneoplastic lesions.

Open Forum

Dr. Robert LeBoeuf: What method did you use to quantitate the degree of inhibition, and was that a fluorescence imaging?

Dr. James Klaunig: With the cell studies, the quantitation is done in terms of plus or minus criteria. If there is dye transfer, then we consider it coupled. If there is no dye transfer, then it is noncoupled.

Dr. LeBoeuf: So when you show an 80% coupling, that means that 80% of the cells are coupling.

Dr. Klaunig: Yes.

Dr. Robert LeBoeuf: My second question is, Do you feel that inhibition of intercellular communication is sufficient for this clonal expansion, or must there be some direct mitogenic effect on the initiated cells by a promoter in addition to inhibition of intercellular communication initiation?

Dr. Klaunig: That is a good question. One question we always ask is, Is inhibition of intercellular communication in vivo related to the stimulation of DNA synthesis and cell division, or is it an event that occurs during and following DNA synthesis and cell division? There is some evidence to suggest that it plays a role in the cell division process in certain cell lines, but we have not shown that in our system.

Dr. Don Stevenson: While your correlation with in vivo data is very good, there are a few things that I think we should be cautious about. For instance, judging by the Haseman, Ashby, and Tennant papers, DDT was negative in the B6 mouse, but DDE was positive. Also you worked with male hepatocytes, and dieldrin was positive in males and negative in females. So I think there are some further correlations that need to be examined. Another point is that when you are looking at levels of toxicant in tissue culture, it is one thing to relate them to the cytotoxicity of the cells in the tissue culture and another thing to relate them to the level of toxicant in the liver in vivo. I think there may be some orders of magnitude difference there.

Dr. Klaunig: That's true, and I think that is something that most toxicologists working in in vitro systems worry about. In tissue culture, the pharmacokinetics of the toxicant are very different than in vivo. Also, the cells that are normally present in the liver are absent (Kupffer cells, endothelial cells, Ito cells). This is one reason we have started to use slices.

Dr. Patrick Eacho: I am curious to know whether you have looked at compounds of different potencies, because I noticed that the two peroxisome proliferators nafenopin and MEHP produced the effect in the same concentration range even though they differ markedly in potency. You can distinguish the potency of the two in cell culture by measuring peroxisome-related parameters.

Dr. Klaunig: I did not expand on those two compounds. Their behavior does differ a little. MEHP inhibited after a 4-h treatment. Nafenopin required peroxisome proliferation to work; in other words, we had to induce peroxisomes during the culture period to have it inhibit intercellular communication. All of these studies were done with normal cells. Of course, our hypothesis is based on the fact that promoters affect the preneoplastic lesions, and that is what we are examining now. There might be a difference in the response in preneoplastic lesions.

Dr. Stevenson: I believe that Jon Mirsalis gets peroxisome proliferation in the whole range of liver cells in the mouse. That makes me wonder whether all mouse liver cells are initiated or there is a general phenomenon of cell proliferation rather than simply a promoting of initiated cells.

Dr. Klaunig: Dr. Hampton in our group is trying to address that now. As for your question about the cells in the lobules that do respond, we know that phenobarbital responds in vivo by hypertrophy and hyperplasia of the centrilobular hepatocytes. We never get zero percent communication in our mixed cultures. What we are attempting to do now is to isolate fractions from the various portions of the lobule and use those isolated centrilobular vs. periportal cells to see if there is a difference in the responsiveness to substances such as phenobarbital.

Dr. Norman Drinkwater: Your data indicated more efficient homotypic communication, between, for example, nodule cells and nodule cells and normal cells and normal cells, than between the two different types of cells. How do you account for that mechanistically?

Dr. Klaunig: You mean there was more communication between the preneoplastic rat hepatocytes than the normal hepaocytes.

Dr. Drinkwater: Or perhaps it was in the mouse slice experiments. You suggested that there is more efficient communication, for example, between preneoplastic cells than between a preneoplastic cell and a surrounding normal cell.

Dr. Klaunig: That is true.

Dr. Drinkwater: How do you account for that mechanistically?

Dr. Klaunig: Based on some of Dr. Beer's work, the explanation may be a decrease in the amount of gap-junction protein or a different type of gap-junction protein in the preneoplastic hepatocytes, for example, than in the normal cells. I should mention that the system we are using right now—Lucifer yellow dye transfer—is very conservative because we are scoring communication on an all or none basis. We will be using image analysis instrumentation to actually measure the amount of dye transferred. I think that will show much more sensitive and specific inhibition.

Dr. Eacho: I am curious about the cell culture again. You said that the intercellular communication in cell culture peaked in 24 h. That is essentially the time it takes for the cells for form confluent monolayers. How much of what you see is attributable to more of the cells being in contact with one another, assuming that you did use confluent monolayers?

Dr. Klaunig: Of course, the cells have to be in contact with one another to form gap junctions. But in our normal situation, the communication stayed at approximately 90% up to 3 or 4 days in culture.

Dr. Eacho: I am talking about at the very early time. You saw less communication prior to 24 h.

Dr. Klaunig: These cells were isolated as individual hepatocytes. They are separate cells that are plated down. I think it just takes time for the cells to form gap junctions and form communication.

Dr. Eugene McConnell: In your slice technique, have you looked at the normal vs. the preneoplastic lesion ultrastructurally? Is there a physical difference in the gap junctions that would allow for this difference in communication?

Dr. Klaunig: We have not looked at this.

References

1. Slaga TJ. Overview of tumor promotion in animals. *Environ Health Perspect* 50:3–14, 1983.
2. Trosko JE, Chang CC. Adaptive and nonadaptive consequences of chemical inhibition of intercellular communication. *Pharmacol Rev* 36:137s–144s, 1984.
3. Hertzberg EL. Antibody probes in the study of gap junctional intercellular communication. *Annu Rev Physiol* 47:305–318, 1985.
4. Paul D. Molecular cloning of cDNA for rat liver gap junction protein. *J Cell Biol* 103:123–134, 1986.
5. Loewenstein WR. Junctional intercellular communication and the control of growth. *Biochim Biophys Acta* 560:1–65, 1979.
6. Loewenstein WR. Junctional intercellular communication: The cell-to-cell membrane channel. *Physiol Rev* 61:829–913, 1981.
7. Meyer DJ, Yancey SB, Revel JP, Peskoff DA. Intercellular communication in normal and regenerating rat liver: A quantitative analysis. *J Cell Biol* 91:505–523, 1981.
8. Spagnoli LG, Pietra GG, Villaschi S, Johns LW. Morphometric analysis of gap junctions in regenerating arterial endothelium. *Lab Invest* 46:139–148, 1982.
9. Freidman EA, Steinberg M. Disrupted communication between late-stage premalignant human colon epithelial cells by 12-O-tetradecanoylphorbol-13-acetate. *Cancer Res* 42:5096–5105, 1982.
10. Flagg-Newton JL, Dahl G, Loewenstein WR. Cell junction and cyclic AMP. I. Upregulation of junctional membrane permeability and junctional membrane particles by administration of cyclic nucleotide or phosphodiesterase inhibitor. *J Membrane Biol* 63:105–121, 1981.
11. Yotti LP, Chang CC, Trosko JE. Elimination of metabolic cooperation in Chinese hamster cells by a tumor promoter. *Science* 206:1089–1091, 1981.
12. Murray AW, Fitzgerald DJ. Tumor promoters inhibit metabolic cooperation in cocultures of epidermal and 3T3 cells. *Biochem Biophys Res Commun* 91:395–401, 1979.
13. Williams GM, Telang S, Tong C. Inhibition of intercellular communication between liver cells by the liver tumor promoter 1,1,1-trichloro-2,2-bis(p-chlorophenyl)ethane. *Cancer Lett* 11:339–344, 1981.
14. Walder L, Lutzelschwab R. Effects of 12-O-tetradecanoylphorbol-13-acetate (TPA), retinoic acid and diazepam on intercellular communication in a monolayer of rat liver epithelial cells. *Exp Cell Res* 152:66–76, 1984.

15. Ruch RJ, Klaunig JE, Pereira MA. Inhibition of intercellular communication between mouse hepatocytes by tumor promoters. *Toxicol Appl Pharmacol* 87:111–120, 1987.
16. Ruch RJ, Klaunig JE. The effects of tumor promoters, genotoxic carcinogens, and hepatocytotoxins on mouse hepatocyte intercellular communication. *Cell Biol Toxicol* 2:469–483, 1986.
17. Klaunig JE, Ruch RJ. Strain and species effects on the inhibition of hepatocyte intercellular communication by liver tumor promoters. *Cancer Lett* 36:161–168, 1987.
18. Ruch RJ, Klaunig JE. Kinetics of phenobarbital inhibition of intercellular communication in mouse hepatocytes. *Cancer Res* 48:2519–2523, 1988.
19. Klaunig JE, Ruch RJ. Role of cyclic AMP in the inhibition of mouse hepatocyte intercellular communication by liver tumor promoters. *Toxicol Appl Pharmacol* 91:159–170, 1987.
20. Klaunig JE, Ruch RJ, Weghorst CM, Hampton JA. Role of inhibition of intercellular communication in hepatic tumor promotion. *In Vitro Toxicol*, in press.
21. Beer DG, Neveu MJ, Paul DL, Rapp UR, Pitot HC. Expression of the c-raf protooncogene, gamma-glutamyltranspeptidase, and gap junction protein in rat liver neoplasms. *Cancer Res* 48:1610–1617, 1988.
22. Janssen-Timmen U, Traub O, Dermietzel R, Rabes HM, Willecke K. Reduced number of gap junctions in rat hepatocarcinomas detected by monoclonal antibody. *Carcinogenesis* 7:1475–1482, 1986.
23. Klaunig JE, Goldblatt PJ, Hinton DE, Lipsky MM, Chacko J, Trump BF. Mouse liver cell culture. I. Hepatocyte isolation. *In Vitro* 17:913–925, 1981.
24. Klaunig JE, Goldblatt PJ, Hinton DE, Lipsky MM, Trump BF. Mouse liver cell culture. II. Primary culture. *In Vitro* 17:926–934, 1981.
25. Smith P, Fisher R, Shubat PJ, Gandolfi AJ, Krumdieck CL, Brendel K. In vitro cytotoxicity of allyl alcohol and bromobenzene in a novel organ culture system. *Toxicol Appl Pharmacol* 87:509–522, 1987.
26. Goethals F, Krack G, Deboyser D, Vossen P, Roberfroid M. Critical biochemical functions of isolated hepatocytes as sensitive indicators of chemical toxicity. *Fundam Appl Toxicol* 4:441–450, 1984.
27. Lescoat G, Pasdeloup N, Kneip B, Guguen-Guillouzo C. Modulation of alphafetoprotein, albumin, and transferrin gene expression by cellular interactions and dexamethasone in cocultures of fetal rat hepatocytes. *Eur J Cell Biol* 44:128–134, 1987.
28. Sugie S, Mori H, Takahashi M. Effect of in vivo exposure to the liver tumor promoters phenobarbital or DDT on the gap junctions of rat hepatocytes: A quantitative freeze fracture analysis. *Carcinogenesis* 8:45–51, 1987.
29. Mesnil M, Fitzgerald DJ, Yamasaki H. Phenobarbital specifically reduces gap junction protein mRNA level in rat liver. *Molecular Carcinogenesis* 1:79–81, 1988.
30. Beer DG, Pitot HC. Biological markers characterizing the development of preneoplastic and neoplastic lesions in rodent liver: Mouse liver tumors. *Arch Toxicol [Suppl]* 10:69–80, 1986.

Mouse Liver Carcinogenesis: Mechanisms
and Species Comparisons, pages 293–309

Proto-oncogene and Gap-Junction Protein Expression in Rodent Liver Neoplasms

David G. Beer and Mark J. Neveu†*
**University of Kansas Medical Center, Department of Pharmacology, Toxicology, and Therapeutics and Center for Environmental and Occupational Health, Kansas City, KS 66103 and †McArdle Laboratory, University of Wisconsin, Madison, WI 53706*

Expression of certain proto-oncogenes is thought to be important to the multistep process of carcinogenesis. In liver tumors from mice, activating mutations in the H-*ras* gene that show carcinogen-specific mutation patterns have been characterized (1,2). These alterations have been suggested to be a direct result of the interaction of carcinogens with the H-*ras* gene. Although similar mutations have not been found consistently in rat liver neoplasms (2–4), elevated expression of the H-*ras* and c-*myc* proto-oncogenes has been reported in liver tumors of rats given various carcinogenic regimens (5–8). Unregulated expression or overexpression of these genes, rather than mutational events, may provide some selective or maintenance role during the development of liver lesions in the rat. To investigate these possibilities, we examined the expression of the H-*ras* and c-*myc* proto-oncogenes in putative preneoplastic focal lesions and primary hepatocellular carcinomas from rat liver. In addition, we explored the expression of both the c-*raf* proto-oncogene and the gap-junction protein in these same types of lesions. The relationship of gene expression to the events occurring during hepatocarcinogenesis is discussed.

Materials and Methods

Female Sprague-Dawley and F-344 rats (Harlan Sprague-Dawley, Madison, WI) weighing 150–170 g were subjected to a 70% hepatectomy according to the method of Higgins and Anderson (9). Eighteen hours later, they were given a single intragastric instillation of 30 mg/kg diethylnitrosamine (DEN) (Eastman Organic Chemicals, Rochester, NY) in distilled water. They were then fed a diet containing 0.05% phenobarbital (PB). Animals were killed after 6–15 months on this diet; some animals were used for preneoplastic cell isolation, and portions of the livers from the remaining rats were either frozen for cryostat sectioning or used for RNA isolation according to the method of Chirgwin et al. (10). Primary liver tumors were dissected free of normal liver, a portion was taken for histologic diagnosis, and the remainder was also used for RNA isolation. For histopathologic analysis, portions of the normal liver and tumors were fixed in 10% buffered formalin, and 5-micron sections were stained with hematoxylin and eosin.

The isolation and purification of γ-glutamyltransferase (GGT)-positive hepatocytes from rat livers containing GGT-positive foci were performed as previously described (11), according to the panning method of Hanigan and Pitot (12). GGT-positive hepatocytes were isolated from animals 6 and 11 months after the start of the treatment protocol and used for RNA isolation. Total cellular RNA was then separated in 1.2% agarose-formaldehyde gels according to the method of Maniatis et al. (13) and transferred to Gene Screen Plus filters (New England Nuclear, North Billerica, MA) for northern gel or dot blot analysis. Mammalian liver 28S and 18S and *Escherichia coli* 23S and 16S ribosomal RNAs were used as size markers for northern gels. The lanes containing the transferred ribosomal markers were cut off and stained with methylene blue as described by Maniatis et al. (13). ^{32}P-labeled DNA probes were prepared by nick translation of the following clones: the 450-bp purified BS-9 fragment of H-*ras* (14); the 3.2-kb *Bam*HI-*Hin*dIII fragment of mouse c-*myc* subcloned into pBR322 (15); the 500-bp *Pst*I pRSA-1 purified fragment of albumin (16); the 2.1-kb human c-*raf*-1 fragment (17); the human H4 histone genomic DNA subcloned into pBR322 (18); and the 790-bp GGT fragment (20). The northern and dot blot filters were prehybridized in 50% deionized formamide, 1% sodium dodecyl sulfate (SDS), 1 M NaCl, and 10% dextran sulfate overnight at 42°C. Two to 5×10^5 dpm of nick-translated, heat-denatured DNA/ml and 200 μg of salmon sperm DNA/ml were then added and the filters hybridized for an additional 18 h at 42°C with 2 × SSC containing 1% SDS. The filters were then washed twice in 0.1 × SSC at room temperature for 30 min and sealed in plastic bags. Autoradiograms were prepared with Kodak XAR 5 x-ray film and exposed at 70°C for 1–5 days with intensifying screens. Autoradiograms were quantified by scanning with an LKB soft laser scanning densitometer (LKB Instruments, Gaithersburg, MD).

In Situ Hybridization

The expression of mRNA for the GGT and c-*myc* genes was examined by a modification of in situ hybridization procedures previously described (21–23) and using a detailed procedure given elsewhere (24).

Immunocytochemistry

The distribution of gap-junctional complexes in normal rat liver and preneoplastic foci was examined with an antibody against the major M_r 27,000 rat liver gap-junction protein (20). Antiserum was used from a rabbit given an injection of a fusion protein expressed in bacteria from a partial cDNA to the rat liver junction protein (20). Preimmune serum from the same rabbit was used as a control. Ten-micron serial frozen sections were fixed in acetone for 10 min at room temperature, hydrated in phosphate-buffered saline (PBS), and incubated with the serum diluted 1:400 in PBS for 45 min at room temperature. Visualization of the antigen-antibody complexes was achieved using the peroxidase-antiperoxidase method of Sternberger et al. (25).

Results and Discussion

We induced preneoplastic and neoplastic lesions in female Sprague-Dawley rats by using a single intragastric administration of 30 mg/kg DEN 18 h after the animals had undergone a 70% partial hepatectomy. This single DEN dose is primarily nonnecrogenic (26). The animals were then fed a diet containing 0.05% PB, which effectively promotes the development of the liver lesions (27). To determine whether focal lesions induced using this protocol showed altered expression of H-*ras* and c-*myc*, we isolated preneoplastic foci cells using a panning procedure developed by Hanigan and Pitot (12). Most preneoplastic foci abundantly express the enzyme GGT, which is located on the outer membrane of the hepatocytes, whereas normal hepatocytes do not express this enzyme. We used an affinity-purified rabbit anti-rat GGT antibody affixed to Petri dishes to cause the hepatocytes that express GGT to bind to the dishes; they were then easily collected. We collected GGT-positive foci cells from rats that had been treated with DEN and then PB for 6 and 11 months and prepared total cellular RNA from these cells as previously described (11). RNA was also prepared from the cells that did not adhere to these plates during the panning procedure (GGT-negative), as well as the starting population. Six months after DEN/PB treatment, approximately 5% of the liver volume was occupied by GGT-positive foci, whereas at 11 months this value was approximately 16% of the liver volume. We obtained approximately a sixfold enrichment of GGT-positive cells relative to the starting cell mixture, but the total yield of GGT-positive cells was only a few million cells (11). Northern blots of mRNA from these cell populations were probed for the expression of the GGT, albumin, H-*ras*, and c-*myc* genes. In GGT-positive cells isolated at either 6 or 11 months after DEN treatment, a large increase in GGT mRNA was observed (Fig. 1). This demonstrates that, as expected, we had enriched for GGT-positive hepatocytes, but also that the increased GGT enzyme activity in these cells was accompanied by increased GGT mRNA levels. Examination of albumin and H-*ras* mRNA levels, however, revealed essentially no differences between GGT-positive cells and -negative cells. In contrast, in the 11-month group (Fig. 1), isolated GGT-positive hepatocytes expressed slightly more c-*myc* than the GGT-negative cells. This increase was not seen in the 6-month group. Because larger nodules and tumors were present in the livers from 11-month animals, this c-*myc* increase may have been due to the contribution of GGT-positive cells from lesions expressing greater amounts of the c-*myc* gene. Alternatively, a slight increase in c-*myc* may actually have occurred in many of the early preneoplastic focal cells. Since we could not distinguish between these two possibilities with the panning procedure, we attempt-

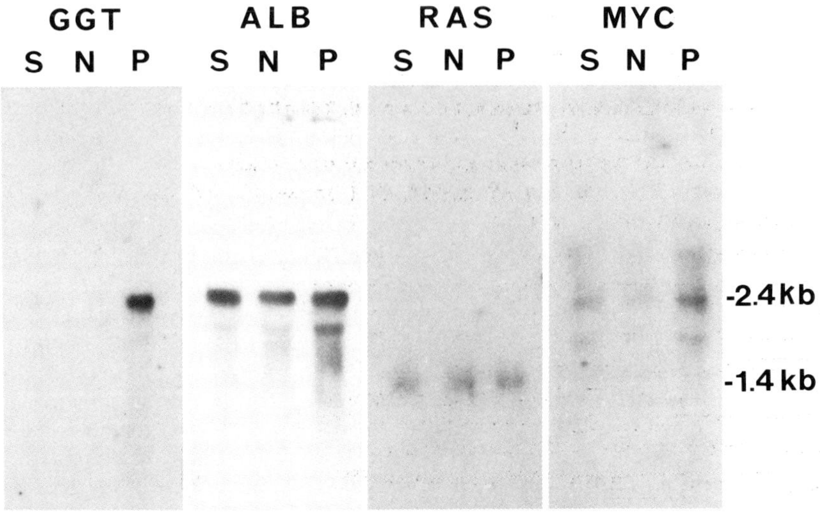

Figure 1. Autoradiograms of GGT, albumin (ALB), H-*ras* (RAS), and c-*myc* (MYC) RNA transcripts in hepatocytes isolated from animals administered a single dose of DEN (30 mg/kg) and fed diets containing 0.05% PB for 11 months. Hepatocytes representing the starting (S), GGT-negative (N), and GGT-positive (P) hepatocyte populations were processed for RNA isolation, and 10 µg of total cellular RNA was separated in agarose-formaldehyde gels and hybridized with ^{32}P-labeled DNA probes as described in Materials and Methods. Reprinted from D. G. Beer et al. *Cancer Res* 46:2435–2441, 1986; used with permission.

ed in situ hybridization of c-*myc* mRNA with frozen sections of livers from these two treatment groups.

In Situ Hybridization of GGT and c-*myc* mRNA

Because the preneoplastic liver lesions abundantly expressed GGT mRNA, we utilized GGT as our positive control to optimize the use of cryostat sections of rat liver for in situ hybridization. Antisense ^{35}S-labeled GGT cRNA was transcribed using the Gemini plasmid (Promega) and then utilized as a probe for GGT mRNA in foci-containing liver. The procedure used for in situ hybridization was previously described (24). As shown in Figure 2, all hepatocytes within preneoplastic foci that express GGT enzyme activity also contain abundant mRNA for GGT, as determined using in situ hybridization. Utilizing the identical in situ hybridization and washing conditions as those used for examining GGT mRNA, antisense c-*myc* ^{35}S-cRNAs were transcribed and the tissue sections analyzed for c-*myc* mRNA. Most GGT-positive foci did not show an increase in expression of c-*myc* mRNA. Areas of bile duct proliferation, however, did show elevated expression (data not shown); this is consistent with previous reports (8). In only a few foci was an apparent increase in c-*myc* observable. This lack of an increase in c-*myc* in most preneoplastic lesions is in contrast to recent work

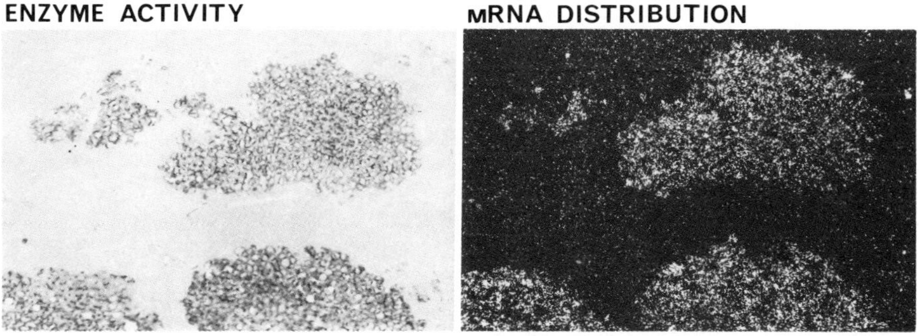

Figure 2. Serial frozen sections of rat liver containing enzyme-altered foci that were either stained for GGT enzyme activity (left) by the histochemical method of Rutenberg et al. (43) or used for in situ hybridization with antisense GGT ^{35}S-labeled RNA (right). In the dark-field images of the autoradiograms after in situ hybridization, the silver grains of the exposed emulsion appear white. These results demonstrate that increased GGT enzyme activity in foci is accompanied by increased GGT mRNA. All magnifications, ×150.

by Nagy et al. (28), who describe elevated expression of c-*myc* in most rat liver lesions induced using the Solt-Farber protocol. The reasons for these differences may reside in the distinct protocols used for inducing them. Analyses of chromosomal alterations in preneoplastic foci cells after a single nonnecrogenic dose of DEN followed by a PB diet (as used in our studies) and after the use of a 20-fold higher, necrogenic dose of DEN and subsequent administration of 2-acetylaminofluorene (as used in the Solt-Farber protocol) show extensive chromosomal alterations in the cells from the Solt-Farber protocol but not in cells initiated with the single low dose of DEN (29). This suggests that additional genetic alterations may be responsible for the increased c-*myc* expression in lesions induced with the Solt-Farber protocol. Additional evidence that the protocol used for inducing the rat liver lesions may influence the observed alterations in gene expression comes from our analyses of H-*ras* and c-*myc* mRNA levels in primary hepatocellular carcinomas (Fig. 3). Most of these tumors showed either abundant or a slight decrease in albumin mRNA levels, but quite variable expression of both H-*ras* and c-*myc* was apparent. Unlike these tumors, which were induced with a single dose of DEN followed by PB, the early lesions and primary hepatocellular carcinomas induced with the Solt-Farber protocol were reported to express elevated levels of c-*myc* (28). This suggests that either the protocol used influences the alterations observed in these liver lesions or that there are distinct pathways to the development of primary hepatocellular carcinomas, one associated with elevated c-*myc* and the other not.

Using a single administration of DEN and a subsequent PB diet as our model system to induce liver lesions, we have examined the expression of genes that are more consistently expressed in primary liver tumors. Although consistently elevated or decreased expression of a gene in tumors does not prove that it is critical to their development, it may indicate that the gene is more closely

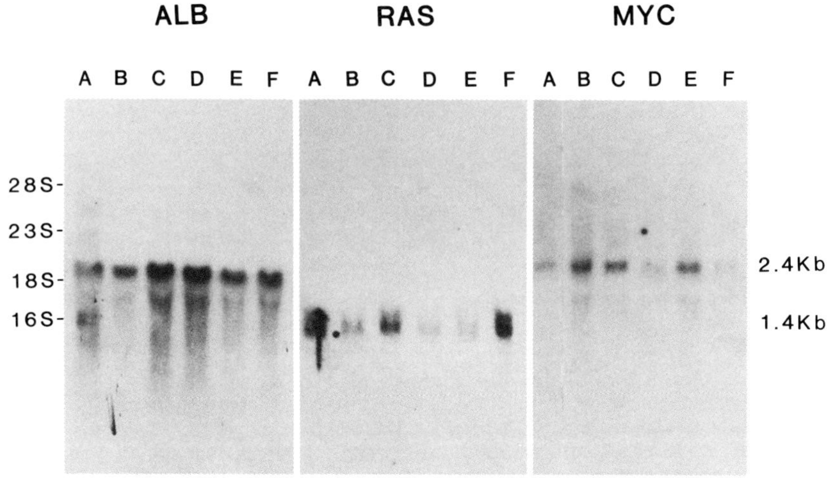

Figure 3. Autoradiograms of albumin (ALB), H-*ras* (RAS), and c-*myc* (MYC) RNA transcripts from normal rat liver (A) and five hepatic lesions (B to F) removed from rats given a single dose of DEN after partial hepatectomy and fed 0.05% PB diets for 12 to 16 months. Ten micrograms total cellular RNA was separated in 1.2% agarose-formaldehyde gels, blotted, and hybridized to ^{32}P-labeled DNA as described in Materials and Methods. The autoradiograms were made after varying exposures for each northern blot. The pathologic diagnosis of each of the lesions was as follows: (A) liver showing focal areas of chronic inflammation and occasional minimal bile duct proliferation; (B) atypical neoplastic nodules with extensive oval cell and bile duct proliferation, multiple hepatocellular carcinomas, and mixed cell types; (C) neoplastic nodules showing markedly atypical cellular hyperplasia with areas compatible with early carcinoma; (D) liver showing hepatocellular carcinoma; (E) several neoplastic nodules showing markedly atypical cellular hyperplasia with foci of early carcinoma; (F) liver showing small neoplastic nodule and hepatocellular carcinoma. Reprinted from D. G. Beer et al. *Cancer Res* 46:2435–2441, 1986; used with permission.

associated with tumor development than a gene that is inconsistently expressed. One gene that is elevated in primary liver tumors from both Sprague-Dawley and F-344 female rats is the c-*raf* proto-oncogene (Fig. 4). Lesions of quite diverse cellular composition and malignant potential show elevated expression of this proto-oncogene. This gene is also expressed during the highly regulated process of liver regeneration after partial hepatectomy. The levels of c-*raf* mRNA increase at 20 h after partial hepatectomy, near the onset of DNA synthesis (30). Histone H4 mRNA, which is tightly coupled to DNA synthesis (18), also is slightly elevated at this time. The increase in c-*raf* in the tumors, however, doesn't show a consistent correlation with increased histone H4 expression, suggesting that the increase in c-*raf* mRNA is not simply due to increased cell proliferation or DNA synthesis in the tumors. We do not know, however, whether this increase in c-*raf* gene expression is accompanied by an increase in the c-*raf* protein. The c-*raf* protein is a protein kinase that catalyzes the phosphorylation of serine and threonine residues (31). In studies in progress using an antibody against the c-*raf* protein, increased c-*raf* immunoreactivity was not observed in most preneoplastic foci (Neveu and Pitot, in preparation).

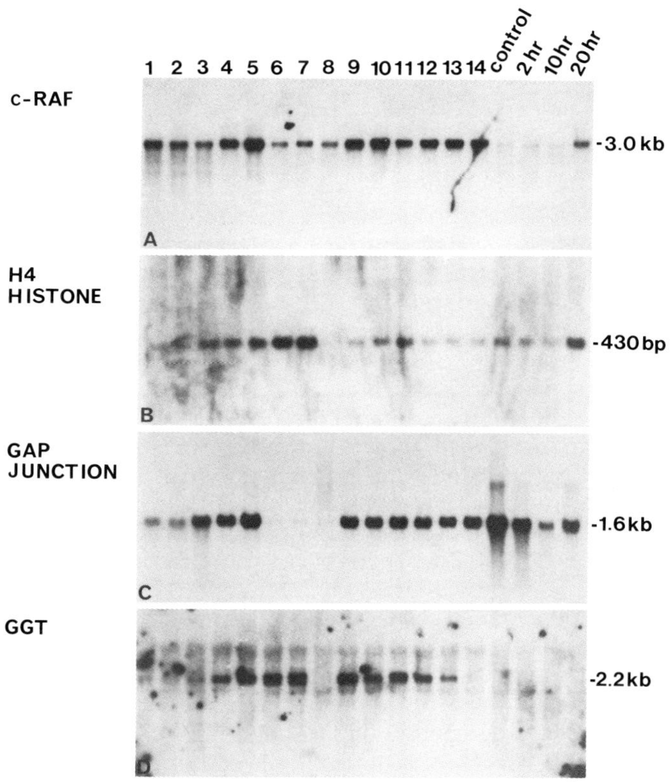

Figure 4. Autoradiograms from northern blots of 10 μg total cellular RNA from 13 primary rat liver tumors (lanes 1–7, 9–14), an age-matched control rat who received a 0.05% PB diet but no DEN treatment (lane 8), and normal rat livers 0, 2, 10, and 20 h after partial hepatectomy. The samples were probed for c-*raf* expression. The same blot was stripped and reprobed for H4 histone, GGT, and the M_r 27,000 major rat liver gap-junction protein. Mammalian and *E. coli* ribosomal RNAs were used as size markers. The histologic diagnoses for the tumors were as follows: lane 1, multiple hepatocellular carcinomas, mixed cell types, and atypical neoplastic nodules with extensive oval cell and bile duct proliferation; lane 2, several neoplastic nodules showing markedly atypical cellular hyperplasia with foci of early carcinoma; lane 3, small neoplastic nodule and hepatocellular carcinoma; lane 4, leukemic infiltration of a hepatic neoplastic nodule showing moderate cellular atypism; lane 5, neoplastic nodule exhibiting extensive spongiosis and moderate cellular atypism; lane 6, neoplastic nodules exhibiting moderate to severe cellular atypism and spongiosis, with focal areas highly suggestive of carcinomatous changes; lane 7, large hepatocellular carcinoma exhibiting trabecular and adenohepatomatous morphology; lane 8, liver showing moderate PB effect, focal fatty metamorphosis, and occasional enzyme-altered foci; lane 9, neoplastic nodules with focal fibrosis and bile duct proliferation—several of the nodules showed moderate to severe atypism with several areas highly suggestive of early carcinomatous changes; lane 10, neoplastic nodules exhibiting moderate to severe atypism, with focal areas of bile duct proliferation; lane 11, mixed cholangiocellular-hepatocellular carcinoma; lane 12, mixed hepatocellular-cholangiocellular carcinoma together with neoplastic nodules exhibiting spongiosis and focal areas of moderate to severe cellular atypism, with some areas compatible with early carcinomatous changes; lane 13, hepatocellular carcinoma; lane 14, hepatocellular carcinoma. Reprinted from D. G. Beer et al. *Cancer Res* 48:1610– 1617, 1988; used with permission.

This suggests that an increase in c-*raf* may not underlie the development of the early lesions but occurs at some stage during promotion or progression. The relevance of the increased c-*raf* mRNA in these tumors is therefore unknown.

Gap-Junction Protein Expression in Rat Hepatocarcinogenesis

Utilizing a cDNA and an antibody to the M_r 27,000 rat liver gap-junction protein that were developed by Paul (20), we have examined the expression of this gene in rat liver lesions induced with our protocol. Gap junctions are known to be intimately involved in regulating cell-to-cell communication (32) and cell proliferation (33) and may also influence the expression of genes characteristic or specific for a cell or tissue (34). Further, inhibition of gap-junctional communication may be important in tumor promotion (35). Decreased expression of the rat liver gap-junction protein has recently been shown in rat hepatocellular carcinomas induced with *N*-methyl-*N*-nitrosourea (36). We observed reduced expression of the rat liver gap-junction protein mRNA and increased GGT mRNA expression in most rat liver lesions induced with DEN (Fig. 4). The gap-junction protein gene is abundantly expressed in normal liver, suggesting a requirement for the continued synthesis of gap-junction proteins. Within 2 h following partial hepatectomy, there is rapid reduction to 50% of control values, and by 10 h there is a maximal decrease to 25% of control mRNA levels for this protein. These results are consistent with work by Yee and Revel (37), who demonstrated a reduction in gap-junctional complexes in rat liver following partial hepatectomy. Interestingly, the dramatic decrease in gap-junction protein mRNA we observed at 2 h suggests rapid reduction in the expression of this gene, probably followed by the rapid turnover of the protein; these may be very early and requisite events leading to compensatory liver growth. The rebounding of the gap-junction protein mRNA is evident by 20 h after hepatectomy. This relatively rapid return of expression suggests that a sustained decrease in gap-junction protein is not necessary to ultimately allow the hepatocytes to divide.

The decrease in gap-junction protein mRNA observed in the DEN-induced liver lesions was not limited to hepatocellular carcinomas but was also observed in samples containing primarily neoplastic nodules (Fig. 4). This suggests that decreased expression was not limited to more advanced lesions or during the later stages of hepatocarcinogenesis. Also there was little correlation between histone H4 mRNA and the reduction in gap-junction protein mRNA, although two samples (lanes 6,7) with very high histone H4 mRNA levels also showed very low gap-junction protein mRNA levels. It was not determined whether these tumor samples actually expressed the gap-junction protein; it is possible that the protein may also be expressed but never incorporated into the cell membrane. Our results, however, appear to be consistent with the very dramatic reductions in the protein observed in hepatocellular carcinomas, as reported by Janssen-Timmen et al. (36). Because the gap-junction protein mRNA is abundant, any normal liver tissue associated with the tumor samples taken for RNA isolation would tend to cause an overestimation of the amount of gap-junction protein mRNA in these tumors. The histologic analysis of each of these tumors indeed confirmed that small areas of normal liver were associated with many of them. Thus in some lesions, the amount of gap-junction protein mRNA may actually have been lower than we estimated. Confirmation that some DEN-induced rat liver lesions do show quite a dramatic reduction in the gap-junction protein was obtained by using an antibody raised against the gap-junction protein (20). Rat livers examined 11 months after partial hepatectomy and DEN and PB treatment contained many preneoplastic foci, some neoplastic nodules, and, more rarely, hepatocellular carcinomas. Antibody analysis of liver sections containing foci and smaller nodules revealed that some lesions contained very reduced levels of this protein (Fig. 5). Because nearly all foci and tumors express GGT (Figs. 3,4), serial frozen sections were histochemically stained for GGT activity to demonstrate the location of each lesion. Surprisingly, many preneoplastic foci showed a decrease in gap-junction protein level. Those that appeared to show the greatest decrease were the lesions with a distinct demarcation from the surrounding normal hepatocytes (Fig. 5). Foci that were less demarcated from the surrounding normal liver did not show as great a reduction in the gap-junction protein. Interestingly, Schulte-Hermann (38) has described these two foci types as nondemarcated or remodeling foci and well-demarcated or non-remodeling foci. It is the latter, the well-demarcated ones, that are promoted to a greater extent with PB. PB is known to be an effective tumor promoter (27) and to inhibit gap-junctional communication in rat and mouse hepatocytes (39,40). It may be no coincidence that the foci that show the greatest reduction in the gap-junction protein are those foci that appear to be most affected by PB (37).

The animals from the 11-month group were given PB continuously until they were killed. The liver lesions from these animals, which show a decrease in the gap-junction protein, therefore may have been more sensitive to PB-induced inhibition of gap-junctional communication than the surrounding normal hepatocytes. Alternatively, because these lesions represent the progeny of initiated cells, their decreased amounts of the protein may have been directly due to carcinogen treatment, without any additional influence from PB. Evidence for the first possibility comes from an examination of livers from the rats in our 6-month group. These animals were given PB in the diet until 10 days before they were killed. At that point, they were placed on a regular diet without PB. Examination of serial frozen sections histochemically stained for GGT activity and immunochemically stained for the gap-junction protein revealed very few foci with reductions in gap-junction protein, similar to results seen with the 11-month group, which had remained on PB. These results suggest that once the promoting stimulus was removed, in this case within 10 days, the foci that may have previously shown significantly reduced gap-junction protein expression regained or began to express this protein.

It therefore appears that the decrease in gap-junction protein expression is not just a function of the preneoplastic state, since a reduction was not observed when PB was removed, although these foci retained GGT expression. It is possible, however, that once the lesions become more advanced, such as in nodules and/or carcinomas, subsequent changes may occur that are associated

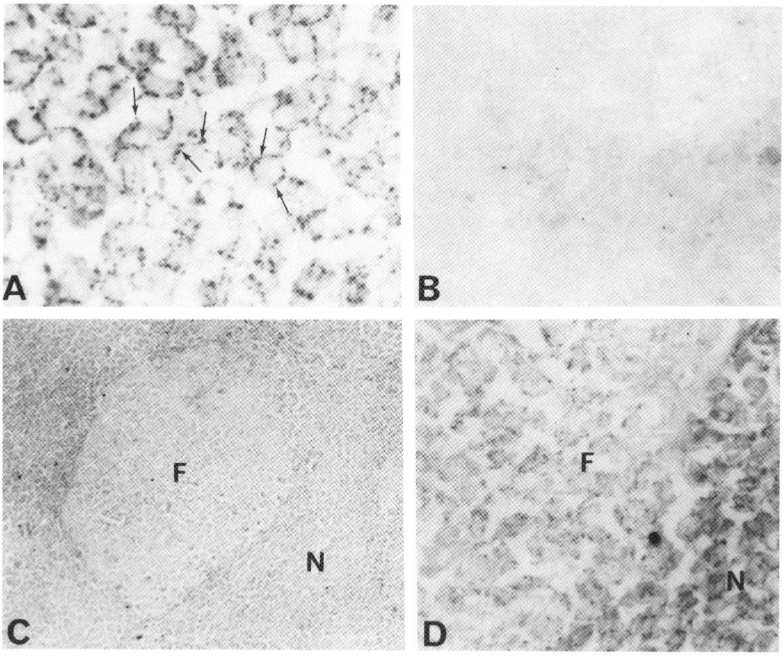

Figure 5. Immunocytochemical localization of the M_r 27,000 rat liver gap-junction protein in normal and preneoplastic liver. Frozen sections of rat liver were incubated with an antibody to the rat liver gap-junction fusion protein (20) at a 1- to 400-fold dilution and visualized by the peroxidase-antiperoxidase method of Sternberger et al. (25). (A) High-power magnification of a normal liver showing the distribution of gap-junction complexes between hepatocytes (arrows); ×450. (B) Normal liver incubated with preimmune serum showing the specificity of the reaction shown in (A); ×450. (C) Low-power image of a foci (F) in normal rat liver (N) demonstrating a reduction in the immunoreactivity of this antibody within the foci cells; ×60. (D) Higher magnification of a portion of the foci in (C) showing the decreased distribution of gap-junction complexes between foci (F) cells compared with the surrounding normal (N) hepatocytes; ×200. Reprinted from D. G. Beer et al. *Cancer Res* 48:1610–1617, 1988; used with permission.

with increased autonomy and result in a gap-junction protein expression decrease that is independent of the promoting stimulus.

Our studies suggest that PB may also affect gap junctions in normal liver. We observed that in the PB-treated animals, the hepatocytes within the centrilobular regions contained fewer gap-junction plaques that were detected with the antibody than any other regions of the liver lobule (Beer and Neveu, unpublished observations). Following PB withdrawal, the hepatocytes regained the gap-junction protein. The dramatic reductions observed in specific foci, however, suggest that most of the normal hepatocytes become refractory to the inhibitory effects of PB, although some foci, primarily the well-demarcated ones, remain very sensitive to PB. Could promotion of foci development by PB be

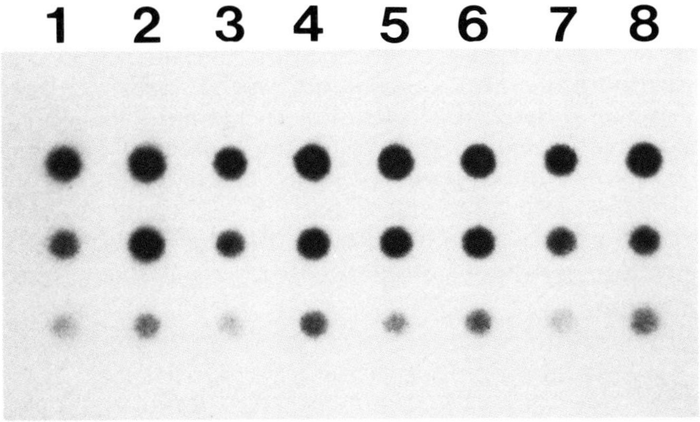

Figure 6. Expression of the gap-junction protein mRNA in mouse liver hepatomas. C3HB6F$_1$ male mice were fed a regular lab chow (lanes 1–3), placed on a choline-deficient diet for 1 year (lanes 4, 5), or fed a choline-deficient diet for 1 year and then a diet containing 0.05% PB (lanes 6–8). Hepatomas arising in the livers either spontaneously in the regular chow group or in the treated groups were collected and used for RNA isolation and dot blot analysis as described in Materials and Methods. The blots were probed using the cDNA to the rat liver gap-junction protein (20). Lane 2 shows a non-tumor-bearing liver from the regular chow group, and lane 8 shows a non-tumor-bearing liver from the group given a choline-deficient diet and then fed a diet containing 0.05% PB.

partly explained by the effects of this compound upon the inhibition of gap-junction protein expression, especially in a specific subset of foci particularly sensitive to such inhibition? Is this phenomenon unique to PB, or is it common to other known liver tumor promoters? The answer may be the latter. Groups of rats given partial hepatectomies, DEN, and then seven structurally different tumor promoters also had decreased gap-junction protein expression in liver tissue (Neveu and Pitot, in preparation). The characterization and quantitation of the foci affected by these treatments, as well as any potential unique phenotypic characteristics of foci showing reductions, are being determined. The diverse compounds used have different effects upon the cell but are all liver tumor promoters and now may also be shown to have a similar action—inhibition of hepatocyte gap junctions. This latter effect may help explain how diversely acting compounds all promote tumors. Since multiple pathways appear to affect gap-junctional communication (41), these compounds may exert a quite similar effect by diverse means—disrupting the communication of the initiated hepatocytes from their neighboring normal hepatocytes. This would free the altered cells from the growth regulation that rules the normal cells and thereby help push them along the path to neoplasia.

The reduction of gap-junctional communication in rat liver lesions is not unique to the rat; some liver tumors from B6C3HF$_1$ mice also show decreased mRNA for the gap-junction protein. Spontaneously arising hepatomas, as well as those induced by a choline-deficient diet, show a reduction in this mRNA (Fig. 6). We have not determined using the antibody whether the early focal liver lesions in the mouse also show a decrease in the gap-junction protein. The mouse hepatomas in these three treatment groups did not all show a decrease in gap-junction protein mRNA, however. We previously characterized these lesions using 10 different histochemical and immunologic markers and observed extensive hetereogeneity in nearly all the proteins and enzyme activities examined (42). Specific areas of these hepatomas may therefore have more significant decreases in gap-junction protein mRNA than these blots indicate, due to the averaging effect of homogenization. Studies with the gap-junction protein antibody should address this possibility.

Conclusions

Using a model system in which a single nonnecrogenic dose of DEN followed by dietary PB is administered to rats, we have examined the potential roles of the H-*ras* and c-*myc* proto-oncogenes in the early stages of hepatocarcinogenesis. Only a few foci and/or nodules and primary hepatocellular carcinomas express elevated levels of c-*myc*. This may suggest that, at least in this model, c-*myc* is unlikely to be a prerequisite for the maintenance or further development of preneoplastic liver lesions, although its involvement in the development of some lesions cannot be excluded. The situation may be similar for H-*ras*, although expression of this gene appears not to be increased in most GGT-positive preneoplastic foci. We did find, however, that the c-*raf* proto-oncogene is abundantly expressed in most primary liver tumors, and these tumors also show a dramatic reduction in mRNA for the 27-kDa major rat liver gap-junction protein. Although the consequences of either an increase in c-*raf* or decreased gap-junction mRNA are not certain, we have explored the expression of the gap-junction protein during the early stages of hepatocarcinogenesis and have found that decreased expression is evident in many GGT-positive preneoplastic foci. This decrease in gap-junction protein expression in foci appears to be directly related to treatment with the tumor promoter PB and may primarily affect well-demarcated foci, which are more sensitive to the promoting effects of phenobarbital. The withdrawal of PB for 10 days appeared to result in the reappearance of gap junctions in most foci, demonstrating the reversible nature of this effect and suggesting that the decrease in gap junctions is not simply due to the fact that these are initiated cells or solely a result of carcinogen treatment.

Acknowledgments

We thank Ms. Hallie Calloway for the preparation of this manuscript. This work was supported by National Cancer Institute grants CA-07652, CA-07175, and CA-22484.

Open Forum

Dr. Stringer Sue Yang: I have a question about your findings with the northern blot using the c-*raf* probe. I have examined more than 30 human hepatomas and do not see any c-*raf* being expressed in their DNA. I am impressed with the blot that you have, showing so many cases of active expression. Did you examine other normal tissues to see whether c-*raf* was expressed?

Dr. David Beer: In that same blot there was a liver from an aged-matched control. Those controls showed less c-*raf* than the tumors. We did not examine any other tissues.

Dr. Yang: Did you look at normal liver?

Dr. Beer: Normal liver was there in the regeneration study. It was quite low.

Dr. Jay Goodman: I would like to comment about your observation regarding the inconsistent association of H-*ras* expression in your DEN-induced rate liver foci. In 1988, Stowers et al. looked for activated H-*ras* in rat liver hepatomas and mouse liver hepatomas induced by DEN. They did not find activated H-*ras* in the rat, but did find it quite consistently in the mouse. I think this points to the fact that you have to look at the chemical, the organ, and the species.

Dr. Beer: I absolutely agree. I think that also shows that some of the steps in mouse tumorigenesis may be different just based on the differences in protocols used for mouse and rat. But I think it is still debatable whether the activation of H-*ras* in the mouse liver is spontaneous or chemically induced. I believe they have a recent paper that addresses that point.

Dr. Gary Williams: I presented data here on the histochemical staining of the P-21 protein in rat liver tumors; we found increased expression of the P-21 protein in tumors. I am puzzled by the discrepancy. The studies that Dr. Goodman just referred to are DNA transfection studies. Isn't there a certain probability of not being able to detect an activated oncogene? In any event, couldn't you have expression of the *ras* oncogene without activation?

Dr. Beer: That is exactly why we did the expression studies. Farber showed that back in 1986; they did 3T3 cell transfections and did not find activated *ras* in DEN-induced tumors. That suggests that activation was not playing a role, but expression could have been. I agree that your results suggest that 70% of the tumors show an increase in *ras* protein. There could be several explanations for this discrepancy. Perhaps the message and protein do not correspond, although that is fairly unlikely. Perhaps our protocols were different. Ours was promoted; perhaps yours was not. First we need to determine whether we are talking about the same thing, and then we can try to find out why it is inconsistent. When you say you find increased expression, what about the other 30%?

Dr. Williams: There was no increase. Our general finding is still consistent with your dilemma that a fraction of the tumors are somehow not expressing the oncogene. We must conclude that this expression is not essential to tumor growth, although it may be important for some tumors.

Dr. James Trosko: Stanbridge, Barrett, and a couple of others now have shown that in some cells, the message as well as the protein product of the *ras* can be expressed, although the phenotype has not changed. It is from these observations that the suppressor gene, or the anti-oncogene concept has come about. Therefore, we have to add that into the possible explanations of potential discrepancies.

Dr. Beer: I think we also must realize that *ras* and *myc* may have no bearing on the phenotypes or the behavior of the tumors.

Dr. Williams: In cell hybridization studies in which a normal cell is hybridized to a neoplastic cell, the neoplastic phenotype in the hybrids is suppressed. The conclusion of all these studies is that the neoplastic phenotype is a recessive characteristic. The studies that have concerned oncogenes so far seem to regard them as a dominant factor. I have been trying to reconcile the genetics of the neoplastic phenotype with the purported role of oncogenes. Do you have any thoughts on that?

Dr. Beer: Nothing that could answer that dilemma.

Dr. Nelson Fausto: In analysis of oncogene expression during carcinogenesis and in tumors, one must distinguish between enhanced expression of a normal proto-oncogene and a mutational event. Proto-oncogenes are generally expressed by dividing normal cells, and alterations of these genes during carcinogenesis may involve increased expression of the normal gene. The nature of the genetic change determines whether the alteration will be dominant or recessive. The term "activation," as used in the context of proto-oncogene expression, is ambiguous. It should also be kept in mind that proto-oncogene mutations do not necessarily lead to large increases in the expression of the gene and that protooncogene mutations in liver carcinomas have been found in some mouse strains but are not common in rats, and their occurrence may depend on the type of carcinogen used. Enhanced protooncogene expression in liver carcinogenesis (such as *myc* and *ras*) may be simply a consequence of increased cell proliferation and not be directly associated with cell transformation. Most of the time, this aspect of the problem is not properly studied. To complicate matters, histochemical methods for the demonstration of p21 ras protein in tissues fail to distinguish between the normal and the mutant protein, although the proteins can be separated in western blots.

Dr. Beer: I agree.

References

1. Wiseman RW, Stowers SJ, Miller EC, Anderson MW, Miller JA. Activating mutations of the c-Ha-ras protooncogene in chemically induced hepatomas of the male B6C3F1 mouse. *Proc Natl Acad Sci USA* 83:5825–5829, 1986.
2. Stowers SJ, Wiseman RW, Ward JM, Miller EC, Miller JA, Anderson MW, Eva A. Detection of activated proto-oncogenes in N-nitrosodiethylamine-induced liver tumors: A comparison between B6C3F1 mice and Fisher 344 rats. *Carcinogenesis* 9: 271–276, 1988.
3. Farber E. Cellular biochemistry of the stepwise development of cancer with chemi-

cals: G. H. A. Clowes Memorial Lecture. *Cancer Res* 44:5463–5474, 1984.
4. McMahon G, Hanson L, Lee J, Wogan GN. Identification of an activated c-Ki-ras oncogene in rat liver tumors induced by aflatoxin B1. *Proc Natl Acad Sci USA* 83: 9418–9422, 1986.
5. Hayashi K, Makino R, Sugimura T. Amplification and over-expression of the c-myc gene in Morris hepatomas. *Gann* 75:475–478, 1984.
6. Ishikawa F, Takaku F, Nagao M, Ochiai M, Hayashi K, Takayama S, Sugimura T. Activated oncogenes in a rat hepatocellular carcinoma induced by 2-amino-3-methyl-imidazo[4,5-f]quinoline. *Jpn J Cancer Res* 76:425–428, 1985.
7. Cote GJ, Lastra BA, Cook JR, Huang DP, Chiu JF. Oncogene expression in rat hepatomas and during hepatocarcinogenesis. *Cancer Lett* 26:121–127, 1985.
8. Yaswen P, Goyette M, Shank P, Fausto N. Expression of c-Ki-ras, c-Ha-ras and c-myc in specific cell types during hepatocarcinogenesis. *Mol Cell Biol* 5:780–786, 1985.
9. Higgins GM, Anderson RM. Experimental pathology of the liver. I. Restoration of the liver of the white rat following partial surgical removal. *Arch Pathol* 12:186–202, 1931.
10. Chirgwin JM, Przybyla AE, MacDonald RJ, Rutter WJ. Isolation of biologically active ribonucleic acid from sources enriched in ribonuclease. *Biochemistry* 18:5294–5299, 1979.
11. Beer DG, Schwarz M, Sawada N, Pitot HC. Expression of H-ras and c-myc proto-oncogenes in isolated γ-glutamyl transpeptidase-positive hepatocytes and hepatocellular carcinomas induced by diethylnitrosamine. *Cancer Res* 46:2435–2441, 1986.
12. Hanigan MH, Pitot HC. Gamma-glutamyltranspeptidase: Its role in hepatocarcinogenesis. *Carcinogenesis* 6:165–172, 1985.
13. Maniatis T, Fritsch EF, Sambrook J. *Molecular Cloning: A Laboratory Manual.* Cold Spring Harbor, NY: Cold Spring Harbor Laboratory, 1982, pp 202–206.
14. Ellis RW, DeFeo D, Maryak JM, Young HA, Shih TY, Chang EH, Lowy DR, Scolnick EM. Dual evolutionary origin for the rat genetic sequences of the Harvey murine sarcoma virus. *J Virol* 36:408–420, 1980.
15. Shen-Ong GLC, Keath EJ, Piccoli SP, Cole MD. Novel myc oncogene RNA from abortive immunoglobulin-gene recombination in mouse plasmacytomas. *Cell* 31: 443–452, 1982.
16. Simmons DL, Lalley PA, Kasper CB. Chromosomal assignments of genes coding for components of the mixed-function oxidase system in mice. *J Biol Chem* 260: 515–521, 1985.
17. Bonner T, O'Brien SJ, Nash WG, Rapp UR, Morton CC, Leder P. The human homologs of the raf (mil) oncogene are located on human chromosomes 3 and 4. *Science* 223:71–74, 1984.
18. Heintz N, Sive HL, Roeder RG. Regulation of human histone gene expression: Kinetics of accumulation and changes in the rate of synthesis and in the half-lives of individual histone mRNAs during the HeLa cell cycle. *Mol Cell Biol* 3:539–550, 1983.
19. Coloma J, Pitot HC. Characterization and sequence of cDNA clone of gamma-glutamyltranspeptidase. *Nucleic Acids Res* 14:1393-1403, 1986.
20. Paul D. Molecular cloning of cDNA for rat liver gap junction protein. *J Cell Biol* 103: 123-134, 1986.
21. Schwarz M, Peres G, Beer DG, Maor M, Buchmann A, Kunz W, Pitot HC. Expression of albumin mRNA detected by in situ hybridization in preneoplastic and neoplastic lesions in rat liver. *Cancer Res* 46:5902–5912, 1986.

22. Angerer LM, Angerer RC. Detection of poly A^{+} RNA in sea urchin eggs and embryos by quantitative in situ hybridization. *Nucleic Acids Res* 9:2819–2840, 1981.
23. Cox KH, DeLeon DV, Angerer LM, Angerer RC. Detection of mRNAs in sea urchin embryos by in situ hybridization using asymmetric RNA probes. *Dev Biol* 101:485–502, 1984.
24. Beer DG, Neveu MJ, Paul DL, Rapp UR, Pitot HC. Expression of the c-raf proto-oncogene, γ-glutamyltranspeptidase, and gap junction protein in rat liver neoplasms. *Cancer Res* 48:1610–1617, 1988.
25. Sternberger LA, Hardy PH, Cuculis JJ, Meyer HG. The unlabeled antibody enzyme method of immunohistochemisty: Preparations and properties of soluble antigen-antibody complex (horseradish peroxidase-antihorseradish peroxidase) and its use in identification of spirochetes. *J Histochem Cytochem* 18:315–333, 1970.
26. Emmelot P, Scherer E. The first relevant cell stage in rat liver carcinogenesis: A quantitative approach. *Biochim Biophys Acta* 605:247–304, 1980.
27. Pitot HC, Sirica AE. The stages of initiation and promotion in hepatocarcinogenesis. *Biochim Biophys Acta* 605:191–215, 1980.
28. Nagy P, Evarts RP, Marsden E, Roach J, Thorgeirsson SS. Cellular distribution of c-myc transcripts during chemical hepatocarcinogenesis in rats. *Cancer Res* 48:5522–5527, 1988.
29. Sargent L, Yi-hua X, Sattler GL, Meisner L, Pitot HC. Ploidy and karyotype of hepatocytes isolated from enzyme-altered foci in two different protocols of multi-stage hepatocarcinogenesis in the rat. *Carcinogenesis* 10:387–391, 1989.
30. Bucher NLR, Malt RA. *Regeneration of Liver and Kidney*. Boston: Little Brown, 1971.
31. Moelking K, Heimann B, Beimling P, Rapp UR, Sander T. Serine- and threonine-specific protein kinase activities of purified gag-mil and gag-raf proteins. *Nature* 312:558–561, 1984.
32. Warner AE, Guthrie SC, Gilula NB. Antibodies to gap-junctional protein selectively disrupt junctional communication in the early amphibian embryo. *Nature* 311:127–131, 1984.
33. Loewenstein WR. Junctional intercellular communication and the control of growth. *Biochim Biophys Acta* 560:1–65, 1979.
34. Fraslin J-M, Kneip B, Vaulont S, Glaise D, Munnich A, Guguen-Guillouzo C. Dependence of hepatocyte-specific gene expression on cell-cell interactions in primary culture. *EMBO J* 4:2487–2491, 1985.
35. Trosko JE, Jones C, Chang CC. The role of tumor promoters on phenotypic alterations affecting intercellular communication and tumorigenesis. *Ann NY Acad Sci* 407:316–327, 1983.
36. Janssen-Timmen U, Traub O, Dermietzel R, Rabes HM, Willecke K. Reduced numbers of gap junctions in rat hepatocarcinomas detected by monoclonal antibody. *Carcinogenesis* 7:1475–1482, 1986.
37. Yee AG, Revel J-P. Loss and reappearance of gap junctions in regenerating liver. *J Cell Biol* 78:554–564, 1978.
38. Schulte-Hermann R, Parzefall W, Bursch W. Role of stimulation of liver growth by chemicals in hepatocarcinogenesis. In *Banbury Report 25: Nongenotoxic Mechanisms in Carcinogenesis*. Cold Spring Harbor, NY: Cold Spring Harbor Laboratory, 1987, pp 91–105.
39. Williams GM. Liver carcinogenesis: The role for some chemicals of an epigenetic mechanism of liver tumor promotion involving modification of the cell membrane. *Food Cosmet Toxicol* 19:577–583, 1981.

40. Ruch RJ, Klaunig JE. Effects of tumor promoters, genotoxic carcinogens and hepatocytotoxins on mouse hepatocyte intercellular communication. *Cell Biol Toxicol* 2:469–483, 1986.
41. Loch-Caruso R, Trosko JE. Inhibited intercellular communication as a mechanistic link between teratogenesis and carcinogenesis. *Crit Rev Toxicol* 16:157–183, 1985.
42. Beer DG, Pitot HC. Biological markers characterizing the development of preneoplastic and neoplastic lesions in rodent liver: Mouse liver tumors. *Arch Toxicol [Suppl]* 10:68–80, 1986.
43. Rutenberg AM, Kim H, Fishbein JW, Hawkes JS, Wasserkurg HL, Seligman AM. Histochemical and ultrastructural demonstration of γ-glutamyltranspeptidase activity. *J Histochem Cytochem* 17:517–526, 1969.

Mouse Liver Carcinogenesis: Mechanisms
and Species Comparisons, pages 311–324

Proto-oncogene Activation in Multistage Murine Hepatocarcinogenesis

Henry C. Pitot
McArdle Laboratory for Cancer Research and Departments of Oncology and Pathology, University of Wisconsin, Madison, WI 53706

Heritable abnormalities in the regulation of gene expression have been found in all neoplasms in which such functions have been carefully studied. Unfortunately, until relatively recently, few if any distinctive patterns for these abnormalities could be identified. Recent studies, a number of which are outlined in this book, suggest at least three general groups of genes whose alteration appears to be a characteristic of at least the majority of hepatic murine neoplasms. These areas include genes involved in xenobiotic metabolism; those of the proteins of the gap-junction complex, a structure necessary for normal cell-to-cell interaction; and those regulating the expression of proto-oncogenes, the latter having the common characteristic of transforming cells following quantitative and/or qualitative alterations in their structure and/or expression. Since the topic of this session relates to the latter of these three areas, this introduction will be restricted to a consideration of the role of the alteration of proto-oncogene expression in hepatocarcinogenesis during initiation, promotion, and progression.

Altered Regulation (Activation) of Proto-oncogenes

The identification of specific cellular DNA sequences in the normal cellular genome that are closely related to the transforming genes (oncogenes of retroviruses [see refs. 1,2]) have been termed proto-oncogenes by Temin and others (3) because of their similarity in gene structure, as well as gene product function. Through mutation, which may involve only one or a few nucleotides or deletions of major portions of the gene, proto-oncogenes can develop the ability to transform cells in culture following their transfection into such cells (4). This type of activation of proto-oncogenes, which has been the focus of considerable interest in recent years, may result from the action of a chemical carcinogen or ionizing radiation on the cellular genome. A number of other potential mechanisms of proto-oncogene activation have been described as well. Many of these are listed in Table 1, which is taken from a recent review (5).

From this table, it can be seen that the activation of proto-oncogenes may occur in cells by a number of theoretical, as well as experimentally proven, mechanisms. In general, such activation may occur by a mutational or a transcriptional mechanism, or both. In the former instance, such activation can be identified by the transfection methodology described above, with subsequent alteration of the phenotype of the transfectant toward that of increased malignancy (6–8). On the other hand, mutations outside exons of the proto-oncogene can result in a transcriptional activation of the gene—as seen with insertional mutagenesis (9,10) or chromosomal translocation (11,12), in which transcription of the affected proto-oncogene is markedly enhanced. Transcriptional activation of proto-oncogenes may also occur as a result of gene amplification (13,14), hypomethylation (15), or an alteration in the environmental or intracellular regulation of transcription of the gene that does not directly result from mechanisms described above or in Table 1 (16,17).

Although the examples listed in Table 1 are general and taken from a number of different types of neoplasms in both animals and humans, most of these potential mechanisms for activation have also been demonstrated in

Table 1. Potential Mechanisms of Proto-Oncogene Activation*

Event	Consequence	Examples
Base mutation in coding sequences	New gene product with altered activity	Rous sarcoma and other v-*onc* genes; bladder carcinoma
Deletion in noncoding sequences	Altered regulation of normal gene product	3T3 cell transformation in vitro
Altered promotion for RNA polymerase	Increased transcription of mRNA (normal gene product)	Murine plasmacytoma; lymphoma in chickens
Insertion/substitution with repetitive DNA elements ("transposons")	Altered regulation of gene product (?normal)	Canine venereal tumor; mouse myeloma
Chromosomal translocation	Altered mRNA, new gene product; altered expression of gene	Non-Hodgkin's lymphoma; mouse plasmacytoma
Gene amplification	Increased expression of normal gene	Human colon carcinoma; human neuroblastomas
Hypomethylation of c-*onc* gene	Altered regulation of gene expression of normal gene product	Human colon and lung cancer

*Adapted from reference 5.

murine hepatocarcinogenesis. Since the process of carcinogenesis involves several stages, however, it is important to determine both the types of activation mechanisms involved and the stage(s) at which such activation occurs, as well as its relationship to the malignant characteristics of the cancer itself.

Stages of Murine Hepatocarcinogenesis—Some Characteristics

For this discussion, the characteristics of the stages of initiation, promotion, and progression that have been found primarily in studies in rodent multistage carcinogenesis systems of skin and liver are listed in Table 2. Some of these characteristics will be briefly expanded on here, but for a more detailed discussion, the reader is referred to the original reference (18).

Table 2. Biologic Characteristics of the Stages of Initiation, Promotion, and Progression in Carcinogenesis*

Initiation	Promotion	Progression
Irreversible, with constant stem cell potential	Reversible increase in replication of progeny of the initiated cell population	Irreversible. Measurable and/or morphologically discernible alteration in cell genome's structure
Efficacy sensitive to xenobiotic and other chemical factors	Reversible alterations in gene expression	Growth of altered cells sensitive to environmental factors during early phase
Spontaneous occurrence of initiated cells can be demonstrated	Promoted cell population existence dependent on continued administration of the promoting agent	Benign and/or malignant neoplasms characteristic
Requires cell division for fixation	Efficacy sensitive to dietary and hormonal factors	Progressor agents act to advance promoted cells into this stage but may not be initiating agents
Dose response does not exhibit a readily measurable threshold	Dose response exhibits measurable threshold and maximal effect dependent on dose of initiating agent	Spontaneous progression can be demonstrated
Relative effect of initiators depends on quantitation of focal lesions following defined period of promotion	Relative effectiveness of promoters depends on their ability with constant exposure to cause an expansion of the progeny of the initiated cell population	

*Adapted from reference 18.

Initiation

The process of initiation, the first stage in the natural history of neoplastic development, is permanent and irreversible in multistage hepatocarcinogenesis, as demonstrated in experiments (19) on the extended time that may lapse between initiation and subsequent promotion with virtually no loss in the appearance of lesions. The process of DNA synthesis itself is critical for the fixation and thus the irreversibility of initiation (20,21). Since spontaneous neoplasms occur ubiquitously in the human and all experimental animal species, spontaneous initiation must occur as well. Spontaneous or "fortuitous" initiated cells have been identified in the tissues of experimental animals (22–24) and in humans (25). The total number of cells developing from such spontaneous initiation in the liver increases with the age of the animal (26,27) and after the administration of promoting agents (28). Thus, it is not surprising that promoting agents, while exhibiting no demonstrable interaction with macromolecules, can on extended chronic administration promote cell growth from spontaneous initiated cells and that this can ultimately result in malignancy.

The stage of initiation can also be altered as a result of the environmental modification of the metabolism of initiating agents to their ultimate forms (29). The presence or absence of a threshold or no-effect level of the action of initiating agents has been determined only by extrapolation. Such investigations rely on the appearance of neoplasms as a measure of the process of initiation in the majority of instances (30). Furthermore, the direct measurement of initiation in systems such as multistage hepatocarcinogenesis, which allow the quantitative evaluation of the early clones of initiated cells, also demonstrates that initiation is a linear, dose-related phenomenon not exhibiting a readily measurable threshold (31,32). In this latter system, it is possible to evaluate quantitatively the relative potency of agents as initiators of multistage hepatocarcinogenesis (32).

Promotion

The characteristic of promotion that clearly distinguishes it from the stages of initiation and progression is its reversibility or instability. This characteristic is evident in the model systems of multistage carcinogenesis studied most extensively (33,34). Furthermore, in multistage hepatocarcinogenesis, focal lesions that disappear when application of the promoting agent is discontinued will reappear on readministration of the promoting agent (35). The exact mechanism of the loss of cell populations during promotion on removal of the promoting agent is not clear, although in multistage hepatocarcinogenesis it may be due to individual cell death (36). Furthermore, studies by Hanigan and Pitot (37) demonstrated that hepatic cells in the promotion stage are actually dependent on the presence of the promoting agent in vivo for their continued existence. This phenomenon is completely analogous to the "dependent" neoplasms of endocrine tissues described by Furth (38).

Unlike initiation, promotion can be continually modulated by a variety of environmental alterations, including how frequently the promoting agent is administered (39), the aging process (40), and the composition and amount of the diet (39,41). In concert with the reversible aspect of the stage of promotion, the dose response of promoting agents in murine hepatocarcinogenesis exhibits a threshold or no-effect level and a maximal response, the latter resulting from the fact that a finite number of cells were initiated and the promoting agent possesses no initiating activity itself (42).

Boutwell (43) has also proposed that the principal molecular action of promoting agents is the alteration of gene expression, an effect also reversed on withdrawal of the promoting agent. Many promoting agents exert their effects on gene expression through the mediation of receptor mechanisms (see ref. 44). Regulation of the expression of many genes is altered by the chronic administration of promoting agents such as ornithine decarboxylase (45), many enzymes of xenobiotic metabolism (see ref. 46), γ-glutamyl transpeptidase (47,48), and others. Finally, the relative potency of promoting agents may be related to their effectiveness in expanding the population of initiated cell clones, as seen in multistage hepatocarcinogenesis (32).

Progression

The term progression was originally employed by Leslie Foulds (49) to signify all periods of development of neoplasia beyond the stage of initiation. With the clear demonstration of the stages of initiation and promotion as critical components in the development of a variety of different histogenetic types of neoplasms, such a concept is no longer tenable. However, it is also clear that the appearance of irreversible, aneuploid malignant neoplasms is a component of a stage distinct from that of initiation and promotion (see refs. 50,51). As can be seen from the characteristics of this irreversible stage (Table 2), readily demonstrable changes in the structure of the genome of the neoplastic cell during progression are evident. Furthermore, such changes are directly related to the increased growth rate, invasiveness, metastatic capability, and biochemical changes in the neoplastic cell during this stage. These latter characteristics of progression were emphasized by Foulds (49). However, with a greater understanding of the molecular nature of changes in the cell genome and their effects on the structure and function of the cell, a relationship as defined above can now be shown. That such changes continue to evolve (progress) during the stage of progression has now been shown in the development of multistage hepatocarcinogenesis (52).

Although the usual course of events in the stage of progression is toward greater karyotypic instability (53), cells in the stage of progression may be induced by treatment with specific chemicals to differentiate terminally, thereby removing them from continued progression to an even more malignant state (54). Other environmental alterations, however, may produce changes in gene expression, growth rate, and functional processes within cells during this stage (55). Chemical agents that act on a cell during promotion to convert it to the stage of progression are complete carcinogens (see ref. 44) and, at least theoretically, chemical agents exhibiting only such "progressor" activity (56). Few if any agents have been definitively characterized as progressors in multistage carcinogenesis, although studies by O'Connell et al. (57) indicate that the free radical generator benzoylperoxide may be such an agent, at least in multistage epidermal carcinogenesis. Theoretically, progressor agents must be capable of inducing the genetic changes characteristic of progression and thus should exhibit some degree of clastogenic activity. That putative progressor agents can act in this manner has now been demonstrated with an initiation-promotion-initiation format such as that proposed earlier by Potter (58). With experimental multistage hepatocarcinogenesis in the rat, a similar regimen has been used to induce foci within foci (59,60); these have been proposed to represent the earliest morphologic, discernible beginning of the stage of progression in this tissue (59). The occurrence of foci within foci in many initiation-promotion protocols for rat liver, such as that described by Pitot et al. (61), is relatively infrequent. However, this would be expected if the alteration actually represented the earliest demonstrable lesion at the interface of promotion and progression, since in this latter model system few malignant neoplasms result despite the large number of altered hepatic foci seen during promotion. Thus, the occasional focus in focus in this system probably represents spontaneous or fortuitous progression, analogous to spontaneous initiation, in both instances resulting in a genetically new population. Furthermore, specific morphologic markers can be correlated during the natural history of neoplastic development with the stages of promotion and progression in many instances, in both animals and humans.

Proto-oncogene Activation during the Stages of Hepatocarcinogenesis

The general assumption made has been that either mutational or transcriptional or both mechanisms of proto-oncogene activation play a significant role in the initiation and development of chemically induced liver neoplasms. Although such a theory is quite tenable in view of our knowledge of the role of oncogenes in retroviral induction of cancer and the proto-oncogene analogues of viral oncogenes, activation of proto-oncogenes may also occur in nonneoplastic processes of growth and development of the liver, indicating that such activation is not unique to neoplasms. A survey of proto-oncogene activation in normal, nonneoplastic, preneoplastic, and neoplastic conditions of the liver is presented in Table 3.

Transcription activation of proto-oncogenes in nonneoplastic tissues appears to represent alterations in the regulation of the expression of these genes. As can be seen from Table 3, both normal alterations in growth, e.g., fetal, neonatal, and regenerative growth, as well as dietary changes, result in increased expression of one or more proto-oncogenes. The dietary change described actually is the transition from the fasting state to the fed state, and the administration of glucagon (71). Cirrhotic liver tissue in the human also demonstrates increased expression of at least two different proto-oncogenes (62,63). During liver regeneration in the rat, there was no consistent change in the expression of c-*abl*, c-*src* (70), or c-d-*raf* (69). Transcripts of c-*mos* were undetectable in either normal or regenerating rat liver (70). In primary cultures of hepatocytes, the responses of at least two proto-oncogenes, c-*fos* and c-Ha-*ras* (72), appear to closely mimic responses seen during liver regeneration in the rat. Attempts to determine whether mutational activation or a combination of mutational and transcriptional activation occurs in nonneoplastic tissues have not been reported in rat hepatocarcinogenesis; however, it is likely that such mechanisms play a significant role in normal development, growth, and physiologic change.

Table 3. Proto-oncogene Activation in Nonneoplastic, Preneoplastic, and Neoplastic Liver in vivo

Liver state	Transcriptional activation	Mutational activation
Nonneoplastic		
Cirrhosis (h)	c-*myc*, c-*erb*B (62,63)	
Fetal and neonatal liver (h)	c-*fms*, c-Ha-*ras*, c-*erb*A and B, c-*fos*, c-*myc* (63)	
Fetal and neonatal liver (r)	c-Ha-*ras*, c-*erb*B, c-*myc*, c-*fos* (64)	
Liver regeneration (r)	c-*fos*, c-*myc*, c-N-*myc*, c-Ha-*ras*, c-Ki-*ras*, p53 (64–70)	
Dietary change (r)	c-*myc* (71)	
Primary hepatocyte cult. (r)	c-*fos*, c-Ha-*ras* (72)	
Preneoplasia		
Altered hepatic foci (r)	c-*myc*, c-*src* (75,76), c-Ha-*ras* (77)	
Neoplastic nodule (r)	c-*myc*, c-*src* (75,76)	
Carcinogen feeding (r)	c-Ki-*ras*, c-Ha-*ras* c-*myc* (78,79,80)	
Neoplasia		
Hepatocellular adenoma (m)		c-Ha-*ras* (83)
Hepatocellular carcin. (m)	c-*myc*, c-Ha-*ras*, c-Ki-*ras* (93)	c-Ha-*ras*, c-K-*ras*, c-*raf* (84,85)
Hepatocellular carcin. (r)	c-*myc*, c-Ki-*ras*, c-Ha-*ras*, c-N-*ras* (94–97), c-*fos*, c-*erb*B (64)	c-*raf*, c-N-*ras*, c-Ha-*ras* (89–91), c-Ki-*ras* (87)
Hepatocellular carcin. (h)	c-*myc*, c-*erb*B (62,63,98)	c-*lca* (92)

Abbreviations: h, human liver; r, rat liver; m, mouse liver; carcin., carcinoma. References are given in parentheses following notation of c-*onc* genes.

In preneoplastic lesions, mostly occurring during promotion in multistage hepatocarcinogenesis in the rat (51), only transcriptional activation, the mechanism for which has not been determined, has been reported thus far (Table 3). While the neoplastic nodule may be considered a neoplasm itself in some respects, the preponderance of evidence indicates that this lesion is reversible and unstable (73,74). Thus far, however, it is only in the rat that preneoplastic lesions during hepatocarcinogenesis have been examined for proto-oncogene activation. While mutational activation has not been found thus far during promotion in rat hepatocarcinogenesis, its role during this stage and that of initiation is not yet clear. On the other hand, several authors have demonstrated increases in the expression of some, but not all proto-oncogenes examined during hepatocarcinogenesis (Table 3). As with liver regeneration, no alteration was seen in the expression of c-*src*, while c-*abl* and c-*mos* transcripts could not be detected in either preneoplastic or neoplastic rat livers (78). Evidence that focal lesions expressed higher levels of the c-*myc* proto-oncogene was demonstrated with in situ hybridization in rats following the Solt-Farber model (81). On the other hand, feeding of a choline-devoid diet, which in itself is carcinogenic, did not lead to alterations in proto-oncogene expression in the nonneoplastic liver (82).

In malignant hepatocytes in the stage of progression, both transcriptional and mutational activation has been reported in hepatocellular carcinomas of mice, rats, and humans (Table 3). Mutational activation of the c-Ha-*ras* proto-oncogene was first described in mice as occurring both in adenomas and hepatocellular carcinomas (83). This work was rapidly followed by others (84,85), showing that while only about a third of hepatocellular adenomas demonstrated mutational activation of proto-oncogenes, especially c-Ha-*ras*, more than 80% of hepatocellular carcinomas demonstrated the same phenomenon. Mutational activation of proto-oncogenes in the rat is relatively uncommon in hepatocellular carcinomas induced by diethylnitrosamine (86) but more common in carcinomas resulting from aflatoxin (87) or methyl(acetoxy-methyl)-nitrosamine (88) administration. In the latter two instances, identification of the activated proto-oncogene has not been accomplished in most cases. In humans, activation by a new proto-oncogene, c-*lca*, has been reported (92).

Transcriptional activation of proto-oncogenes in hepatocellular carcinomas has been described in humans, mice, and rats (Table 3). At least for c-*myc* proto-oncogenes in Morris hepatomas, activation appears to be due in part to gene amplification (99). In addition, in at least one human hepatocellular carcinoma, the c-*myc* proto-oncogene was hypomethylated as compared with that in normal liver (100), suggesting another mechanism of transcriptional activation to account for the increased expression of this proto-oncogene seen in some human hepatocellular carcinomas.

Unfortunately, there appears to be no specific pattern to proto-oncogene activation in hepatocellular carcinomas, with the possible exception of the very high incidence of mutational activation of c-Ha-*ras* seen in mouse hepatocellular carcinomas (83–85). Since this does not appear to be a general effect across species, such an activation may be unique to the mouse but still may have a critical role in the carcinogenic process itself. One would also assume that mutational activation of the neoplasms as noted in Table 3 occurred during initiation. The relatively lower incidence of such mutational activation in hepatocellular adenomas, as compared with carcinomas, does not favor this explanation, but in view of the relatively small number of studies that have been carried out as yet, I certainly make no generalization in this area. Thus on the basis of the available data, it is apparent that proto-oncogene activation, both mutational and transcriptional, is seen somewhat more frequently in the stage of progression. Transcriptional but probably not mutational activation of proto-oncogenes occurs during promotion, and mutational activation likely occurs during initiation in some instances. The role of such activation in the carcinogenic process is presumed to be important, although how it fits into carcinogenesis is not yet clear.

References

1. Ratner L, Josephs SF, Wong-Staal F. Oncogenes: Their role in neoplastic transformation. *Annu Rev Microbiol* 39:419–449, 1985.
2. Garrett CT. Oncogenes. *Clin Chim Acta* 156:1–40, 1986.
3. Temin HM. On the origin of RNA tumor viruses. *Annu Rev Genet* 8:155–177, 1974.
4. Cooper GM, Lane M-A. Cellular transforming genes and oncogenesis. *Biochim Biophys Acta* 738:9–20, 1984.
5. Beer DG, Pitot HC. Proto-oncogene activation during chemically induced hepatocarcinogenesis in rodents. *Mutat Res*, 220:1–10, 1989.
6. Kato J-Y, Takeya T, Grandori C, Iba H, Levy JB, Hanafusa H. Amino acid substitutions sufficient to convert the nontransforming $p60^{c\text{-}src}$ protein to a transforming protein. *Mol Cell Biol* 6:4155–4160, 1986.
7. Meijlink F, Curran T, Miller AD, Verma IM. Removal of a 67-base-pair sequence in the noncoding region of protooncogene *fos* converts it to a transforming gene. *Proc Natl Acad Sci USA* 82:4987–4991, 1985.
8. Tabin CJ, Bradley SM, Bargmann CI, Weinberg RA, Papageorge AG, Scolnick EM, Dhar R, Lowy DR, Chang EH. Mechanism of activation of a human oncogene. *Nature* 300:143–149, 1982.
9. Katzir N, Rechavi G, Cohen JB, Unger T, Simoni F, Segal S, Cohen D, Givol D. "Retroposon" insertion into the cellular oncogene c-*myc* in canine transmissible venereal tumor. *Proc Natl Acad Sci USA* 82:1054–1058, 1985.
10. Rechavi G, Givol D, Canaani E. Activation of a cellular oncogene by DNA rearrangement: Possible involvement of an IS-like element. *Nature* 300:607–611, 1982.
11. Fahrlander PD, Sümegi J, Yang J-Q, Wiener F, Marcu KB, Klein G. Activation of the c-myc oncogene by the immunoglobulin heavy-chain gene enhancer after multiple switch region-mediated chromosome rearrangements in a murine plasmacytoma. *Proc Natl Acad Sci USA* 82:3746–3750, 1985.
12. Hayday AC, Gillies SD, Saito H, Wood C, Wiman K, Hayward WS, Tonegawa S. Activation of a translocated human c-*myc* gene by an enhancer in the immunoglobulin heavy-chain locus. *Nature* 307:334–340, 1984.

13. Brodeur GM, Seeger RG. Gene amplification in human neuroblastomas: Basic mechanisms and clinical implications. *Cancer Genet Cytogenet* 19:101–111, 1986.
14. Lin CC, Alitalo K, Schwab M, George D, Varmus HE, Bishop JM. Evolution of karyotypic abnormalities and c-*myc* oncogene amplification in human colonic carcinoma cell lines. *Chromosoma* 92:11–15, 1985.
15. Feinberg AP, Vogelstein B. Hypomethylation of *ras* oncogenes in primary human cancers. *Biochem Biophys Res Commun* 111:47–54, 1983.
16. Yang JQ, Bauer SR, Mushinski JF, Marcu KB. Chromosome translocations clustered 5′ of the murine c-*myc* gene qualitatively affect promoter usage: Implications for the site of normal c-*myc* regulation. *EMBO J* 4:1441-1447, 1985.
17. Marth JD, Overell RW, Meier KE, Krebs EG, Perlmutter RM. Translational activation of the *lck* proto-oncogene. *Nature* 332:171–173, 1988.
18. Pitot HC. The stages in neoplastic development. In: Schottenfeld D, Fraumeni J (eds): *Cancer Epidemiology and Prevention*, 2nd Ed. New York: Oxford University Press, in press.
19. Pitot HC. Drugs as promoters of carcinogenesis. In: Estabrook RW, Lindenlaub E (eds): *The Induction of Drug Metabolism Symposium*, Ashford Castle, Ireland, May 24–27, 1978. Stuttgart, New York: FK Schauttauer Verlag, 1979, pp 471–483.
20. Warwick GP. Effect of the cell cycle on carcinogenesis. *Fed Proc* 30:1760–1765, 1971.
21. Borek C, Sachs L. The number of cell generations required to fix the transformed state in X-ray-induced transformation. *Proc Natl Acad Sci USA* 59:83–85, 1968.
22. Emmelot P, Scherer E. The first relevant cell stage in rat liver carcinogenesis: A quantitative approach. *Biochim Biophys Acta* 605:247–304, 1980.
23. Ethier SP, Ullrich RL. Detection of ductal dysplasia in mammary outgrowths derived from carcinogen-treated virgin female BALB/c mice. *Cancer Res* 42:1753–1760, 1982.
24. Popp JA, Scortichini BH, Garvey LK. Quantitative evaluation of hepatic foci of cellular alteration occurring spontaneously in Fischer-344 rats. *Fundam Appl Toxicol* 5:314–319, 1985.
25. van Bogaert L-J. Mammary hyperplastic and preneoplastic changes: Taxonomy and grading. *Breast Cancer Res Treat* 4:315–322, 1984.
26. Schulte-Hermann R, Timmermann-Trosiener I, Schuppler J. Promotion of spontaneous preneoplastic cells in rat liver as a possible explanation of tumor production by nonmutagenic compounds. *Cancer Res* 43:839–844, 1983.
27. Ogawa K, Onoe T, Takeuchi M. Spontaneous occurrence of gamma-glutamyl transpeptidase-positive hepatocytic foci in 105-week-old Wistar and 72-week-old Fischer 344 male rats. *JNCI* 67:407–412, 1981.
28. Pitot HC, Grosso LE, Goldsworthy T. Genetics and epigenetics of neoplasia: Facts and theories. *Carcinogenesis* 10:65–79, 1985.
29. Wattenberg LW. Chemoprevention of cancer. *Cancer Res* 45:1–8, 1985.
30. Ehling UH, Averbeck D, Cerutti PA, Friedman J, Greim H, Kolbye AC Jr, Mendelsohn ML. Review of the evidence for the presence or absence of thresholds in the induction of genetic effects by genotoxic chemicals. *Mutat Res* 123:281–341, 1983.
31. Scherer E, Emmelot P. Kinetics of induction and growth of precancerous liver-cell foci, and liver tumour formation by diethylnitrosamine in the rat. *Eur J Cancer* 11:689–696, 1975.
32. Pitot HC, Goldsworthy TL, Moran S, Kennan W, Glauert HP, Maronpot RR, Campbell HA. A method to quantitate the relative initiating and promoting potencies of hepatocarcinogenic agents in their dose-response relationships to

altered hepatic foci. *Carcinogenesis* 8:1491–1499, 1987.
33. Glauert HP, Schwarz M, Pitot HC. The phenotypic stability of altered hepatic foci: Effect of the short-term withdrawal of phenobarbital and of the long-term feeding of purified diets after the withdrawal of phenobarbital. *Carcinogenesis* 7:117–121, 1986.
34. Tatematsu M, Nagamine Y, Farber E. Redifferentiation as a basis for remodeling of carcinogen-induced hepatocyte nodules to normal appearing liver. *Cancer Res* 43:5049–5058, 1983.
35. Hendrich S, Glauert HP, Pitot HC. The phenotypic stability of altered hepatic foci: Effects of withdrawal and subsequent readministration of phenobarbital. *Carcinogenesis* 7:2041–2045, 1986.
36. Bursch W, Lauer B, Timmermann-Trosiener I, Barthel G, Schuppler J, Schulte-Hermann R. Controlled death (apoptosis) of normal and putative preneoplastic cells in rat liver following withdrawal of tumor promoters. *Carcinogenesis* 5:453–458, 1984.
37. Hanigan M, Pitot HC. Growth of carcinogen-altered rat hepatocytes in the liver of syngeneic recipients promoted with phenobarbital. *Cancer Res* 45:6063–6070, 1985.
38. Furth J. Hormones and neoplasia. *Thule International Symposia on Cancer and Aging*. Stockholm: Nordiska Bokhandelns Forlag, 1968, pp 131–151.
39. Boutwell RK. Some biological aspects of skin carcinogenesis. *Prog Exp Tumor Res* 4:207–250, 1964.
40. Van Duuren BL, Sivak A, Katz C, Seidman I, Melchionne S. The effect of aging and interval between primary and secondary treatment in two-stage carcinogenesis on mouse skin. *Cancer Res* 35:502–505, 1975.
41. Hendrich S, Glauert HP, Pitot HC. Dietary effects on initiation and promotion of hepatocarcinogenesis in rat. *J Cancer Res Clin Oncol* 114:149–157, 1988.
42. Goldsworthy T, Campbell HA, Pitot HC. The natural history and dose-response characteristics of enzyme-altered foci in rat liver following phenobarbital and diethylnitrosamine administration. *Carcinogenesis* 5:67–71, 1984.
43. Boutwell RK. The function and mechanism of promoters of carcinogenesis. *CRC Crit Rev Toxicol* 2:419–443, 1974.
44. Pitot HC. *Fundamentals of Oncology*, 3rd Ed. New York: Marcel Dekker, 1986.
45. O'Brien TG, Simsiman RC, Boutwell RK. Induction of the polyamine-biosynthetic enzymes in mouse epidermis and their specificity for tumor promotion. *Cancer Res* 35:2426–2433, 1975.
46. Williams GM. Modulation of chemical carcinogenesis by xenobiotics. *Fundam Appl Toxicol* 4:325–344, 1984.
47. Kitagawa T, Watanabe R, Sugano H. Induction of γ-glutamyl transpeptidase activity by dietary phenobarbital in "spontaneous" hepatic tumors of C3H mice. *Gann* 71:536–542, 1980.
48. Sirica AE, Jicinsky JK, Heyer EK. Effect of chronic phenobarbital administration on the gamma-glutamyl transpeptidase activity of hyperplastic liver lesions induced in rats by the Solt/Farber initiation: Selection process of hepatocarcinogenesis. *Carcinogenesis* 5:1737–1740, 1984.
49. Foulds L. Multiple etiologic factors in neoplastic development. *Cancer Res* 25:1339–1347, 1965.
50. Schulte-Hermann R. Tumor promotion in the liver. *Arch Toxicol* 57:147–158, 1985.
51. Pitot HC, Beer D, Hendrich S. Multistage carcinogenesis: The phenomenon under-

lying the theories. In Iversen O (ed): *Theories of Carcinogenesis*. Washington, DC: Hemisphere Press, 1988, pp 159–177.
52. Sargent L, Xu Y-H, Sattler GL, Meisner L, Pitot HC. Ploidy and karyotype of hepatocytes isolated from enzyme-altered foci in two different protocols of multistage hepatocarcinogenesis in the rat. *Carcinogenesis* 10:387–391, 1989.
53. Kraemer PM, Deaven LL, Crissman HA, Steinkamp JA, Petersen DF. On the nature of heteroploidy. *Cold Spring Harbor Symp Quant Biol* 38:133–144, 1974.
54. Reiss M, Gamba-Vitalo C, Sartorelli AC. Induction of tumor cell differentiation as a therapeutic approach: Preclinical models for hematopoietic and solid neoplasms. *Cancer Treat Rep* 70:201–218, 1986.
55. Noble RL. Hormonal control of growth and progression in tumors of Nb rats and a theory of action. *Cancer Res* 37:82–94, 1977.
56. Pitot HC, Beer DG, Hendrich S. Multistage carcinogenesis of the rat hepatocyte. In *Banbury Report 25: Nongenotoxic Mechanisms in Carcinogenesis*. Cold Spring Harbor, NY: Cold Spring Harbor Laboratory, 1987, pp 41–53.
57. O'Connell JF, Klein-Szanto AJP, DiGiovanni DM, Fries JW, Slaga TJ. Enhanced malignant progression of mouse skin tumors by the free-radical generator benzoyl peroxide. *Cancer Res* 46:2863–2865, 1986.
58. Potter VR. A new protocol and its rationale for the study of initiation and promotion of carcinogenesis in rat liver. *Carcinogenesis* 2:1375–1379, 1981.
59. Scherer E. Relationship among histochemically distinguishable early lesions in multistep-multistage hepatocarcinogenesis. *Arch Toxicol [Suppl]* 10:81–94, 1987.
60. Farber E. Hyperplastic liver nodules. *Methods Cancer Res* 7:345–375, 1973.
61. Pitot HC, Barsness L, Goldsworthy T, Kitagawa T. Biochemical characterization of stages of hepatocarcinogenesis after a single dose of diethylnitrosamine. *Nature* 271:456–458, 1978.
62. Himeno Y, Fukuda Y, Hatanaka M, Imura H. Expression of oncogenes in human liver disease. *Liver* 8:208–212, 1988.
63. Zhang X-K, Huang D-P, Chiu D-K, Chiu J-F. The expression of oncogenes in human developing liver and hepatomas. *Biochem Biophys Res Commun* 142:932–938, 1987.
64. Zhang X-K, Wang Z, Lee A, Huang D-P, Chiu J-F. Differential expression of cellular oncogenes during rat liver development. *Cancer Lett* 41:147–155, 1988.
65. Makino R, Hayashi K, Sugimura T. c-*myc* transcript is induced in rat liver at a very early stage of regeneration or by cycloheximide treatment. *Nature* 310:697–698, 1984.
66. Corral M, Tichonicky L, Guguen-Guillouzo C, Corcos D, Raymondjean M, Paris B, Kruh J, Defer N. Expression of c-*fos* oncogene during hepatocarcinogenesis, liver regeneration and in synchronized HTC cells. *Exp Cell Res* 160:427–434, 1985.
67. Thompson NL, Mead JE, Braun L, Goyette M, Shank PR, Fausto N. Sequential protooncogene expression during rat liver regeneration. *Cancer Res* 46:3111–3117, 1986.
68. Corral M, Paris B, Guguen-Guillouzo C, Corcos D, Kruh J, Defer N. Increased expression of the N-*myc* gene during normal and neoplastic rat liver growth. *Exp Cell Res* 174:107–115, 1988.
69. Hsieh LL, Hsiao W-L, Peraino C, Maronpot RR, Weinstein IB. Expression of retroviral sequences and oncogenes in rat liver tumors induced by diethylnitrosamine. *Cancer Res* 47:3421–3424, 1987.
70. Goyette M, Petropoulos CJ, Shank PR, Fausto N. Regulated transcription of c-Ki-

ras and c-*myc* during compensatory growth of rat liver. *Mol Cell Biol* 4:1493–1498, 1984.

71. Corcos D, Vaulont S, Denis N, Lyonnet S, Simon M-P, Kitzis A, Kahn A, Kruh J. Expression of c-*myc* is under dietary control in rat liver. *Oncogene Res* 1:193–199, 1987.
72. Kruijer W, Skelly H, Botteri F, van der Putten H, Barber JR, Verma IM, Leffert HL. Proto-oncogene expression in regenerating liver is simulated in cultures of primary adult rat hepatocytes. *J Biol Chem* 261:7929–7933, 1986.
73. van der Heijden CA, Dormans JAMA. Short-term induction of neoplastic nodules in the rat liver. II. Study of their development and the effects of withdrawal of 2-acetylaminofluorene. *Carcinogenesis* 2:147–156, 1981.
74. Bannasch P, Moore MA, Hacker HJ, Klimek F, Mayer D, Enzmann H, Zerban H. Potential significance of phenotypic instability in focal and nodular liver lesions induced by hepatocarcinogens. In Brunner H, Thaler H (eds): *Hepatology: A Festschrift for Hans Popper.* New York: Raven Press, 1985, pp 191–209.
75. Ito S, Watanabe T, Abe K, Yanaihara N, Tateno C, Okuno Y, Yoshitake A, Miyamoto J. Immunohistochemical demonstration of the c-*myc* oncogene product in rat chemical hepatocarcinogenesis. *Biomed Res* 9:177–180, 1988.
76. Richmond RE, Pereira MA, Carter JH, Carter HW, Long RE. Quantitative and qualitative immunohistochemical detection of myc and src oncogene proteins in normal, nodule, and neoplastic rat liver. *J Histochem Cytochem* 36:179–184, 1988.
77. Galand P, Jacobovitz D, Alexandre K. Immunohistochemical detection of c-Ha-*ras* oncogene p21 product in pre-neoplastic and neoplastic lesions during hepatocarcinogenesis in rats. *Int J Cancer* 41:155–161, 1988.
78. Yaswen P, Goyette M, Shank PR, Fausto N. Expression of c-Ki-*ras*, c-Ha-*ras*, and c-*myc* in specific cell types during hepatocarcinogenesis. *Mol Cell Biol* 5:780–786, 1985.
79. Cote GJ, Lastra BA, Cook JR, Huang D-P, Chiu J-F. Oncogene expression in rat hepatomas and during hepatocarcinogenesis. *Cancer Lett* 26:121–127, 1985.
80. Nagy P, Evarts RP, Marsden E, Roach J, Thorgeirsson SS. Cellular distribution of c-*myc* transcripts during chemical hepatocarcinogenesis in rats. *Cancer Res* 48:5522–5527, 1988.
81. Solt DB, Farber E. New principle for the analysis of chemical carcinogenesis. *Nature* 263:702–703, 1976.
82. Chandar N, Lombardi B, Schulz W, Locker J. Analysis of *ras* genes and linked viral sequences in rat hepatocarcinogenesis. *Am J Pathol* 129:232–241, 1987.
83. Fox TR, Watanabe PG. Detection of a cellular oncogene in spontaneous liver tumors of $B6C3F_1$ mice. *Science* 228:596–597, 1985.
84. Reynolds SH, Stowers SJ, Maronpot RR, Anderson MW, Aaronson SA. Detection and identification of activated oncogenes in spontaneously occurring benign and malignant hepatocellular tumors of the $B6C3F_1$ mouse. *Proc Natl Acad Sci USA* 83:33–37, 1986.
85. Wiseman RW, Stowers SJ, Miller EC, Anderson MW, Miller JA. Activating mutations of the c-Ha-*ras* protooncogene in chemically induced hepatomas of the male $B6C3F_1$ mouse. *Proc Natl Acad Sci USA* 83:5825–5829, 1986.
86. Stowers SJ, Wiseman RW, Ward JM, Miller EC, Miller JA, Anderson MW, Eva A. Detection of activated proto-oncogenes in N-nitrosodiethylamine-induced liver tumors: A comparison between $B6C3F_1$ mice and Fischer 344 rats. *Carcinogenesis* 9:271–276, 1988.

87. McMahon G, Hanson L, Lee J-J, Wogan GN. Identification of an activated c-Ki-*ras* oncogene in rat liver tumors induced by aflatoxin B1. *Proc Natl Acad Sci USA* 83:9418–9422, 1986.
88. Goyette M, Dolan M, Kaufmann W, Kaufman D, Shank PR, Fausto N. Transforming activity of DNA from rat liver tumors induced by the carcinogen methyl(acetoxymethyl)nitrosamine. *Mol Carcinogenesis* 1:26–32, 1988.
89. Ishikawa F, Takaku F, Hayashi K, Nagao M, Sugimura T. Activation of rat c-*raf* during transfection of hepatocellular carcinoma DNA. *Proc Natl Acad Sci USA* 83:3209–3212, 1986.
90. Funato T, Yokota J, Kasamoto H, Kameya T, Fukushima S, Ito N, Terada M, Sugimura T. Activation of N-*ras* gene in a rat hepatocellular carcinoma induced by dibutylnitrosamine and butylated hydroxytoluene. *Jpn J Cancer Res* 78:689–694, 1987.
91. Ishikawa F, Takaku F, Nagao M, Ochiai M, Hayashi K, Takayama S, Sugimura T. Activated oncogenes in a rat hepatocellular carcinoma induced by 2-amino-3-methylimidazo[4,5-f]quinoline. *Jpn J Cancer Res* 76:425–428, 1985.
92. Ochiya T, Fujiyama A, Fukushige S, Hatada I, Matsubara K. Molecular cloning of an oncogene from a human hepatocellular carcinoma. *Proc Natl Acad Sci USA* 83:4993–4997, 1986.
93. Beer DG, Pitot HC. Biological markers characterizing the development of preneoplastic and neoplastic lesions in rodent liver. *Arch Toxicol [Suppl]* 10:68–80, 1987.
94. Corcos D, Defer N, Raymondjean M, Paris B, Corral M, Tichonicky L, Kruh J, Glaise D, Saulnier A, Guguen-Guillouzo C. Correlated increase of the expression of the c-*ras* genes in chemically induced hepatocarcinomas. *Biochem Biophys Res Commun* 122:259–264, 1984.
95. Tashiro F, Morimura S, Hayashi K, Makino R, Kawamura H, Horikoshi N, Nemoto K, Ohtsubo K, Sugimura T, Ueno Y. Expression of the c-Ha-*ras* and c-*myc* genes in aflatoxin B1-induced hepatocellular carcinomas. *Biochem Biophys Res Commun* 138:858–864, 1986.
96. Cote GJ, Chiu J-F. The expressions of oncogenes and liver-specific genes in Morris hepatomas. *Biochem Biophys Res Commun* 143:624–629, 1987.
97. Makino R, Hayashi K, Sato S, Sugimura T. Expressions of the c-Ha-*ras* and c-*myc* genes in rat liver tumors. *Biochem Biophys Res Commun* 119:1096–1102, 1984.
98. Su T-S, Lin L-H, Lui W-Y, Chang CM, Chou C-K, Ting L-P, Hu C-P, Han S-H, P'eng F-K. Expression of c-*myc* gene in human hepatoma. *Biochem Biophys Res Commun* 132:264–268, 1985.
99. Hayashi K, Makino R, Sugimura T. Amplification and over-expression of the c-*myc* gene in Morris hepatomas. *Gann* 75:475–478, 1984.
100. Nambu S, Inoue K, Sasaki H. Site-specific hypomethylation of the c-*myc* oncogene in human hepatocellular carcinoma. *Jpn J Cancer Res* 78:695–704, 1987.

Mouse Liver Carcinogenesis: Mechanisms
and Species Comparisons, pages 325–334

Oval Cells and Liver Carcinogenesis: An Analysis of Cell Lineages in Hepatic Tumors Using Oncogene Transfection Techniques

Nelson Fausto
Department of Pathology and Laboratory Medicine,
Brown University, Providence, RI 02916

A major goal of present-day cancer research is the development of methods to detect tumors at an early stage of growth, before local invasion and metastasis occur. To accomplish this objective, it is necessary to identify and characterize cells that can function as tumor progenitors long before a tumor is formed in tissues of carcinogen-treated animals. The identification of tumor stem cells makes it possible to trace the natural history of a tumor and to study the molecular events involved in tumorigenesis in a defined cell population that is probably the primary target for carcinogen action. Furthermore, a cell that serves as a tumor progenitor may also be expected to function as a stem cell in the normal tissue. The recognition of carcinogen-induced alterations in such cells may eventually lead to the establishment of techniques aimed at selectively eliminating abnormal stem cells from animal tissues soon after carcinogen exposure.

In a general sense, initiated cells can be considered tumor progenitors. The major difficulty with this concept as it relates to liver carcinogenesis is that initiation remains an operational definition, without a clear molecular correlate. Furthermore, in most of the experimental models of chemical carcinogenesis in the rat liver, tumorigenesis is preceded by the development of hyperplastic nodules presumed to be the sites where neoplasms originate. However, no more than 5%–10% of these nodules persist, and there are no reliable markers to distinguish nodules that will become persistent from those that regress (1–3).

Studies of cell lineages in liver tumors do not have to rely on concepts of initiation and promotion. At least two types of strategies can be used: the study of the expression of relevant genes in specific cell types by in situ hybridization and immunohistochemical techniques and the analysis of tumors produced by cell populations isolated from livers of normal or carcinogen-fed rats and transformed in culture. These general methods have been used by various laboratories for lineage studies of hepatic tumors (4–8). Our own work has involved isolating liver epithelial cells (commonly referred to as oval cells) from livers of rats fed a carcinogenic diet for a brief period, establishing the cells in culture, and transfecting them with an oncogene. We then compare (histologically and by electron microscopy, monoclonal antibody reactivity, and nucleic acid hybridization) the tumors produced by these cells after injection into nude mice or weanling rats with the tumors that develop in the liver of rats fed the carcinogenic diet for 35–50 weeks (4,8, and M. Goyette, R. Faris, L. Braun, D. Hixon, and N. Fausto, submitted).

Oval Cells and the Analysis of Cell Lineages in Hepatocarcinogenesis

Oval cell proliferation (under various names) was first described more than 50 years ago (see ref. 9 for review). Farber, who coined the term oval cell in 1956, noted these cells in the liver of rats exposed to three different chemical carcinogens (10). This and subsequent work established that oval cells play a prominent part in rat liver carcinogenesis induced by azo dyes. Several investigators showed that in livers of rats exposed to these chemicals, oval cells can differentiate into hepatocytes and that intermediate forms between these two cell types may exist (11). During the last few years, studies on oval cells have been carried out in many different models of liver carcinogenesis, as well as in noncarcinogenic hepatic injury (12,13). Progress has been rapid, in great part due to the development of cell isolation techniques, the availability of monoclonal antibodies that recognize different cell types in the liver, the establishment of methods to maintain liver epithelial cells in culture in an immortalized but untransformed state, and the use of chemicals or oncogene transfection to successfully transform these cells and render them tumorigenic (6,7,9,14–17).

All these advances make it possible to use oval cells in lineage studies not yet achievable with hepatocytes and to learn a great deal about the developmental potential of these cells. Studies from our and other laboratories show that transformed liver epithelial cells can form well-differentiated hepatocellular carcinomas and thus serve as stem cells for these neoplasms (4,6). Furthermore, we and others have shown that oval cells possess dual phenotypic traits; that is, they have antigens and enzymes present in fetal liver and adult hepatocytes (9,18–20). Cells with similar features exist at some stages of embryonic development of rat and mouse liver and in adult livers of rats, mice, and humans (21–24).

Oval Cells: A Specific Cell Type or a Cell Compartment?

"Oval cells" do not actually constitute a homogeneous population but are instead a compartment consisting of multiple epithelial cell types possessing different developmental potential that probably derive from different segments of the intrahepatic biliary duct system (9). The majority of these cells are well-differentiated bile duct cells, strictly committed to the bile duct lineage. The oval cell compartment contains another cell type, most likely originating from bile ductules (Hering canals), that has broader developmental potential and can evolve either into the hepatocyte or ductular cells. Furthermore, liver ductular cells can undergo metaplasia and become intestinal epithelium (25). Thus, the term "oval cell" is not meaningful if only a simple morphologic description is given.

It is not surprising that cells with broad developmental potential can be detected in adult liver during carcinogenesis. It has become evident that even adult differentiated tissues possess a great plasticity in their cellular composition. A very interesting illustration of this type of cell modulation is the finding that fully differentiated hepatocytes can be generated from pancreatic ducts in adult rats (26).

Epithelial Stem Cells in Normal Liver

The epithelial stem cells detected in the livers of carcinogen-treated rats must either exist as such in the normal liver or constitute a population that gained new developmental properties as a consequence of carcinogen action. Although not entirely proven, several arguments suggest that stem-like cells exist in normal adult rat liver. Nonparenchymal cells containing α-fetoprotein mRNA can be detected in normal adult rat liver with sensitive nucleic acid hybridization probes (24). These cells proliferate after liver injury produced by noncarcinogenic agents, as shown by nucleic acid hybridization and immunohistochemical techniques (J. Lemire, N. Shiojiri, N. Fausto, submitted). Also, nonparenchymal cells from normal livers can produce a variety of tumors when transformed in culture, including hepatoblastomas, hepatocellular carcinomas, and cholangiocarcinomas (4,6,8, and M. Goyette, R. Faris, L. Braun, D. Hixson, N. Fausto, submitted).

However, although the liver may contain cells with stem cell potential, such cells apparently function as hepatocyte progenitors only under special situations. Indeed, in liver regeneration after partial hepatectomy, mature hepatocytes divide and replenish the organ without the help of a stem cell compartment. Liver stem cells seem to form a reserve or facultative compartment that becomes active only when hepatocyte damage is relatively extensive (27). Obviously, the proliferation of this cell type, although common, is not an obligatory event in carcinogenesis; the extent of proliferation probably depends on the nature of the agent and its effect on the liver.

Oval Cells in Mouse and Human Liver

If one accepts the definition of oval cells in rat liver as cells that derive from bile ductules and retain the broad developmental capabilities they displayed during development, it is logical to ask whether similar cells also exist in human and mouse liver. Studies with human liver using cytokeratin markers show that oval cells proliferate in many diseases in humans and might perhaps be a component of human hepatocellular carcinomas (28). The most interesting demonstration of epithelial stem cells in human liver is in patients with massive hepatic failure. In this condition, hepatocytes are unable (because of low numbers, extensive injury, or lack of hepatic architectural framework) to replenish the tissue. Instead, new hepatocytes appear to be generated from a cell type that has many of the features of oval cells, including the presence in the same cell of antigens found in fetal and adult liver (23).

There is clear evidence that ductular cells with the potential to differentiate into hepatocytes are also present in mouse liver. Some of the early, basic studies on the modulation of differentiation of ductular cells were done in mice subjected to nutritional injury (22,29). Oval cells similar to those of rat liver have also been described in ethionine-induced mouse liver hepatocarcinogenesis and in mice given CCl_4 (30,31). A recent report shows that epithelial cells isolated from normal C3H mice can be immortalized and transformed in culture (32). These cells, which have similar characteristics to the facultative stem cells isolated from livers of normal and carcinogen-treated rats, produced trabecular hepatocellular carcinomas when injected into nude mice. Finally, current work shows that oval cells are present in livers of transgenic mice carrying hepatitis virus, at the time that neoplastic transformation occurs (H. Dunsford, personal communication).

Evidence That Transformed Liver Epithelial Cells Can Generate Hepatocellular Carcinomas

To prove that certain types of liver epithelial cells function as stem cells, it is essential to show that such cells can generate differentiated hepatocellular carcinomas. For our studies examining this assumption, we fed rats a choline-deficient diet containing 0.1% ethionine for 2–6 weeks to elicit the proliferation of liver epithelial cells (9,18,33). We then isolated liver epithelial cells using collagenase/pronase/DNase digestion and then by centrifugal elutriation. The isolated epithelial cell fraction was not contaminated by hepatocytes but included a small proportion (~10%) of Kupffer cells (9,18,33). We established cultures of liver epithelial cells and analyzed their biologic properties (proto-oncogene and growth factor expression, cell cycle regulation, karyotypes, growth characteristics). The cells became immortalized but did not grow in soft agar or produce tumors when injected into nude mice. Cultures kept for more than 50 passages grew in soft agar if epidermal growth factor was present, but cells from these colonies were also not tumorigenic (4,8, and M. Goyette, R. Faris, L. Braun, D. Hixson, N. Fausto, submitted).

We transfected liver epithelial cells at the 25–30th passage with various plasmids (4, and M. Goyette, R. Faris, L. Braun, D. Hixson, N. Fausto, submitted) containing the normal proto-oncogene c-Ha-*ras*, the activated form of this oncogene (EJ gene), and a drug selection gene (*neo*). pSV2 *neo*/Ha-*ras* is a plasmid that contains the *neo* gene and the normal human cellular proto-oncogene c-

Ha-*ras*. The *neo* gene codes for aminoglycoside transferase, which enables the cell to survive in medium containing the antibiotics neomycin or G418; this gene was introduced into the cells to provide a way to select for cells that contain this marker gene linked to the gene of interest. Since in the pSV2 *neo*/Ha-*ras* plasmid, *neo* is linked to c-Ha-*ras*, cells that have incorporated human c-Ha-*ras* can be selected by growing transfected cultures in medium containing the antibiotic G418. We also used the plasmid pSV2 *neo*/EJ, which contains *neo* and the EJ gene (an oncogene derived from the normal human c-Ha-*ras* proto-oncogene). While the proto-oncogene does not transform NIH 3T3 fibroblasts upon transfection, the EJ oncogene is capable of inducing transformation in these cells. The EJ oncogene was isolated from bladder carcinoma cell lines and differs from the normal allele by a single point mutation in codon 12, which converts the 12th amino acid of the protein coded by the gene from glycine to valine. This involves a change in the codon from GGC to GTC. Proteins derived from the normal or mutant genes are referred to as p21 *ras* proteins to indicate their molecular weight. The construction of pSV2 *neo*/Ha-*ras* and pSV2 *neo*/EJ is identical, with the exception that the plasmids contain either the normal cellular proto-oncogene or the transforming allele. Both contain the *neo* marker gene.

We also used the plasmid pGV-16 cotransfected with pUC/EJ. In this cotransfection system, the *neo* and *ras* genes are in separate plasmids. pGV-16 contains the *neo* gene but no *ras* genes; pUC/EJ contains the EJ gene but not the *neo* gene. We used this system to check the efficiency of transfection of each of these genes separately and to determine whether the presence of the *neo* gene altered the tumorigenicity of EJ. The pGV-16 plasmid by itself was transfected into liver epithelial cells as a control to determine whether the transfection procedure itself, in the absence of normal or abnormal *ras* genes, might cause cell transformation.

Selection of transfected cells was accomplished in several steps. Liver epithelial cells transfected with the various plasmids were maintained for two weeks in medium containing the antibiotic G418 to select the cells that contained the transfected genes. Cells not containing *neo* die when grown in this medium. DNA was extracted from these cells and analyzed by the Southern blot technique to ensure that they contained copies of the normal human c-Ha-*ras* or the oncogenic homologue EJ. These genes can be distinguished from the normal endogenous c-Ha-*ras* gene by Southern analysis.

For tumor production (4,8, and M. Goyette, R. Faris, L. Braun, D. Hixson, N. Fausto, submitted), cells containing either human c-Ha-*ras* or EJ were injected subcutaneously into nude mice. After approximately 45 days, tumors developed in all eight injected mice. Histologic analysis of the tumors showed that seven of the eight tumors were differentiated trabecular hepatocellular carcinomas (Fig. 1). The other tumor was a poorly differentiated hepatocellular carcinoma. None of the mice injected with plasmids containing the normal c-Ha-*ras* or with plasmids containing the neo gene alone developed tumors (3–5 months' observation).

Further analysis of the tumors and of tumor-derived cell lines (4,8, and M. Goyette, R. Faris, L. Braun, D. Hixon, N. Fausto, submitted) revealed that all the

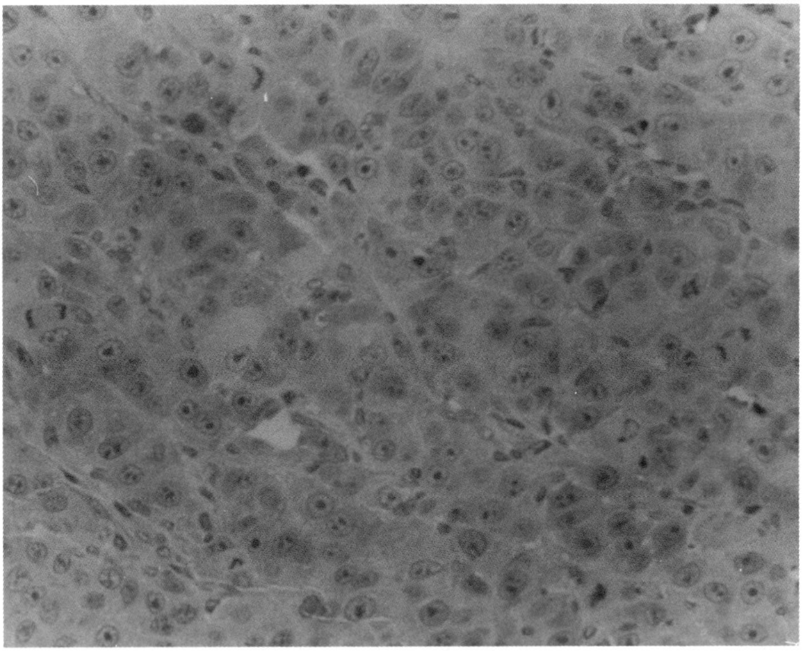

Figure 1. Trabecular hepatocellular carcinoma produced in nude mouse by subcutaneous injection of oval cells transfected with the activated c-Ha-*ras* oncogene (EJ gene).

tumors and clonal cell lines derived from them contained α-fetoprotein mRNA and that one clonal line produced albumin mRNA. Although all tumor-derived clonal cell lines were tumorigenic, GGT expression was variable and could not be detected in 2 of 10 lines. Two clonal lines and the tumors they produced in either nude mice or weanling rats were studied in detail using a panel of monoclonal antibodies that recognize oval cells and mature hepatocytes. The tumors reacted with antibodies that recognize either cell type. Serial sections indicated that oval cell and hepatocyte antigens recognized by the antibodies were present in the same cell.

Conclusions

New techniques have made it possible to analyze the expression of various genes in different cell types, as well as to study cell lineages in hepatocarcinogenesis. Some of the results obtained with methods such as in situ hybridization and oncogene transfection clearly establish that rat, mouse, and probably human liver contain a facultative stem cell compartment, which may be activated during carcinogenesis induced by certain agents, as well as in various types of noncarcinogenic liver injury. As more data are gathered about these cells, their relationships to counterparts that exist during liver development and their role in normal and abnormal hepatic growth will become better understood. Of great interest is the finding that oval cells not only may generate differentiated hepatocellular carcinomas but also are a major source for growth factors during chemically induced rat liver carcinogenesis (L. Braun, P. Gruppuso, R. Mikumo, N. Fausto, in preparation) and in woodchuck carcinogenesis mediated by hepatitis virus (34).

Acknowledgments

I thank Mrs. Anna Louise Baxter for her help in preparing the manuscript. Work from my laboratory was supported by grants CA23226 and CA35249 from the National Cancer Institute.

Open Forum

Dr. James Trosko: Dr. Fausto, Jim Klaunig, Dr. Goodman, and I have generated data essentially repeating much of what you have done, starting with the rat epithelial cells and the Harvey-*ras* oncogene. However, we were primarily trying to determine whether cell-to-cell communication would be affected by expression of *ras*, and whether it would be correlated with tumorigenesis when the cells were put into the rat. We found that we were able to get these *ras*-transfected cells to show diminished cell-to-cell communication and increased tumorigenicity when they were put back into the portal vein of the rat.

Dr. Samuel French: Desmet in Belgium has studied this phenomenon in the embryo in the development of bile ducts from normal liver cells. Have you considered this phenomenon, particularly in terms of trying to use methods other than transfection with an oncogene (e.g., retinoic acid exposure) to produce differentiation?

Dr. Nelson Fausto: In early development of both the mouse and rat, α-fetoprotein (AFP)- and albumin-positive cells are detected before there is a separation of lineages between ductular cells and hepatocytes. Surprisingly, some AFP- and albumin-containing precursor cells have the capacity to enter two different paths of differentiation, to generate either bile ducts or mature hepatocytes. This has been well described by Dr. V. Desmet in Belgium and Dr. N. Shiojiri in Japan. Dr. N. Marceau has done elegant work using cultures from fetal liver cells and has shown that it is possible to modulate the differentiation path of these cells in culture. We are vigorously trying to find a system in which full hepatocyte differentiation of epithelial cells occurs without their transformation. We have not yet been successful.

Dr. Stringner Sue Yang: I have a question about the pSV *neo* vector. Does it have an SV40 promoter carrying the T antigen coding sequence—at least partially if not completely? Did you try any control?

Dr. Fausto: That is why we used various types of plasmids in the transfections. Cells were transfected with identical plasmids carrying either the normal c-Ha-*ras* or the mutant allele. Cells transfected with the normal gene were not tumorigenic. In addition, we did other controls using plasmids containing the *neo* gene alone, and these cells also were not tumorigenic. Regardless of the plasmids used for transfection, the only cells that were tumorigenic were those that received the mutant gene (EJ gene).

Dr. Yang: I noticed that, too, because I also use that particular vector carrying a human gene. The difference is the time course in inducing the tumor. I noticed that solid tumors form very fast with the transfected cells. However, if the tumor goes a long time, other expression is induced within the animal. I don't know if you looked at that. I have seen a very dramatic lymphadenopathy.

Dr. Fausto: No, we have not. We remove the tumors when they are 1–2 cm in diameter and do not wait for further growth or tumor spread.

Dr. Yang: My point was that obviously the T antigen gene, as well as the SV40 promoter, is working there also. Of course, it depends on how you link the gene together.

Dr. Fausto: That's right. I think that we obtained differentiated tumors because we used an inefficient method of transfection that resulted in the incorporation of 1–10 copies of the normal gene or the oncogene in the cell genome. Had we used retrovirus vectors, we might have obtained different results.

References

1. Farber E, Sarma DSR. Hepatocarcinogenesis: A dynamic cellular perspective. *Lab Invest* 56:4–22, 1987.
2. Pitot HC, Sirica AE. The stages of initiation and promotion in hepatocarcinogenesis. *Biochim Biophys Acta* 605:191–215, 1980.
3. Kaufmann WK, Mackenzie SA, Kaufman DG. Quantitative relationship between hepatocytic neoplasms and islands of cellular alteration during hepatocarcinogenesis in the male F344 rat. *Am J Pathol* 119:171–174, 1985.
4. Braun L, Goyette M, Yaswen P, Thompson NL, Fausto N. Growth in culture and tumorigenicity after transfection with the ras oncogene of liver epithelial cells from carcinogen-treated rats. *Cancer Res* 47:4116–4124, 1987.
5. Evarts RP, Nagy P, Marsden E, Thorgeirsson SS. In situ hybridization studies on expression of albumin and α-fetoprotein during the early stage of neoplastic transformation in rat liver. *Cancer Res* 47:5469–5475, 1987.
6. Tsao MS, Grisham JW. Hepatocarcinomas, cholangiocarcinomas and hepatoblastomas produced by chemically transformed cultured rat epithelial cells: A light- and electron-microscopic analysis. *Am J Pathol* 127:168–181, 1987.
7. Germain L, Blouin MJ, Marceau N. Biliary epithelial and hepatocytic cell lineage relationships in embryonic rat liver as determined by the differential expression of cytokeratins, α-fetoprotein, albumin and cell surface-exposed components. *Cancer Res* 48:4909–4918, 1988.
8. Braun L, Mikumo R, Fausto N. Production of hepatocellular carcinoma by oval

cells: Cell cycle expression of c-myc and p53 at different stages of oval cell transformation. *Cancer Res* 49:1554–1561, 1989.

9. Fausto N, Thompson NL, Braun L. Purification and culture of oval cells from rat liver. In Pretlow TG II, Pretlow TP (eds): *Cell Separation: Methods and Selected Applications*, Vol 4. Orlando, FL: Academic Press, 1987, pp 45–77.
10. Farber E. Similarities in the sequence of early histological changes induced in the liver of the rat by ethionine, 2-acetylaminofluorene and 3′-methyl-4-dimethylaminoazobenzene. *Cancer Res* 16:142–149, 1956.
11. Ogawa K, Minase T, Onoe T. Demonstration of glucose-6-phosphatase activity in the oval cells of rat liver and the significance of the oval cells in azo dye carcinogenesis. *Cancer Res* 34:3379–3386, 1974.
12. Evarts RP, Nagy P, Marsden E, Thorgeirsson SS. A precursor-product relationship exists between oval cells and hepatocytes in rat liver. *Carcinogenesis* 8:1737–1740, 1987.
13. Tournier I, Legres L, Schoevaert D, Feldmann G, Bernuau D. Cellular analysis of α-fetoprotein gene activation during carbon tetrachloride and D-galactosamine-induced acute liver injury in rats. *Lab Invest* 59:657–665, 1988.
14. Sirica AE, Cihla HP. Isolation and partial characterizations of oval and hyperplastic bile ductular cell-enriched populations from the liver of carcinogen and non-carcinogen-treated rats. *Cancer Res* 44:3454–3466, 1984.
15. Hixson DC, Allison JP. Monoclonal antibodies recognizing oval cells induced in the liver of rats by N-2-fluorenylacetamide or ethionine in a choline-deficient diet. *Cancer Res* 45:3750–3760, 1985.
16. Sell S, Leffert HL. An evaluation of cellular lineages in the pathogenesis of experimental hepatocellular carcinoma. *Hepatology* 2:77–86, 1982.
17. Lombardi B. On the nature, properties and significance of oval cells. In Pani P, Feo F, Columbano A (eds): *Recent Trends in Chemical Carcinogenesis*, Vol 1. Cagliary, Italy: ESA, 1982, pp 36–56.
18. Hayner NT, Braun L, Yaswen P, Brooks M, Fausto N. Isozyme profiles of oval cells, parenchymal cells and biliary cells isolated by centrifugal elutriation from normal and preneoplastic livers. *Cancer Res* 44:332–338, 1984.
19. Tsao MS, Smith JD, Nelson KG, Grisham JW. A diploid epithelial cell line from normal adult rat liver with phenotypic properties of "oval" cells. *Exp Cell Res* 154: 38–52, 1984.
20. Desmet VJ. Modulation of biliary epithelium. In Reutter W, Heinrich PC, Popper H, Keppler D, Arias IM (eds): *Modulation of Liver Cell Expression*. Falk Symposium 43. Lancaster: MTP Press, pp 195–214, 1987.
21. Shiojiri N. Enzymo- and immunocytochemical analyses of the differentiation of liver cells in the prenatal mouse. *J Embryol Exp Morphol* 62:139–152, 1981.
22. Leduc EH. Cell modulation in liver pathology. *J Histochem Cytochem* 7:253–255, 1959.
23. Gerber MA, Thung SN, Shen S, Stromeyer FW, Ishak KG. Phenotypic characterization of hepatic proliferation: Antigen expression by proliferating epithelial cells in fetal liver, massive hepatic necrosis and nodular transformation of the liver. *Am J Pathol* 110:70–74, 1983.
24. Petropoulos CJ, Yaswen P, Panzica M, Fausto N. Cell lineages in liver carcinogenesis: Possible clues from studies of the distribution of α-fetoprotein RNA sequences in cell populations isolated from normal, regenerating and preneoplastic rat liver. *Cancer Res* 45:5762–5768, 1985.

25. Tatematsu M, Kaku T, Medline A, Farber E. Intestinal metaplasia as a common option of oval cells in relation to cholangiofibrosis in liver of rats exposed to 2-acetylaminofluorene. *Lab Invest* 52:354–362, 1985.
26. Rao MS, Dwivedi RS, Subbarao V, Usman MI, Scarpelli DG, Nemali MR, Yeldandi A, Thangada S, Kumar S, Reddy JK. Almost total conversion of pancreas to liver in the adult rat: A reliable model to study transdifferentiation. *Biochem Biophys Res Commun* 156:131–136, 1988.
27. Grisham JW. Cell types in long-term propagable cultures of rat liver. *Ann NY Acad Sci* 349:128–137, 1980.
28. Van Eyken P, Sciot R, Paterson A, Callea F, Kew MC, Desmet VJ. Cytokeratin expression in hepatocellular carcinoma: An immunohistochemical study. *Hum Pathol* 19:562–568, 1988.
29. Wilson JW, Leduc EH. Role of cholangioles in the restoration of the liver of the mouse after dietary injury. *J Pathol Bacteriol* 76:441–449, 1958.
30. Hoover KL, Hyde CL, Wenk ML, Poirier LA. Ethionine carcinogenesis in CD-1, BALB/c and C3H mice. *Carcinogenesis* 7:1143–1148, 1986.
31. Engelhardt NV, Baranov VN, Lazareva MN, Goussev AI. Ultrastructural localization of alpha-fetoprotein (AFP) in regenerating mouse liver poisoned with CCl_4. *Histochemistry* 80:401–407, 1984.
32. Lee GH, Sawada N, Mochizuki Y, Nomura K, Kitagawa T. Immortal epithelial cells of normal C3H mouse liver in culture: Possible precursor populations for spontaneous hepatocellular carcinoma. *Cancer Res* 49:403–409, 1989.
33. Yaswen P, Hayner N, Fausto N. Isolation of oval cells by centrifugal elutriation and comparison with other cell types purified from normal and preneoplastic livers. *Cancer Res* 44:324–331, 1984.
34. Fu X-X, Su CY, Lee Y, Hintz R, Biempica L, Snyder R, Rogler CE. Insulinlike growth factor II expression and oval cell proliferation associated with hepatocarcinogenesis in woodchuck hepatitis virus carriers. *J Virol* 62:3422–3430, 1988.

Mouse Liver Carcinogenesis: Mechanisms
and Species Comparisons, pages 335–343

Alterations in the Methylation Status of *ras* Oncogenes in B6C3F$_1$ Mouse Liver Tumors

Roseann L. Vorce and Jay I. Goodman
Department of Pharmacology and Toxicology, Center for Environmental Toxicology, Michigan State University, East Lansing, MI 48824

Under the guidelines of the National Toxicology Program's standard carcinogen bioassay, the B6C3F$_1$ mouse has been used extensively in carcinogenicity testing of chemicals. The male and female of this mouse strain are sensitive to induction of hepatocellular carcinomas by a wide variety of carcinogens, including both initiators and promoters. However, evaluation of carcinogen bioassay data is difficult, particularly in light of the high incidence of spontaneous hepatomas in the B6C3F$_1$ mouse (approximately 30% in the male at 18 months of age) (1). This is especially evident when one reflects that a variety of compounds are uniquely tumorigenic in the liver of this animal (2). This group of chemicals includes established promoters of carcinogenesis, e.g., chlordane (3), heptachlor (3), and phenobarbital (4), as well as trichloroethylene (5), all of which produce hepatocellular carcinomas in the B6C3F$_1$ mouse when administered chronically in the absence of any other chemical treatment. The high degree of sensitivity of the B6C3F$_1$ mouse toward hepatoma development makes the liver of this animal a valuable organ in which to study mechanisms of carcinogenesis in vivo. In addition, it is vital to elucidate the molecular mechanism(s) by which hepatomas arise in the B6C3F$_1$ mouse to generate the type of information that will allow a more rational approach toward the evaluation of carcinogen bioassay data.

DNA derived from spontaneous hepatocellular neoplasms of the B6C3F$_1$ mouse has been found to contain an activated Ha-*ras* oncogene (6,7). Mutations in the 13th, 61st, and 117th codons of the Ha-*ras* oncogene have been detected in both spontaneous and chemically induced B6C3F$_1$ mouse hepatic tumors (8,9), and it appears that the resulting Ha-*ras* gene product possesses altered biochemical activity that contributes to phenotypic changes. Thus, it is evident that specific mutations are involved in the development of hepatomas in the B6C3F$_1$ mouse. However, it must be kept in mind that a mutation in the Ha-*ras* oncogene can only produce phenotypic alterations if the gene is also expressed. For this reason, our laboratory has become interested in possible mechanisms by which the expression of oncogenes is regulated. One factor that appears to play a role in the control of gene expression is gene methylation state (10,11). Hypomethylation of a gene appears to be necessary, but not sufficient, for expression, so it follows that a hypomethylated gene has a greater potential for transcriptional activity than does a hypermethylated gene. There is convincing evidence that DNA methylation is involved in the regulation of differentiation (12). In addition, carcinogen-induced alterations in DNA methylation have been suggested as one possible mechanism underlying the aberrant gene expression that is thought to contribute to carcinogenesis (13). We, therefore, hypothesized that the Ha-*ras* oncogene may be hypomethylated in the nascent liver of the B6C3F$_1$ mouse, and this condition may facilitate inappropriate Ha-*ras* expression. Hepatoma development may then proceed through a combination of a critical mutation in and inappropriate expression of the Ha-*ras* gene, along with other factors.

We have studied the methylation state of the Ha-*ras*, Ki-*ras*, and *myc* oncogenes assessed in spontaneous and chemically induced hepatomas, and in non-tumor liver tissue, of B6C3F$_1$ mice as a means to determine their potential for expression. Because the predisposition of the B6C3F$_1$ mouse to develop hepatomas is a heritable trait and because the paternal C3H/He strain has a very high (60%) (1) and the maternal C57BL/6 strain has a very low (<1%) spontaneous hepatoma incidence, the methylation state of the oncogenes of interest in both parental strains was also assessed.

Methods

Animals

B6C3F$_1$, C3H/He, and C57BL/6 mice were obtained from Charles River Laboratories, housed in constant temperature and humidity conditions with a 12-h light/dark cycle, and provided with food and water ad libitum. Hepatomas were harvested from four different groups of male B6C3F$_1$ mice. For the benzidine-treated group, animals were given 120 ppm benzidine dihydrochloride in their drinking water starting when they were 10–12 weeks old and continuing for 1 year before they were killed; this treatment resulted in an 85% tumor incidence. For the phenobarbital-treated group, animals received 0.05% (wt/vol) phenobarbital in drinking water for one year and were killed 6 months later. Tumors were apparent in 68% of these animals. Tumors were induced with chloroform by administering 200 mg/kg of this chemical in a corn oil vehicle by gavage twice weekly for 1 year, a regimen resulting in a tumor incidence of 80%. A third group of animals was allowed 24 months for the development of spontaneous tumors in the absence of any chemical treatment. The control animals were killed at approximately 6 weeks of age and their livers were immediately frozen in liquid nitrogen. Tumor and non-tumor liver tissue was excised and used either immediately after freezing or stored at –85°C.

DNA Isolation

DNA was isolated by an adaptation of the method described by Chirgwin et al. (14). Frozen tissue was ground to a fine powder in liquid nitrogen with a mortar and pestle. The frozen powder was then stirred into 24 vol of 4.0 M guanidinium isothiocyanate, 0.5% sarcosyl, 1 M sodium citrate (pH 7.0), 0.5% β-mercaptoethanol until a particle-free viscous liquid was produced. This was then layered onto a CsCl step gradient (3 ml 5.7 M, 2 ml 3.0 M CsCl; CsCl made in 0.1 M EDTA) and ultracentrifuged at 29,000 rpm for 22 h at 20°C in a Beckman SW41 rotor (Beckman Instruments, Brea, CA). At the end of the centrifuge run, the upper guanidium-containing layer was removed and discarded. The DNA, which migrates to just below the 5.7 M/3.0 M CsCl interface, was removed with a wide-mouth pipette. RNAse A (preboiled 10 min) was added to a final concentration of 50 μg/ml, and the DNA was dialyzed against 1 liter of 10 mM Tris/1 mM EDTA/0.1% SDS (pH 7.5) for 1 h at room temperature. Proteinase K was added to a final concentration of 50 μg/ml, the dialysis buffer was changed, and dialysis was continued for 1 h. The DNA was then dialyzed against 2 liters of 10 mM Tris/1 mM EDTA (pH 7.5) at 4°C for 16–20 h. An equal volume of preequilibrated phenol was used to extract proteins, and this was followed by a wash with an equal volume of chloroform/3% isoamyl alcohol. The DNA was then precipitated with 1/2 vol of 7.5 M NH_4OAc and 2 vol of ethanol. After the DNA was dissolved in TE (10 mM Tris/1 mM EDTA, pH 8.0), it was reprecipitated with NH_4OAc and ethanol and again dissolved in TE. DNA was quantitated by the A_{260} absorbance, and the A_{260}/A_{280} ratio routinely approached 1.8.

Restriction Enzyme Digestion, Electrophoresis, and Southern Transfer

Twenty-microgram aliquots of DNA were digested to completion with 5 units/μg *Msp*I, *Hpa*II, and *Hha*I. Reaction buffers were supplied as 10× concentrates by BRL, and incubation was carried out for 2 h at 37°C. After the addition of marker dye (5× = 15% Ficoll 400, 0.125% each bromophenol blue and xylene cyanole ff in 5 × running buffer), samples were heat treated for 10 min at 65°C and cooled on ice. The DNA was electrophoresed on a 0.95% agarose submerged gel, using TBE (89 mM Tris, 89 mM EDTA, 2.5 mM EDTA, pH 8.3) as the running buffer. Electrophoresis was carried out at 50 V for approximately 16 h.

After the gel was stained with ethidium bromide and photographed, it was prepared for Southern transfer (15). To this end, the gel was soaked in 0.25 N HCl for 10 min, 0.4 N NaOH/0.6 M NaCl for 30 min, and 1.5 M NaCl/0.5 M Tris (pH 7.5) for 30 min. Transfer of DNA to Gene Screen Plus (New England Nuclear/DuPont, North Billerica, MA) was allowed to proceed for approximately 20 h using 10 × SSC (1.5 M NaCl/0.5 M sodium citrate, pH 7.0) as the transfer buffer. Following transfer, the membrane was briefly rinsed in 0.4 N NaOH, then in 0.2 M Tris/2 × SSC (pH 7.5), air dried, and baked in a vacuum oven at 80°C for 2 h.

Probes and Hybridization

Probes for Ha-*ras* and Ki-*ras* were purchased from Oncor (Gaithersburg, MD). Probes were labeled with ^{32}P using a random primers kit from BRL, and separation of newly labeled probe from unincorporated nucleotides was accomplished by spin column chromatography (16).

Prehybridization and hybridization were carried out in Scotch-pak (kapak) heat-sealable bags placed in a 65°C waterbath with constant agitation. Membranes were prehybridized for at least 15 min with approximately 50 µl/cm^2 of a solution containing 10% dextran sulfate, 1% SDS, 1 M NaCl, and 50 mM Tris (pH 7.5), preheated to 65°C. The denatured probe (1×10^5 cpm per ml hybridization solution) and sonicated salmon sperm DNA (100 µg/ml) were then added and mixed thoroughly with the prehybridization solution. Hybridization was carried out for approximately 16 h. Membranes were washed twice with $2 \times$ SSC for 5 min at room temperature, twice with $2 \times$ SSC/1% SDS for 15 min each at 65°C, and twice with $0.1 \times$ SSC for 15 min each at room temperature (250 ml solution per wash), all with constant agitation. Membranes were then blotted with 3-mm paper (Whatman, Clifton, NJ) and wrapped in Handi-Wrap (Dow Chemical, Midland, MI). Autoradiography was performed for 1–4 days using Kodak X-OMAT AR film and Cronex Lightning Plus intensifying screens.

Results and Discussion

Our results indicate that the B6C3F$_1$ mouse, C3H/He mouse, and C57BL/6 mouse exhibit differential degrees of methylation of the Ha-*ras* oncogene. Restriction enzyme analysis (with *Msp*I, *Hpa*II, and *Hha*I) was used to assess the extent of DNA methylation. *Msp*I digestion of both male and female B6C3F$_1$ and male C3H/He DNA revealed the absence of a 15-kb Ha-*ras* band present in *Msp*I-digested female C57BL/6 DNA, suggesting that the Ha-*ras* oncogene of B6C3F$_1$ and C3H/He mouse liver lacks a methylated site (methylation at the external cytosine in a 5′-CCGG-3′ sequence. In other respects, the Ha-*ras* and Ki-*ras* oncogenes are methylated to a similar degree (R. L. Vorce, J. I. Goodman, submitted [a]).

The methylation state of the Ha-*ras* and Ki-*ras* oncogenes was also assessed in benzidine-induced hepatomas and adjacent nontumor tissue from B6C3F$_1$ mice. In 4/4 cases, the Ha-*ras* oncogene was hypomethylated in tumor as compared to non-tumor tissue and increased expression of the gene was detected in 3/4 hepatomas; the Ki-*ras* oncogene was hypomethylated in 2/4 cases (R.L. Vorce, J.I. Goodman, submitted [a]). The methylation status of the *ras* genes in the non-tumor liver tissue obtained from tumor-bearing animals was the same as that observed in control animals. Therefore, the hypomethylation observed in hepatoma tissue was not simply a phenomenon associated with aging. In addition, the methylation status of these genes was examined in spontaneous liver tumors and in tumors induced by phenobarbital and chloroform (R. L.

Vorce, J. I. Goodman, submitted [b]). Ha-*ras* was found to be hypomethylated in all tumors examined, whereas Ki-*ras* was sometimes hypomethylated; such hypomethylation might play a role in the promotion stage of carcinogenesis. Modifications in the methylation of DNA provide a possible epigenetic mechanism leading to heritable alterations in the regulation of gene expression (17, 18). These results suggest that oncogene activation in $B6C3F_1$ mouse liver tumors may involve an increased potential for expression. The data support the assertion that tumors induced by different classes of carcinogens or arising spontaneously share common biochemical pathways of oncogene activation during tumorigenesis.

When considering these three strains of mice, we have discerned a positive correlation between relative hypomethylation of the Ha-*ras* oncogene and susceptibility towards tumorigenesis. The Ha-*ras* oncogene is relatively hypomethylated in $B6C3F_1$ and C3H/He mice, and both of these strains have a high propensity for hepatoma development. In contrast, the Ha-*ras* oncogene in the C57BL/6 mouse is relatively hypermethylated, and this strain exhibits a relatively low tendency to develop hepatomas. We consider methylation to be one of a variety of factors governing gene transcription (19). However, if activation of the Ha-*ras* oncogene is indeed necessary for the development of mouse liver tumors, it is possible that a state of relative hypomethylation of this gene facilitates its aberrant expression and thereby contributes to neoplastic growth. Supporting this view is the finding that an unmethylated Ha-*ras* oncogene is capable of transforming NIH3T3 cells following transfection, whereas in vitro methylation of the Ha-*ras* oncogene markedly decreases this action (20). Furthermore, the Ha-*ras* oncogene is hypomethylated in several human carcinomas (21), and Ha-*ras* gene expression is elevated during hepatocarcinogenesis (22) and skin carcinogenesis (23), where gene amplification was not observed. Thus, experimental evidence supports the possibility that deregulation of Ha-*ras* oncogene expression is involved in carcinogenesis.

This view is compatible with evidence that suggests that a mutation is involved in Ha-*ras* activation (8,9,24), especially in light of the fact that carcinogenesis is a multistep process (25). If a gene acquires a crucial mutation in the Ha-*ras* oncogene, no phenotypic effect will be seen as long as the gene remains quiescent. The critical step now becomes deregulation of gene expression, and this stage may involve DNA hypomethylation. If, however, the Ha-*ras* oncogene is already hypomethylated, a state permissive for gene expression, a crucial mutation is much more likely to result in phenotypic alterations consistent with the malignant phenotype. This scenario is supported by the finding that the Ha-*ras* oncogene is hypomethylated in mouse epidermis, a cell type known to be susceptible to oncogenic conversion by a single point mutation (26). In this context, our results indicate that hepatocytes of the nascent liver of the $B6C3F_1$ mouse might not be normal but rather a step toward the malignant phenotype.

Because it is prone to develop hepatomas, the $B6C3F_1$ mouse is an appropriate model for the study of hepatocarcinogenesis in research aimed at discerning mechanisms involved in the transformation of a normal cell into a cancer cell. This knowledge is vital for developing a rational approach to assessing the level of risk associated with exposure to chemicals. Our results indicate that an oncogene believed to be involved in hepatocarcinogenesis might be primed for expression in the liver of the B6C3F$_1$ mouse. This could, in part, explain the high propensity of this animal to develop hepatomas. The data presented in this report support the contention that positive results of carcinogen bioassays in B6C3F$_1$ mouse liver need to be critically evaluated. This is particularly true when we are attempting to assess the significance of positive bioassay results for compounds that are uniquely tumorigenic in the liver of this animal.

Acknowledgments

Support for this research was provided by a grant from the International Life Sciences Institute Risk Science Institute. R. L. V. was the recipient of an Advanced Predoctoral Fellowship awarded by the Pharmaceutical Manufacturers Association Foundation, Inc.

Open Forum

Dr. Jon Mirsalis: Is there really good evidence that, in going through semiconservative replication, once you lose a methyl group, it never picks it up again? I am not aware of literature that shows that. Somewhere along the line, the DNA was methylated. What you are proposing is that if the methyl group is lost, and the cells are turning over very quickly, it might be permanently lost if that one strand gets replicated again before it is methylated. I am not sure that I buy that. I think it is probably more likely that there are site-specific methylases that recognize certain sequences and put methyl groups there. I am not sure there is evidence for either.

Dr. Jay Goodman: There are two schools of thought. There does seem to be some indication that differentiation may start with a hypermethylated DNA and then begin to hypomethylate selected regions and allow for expression as the cells move to differentiate. There are some indications that there is loss of methyl groups, for example, during estrogen stimulation of differentiation. On the other hand, the possibility you raise does need to be considered. Some aspects of differentiation might involve starting with hypomethylated DNA and then methylating to quiet certain regions. But it does seem that differentiated patterns of methylation are maintained as the cells replicate, and that isolated maintenance methylase can methylate semimethylated DNA quite well and quite faithfully. For example, if we look at regenerating liver (and we looked at it only at 24–72 h posthepatectomy), we get the same restriction pattern as in the normal liver. But here, at the 24 h time point, we are really seeing that the restriction nuclease is also inhibited by the semimethylated state.

Dr. Mirsalis: I noticed in your phenobarbital study that you treated the animals for a year and then let them go for 6 months. Did you look at the methylation status in the normal tissue?

Dr. Goodman: We did not have surrounding tissue for all of them. We did look at the methylation status in the surrounding non-tumor tissue in the benzidene group, which would have been approximately the same age vs. the relatively young adults that we typically use. There were no differences in liver from the young animals and non-tumor tissue from the older ones.

Dr. Mirsalis: That would argue against the heritable methylation theory. In other words, at the end of that first year, when the cells are turning over rapidly, you might have seen hypomethylation. But give them 6 months to recover and give the various methylases a chance to remethylate this DNA, and you would expect to see the methylation return to control.

Dr. Goodman: But not in the tumor tissue. In the normal tissue, we feel that maintenance methylation proceeds with a high degree of fidelity.

Dr. Mirsalis: The tumor tissue is a whole different story. I am saying in the normal tissue.

Dr. Gary Williams: Did you speculate on the basis for the hypomethylation in the $B6C3F_1$? Could it be a deficiency in the maintainence methylase?

Dr. Goodman: I presented an empirical observation. It has been known for a while that there is some general hypomethylation in tumor tissue, as compared with normal tissue, in tumors produced by a variety of regimens. We were interested in asking whether there is hypomethylation in some selected genes of interest, for example H-*ras*. Second, we asked whether the frequency or extent of hypomethylation in a gene of interest that seemed to frequently be involved in tumorigenesis in the tissue of interest would differ from that in a gene with relatively infrequent involvement in tumorigenesis. That is why we used H-*ras* vs. K-*ras*. We are interested in pursuing whether there is any defect in the maintainence methylase in terms of fidelity, for example, in $B6C3F_1$ vs. C57 black. We are also interested in whether, under bioassay conditions, compounds like benzidene, phenobarbital, or chloroform can interfere with maintenance methylation.

Dr. Williams: If that is your hypothesis, you should really look at the effect of ethionine.

Dr. Goodman: I agree.

References

1. Becker FF. Morphological classification of mouse liver tumors based on biological characteristics. *Cancer Res* 42:3918–3923, 1982.
2. Clayson DB. The need for biological risk assessment in reaching decisions about carcinogens. *Mutat Res* 185:243–269, 1987.
3. Williams GM, Numoto S. Promotion of mouse liver neoplasms by the organochlorine pesticides chlordane and heptachlor in comparison to dichlorodiphenyltrichloroethane. *Carcinogenesis* 5:1689–1696, 1984.

4. Peraino C, Staffeldt EF, Haugen DA, Lombard LS, Stevens FJ, Fry RJM. Effects of varying the dietary concentration of phenobarbital on its enhancement of 2-acetylaminofluorene-induced hepatic tumorigenesis. *Cancer Res* 40:3268–3273, 1980.
5. Kimbrough RD, Mitchell FL, Houk VN. Trichloroethylene: An update. *J Toxicol Environ Health* 15:369–383, 1985.
6. Fox TR, Watanabe PG. Detection of a cellular oncogene in spontaneous liver tumors of B6C3F$_1$ mice. *Science* 228:596–597, 1985.
7. Reynolds SH, Stowers SJ, Maronpot RR, Anderson MW, Aaronson SA. Detection and identification of activated oncogenes in spontaneously occurring benign and malignant hepatocellular tumors of the B6C3F$_1$ mouse. *Proc Natl Acad Sci USA* 83:33–37, 1986.
8. Wiseman RW, Stowers SJ, Miller EC, Anderson MW, Miller JA. Activating mutations of the c-Ha-ras protooncogene in chemically induced hepatomas of the male B6C3F$_1$ mouse. *Proc Natl Acad Sci USA* 83:5825–5829, 1986.
9. Reynolds SH, Stowers SJ, Patterson RM, Maronpot RR, Aaronson SA, Anderson MW. Activated oncogenes in B6C3F$_1$ mouse liver tumors: Implications for risk assessment. *Science* 237:1309–1316, 1987.
10. Riggs AD, Jones PA. 5-Methylcytosine, gene regulation, and cancer. *Adv Cancer Res* 40:1–30, 1983.
11. Jones PA. DNA methylation and cancer. *Cancer Res* 46:461–466, 1986.
12. Michalowsky LA, Jones PA. DNA methylation and differentiation. *Environ Health Perspect* 80:189–197, 1989.
13. Boehm TLJ, Drahovsky D. Alteration of enzymatic methylation of DNA cytosines by chemical carcinogens: A mechanism involved in the initiation of carcinogenesis. *JNCI* 71:429–433, 1983.
14. Chirgwin JM, Przybyla E, MacDonald RJ, Rutter WJ. Isolation of biologically active ribonucleic acid from sources enriched in ribonuclease. *Biochemistry* 18:5294–5299, 1979.
15. Southern EM. Detection of specific sequences among DNA fragments separated by gel electrophoresis. *J Mol Biol* 98:503–517, 1975.
16. Maniatis T, Fritsch EF, Sambrook J. *Molecular Cloning: A Laboratory Manual.* Cold Spring Harbor, NY: Cold Spring Harbor Laboratory, 1982.
17. Holliday R. The inheritance of epigenetic defects. *Science* 238:163–170, 1987.
18. Holliday R. DNA methylation and epigenetic defects in carcinogenesis. *Mutat Res* 181:215–217, 1987.
19. Vorce RL, Goodman JI. Investigation of parameters associated with activity of the Kirsten-ras, Harvey-ras, and myc oncogenes in normal rat liver. *Toxicol Appl Pharmacol* 90:86–95, 1987.
20. Borrello MG, Pierotti MA, Bongarzone I, Donghi R, Mondellini P, Porta GD. DNA methylation affecting the transforming activity of the human Ha-ras oncogene. *Cancer Res* 47:75–79, 1987.
21. Feinberg AP, Vogelstein B. Hypomethylation of ras oncogenes in primary human cancers. *Biochem Biophys Res Commun* 111:47–54, 1983.
22. Cote GJ, Lastra BA, Cook JR, Huang D-P, Chiu J-F. Oncogene expression in rat hepatomas and during hepatocarcinogenesis. *Cancer Lett* 26:121–127, 1985.
23. Pelling JC, Ernst EM, Strawhecker JM, Johnson JA, Nairn RS, Slaga TJ. Elevated expression of Ha-ras is an early event in two-stage skin carcinogenesis in SENCAR mice. *Carcinogenesis* 7:1599–1602, 1986.

24. Zarbl H, Sukumar S, Arthur AV, Martin-Zanca D, Barbacid M. Direct mutagenesis of Ha-ras-1 oncogenes by N-nitroso-N-methylurea during initiation of mammary carcinogenesis in rats. *Nature* 315:382–385, 1985.
25. Farber E. The multistep nature of cancer development. *Cancer Res* 44:4217–4223, 1984.
26. Ramsden M, Cole G, Smith J, Balmain A. Differential methylation of the c-H-ras gene in normal mouse cells and during skin tumour progression. *EMBO J* 4:1449–1454, 1985.

Mouse Liver Carcinogenesis: Mechanisms
and Species Comparisons, pages 345–365

Mouse Liver Tumors and Microsomal Enzyme–Inducing Drugs: Experimental and Clinical Perspectives With Phenobarbital

R. Michael McClain
Department of Toxicology and Pathology, Hoffmann-La Roche Inc., Nutley, NJ 07110

Phenobarbital (PB) is an extensively studied prototype for a class of hepatic microsomal enzyme inducers that induce a similar spectrum of isoenzymes (1). When chemicals are tested at high dosages in subchronic and chronic toxicity tests in rodents, histologic evidence for microsomal enzyme induction—i.e., increased hepatic weight and centrilobular hepatocytic hypertrophy—is commonly observed. In a previous review of chemicals that cause liver enlargement, Golberg (2) stated that 75% of chemicals induce hepatomegaly in chronic toxicity studies. In a recent survey by the Pharmaceutical Manufacturers Association, it was estimated that 171/399 (43%) of compounds put into development gave some indication that enzyme induction occurred, as evidenced by either histological observations or by in vivo or in vitro tests for the assessment of microsomal enzyme induction.

The short-term consequences of hepatic microsomal enzyme induction in laboratory animals are relatively minor from a toxicologic point of view. One may observe changes in hepatic function and, since microsomal enzymes metabolize a variety of endogenous hormones, vitamins, and cofactors, functional changes in various endocrine organs occur. The long-term consequences of microsomal enzyme induction are of greater concern, since an alteration in tumor incidence, most commonly in the liver and thyroid gland, may be observed. There is a relatively well recognized association between microsomal enzyme–inducing compounds and hepatic tumor promotion in rats and mice in various initiation-promotion models (3–5). PB is also a promoter of thyroid tumors when administered after various nitrosamines (6–9).

Many compounds that induce hepatic tumors in responsive strains of mice (such as the $B6C3F_1$ and C3H strains) but do not possess mutagenic or DNA-damaging activity, are microsomal enzyme inducers. With respect to hepatocarcinogens, as stated in numerous articles by Schulte-Hermann (10–12) the "nongenotoxic" hepatocarcinogens (or the hepatic tumor promoters) are, for the most part, compounds that stimulate liver growth. The two main categories of drugs that stimulate liver growth are the peroxisome proliferators, such as a clofibrate, and the microsomal enzyme inducers, such as phenobarbital. Examples of three groups of microsomal enzyme–inducing chemicals associated with the induction of hepatic tumors in rats and/or mice would include the barbiturates, the chlorinated hydrocarbon pesticides, and the polyhalogenated biphenyls (13).

The barbiturates are considered to be promoters of hepatocarcinogenesis, based on a large body of evidence accumulated over the last 20 years. There is no evidence to support the conclusion that PB, the most extensively studied barbiturate, possesses any initiating activity. Tests of this nature have all been negative (14,15).

The association of tumor-promoting activity in the liver with the ability to induce microsomal enzymes comes from various studies with inducing and noninducing structural analogues. In the case of the barbiturates, there is a very significant correlation between microsomal enzyme–inducing capabilities and hepatic tumor–promoting properties in the rat (16). For example, the enzyme inducers phenobarbital and barbital are hepatic tumor promoters, whereas noninducing analogues such as hexobarbital and monoethyl- or monophenylbarbital are not hepatic tumor promoters. Furthermore, within the barbiturate series, the potency for enzyme induction correlates quite well with the potency for hepatic tumor promotion (16).

Although there is a good association between the ability to induce microsomal enzymes and hepatic tumor promotion, it is unlikely that microsomal enzyme induction per se causes tumor promotion. PB and many other inducers also increase cell proliferation via a mitogenic effect that may be important for hepatic tumor promotion (10,12,17).

Since such a large portion of chemicals are microsomal enzyme inducers when tested in rodents at high dosages and since the liver is the most common site of carcinogenic response (18), the relationship of microsomal enzyme induction and tumor promotion is an important one. Many microsomal enzyme–inducing substances that induce hepatocellular tumors do so in strains of mice such as the $B6C3F_1$ and C3H, which have a high incidence of spontaneous liver tumors.

The scope of this chapter is limited to a discussion of the response to PB in rodents not treated with an initiating chemical prior to PB exposure.

Phenobarbital

Phenobarbital (5-ethyl,5-phenyl barbituric acid) was introduced into clinical medicine in 1912 as a sedative-hypnotic and anticonvulsant. Its use as a sedative-hypnotic has declined with the advent of safer and more effective drugs. Although other anticonvulsant drugs have been introduced, PB is still an important anticonvulsant drug used alone or in combination with other drugs for the prevention of grand mal seizures.

Metabolism

PB is slowly but well absorbed from the gastrointestinal tract, is moderately bound to plasma proteins and has a relatively long half-life in animals and man (19–21). Approximately 50%–60% of PB is metabolized, and 20%–40% is excreted into the urine unchanged (22). Phenobarbital is extensively metabolized, primarily by *p*-hydroxylation of the phenyl ring (23) and is excreted as the sulfate or glucuronide conjugate (24) of the *para*hydroxyphenyl metabolite. The compound is also *N*-glucosidated (25), and limited hydroxylation of the side chain occurs (26). Phenobarbital may be metabolized to a limited extent through a 3,4-arene oxide, which, via epoxide hydratase, would be metabolized to the dihydrodiol (27) or oxidized to form a catechol (26). These products have only been tentatively identified, and how operative these pathways are is not known. There is no evidence, however, that phenobarbital forms reactive metabolites that would bind to DNA, since tests for DNA binding and damage in vitro and in vivo have been uniformly negative.

Phenobarbital Short-Term Tests

Tests for Mutagenicity and Clastogenicity

Phenobarbital has been extensively tested for mutagenic, clastogenic, cell-transforming, and DNA-damaging activity. It has been considered generally negative for genotoxicity, but mixed and conflicting results have been obtained in both bacterial and mammalian systems.

PB was not mutagenic in bacteria in tests conducted by several investigators (28–32) or in a host-mediated assay (33) and did not induce mutation in *Drosophila* (34). It produced sister chromatid exchange (SCE) in Chinese hamster cells (35) but no SCE in vivo (36). PB treatment did not lead to transformation of rat liver cells (36) but did produce chromosome aberrations in cultured human lymphocytes (38,39). Additionally, it produced negative results in the mouse micronucleus test (40).

An extensive evaluation of PB and seven other "carcinogens" was performed by a group established by the International Program on Chemical Safety (IPCS) for the Collaborative Study on Short Term Tests for Genotoxicity and Carcinogenicity, the results of which were published in 1985 (41). Although PB had produced generally negative results in previous tests, mixed and conflicting results were observed in various IPCS tests as follows:

Mixed results were observed in the bacterial *Salmonella typhimurium* test. In three labs, a very weak response was observed in either TA100 or TA1535 cells only in the absence of the S9-activating system, whereas two labs reported negative results. PB generally had no effect in fungi, as assessed by end points including gene mutation, mitochondrial mutation, gene conversion, and mitotic crossing over, but did produce aneuploidy. In *Drosophila*, tests for PB-caused mutations, deletions, chromosome breakage, mitochondrial recombination, and aneuploidy were negative.

In mammalian cell culture, PB treatment yielded clastogenic activity in three of five assays. Two were weakly positive, one in Chinese hamster liver (CHL), only in the absence of S9, and one in Chinese hamster ovary cells (CHO), only in the presence of S9. The third was positive with and without S9 in CHO cells. Phenobarbital treatment gave negative results in four assays for polyploidy and aneuploidy and weakly positive results for SCE in one of three assays using CHO cells.

The results of PB treatment in tests for gene mutation in mammalian cell culture were also mixed (five negative, four positive, and two equivocal). The effects were weakly positive in human lymphoblasts, positive in Chinese hamster V79 cells (only in the absence of S9), and positive in mouse lymphoma cells at highly cytotoxic doses (only in the presence of S9). Two equivocal responses were observed at cytotoxic doses in mouse lymphoma cells; one only in the presence of and another only in the absence of S9. Three additional tests in mouse lymphoma cells and two tests with Chinese hamster V79 cells, all with and without S9, had negative results.

Cell Transformation

PB gave both positive and negative results for cell transformation using the primary Syrian hamster embryo (SHE) assay; negative results in C3H/10T1/2 cells, with and without activation, and weakly positive results for transformation, but not mutagenicity, in BALB/c 3T3 cells with S9.

DNA Damage

Tests for DNA damage in hepatocytes would be considered to be the most relevant short-term tests with respect to hepatocarcinogenesis. PB has been uniformly negative in tests for DNA binding and DNA repair. It does not bind to DNA, as evidenced by the lack of strand breaks in rat hepatocytes (42–44). In the IPCS studies, negative results for strand breaks were obtained in two labs that performed the test. Phenobarbital did not increase unscheduled DNA synthesis (UDS) in rat primary hepatocyte cultures in studies by Williams (45) or in tests from 4/4 labs in the IPCS study.

Tests for DNA damage in the mouse liver are particularly important. Doolittle and coworkers (46), reported that PB did not induce UDS in hepatocytes cultured from male CD-1 mice. We have performed additional tests for DNA repair using $B6C3F_1$ mice. Phenobarbital did not increase UDS in primary hepatocyte cell culture from male and female $B6C3F_1$ mice or in the in vivo–in vitro UDS assay using both male and female $B6C3F_1$ mice.

Cell-to-Cell Communication

Results of in vitro tests for metabolic cooperation have correlated with the tumor-promoting activity of certain chemicals. PB has been shown to inhibit metabolic cooperation at noncytotoxic doses (45,47–50). In the IPCS studies, phenobarbital was weakly positive in one study and negative in another.

Evaluation

In the IPCS studies, the results of both bacterial and mammalian tests for mutagenicity and clastogenicity are mixed. For several tests, both positive and negative results have been obtained for the same tests, as well as in different tests measuring the same genetic end point. Both positive and negative results have been obtained in the same lab, and an overall result was reported in these cases. In some cases, a test was positive in one laboratory only with metabolic activation, whereas in another lab the same test was positive only without metabolic activation. In most cases, positive results, when they have been observed, have been only weakly positive, and they have occasionally been observed at highly cytotoxic concentrations.

It is important to further explore the tests with genotoxic end points that have yielded mixed, conflicting, or equivocal results. For example, we were interested in the results of the *S. typhimurium* test, since the authors concluded that PB was a bacterial mutagen. In the IPCS studies, the response to phenobarbital was observed in TA100 or TA1535, only in the absence of S9. Although PB did not increase the number of revertants to more than twice the background level (a commonly used criteria for this test), the authors considered the tests positive because the response was dose related and reproducible in several tests in three labs. We conducted a series of tests with both phenobarbital and sodium phenobarbital using the TA100 and TA1535 strains with and without S9 activation and with and without preincubation. This combination resulted in 16 tests (8 tests without S9), with three replicates per test. None of the tests or replicates showed an increase in the number of revertants with PB treatment. Furthermore, since PB is only slowly metabolized in vivo and in vitro (51), the amount of phenobarbital at high test concentrations should be similar with or without S9, thus the positive responses observed in tests only in the absence of S9 need further examination. Because of the importance of PB as a prototype, the conflicting responses in various tests need to be confirmed one way or the other.

Phenobarbital Hepatic Neoplasia in Mice

Long-term treatment of mice with phenobarbital has resulted in a marked increase in both hepatocellular adenomas and carcinomas, as detailed in the following studies.

Peraino et al. (52) reported in 1973 that PB treatment of male and female C3H mice for 12 months at a dose of 500 ppm in the diet resulted in a marked increase in the incidence (control vs. treated) of hepatomas in both male (67% vs. 98%) and female (13% vs. 100%) mice. The histologic type of the tumors was not indicated, and opinions varied on whether benign hepatoma or hyperplastic nodule was the most appropriate diagnosis.

Thorpe and Walker (53) treated CF1 mice of both sexes with 500 ppm of phenobarbital in the diet and observed hepatic tumors in about 75%–80%, as compared with 23%–24% of control mice. Tumors were classified as either type A or type B; type B tumors represented invasive neoplasms expected to be metastatic and transplantable. The number of both type A and B tumors markedly increased with PB treatment. Although no metastatic lesions were observed at 500 ppm, the authors found lung metastases in mice treated with higher dosages of PB (1,000 and 3,000 ppm) when type B tumors were found.

In a similar experiment, Ponomarkov et al. (54) treated CF1 mice with 0.05% PB in their drinking water for their life spans (120 weeks). Control male and female mice exhibited a 27% and 0% incidence of hepatomas, respectively, as compared with incidences of 78% and 62% in male and female mice treated with PB. The authors stated that of 122 hepatomas observed in mice treated with phenobarbital, 58 (48%) could be classified, without hesitation, as hepatocellular carcinoma; however, no metastases were observed.

Results of treatment of smaller groups of mice have been reported by Jones and Butler (55) and Butler and Jones (56). An increased incidence of nodules (70%–80%) in C3H male mice treated with 1,000 ppm PB was seen at 80 weeks, as compared with 20% in controls. Although the nodules were morphologically similar to type A and B lesions (57), the authors questioned whether the lesions represented malignant neoplasia because they could not identify invasion or metastases. In a subsequent report (58), a small number of male C3H mice treaed for 2 years were studied; 11 of 13 macroscopic nodules were similar histologically—simple trabeculae of hepatic parenchymal cells with rather abundant eosinophilic cytoplasm. Two other PB-treated mice with nodules showed more atypical areas with thickened irregular trabeculae and wide vascular spaces; both mice had lung metastases.

Becker (59) treated male C3H and B6C3F$_1$ mice with 0.05% phenobarbital in their drinking water for 12 months and produced a marked increase in hepatic tumors classified as MLT (mouse liver tumor) I or II, which are adenomas. Both the number and the size of tumors were increased in PB-treated mice.

In two subsequent studies, in male C3H/He mice treated with phenobarbital at a dose of 500 ppm for 80 weeks (60) or 104 weeks (61), the histologic appearance of liver tumors was recorded as tumors composed of either small basophilic cells or large eosinophilic cells. In control mice, only tumors containing small basophilic cells were observed. In PB-treated mice, the incidence of basophilic tumors was similar to that of controls; however, the number of eosinophilic tumors, a type not usually observed in controls, was markedly increased. There was no increase in carcinomas; the eosinophilic tumors observed were considered to be benign. The authors concluded that PB treatment of a susceptible strain of mouse such as the C3H/He leads to the formation of eosinophilic nodules distinct from the basophilic nodules that develop spontaneously and that the histogenesis of the spontaneous and PB-induced tumors may be different.

Ward et al. (62) found similar results when they treated C3H/He mice with 500 ppm PB in drinking water. It was observed that aging male mice were more susceptible to the effects of PB than younger mice and that PB primarily increased the incidence of unique eosinophilic proliferative lesions; it had no effect on the normal progression of natural liver tumors, especially those of the basophilic type.

Phenobarbital treatment (500 ppm in the diet) of male mottled yellow A^{vy}/A and agouti A/a (C3H x vy) F_1 hybrid mice (63) increased the incidence of hepatocellular adenomas in these strains but appeared to substantially decrease the incidence of carcinoma.

Phenobarbital has not been tested for a full 2-year treatment period in both sexes of the two mouse strains most commonly used for carcinogenicity testing (the CD-1 strain used primarily by industry and the $B6C3F_1$ strain, which has been used from the inception of the National Cancer Institute/National Toxicology Program bioassay program). Matsuyama and Amo (64), however, treated male and female $B6C3F_1$ mice with barbital, a barbiturate closely related to phenobarbital, at a dose of 5,000 ppm in the diet for 2 years. Barbital produced a marked increase in hepatomas in both male (19% vs. 57%) and female (3% vs. 50%) mice. In the barbital-treated mice, 22% of the tumors were classified as hepatocellular carcinoma; no metastases were observed.

To summarize the effects of PB in the mouse, both an increase in hepatocellular adenomas and carcinomas were observed in the two 2-year studies conducted in the CF1 mouse (53,54) as was the case for barbital (64) in the $B6C3F_1$ mouse. In the two 12-month studies using C3H (52) and male $B6C3F_1$ mice (59), only an increase in adenomas was observed. The two long-term studies of the C3H/He mice (60,61) appear to conflict with the CF1 studies in that only an increase in adenomas was observed. Similarly, an increase only in adenomas with a decrease in carcinomas was observed in treated yellow A^{vy}/A and agouti A/a (C3H × vy) F_1 male mice (63). Overall, in some mouse strains, phenobarbital can increase the incidence of both hepatocellular adenomas and carcinomas. With time, some carcinomas metastasize. It is important to note, however, that the incidence of and potential for metastasis of PB-induced tumors are very low—lower than would be expected from the histologic appearance of the tumors in the studies in which hepatocellular carcinomas were diagnosed. In addition, there is very little increase in mortality in PB-treated mice, even though the tumor incidence approaches 100%. This has led several of the authors cited to question the metastatic potential of PB-induced tumors as compared with those of spontaneous tumors or of tumors induced by direct-acting hepatocarcinogens.

It is also clear that in the C3H male mouse, PB primarily increases the incidence of a specific tumor type composed of large eosinophilic cells without altering the incidence of tumors commonly observed in control mice; the latter are mainly composed of small basophilic cells (56,60–62). This observation indicates that the phenotype of the PB-induced tumor may be different from that of spontaneous tumors and suggests that PB either alters the phenotype of spontaneously occurring tumors or promotes a tumor with a phenotype not commonly expressed in control mice. The difference in histologic appearances of the PB-induced and spontaneous tumors has been noted in several studies. Jones and Butler (55) thought that spontaneously occurring tumors were responding to the inducing effects of PB, since they closely resembled the hypertrophied centrilobular hepatocytes, containing abundant eosinophilic cytoplasm, observed with various microsomal enzyme inducers. This does not appear to be the case, since the appearance and number of basophilic tumors commonly observed in untreated mice do not appear to be altered by PB treatment. Electron microscopy of the large eosinophilic cells shows the presence of markedly proliferated smooth endoplasmic reticulum and/or mitochondria, whereas the small basophilic tumor cells show a marked increase in rough endoplasmic reticulum (65). In addition, studies by Ward (66) have shown that mice treated with a direct-acting carcinogen such as diethylnitrosamine (DEN) exhibit an increase in tumors that resemble the spontaneous small cell basophilic tumors; however, in initiation-promotion studies PB increases the number of eosinophilic tumors after the administration of DEN to both mice and rats (66,67).

Species and Strain Differences

The strain and species variation in susceptibility to phenobarbital-induced liver tumors roughly parallels the spontaneous incidence of hepatic tumors in the test animals. Those species or strains with high incidences of spontaneous liver tumors appear to be more sensitive than species with low tumor incidences. Many mouse strains exhibit high spontaneous incidence of liver tumors, which can range from 0% to almost 100%; male mice generally have a higher spontaneous incidence of liver tumors than do female mice (68). The $B6C3F_1$ mouse, commonly used in carcinogenicity studies, has a relatively high spontaneous incidence of hepatomas at 2 years of age (69,70). Strain differences in response to phenobarbital were nicely demonstrated in studies by Becker (59) in which PB-treated C3H mice had a marked increase in liver tumors after 12 months of treatment but no tumors were observed in the C57BL mouse treated with PB for 18 months.

Most rat strains, on the other hand, have relatively low incidences of spontaneous liver tumors. The Charles River CD rats and the Fischer 344 rat, both of which are commonly used in chronic toxicity/carcinogenicity studies, have spontaneous liver tumor rates of from 2% to 3% (69,70). Although it is possible to produce hepatic tumors in rats with phenobarbital and other microsomal enzyme inducers, rats are relatively resistant, as compared to mice, and in many instances positive findings may not be observed.

Sex Differences

Male mice are generally considered more susceptible to the induction of liver tumors than are female mice; however, an examination of the sex difference in response to PB treatment reveals that this is not usually the case. The data from Thorpe and Walker (53) show that the increased incidences of hepatomas in CF1 male mice (from 24% to 80%) and in female mice (from 23% to 75%) were equivalent. In the study of Ponomarkov et al. (54), the hepatic tumor incidence in CF1 male mice increased from 27% to 78%, whereas in female mice, it increased from 0% to 62%. Although the absolute percentage of hepatomas is somewhat greater in treated male than in treated female mice, the response in terms of the percentage increase over the control incidence is greater in female mice. Similarly, in the study of Peraino et al. (52), the incidence of hepatomas in C3H male mice increased from 68% to 97%, as compared with an increase from 13% to 100% in female mice. Although the magnitude of the sex difference cannot be determined from the last study since both sexes showed the maximum possible response, the percentage increase over control tumor incidence was greater in female mice.

With respect to the sex difference in response to treatment with phenobarbital, one must conclude that the response of CF1 female mice is at least equal to male, and some data suggest that the female is more responsive. Ponomarkov et al. (54), in discussing the sex difference in response to PB treatment, stated, "This high effect was more evident in female than in male mice, since in females of this strain, hepatomas are a very rare spontaneous event, and thus the difference in incidence in untreated and phenobarbital treated animals was more conspicuous."

Phenobarbital Rat Studies

Butler (71) treated F344 male rats with 1,000 ppm PB in the diet for 103 weeks and did not observe an increase in the incidence of hepatic neoplasms. Similarly, no increase in hepatic tumors was observed in Sprague-Dawley rats injected intraperitoneally with phenobarbital (72). In contrast, Rossi et al. (73) observed an increase in benign adenomas to about 30% in male and female Wistar rats treated with 0.05% PB in their drinking water for their life spans. Ward (74) treated young and aged F344 rats with 500 ppm PB in their drinking water for up to 233 days. There was an increase in the number of GGT-positive eosinophilic hepatocellular adenomas in aged rats, which were more susceptible to the effects of phenobarbital. Hepatocellular neoplasms in control rats were basophilic and generally GGT-negative. Tumors of this type did not appear to respond to phenobarbital (67).

Human Epidemiology

Phenobarbital has been in clinical use for more than 80 years, and since such anticonvulsants are used at relatively high, microsomal enzyme–inducing dosages for up to lifetime exposure, a sizeable patient population has been continuously exposed to known dosages of PB. Although it is difficult to obtain negative data, considered to be conclusive, from human epidemiologic studies, the patient population studied by Clemmesen is extensive (75–79).

Clemmesen followed more than 8,000 patients admitted to a Danish epilepsy center from 1933 to 1962, and at later time periods by using registers of population, of deaths, and of cancer cases. In this patient population, excess tumors occurred at two sites, the liver and the central nervous system; however, neither of these was attributed to PB therapy. In the case of hepatic tumors, 11 cases were observed against an expected incidence of 2.8. Eight of the 11 patients, however, had received treatment with thorium oxide (thorotrast) which was used at one time as a radiopaque agent. Thorotrast is a known human hepatocarcinogen, and its presence can be observed histologically as deposits in the liver. In addition, thorotrast-caused tumors are hemangiosarcomas, not hepatocellular tumors. After subtracting the 8 patients that received thorotrast from the total of 11, 3 patients had hepatic tumors in the anticonvulsant group, against an expected incidence of 2.8. The authors concluded, therefore, that there was no evidence for an increase in hepatic tumors in PB-treated patients. A recently published update of this patient data base (80) confirms and extends the observations of Clemmesen and coworkers. The results of a smaller study of 2,099 patients admitted with severe epilepsy from 1931 to 1971, reported by White et al. (81), also confirmed the lack of a hepatic effect, in that no hepatic tumors were observed at all.

With respect to tumors of the central nervous system, an excess of intracranial tumors was observed in the Clemmesen studies (45 vs. an expected 3.87) and those reported by others (80–82). In addition, Gold et al. (83) determined that there was an excess of brain tumors in children exposed to PB and suggested that the tumors were caused by the drug. Clemmesen concluded, however, that the increase in central nervous system tumors in a patient population treated with anticonvulsants would be expected, since brain tumors can cause convulsions, which would be an indication for the drug. Furthermore, a clear pattern of decreased incidence of central nervous system tumors over time in the Clemmesen and other studies disproves an association between drug treatment and central nervous system tumors (79,80,82).

In several studies, a slight increase in lung tumors was observed (79–82,84). In several cases, this increase was accounted for by cigarette smoking. Shirts et al. (81) concluded that the observed pattern of lung tumors was inconsistent with an etiologic role for anticonvulsants; however, Olsen et al. (80) stated that further research on this link appears warranted.

Although it is seldom possible to obtain negative data from epidemiologic studies that is considered conclusive by regulatory bodies, the incidence of hepatic cancer in humans is rare; thus an increase should be readily apparent. There is no evidence of an increase in hepatic tumors in patient populations treated with phenobarbital and no indication of an increase in the total cases of cancer or in cancers at other sites except for the central nervous system; this increase is associated with the disease process rather than anticonvulsant therapy.

IARC Classification for Phenobarbital

The current classification of phenobarbital by the International Agency for Research on Cancer (85) is group 2B, meaning the agent is possibly carcinogenic to humans. For phenobarbital, the evidence for carcinogenicity in humans is considered inadequate (i.e., the studies are insufficient to draw a conclusion regarding a causal relationship). In animals, the evidence for carcinogenicity is considered sufficient (i.e., a causal relationship to treatment has been established).

The evidence for animals is based on the production of hepatic adenomas in rats and hepatic adenomas and carcinomas in mice. Supporting evidence from initiation-promotion experiments includes enhancement of hepatocellular tumors in rats and mice and thyroid follicular tumors in rats after the administration of initiating agents. Although the available animal data have not changed substantially since supplement 4 in 1982, the animal data classification was upgraded from limited to sufficient. The effect of this was an upgrade from IARC's group 3 (the agent is not classifiable with respect to its carcinogenicity in humans) to group 2B (the agent is possibly carcinogenic to humans) in supplement 7, 1987 (85).

One of qualifications for "limited" evidence of carcinogenicity under the IARC scheme is that chemicals "increase the incidence of certain neoplasms which may occur spontaneously in high incidence in certain strains." In the case of phenobarbital, evidence could be considered "limited," even though there was a clear increase in hepatocellular tumors in mice treated with the drug.

Other agents with activities in mice similar to that of phenobarbital that have beén less extensively studied remain in the "limited" evidence for animals and are in category 3. As a compound becomes more extensively studied, it is likely to move up in the IARC classification scheme regardless of mechanistic considerations. Phenobarbital remains an important prototype, and a full understanding of how it works may lead to more appropriate ways of classifying chemicals.

Risk Assessment

The relevance of mouse liver tumors to humans has been the subject of widespread scientific debate, more so than the relevance of any other tumor response observed in the rodent bioassay. The proceedings of a conference on mouse hepatic neoplasia published in 1975 (87) addressed this problem. Newberne (88) discussed the significance and extrapolation to man of mouse liver tumor data and concluded on the basis of the particular problems in the mouse that the utility of the mouse for the purpose of routine screening of chemicals for carcinogenic potential is highly questionable. It was suggested that strong consideration be given to deleting the mouse as a routine test animal for this purpose. On the other hand, Ruebner et al. (65) and others have concluded that an ideal mouse strain should have a low incidence of tumors but a high susceptibility to carcinogenesis and considered that the incidence of background tumors in the B6C3F$_1$ mouse was sufficiently low, coupled with its high sensitivity to carcinogens, to make this mouse quite suitable for such tests. In any event, the mouse liver tumor cannot be lightly dismissed, since it can represent a valid end point in carcinogenicity studies (89,90). Pereira (91) concluded that mouse liver tumor data are important for detecting nongenotoxic and genotoxic carcinogens that might be missed in the rat and that although it is important to distinguish between genotoxic and nongenotoxic mechanisms, further study is required to determine the relationship between their activity in mouse liver and their activities in humans.

A major problem with the rodent bioassay is that clear carcinogenic responses can be elicited by both initiators and phenobarbital-like promoters; as such, the bioassay does not distinguish between these two types of activities. Classification schemes and risk assessment are based primarily on the bioassay results. Ancillary information, such as positive tests for mutagenicity or DNA damage, are used to corroborate a conclusion of carcinogenic activity more often than the lack of evidence for genotoxicity or known promoting activity is taken into account to qualify a conclusion of "carcinogenic activity." Thus in the case of the occurrence of spontaneously initiated eosinophilic foci, Schulte-Hermann (11) has stated that carcinogenicity studies with PB-like compounds may lead to false-positive results with respect to the initiating potential of the test compound.

A substance such as phenobarbital, which induces both hepatocellular adenomas and carcinomas in mice in a bioassay, will appear to be a "complete carcinogen" because the animals were not knowingly exposed to an initiating carcinogen. Since various tests are available that can distinguish between chemicals that have direct (initiating) carcinogenic activity from those that act as tumor promoters, initiation can be sufficiently distinguished from promotion for purposes of risk assessment and management. Williams and Weisburger (92) have classified chemicals with respect to their mechanism of action into DNA-reactive, genotoxic carcinogens and compounds that act through a variety of epigenetic mechanisms. They maintain that since current evidence shows that many classes of epigenetic agents, in particular promoters, have reversible and highly dose-dependent effects, sound theoretical and empirical reasons exist for treating such agents in an entirely distinct manner with respect to human risk assessment.

Several expert committees have addressed the issue of mouse liver tumors. In 1982, an ECETOC task force on hepatocarcinogenicity noted that some strains of mice are especially sensitive to the development of hepatic tumors and concluded that it is possible to distinguish between genotoxic and nongenotoxic hepatocarcinogens (93). The former compounds are generally of great concern; however, enhanced tumor incidence by nongenotoxic mechanisms under conditions of high-dose exposure should not present a risk under conditions of low-dose exposure.

In 1983, in a report by an International Expert Advisory Committee to the Nutrition Foundation (94), the committee concluded, "There could be little doubt that it would be prudent to severely limit or even eliminate exposure to potentially genotoxic chemicals which induce malignant tumors at multiple sites or in multiple species, and at low exposure levels. Less concern is warranted in the case of chemical induction of tumors only in mouse liver, particularly if the tumors are primarily benign neoplasms, that are associated only with high exposure levels that produce additional biologic effects such as chronic tissue injury. Further evidence that would alleviate concern would include evidence that the substance fails to bind covalently to cellular macromolecules and negative results in well validated short-term in vivo and in vitro tests for DNA damage. In such an instance there would be little basis for severely restricting exposure to a chemical."

Regulatory agencies also have to weigh the relevance of the mouse liver tumor to human risk. The Environmental Protection Agency (95) in its 1986 guidelines for cancer risk assessment (although recognizing that there are widely diverging scientific views regarding the significance of mouse liver neoplasms) states: "...an increased incidence of neoplasms that occur with high spontaneous background incidence (e.g. mouse liver tumors and rat pituitary tumors in certain strains) generally constitutes 'sufficient' evidence of carcinogenicity but may be changed to 'limited' when warranted by the specific information available in the agent."

In the IARC classification scheme, "limited evidence" for animal carcinogenicity would include the situation in which the data suggest that a carcinogenic effect is limited for making a definitive evaluation because, e.g., "...the agent increases the incidence of only benign neoplasms or lesions of uncertain neoplastic potential, or of certain neoplasms which may occur spontaneously in high incidences in certain strains."

In an IARC monograph on a critical appraisal of long-term assays for carcinogens (96), the issue of tumors that occur at a high spontaneous incidence is addressed: "Some strains of animals have extremely high spontaneous tumor rates; for example, most Fisher F344 rats develop testicular tumors within two years, and some strains of mice likewise have very high spontaneous rates of liver, lung or mammary tumors. Alterations in gross or even apparently minor aspects of diet can easily double or halve these spontaneous rates. It is thus uncertain how to interpret chemical carcinogenesis experiments in which, although there is a highly statistically significant effect of treatment, the chemical only affects the incidence of such tumors. In the absence of any further data, there is a general consensus that such a result does not provide sufficient evidence of carcinogenicity. Whether or not any objective evidence exists to support this consensus view, the modes of action of those agents that seem to affect only tumor types with a high spontaneous frequency should perhaps be studied, and perhaps further carcinogenicity tests on other species or strains should also be done."

Although some regulatory bodies have placed qualifications on the significance of the mouse liver tumor, as indicated above, the problem has yet to be meaningfully resolved for regulatory purposes as an issue for carcinogen classification and cancer risk assessment. Some of the factors that the EPA (95) would take into account on a case-by-case basis for considering mouse liver tumors "limited" rather than "sufficient" evidence of carcinogenicity include: the mouse liver is the only response site; evidence is negative or inconclusive in short-term tests for mutagenic activity; the response occurs only at the high dose and/or only at the end of the study; excess tumors occur only in a single sex; there is no shortened time to appearance of tumors; and the tumors are predominantly benign, with no substantial increase in the proportion of tumors that are malignant.

Few of these criteria would be met by phenobarbital, since the response is observed in both sexes of several strains of mice, and tumors in rat liver have also been observed. Depending upon dose selection, the tumor response could be observed at lower dosages in addition to the high dose, and in a 2-year study, an increase in tumor incidence would be observed at least as early as 12 months. A substantial increase in hepatocellular carcinoma (with a low malignancy rate), in addition to benign tumors (adenomas), has been observed in several studies.

The factors taken into account in the EPA cancer risk assessment guidelines are limited to descriptive characteristics of the tumor response and as such do not represent a consideration of mechanisms. The phenobarbital example suggests that mechanistic considerations in terms of whether a compound is a nongenotoxic tumor promoter vs. a genotoxic initiating carcinogen are more important for risk assessment than descriptive characteristics of the tumor response, which, in the case of phenobarbital can be observed in multiple species, strains, and sexes and potentially at multiple sites.

Tumor Site Multiplicity

Several microsomal enzyme inducers that produce mouse liver tumors also increase the incidence of thyroid follicular neoplasia in rats. Since this is not an uncommon tumor combination (97), tumors at multiple sites and in multiple species need to be examined, as discussed above, and considered on a mechanistic basis.

In addition to promoting hepatic tumors in mice and rats, PB promotes thyroid gland tumors in rats after the administration of various nitrosamines (6–9). Phenobarbital is known to alter thyroid function in rodents (98) as a result of enhanced biliary excretion of thyroid hormone (99). McClain and coworkers (100) have demonstrated that the increase in the biliary excretion of thyroid hormone is accounted for by an increase in the biliary excretion of thyroxine glucuronide as a result of an induction of hepatic microsomal thyroxine-glucuronyltransferase. We also demonstrated that supplemental administration of thyroxine blocked the promoting effect of PB and that the tumor-promoting effect was directly proportional to the plasma level of thyroid-stimulating hormone (TSH) (9,97). The tumor-promoting effect of PB in the thyroid gland thus appears to be mediated by pituitary TSH as a compensatory response to the increased hepatic disposition of thyroid hormone. In the case of thyroid follicular neoplasia in rat carcinogenicity studies with microsomal enzyme inducers, thyroid function should be assessed, and a secondary mechanism of hormone imbalance should be considered in the interpretation of these findings in rodents.

Research Needs

Further research on phenobarbital and a better understanding of the pathogenesis of hepatic tumor promotion should provide a better scientific basis for distinguishing hepatocarcinogens by their mechanisms of action. Even though phenobarbital has been extensively studied, the pathogenesis and biologic characteristics of the induced tumors are not well understood. In the first place, it is uncertain what the response in mouse strains such as the $B6C3F_1$ and C3H actually represents. Most studies support the conclusion that PB is promoting preexisting, spontaneously initiated lesions; others maintain that this represents a true carcinogenic response. Since the phenotype of the PB-induced tumor appears to be different from the commonly expressed tumors of control mice and from tumors induced by many direct-acting (initiating) carcinogens, further characterization of the biologic characteristics of this phenotype during treatment and after PB withdrawal are necessary. The nature of the cells of the PB-induced tumor is of particular importance. Do these arise from an alteration in the phenotype of the spontaneously occurring tumor, or do they represent the promotion of spontaneously initiated cells not commonly expressed in controls? Although most studies are conducted in male mice, there may be some advantage to studies in the female mouse. The spontaneous tumor incidence is lower in females than males, and since the female mouse response is at least equal to that in the male, the PB-induced tumor should be the predominant type observed.

Since phenobarbital is an important prototype for a class of hepatic tumor promoters, the physiologic, biochemical, and molecular biology events that link hepatic microsomal enzyme induction, cell proliferation, and hepatic tumor promotion need to be understood. In addition, since PB is a promoter of both hepatocellular and thyroid follicular tumors, the relationship between this particular combination of tumors should be considered on a mechanistic basis.

The conflicting, nonconcordant, and equivocal responses to phenobarbital in short-term tests for mutagenicity and clastogenicity must be resolved. In addition to a complete assessment for DNA damage in the species and strains used for carcinogenicity testing, a more thorough evaluation for DNA binding and adduct formation would be of value.

The available 2-year studies with PB using both male and female mice are limited to the CF1 and C3H strains. A full characterization of the response to phenobarbital in the strains commonly used for bioassay (CD-1 and $B6C3F_1$) should be conducted using both sexes treated for 2 years for comparison to compounds with similar activity. Any differences in response, qualitatively or quantitatively, between the latter two strains or between the response observed in the C3H/He strain would be of considerable importance.

Conclusions

In summary, phenobarbital treatment induces hepatocellular adenomas and carcinomas in male and female mice and adenomas in rats. Although the response in mice is observed predominantly in strains with a high spontaneous background of hepatic tumors, PB appears to enhance a tumor with a phenotype not commonly expressed in either male or female mice. PB is generally considered nongenotoxic; however, mixed and conflicting results have been observed in both bacterial and mammalian systems. There is no evidence for DNA binding or DNA damage by PB in various in vivo and in vitro tests using rats and mice. Phenobarbital is a hepatic tumor promoter and has yielded no evidence for initiating activity. Relatively extensive human epidemiologic studies at enzyme-inducing doses have indicated no increased cancer risk; thus the relevance of the hepatic tumor response in mice is doubtful.

Phenobarbital is an extensively studied and particularly important prototype for microsomal enzyme–inducing compounds that produce similar effects. The mechanisms involved need to be fully understood so that this knowledge can be applied to the schemes used for carcinogen classification and the procedures followed for cancer risk assessment.

References

1. Conney AH. Pharmacologic implications of microsomal enzyme induction. *Pharmacol Rev* 19:317–366, 1967.
2. Golberg L. Liver enlargement produced by drugs: Its significance. In *Proceedings of the European Society for the Study of Drug Toxicity*, Vol. 7. Amsterdam: Excerpta Medica Foundation, 1966, p 171.
3. Peraino C, Fry RJ, Staffeldt E. Reduction and enhancement by phenobarbital of hepatocarcinogenesis induced in the rat by 2-acetyl-aminofluorene. *Cancer Res* 31:1506–1512, 1971.
4. Pitot HC, Barsness L, Goldsworthy T, Kitagawa T. Biochemical characterisation of stages of hepatocarcinogenesis after a single dose of diethylnitrosamine. *Nature* 271:456–458, 1978.

5. Solt DB, Farber E. New principle for the analysis of chemical carcinogenesis. *Nature* 263:701–703, 1976.
6. Hiasa Y, Kitahori Y, Ohshima M, Fujita T, Yuasa T, Konishi N, Miyashiro A. Promoting effects of phenobarbital and barbital on development of thyroid tumors in rats treated with N-bis(2-hydroxypropyl) nitrosamine. *Carcinogenesis* 3:1187–1190, 1982.
7. Tsuda H, Fukushima S, Imaida K, Kurata Y, Ito N. Organ-specific promoting effect of phenobarbital and saccharin in induction of thyroid, liver, and urinary bladder tumors in rats after initiation with N-nitrosomethylurea. *Cancer Res* 43:3292–3296, 1983.
8. Diwan BA, Palmer AE, Ohshima M, Rice JM. N-Nitroso-N-methylurea initiation in multiple tissues for organ-specific tumor promotion in rats by phenobarbital. *JNCI* 75:1099–1105, 1985.
9. McClain RM, Posch RC, Bosakowski T, Armstrong JM. Studies on the mode of action for thyroid gland tumor promotion in rats by phenobarbital. *Toxicol Appl Pharmacol* 94:254–265, 1988.
10. Schulte-Hermann R. Induction of liver growth by xenobiotic compounds and other stimuli. *CRC Crit Rev Toxicol* 3:97–158, 1974.
11. Schulte-Hermann R. Tumor promotion in the liver. *Arch Toxicol* 57:147–158, 1985.
12. Schulte-Hermann R, Schuppler J, Timmermann-Trosiener I, Ohde G, Bursch W, Berger H. The role of growth of normal and preneoplastic cell populations for tumor promotion in rat liver. *Environ Health Perspect* 50:185–194, 1983.
13. Moore MA, Kitagawa T. Hepatocarcinogenesis in the rat: The effect of promoters and carcinogens in vivo and in vitro. *Int Rev Cytol* 101:125–172, 1986.
14. Pitot HC, Sirica AE. The stages of initiation and promotion in hepatocarcinogenesis. *Biochim Biophys Acta* 605:191–215, 1980.
15. Peraino C, Staffeldt E, Haugen DA, Lombard LS, Stevens FJ, Fry RJ. Effects of varying the dietary concentration of phenobarbital on its enhancement of 2-acetylaminofluorene-induced hepatic tumorigenesis. *Cancer Res* 40:3268–3273, 1980.
16. Lubet RA, Nims RW, Ward JM, Rice JM, Diwan BA. Induction of cytochrome P450B and its relationship to liver tumor promotion. *J Am Coll Toxicol* 8:253–262, 1989.
17. Schulte-Hermann R, Parzefall W, Bursch W. Role of stimulation of liver growth by chemicals in hepatocarcinogenesis. In Butterworth BE, Slaga TJ (eds): *Banbury Report No 25: Nongenotoxic Mechanisms in Carcinogenesis.* Cold Spring Harbor, NY: Cold Spring Harbor Laboratory, 1987, pp 91–106.
18. Maronpot RR, Haseman JK, Boorman GA, Eostin SE, Rao GN, Hoft JE. Liver lesions in B6C3F$_1$ mice: The National Toxicology Program experience and position. In Chambers PL, Henschler D, Oesch F (eds): *Mouse Liver Tumors: Relevance to Human Cancer Risks.* Berlin: Springer-Verlag, 1987, pp 10–26.
19. Butler TC, Mahatee C, Waddell WJ. Phenobarbital: Studies of elimination, accumulation, tolerance and dosage schedules. *J Pharmacol Exp Ther* 111:425–435, 1954.
20. Waddell WJ, Butler TC. The distribution and excretion of phenobarbital. *J Clin Invest* 36:1217–1226, 1957.
21. Alvin J, McHorse T, Hoyumpa A, Bush MT, Schenker S. The effect of liver disease in man on the disposition of phenobarbital. *J Pharmacol Exp Ther* 192:224–235, 1975.
22. Morselli PL, Franco-Morselli R. Clinical pharmacokinetics of antiepileptic drugs in adults. *Pharmacol Ther* 10:65–101, 1980.

23. Butler TC. The metabolic hydroxylation of phenobarbital. *J Pharmacol Exp Ther* 116:326–336, 1956.
24. Whyte MP, Dekaban AS. Metabolic fate of phenobarbital: A quantitative study of p-hydroxyphenobarbital elimination in man. *Drug Metab Dispos* 5:63–70, 1977.
25. Tang BK, Inaba T, Kalow W. N-Hydroxylation of pentobarbital in man. *Drug Metab Dispos* 5:71–74, 1977.
26. Horning EC, Horning MG. Metabolic profiles: Gas-phase methods for analysis of metabolites. *Clin Chem* 17:802–809, 1971.
27. Harvey DJ, Glazener L, Stratton C, Nowlin J, Hill RM, Horning MG. Detection of a 5-(3,4-dihydroxy-1,5-cyclohexadien-1-yl) metabolite of phenobarbital and mephobarbital in rat, guinea pig and human. *Res Commun Chem Pathol Pharmacol* 3:557–565, 1972.
28. McCann J, Choi E, Yamasaki E, Ames BN. Detection of carcinogens as mutagens in the Salmonella/microsome test: Assay of 300 chemicals. *Proc Natl Acad Sci USA* 72:5135–5139, 1975.
29. Rao TK, Hardigreo AA, Young JA, Lijinsky W, Epler JL. Mutagenicity of N-Nitrosopiperdins with Salmonella typhimurium/microsomal activating system. *Mutat Res* 56:131–145, 1977.
30. Hollstein M, McCann J, Angelosanto FA, Nichols WW. Short term tests for carcinogens and mutagens. *Mutat Res* 65:133–226, 1979.
31. Purchase IF, Richardson CR, Anderson D, Paddle GM, Adams WG. Chromosomal analyses in vinyl chloride-exposed workers. *Mutat Res* 57:325–334, 1978.
32. Anderson D, Styles JA. The bacterial mutation test: Six tests for carcinogenicity. *Br J Cancer* 37:924–930, 1978.
33. King MT, Beikirch H, Eckhardt K, Gocke E, Wild D. Mutagenicity studies with x-ray-contrast media, analgesics, antipyretics, antirheumatics and some other pharmaceutical drugs in bacterial, Drosophila and mammalian test systems. *Mutat Res* 66:33–43, 1979.
34. Donner M, Sorsa M, Vainio H. Recessive lethals induced by styrene and styrene oxide in Drosophila melanogaster. *Mutat Res* 67:373–376, 1979.
35. Ray-Chaudhuri R, Currens M, Iype PT. Enhancement of sister-chromatid exchanges by tumour promoters. *Br J Cancer* 45:769–777, 1982.
36. Tice RR, Costa DL, Drew RT. Cytogenetic effects of inhaled benzene in murine bone marrow: Induction of sister chromatid exchanges, chromosomal aberrations, and cellular proliferation inhibition in DBA/2 mice. *Proc Natl Acad Sci USA* 77:2148–2152, 1980.
37. Traul KA, Hink RJ Jr, Kachevsky V, Wolff JS III. Two-stage carcinogenesis in vitro: Transformation of 3-methylcholanthrene-initiated Rauscher murine leukemia virus-infected rat embryo cells by diverse tumor promoters. *JNCI* 66:171–176, 1981.
38. Foerst D. Chromosome studies on the effect of primidone (mylepsinum) and its metabolites phenobarbital and phenylethylmalondiamide in vitro. *Acta Genet Med Gemellol* 21:305–318, 1972.
39. Nandan SD, Rao MS. Action of phenobarbitone on human chromosomes in vitro. *IRCS Med Sci Libr Compend* 10:226–227, 1982.
40. Bruce WR, Heddle JA. The mutagenic activity of 61 agents as determined by the micronucleus, Salmonella, and sperm abnormality assays. *Can J Genet Cytol* 21: 319–334, 1979.
41. Ashby J, de Serres FJ, Draper M, Ishidate M, Margolin BH, Matter BE, Shelby MD.

Evaluation of short term tests for carcinogens: Report of the International Programme on Chemical Safety's Collaborative Study on In-Vitro Assays. *Progress in Mutation Research* Vol. 5, 1985.
42. Schwarz M, Hummel KE, Appel R, Rickart R, Kunz W. DNA damage induced in vivo evaluated with a non-radioactive alkaline elution technique. *Cancer Lett* 6:221–226, 1979.
43. Swenberg JA, Petzold GL, Harback PR. In vitro DNA damage/alkaline elution assay for predicting carcinogenic potential. *Biochem Biophys Res Commun* 72:732–738, 1976.
44. Stout DL, Becker FF. Occurrence of progressive DNA damage coincident with the appearance of foci of altered hepatocytes. *Carcinogenesis* 3:599–602, 1982.
45. Williams GM. Classification of genotoxic and epigenetic hepatocarcinogens using liver culture assays. *Ann NY Acad Sci* 349:273–282, 1980.
46. Doolittle DJ, Muller G, Scribner HE. The in vivo–in vitro hepatocyte assay for assessing DNA repair and DNA replication: Studies in the CD-1 mouse. *Food Chem Toxicol* 25:399–405, 1987.
47. Trosko JE, Yotti LP, Dawson B, Chang CC. In vitro assays for tumor promoters. In Stitch HF, Sam RHC (eds): *Short-Term Tests for Chemical Carcinogens*. New York: Springer, 1981.
48. Jone CM, Erickson LM, Trosko JE, Netzloff ML, Chang CC. Inhibition of metabolic cooperation by the anticonvulsants, diphenylhydantoin and phenobarbital. *Teratogenesis Carcinog Mutagen* 5:379–391, 1985.
49. Ruch RJ, Klaunig JE, Pereira MA. Inhibition of intercellular communication between mouse hepatocytes by tumor promoters. *Toxicol Appl Pharmacol* 87:111–120, 1987.
50. Klaunig JE, Ruch RJ. Strain and species effects on the inhibition of hepatocyte intercellular communication by liver tumor promoters. *Cancer Lett* 36:161–168, 1987.
51. Wright AS. The role of metabolism in chemical mutagenesis and chemical carcinogenesis. *Mutat Res* 75:215–241, 1980.
52. Peraino C, Fry RJ, Staffeldt E. Enhancement of spontaneous hepatic tumorigenesis in C3H mice by dietary phenobarbital. *JNCI* 51:1349–1350, 1973.
53. Thorpe E, Walker AI. The toxicology of dieldrin (HEOD). II. Comparative long-term oral toxicity studies in mice with dieldrin, DDT, phenobarbitone, β-BHC and γ-BHC. *Food Cosmet Toxicol* 11:433–442, 1973.
54. Ponomarkov V, Tomatis L, Turusov V. The effect of long term administration of phenobarbitone in CF-1 mice. *Cancer Lett* 1:165–172, 1976.
55. Jones G, Butler WH. Morphology of spontaneous and induced neoplasia. In Butler WH, Newberne PM (eds): *Mouse Hepatic Neoplasia*. Amsterdam: Elsevier, 1975, pp 21–59.
56. Butler WH, Jones G. Pathological and toxicological data on chlorinated pesticides and phenobarbital. *Ecotoxicol Environ Safety* 1:503–509, 1978.
57. Walker AI, Thorpe E, Stevenson DE. The toxicology of dieldrin (HEOD). I. Long-term oral toxicity studies in mice. *Food Cosmet Toxicol* 11:415–432, 1972.
58. Butler WH, Hempsall V. Histochemical observations on nodules induced in the mouse liver by phenobarbitone. *J Pathol* 125:155–161, 1978.
59. Becker FF. Characterization of spontaneous and chemically induced mouse liver tumors. In Popp JA (ed): *Mouse Liver Neoplasia: Current Perspectives*. Washington, DC: Hemisphere, 1984, pp 145–159.

60. Gangolli SD, Lake BG, Evans JG. The histopathology and biochemistry of phenobarbitone-induced liver nodules in C3H mice. *Arch Toxicol [Suppl]* 10:95–107, 1987.
61. Evans JG, Collins MA, Savage SA, Lake BG, Butler WH. The histology and development of hepatic nodules in C3H/He mice following chronic administration of phenobarbitone. *Carcinogenesis* 7:627–631, 1986.
62. Ward JM, Lynch P, Riggs C. Rapid development of hepatocellular neoplasms in aging male C3H/HeNCr mice given phenobarbital. *Cancer Lett* 39:9–18, 1988.
63. Wolff GL, Morrissey RL, Chen JJ. Amplified response to phenobarbital promotion of hepatotumorigenesis in obese yellow A^{vy}/A (C3H × VY) F-1 hybrid mice. *Carcinogenesis* 7:1895–1898, 1986.
64. Matsuyama M, Amo H. Promoting effects of long-term administration of barbital on spontaneous hepatic tumorigenesis in $B6C3F_1$ mice. *Jpn J Hygiene* 37:892–896, 1983.
65. Ruebner BH, Gershwin ME, French SW, Meierhenry E, Dunn D, Hsieh LS. Mouse hepatic neoplasia: Differences among strains and carcinogens. In Popp JA (ed): *Mouse Liver Neoplasia: Current Perspectives.* Washington, DC: Hemisphere, 1984, pp 115–143.
66. Ward JM. Morphology of potential preneoplastic hepatocyte lesions and liver tumors in mice and a comparison with other species. In Popp JA (ed): *Mouse Liver Neoplasia: Current Perspectives.* Washington, DC: Hemisphere, 1984, pp 1–19.
67. Ward JM, Ohshima M. Evidence for lack of promotion of the growth of the common naturally occurring basophilic focal hepatocellular proliferative lesions in aged F344/NCr rats by phenobarbital. *Carcinogenesis* 6:1255–1259, 1985.
68. Grasso P, Hardy J. Strain differences in natural incidence and response to carcinogens. In Butler WH, Newberne PM (eds): *Mouse Hepatic Neoplasia.* Amsterdam: Elsevier, 1974, pp 111–142.
69. Tarone RE, Chu KC, Ward JM. Variability in the rates of some common naturally occurring tumors in Fischer 344 rats and $(C57BL/6N \times C3H/HeN)F_1$ ($B6C3F_1$) mice. *JNCI* 66:1175–1181, 1981.
70. Sher SP, Jensen RD, Bokelman DL. Spontaneous tumors in control F344 and Charles River-CD rats and Charles River CD-1 and $B6C3HF_1$ mice. *Toxicol Lett* 11: 103–110, 1982.
71. Butler WH. Long-term effects of phenobarbitone-Na on male Fischer rats. *Br J Cancer* 37:418–423, 1978.
72. Schmahl D, Habs M. Lifespan investigations of some immunostimulating immunosuppressive and neurotropic substances in Sprague-Dawley rats. *Z Krebsforsch* 86:77–84, 1976.
73. Rossi L, Ravera M, Repetti G, Santi L. Long-term administration of DDT or phenobarbital-Na in Wistar rats. *Int J Cancer* 19:179–185, 1977.
74. Ward JM. Increased susceptibility of livers of aged F344/NCr rats to the effects of phenobarbital on the incidence, morphology, and histochemistry of hepatocellular foci and neoplasms. *JNCI* 71:815–823, 1983.
75. Clemmesen J, Fuglsang-Frederiksen V, Plum CM. Are anticonvulsants oncogenic? *Lancet* 1:705–707, 1974.
76. Clemmesen J. Phenobarbitone, liver tumors and thorotrast. *Lancet* 1:37–38, 1975.
77. Clemmesen J, Hjalgrim-Jensen S. On the absence of carcinogenicity to man of phenobarbital. *Acta Pathol Microbiol Scand* 261(Suppl):38–50, 1977.
78. Clemmesen J, Hjalgrim-Jensen S. Is phenobarbital carcinogenic? A follow-up of 8078 epileptics. *Ecotoxicol Environ Safety* 1:457–470, 1978.

79. Clemmesen J, Hjalgrim-Jensen S. Does phenobarbital cause intracranial tumors? A follow-up through 35 years. *Ecotoxicol Environ Safety* 5:225–260, 1981.
80. Olsen JH, Boice JD, Jensen PA, Fraumeni JF. Cancer among epileptic patients exposed to anticonvulsant drugs. *JNCI* 81:803–808, 1989.
81. White SJ, McLean AE, Howland C. Anticonvulsant drugs and cancer: A cohort study in patients with severe epilepsy. *Lancet* 2:458–461, 1979.
82. Shirts SB, Annegers JF, Hauser WA, Kurland LT. Cancer incidence in a cohort of patients with seizure disorders. *JNCI* 77:83–87, 1986.
83. Gold E, Gordis L, Tonascia J, Szklo M. Increased risk of brain tumors in children exposed to barbiturates. *JNCI* 61:1031–1034, 1978.
84. Friedman GD. Barbiturates and lung cancer in humans. *JNCI* 67:291–295, 1981.
85. International Agency for Research on Cancer. Overall evaluation of carcinogenicity: An updating of IARC monographs, Vols. 1–42. *IARC Monogr Eval Carcinog Risk Chem Hum* Suppl 7:17–34 & 313–316, 1987.
86. International Agency for Research on Cancer. Chemicals, industrial processes and industries associated with cancer in humans. *IARC Monogr Eval Carcinog Risk Chem Hum* Vol. 1–29, Suppl 4, 1982.
87. Butler WH, Newberne PM. *Mouse Hepatic Neoplasia*. Amsterdam: Elsevier, 1975.
88. Newberne JW. Mouse hepatic neoplasia: Significance and extrapolation to man. In Butler WH, Newberne PM (eds): *Mouse Hepatic Neoplasia*. Amsterdam: Elsevier, 1975.
89. Tomatis L, Partensky C, Montesano R. The predictive value of mouse liver tumor induction in carcinogenicity testing: Literature survey. *Int J Cancer* 12:1–20, 1973.
90. Ward JM, Griesemer RA, Weisburger EK. The mouse liver tumor as an endpoint in carcinogenesis tests. *Toxicol Appl Pharmacol* 51:389–397, 1979.
91. Pereira MA. Mouse liver tumor data: Assessment of carcinogenic activity. *Toxicol Indus Health* 1:311–333, 1985.
92. Williams GM, Weisburger JH. Carcinogenicity testing of drugs. *Prog Drug Res* 29: 155–203, 1985.
93. European Chemical Industry Ecology and Toxicology Center. *Hepatocarcinogenesis in Laboratory Rodents: Relevance for Man*. Brussels: ECETOC, 1982.
94. Nutrition Foundation. *The Relevance of Mouse Liver Hepatoma to Human Carcinogenic Risk: A Report of the International Expert Advisory Committee to the Nutrition Foundation*. Washington, DC, 1983.
95. Environmental Protection Agency. Guidelines for carcinogen risk assessment. Part II. *Federal Register* 51:33992–34003, 1986.
96. International Agency for Research on Cancer. Long-term and short-term screening assays for carcinogens: A critical appraisal. *IARC Monogr Eval Carcinog Risk Chem Hum* Suppl 2:73, 1980.
97. McClain RM. The significance of hepatic microsomal enzyme induction and altered thyroid function in rats: Implications for thyroid gland neoplasia. *Toxicol Pathol* 17:294–306, 1989.
98. Japundzic MM. The goitrogenic effect of phenobarbital-Na on the rat thyroid. *Acta Anat* 74:88–96, 1969.
99. Oppenheimer JH, Bernstein G, Surks MI. Increased thyroxine turnover and thyroidal function after stimulation of hepatocellular binding of thyroxine by phenobarbital. *J Clin Invest* 47:1399–1406, 1968.
100. McClain RM, Levin A, Posch RC, Downing JC. The effect of phenobarbital on the metabolism and excretion of thyroxine in rats. *Toxicol Appl Pharmacol* 99:216–228, 1989.

Mouse Liver Carcinogenesis: Mechanisms
and Species Comparisons, pages 367–383

Organochlorine Insecticides and Mouse Liver Tumors: An Analysis of Responses and Mechanisms

Donald E. Stevenson
Toxicology; Health, Safety and Environment, Shell Oil Company, Houston, TX 77210 and Scientific Committee, American Industrial Health Council, Washington, DC 20036

The era of the Second World War marked a major change in the use of insecticides. Until that time the products used were mainly compounds of natural origin such as salts of lead, copper, and arsenic and mercury, sulphur, white oils, nicotine, rotenone, and pyrethrins. While dichlorodiphenyltrichloroethane (DDT) had been synthesized many years previously, it was rapidly developed in order to deal with the immediate health threat of insect-carried plague to American troops in Italy in 1943–44. Large numbers of troops were literally dusted with DDT, with no observed side effects. Other public health and agricultural uses were developed in the following 3 or 4 years. While DDT had proved immediately safe and effective for some specific uses, it was not sufficiently active for all situations. To identify other active compounds, researchers focused on two major types of chemicals—other organochlorine compounds and the organophosphates. The search emphasized the need for broad-spectrum materials that would kill a range of pests and would be active for an entire growing season. The application rates for the older metal-based insecticides could be as high as 50–100 lbs/acre, but were in ounces or a few pounds/acre for the new insecticides.

The development and use of the organochlorine insecticides (OCI) resulted in a major evolution of the process regulating pesticides. The Insecticide Act of 1910 was aimed at preventing the manufacture, sale, or transportation of adulterated or misbranded Paris greens, lead arsenates, and other pesticides. In 1947 this Act was replaced by the Federal Insecticide, Fungicide, and Rodenticide Act (FIFRA), which broadened the range of pesticides covered and provided for registration of products. In 1959, an amendment included nematocides (such as 1,2-dibromo-3-chloropropane), plant regulators, and defoliants.

FIFRA was first administered by the Department of Agriculture, and its basic purpose was to protect the public from personal or economic injury. Among the requirements was label information that, if complied with, would be sufficient to prevent injury to living man and animals and useful vegetation. The criteria for toxicity related to short-term exposure, e.g., applicators. The tests required were acute oral and dermal toxicity and subacute studies when warranted. In 1954, an amendment known as the Pesticide ("Miller") Amendment to the Federal Food, Drug and Cosmetic Act (FDCA) established the requirement for a tolerance on or in a raw agricultural commodity for pesticide residues. The Food Additives Amendment of the FDCA was used to regulate residues in processed foods. Under these amendments, the FDA required long-term animal toxicology information in order to assess the potential risk arising from pesticide residues. Thus, testing requirements increased as analytical methods became more sensitive. The first monograph outlining FDA testing guidelines under the FDCA was published in 1959 (1). Initially, long-term studies were conducted in rats and dogs; it was not until the mid-1960s that mouse studies became commonplace.

Nearly 30 years ago, the FDA conducted a series of rat and mouse studies with OCI that showed that a common response was an increased incidence of hepatomas in C3H mice. It was their conclusion that the results were consistent with a tumorigenic rather than a carcinogenic response (2). In a transplantation study conducted by Andervont (K. J. Davis, unpublished observations) on animals from the FDA, the control hepatic tumors transplanted more readily than those from the treated animals. DDT, heptachlor, heptachlor epoxide, and chlordane were also found to lead to adenoma formation in mice.

Even before the FDA studies, Annau (3) had reported that levels of DNA and RNA were increased in both liver tumors and parenchyma following treatment with aldrin.

A pattern of results has emerged with the OCI that indicates that most members of the class are associated with mouse liver tumors—endrin and methoxychlor are exceptions. Some compounds such as DDT and mirex may also be associated with liver tumors in rats, although the response is not so marked as in mice. Studies in the hamster with DDT and dieldrin have been negative (4,5), while a near lifetime study in monkeys with DDT has also been negative (6). Following exposure of monkeys to dieldrin for 6 years, there was no evidence for any liver damage or tumors (7). Lastly, several epidemiologic studies (8,9) have been essentially negative even though exposure occurred for 30 or more years. There have also been human clinical studies that have not demonstrated abnormal changes in the human liver (10,11).

Studies that have examined the interaction between OCI have been negative in rats (12) but have demonstrated an additive response in mice with respect to liver tumors (13,14).

The strain and sex response of mice to OCI is not entirely uniform. For instance DDT was positive in CF1 mice (8), but negative in the National Toxicology Program (NTP) bioassays that utilized the $B6C3F_1$ mouse, although the metabolite DDE was positive (15). Dieldrin, which is also positive in both sexes of the CF1 strain was negative in the female $B6C3F_1$ mouse (15). Thus, there are some potentially interesting comparisons for mechanistic studies.

It may be concluded, therefore, that the tumorigenic response of the mouse liver to OCI does not represent a response that is consistent with the observations in other species, although some members of the class have been associated with a less marked response in rats. The remainder of this chapter reviews the possible underlying mechanisms and the reasons for this apparent species difference. An understanding of this difference may be important in relation to our total knowledge on cancer causation.

In our own studies (13,14), it was possible to rule out the potential question as to the role of active contaminants because the pure active component of dieldrin, HEOD, was used. The clinical course of high degree of liver enlargement and also the formation of discrete masses was followed by abdominal palpation. Animals were killed when the experienced technician believed that the tumor was sufficiently advanced to present a life-threatening risk. However, as we gained experience we realized that the judgment was conservative and that mice could survive a full life span in many cases with large tumors that were palpable for several months. Histologic examination of the tumors led to a classification of two types of tumor that were designated A and B, based on the cell pattern (13,14,16). Lung emboli and metastases were occasionally associated with the latter type. One of the intriguing aspects of these was the comparatively normal appearance of the cells. The clinical course was very different from that associated with human hepatocellular carcinoma.

There was concern that the ethylene oxide method of diet sterilization that was used might have modulated the response, since it left residues of chloroethanol in the diet. However, a comparison between unsterilized diets and diets sterilized by irradiation or ethylene oxide did not suggest any differences in response. More recent studies (17) have shown that the response was not affected by the use of semisynthetic diets or bedding other than sawdust.

Since the histologic response in mice and rats suggested a marked effect on the endoplasmic reticulum and enzyme induction, it was postulated that mice would respond to all enzyme inducers in a similar manner (18). Accordingly, DDT, beta- and gamma-benzene hexachloride (BHC), phenobarbitone, and butylated hydroxytoluene (BHT) were tested; and with the exception of BHT similar results were obtained (14). However the doses chosen for the latter were toxic, and the study was not repeated. Dix (19) reinvestigated the detail of histologic changes and confirmed that the most prominent changes occurred in the smooth endoplasmic reticulum but also noted changes in the rough endoplasmic reticulum and to a lesser extent in other organelles such as the Golgi bodies and mitochondria. An increased incidence of Mallory bodies was also noted (see Cadrin et al. in this book).

Tennekes et al. (20) and Kunz et al. (21) have reexamined the data from the Shell dieldrin experiments. There were several implications from their studies for hazard evaluation, and their approach should be extended to other compounds. They utilized the approach of Druckery (22), who investigated the dose-time response relationship for several nitrosamines in rats and concluded dose multiplied by a power of time (often 4–5) was a constant and that there was a linear log/log relationship between dose and induction time for the 50% response rate. The same relationships did not hold for dieldrin in the CF1 mouse or for the promotion by 12-*O*-tetradecanoylphorbol-13-acetate of 7,12-dimethylbenz[*a*]anthracene (DMBA)-initiated skin tumors in mice. However, for both the nitrosamines in rats and dieldrin in mice, there was a linear relationship on a log-probit plot of the cumulative frequency of tumors for each dose rate with time, suggesting a stochastic response in a homogeneous population. The lines are parallel for each dose group and the controls, suggesting that it is reasonable to make an estimate of the median time to tumor, even for incidences well below 50%.

Utilizing the median time to tumor information, Tennekes et al. (23) determined that the velocity of tumor formation in relation to dose was linear. It was possible to estimate the compound response and the amount of dieldrin that would have produced the response in the control animals if it had been caused entirely by dieldrin. This amount was approximately equivalent to 10 ppm in the CF1 mouse—a very high dose—and is an indication of the high background pressure on the mouse liver with respect to tumor formation.

Wright and et al. (24,25) have shown that the mouse liver responds to OCI such as dieldrin by both hypertrophy and hyperplasia, but the response in the rat is mainly hypertrophy. In the monkey, there is little evidence of a hypertrophic or hyperplastic response (8).

Consideration of Mechanisms

While there is a wealth of information on the mechanism of action of the OCI, it often lacks consistency in terms of compound, species, and type of study, so that direct comparisons between strain or species are not possible for every end point. Furthermore, much of the work was conducted during the 1960s and 1970s using techniques that were not as discriminating as those now available. However, a review of the information allows the development of hypotheses for further testing and confirmation.

Miller and Miller (26) proposed that carcinogens were electrophils that could react covalently with DNA and induce mutations. The reactive moiety might either be the parent compound or a metabolite. These reactions can be detected by examining mutagenic activity or by direct determination of adducts with DNA bases or the investigation of the products of DNA repair.

Mutagenicity

Maslansky and Williams (27) summarized the evidence for an epigenetic mode of action for OCI hepatocarcinogenicity. There was no evidence of genotoxic activity in primary culture/DNA repair assays using rat, mouse, or hamster liver cells exposed to chlordane, endrin, heptachlor, mirex, DDT, dichlorodiphenyldichloroethane (DDD), and DDE.

The mutagenicity studies on aldrin and dieldrin were summarized by Ashwood-Smith (28). The International Commission for Protection against Environmental Mutagens and Carcinogens has also summarized the data on

DDT, dieldrin, and phenobarbitone (29). The majority of studies have been negative, and the general conclusion is that these compounds are nongenotoxic and that they are probably acting as promoters rather than complete carcinogens in mice (27). In some cytogenicity studies, positive results have been obtained at high concentrations, i.e., above the levels that would be compatible with life in the whole animal.

Interactions with DNA and Other Macromolecules

Studies to define the potential for covalent binding between OCI and DNA have not demonstrated that this occurs, although Wright et al. (25) have reported very low levels of an unidentified biotransformation product of dieldrin that became bound to the liver DNA of CFE rats and CF1 and LACG mice. The degree of binding correlated with the rate of metabolism of dieldrin in the three animal types rather than with any biologic effect. The level of binding was very low, and the technique used did not distinguish between nuclear and mitochondrial DNA.

Kupfer et al. (30) reported that methoxychlor produced a reactive intermediate that bound covalently to the microsomal protein of rats and appeared to be catalyzed by cytochrome P-450. The reaction was inhibited by antioxidants. Interestingly, methoxychlor is not regarded as being carcinogenic in rats or mice.

Charles et al. (31) conducted a somewhat similar study with mirex, which causes liver tumors in both rats and mice. Mirex is associated in both rats and mice with liver enlargement, increases in DNA, protein, and microsomal enzyme synthesis and proliferation of the smooth endoplasmic reticulum. Labeled mirex becomes predominately distributed in mitrochondria and microsomes in a covalently bound form. Mirex was found in high concentrations in the nuclear and plasma membranes but could be dissociated from these by trichloroacetic acid, i.e., not in a covalently bound form. Polyploid hepatocytes selectively accumulated higher concentrations (two- to threefold) compared with diploid hepatocytes. No studies have been identified that would extend these interesting observations to other compounds.

It is possible to monitor the degradation products of DNA in urine as a way of measuring the rates of DNA repair. Saul et al. (32) have has shown that there are substantial species differences in the excretion rates. No studies have been identified that have used this approach to investigate whether OCI induce a higher level of oxidative damage in the livers of mice (see the discussion of lipid peroxidation below).

DNA Synthesis

Many studies have indicated a potential correlation between the stimulation of DNA synthesis and the development of mouse liver tumors with nongenotoxic compounds. For instance, the studies of Wright and his coworkers (24,25) showed that with dieldrin, the mouse responded by hyperplasia and an increase in liver DNA, whereas the liver response in the rat, dog, and monkey was mainly hypertrophy. Busser and Lutz (33) investigated the stimulation of DNA synthesis in rat and mouse liver for several compounds, including phenobarbital, DDT, and aldrin. There was generally a good correlation between an increased level of DNA synthesis and the corresponding bioassay data. Of particular interest was the result with aldrin, for which the sex difference in the NTP bioassay was consistent with the DNA synthesis data. Busser and Lutz distinguished the effects of genotoxic and nongenotoxic carcinogens in terms of their ability to induce or reduce DNA synthesis. These observations are consistent with those of Mirsalis and Steinmetz in these proceedings, which suggest that a sustained increase in cell division is an important component in the total tumorigenic process. A key issue is how cell turnover is initiated and sustained.

The work of Moolgavkar and his associates (34) suggested that it might be possible to determine the predominant mechanisms of action from modeling considerations. This approach suggests that increased cell turnover alone (cocarcinogenesis or proliferation) will lead to an increased incidence of tumors by amplifying the base level of mutations. On the other hand, a promoter increases the cell turnover of initiated cells.

Pharmacokinetics and Metabolism

One possible explanation of the species differences in response could be that the mouse acquires much higher concentrations of a compound in the liver than other species. However, data summarized by the National Institute for Occupational Safety and Health (35) indicate that this is not the case for dieldrin and that tissue levels of HEOD (dieldrin) achieved in other species would be associated with a large increase in liver tumors if the species susceptibility were similar. Analogous data is also available for DDT (35). The half-life of dieldrin in the mouse is in between that in man and the rat.

Another possibility is that there are important differences in metabolites that could explain the species differences in tumor response. In brief, while there are some differences in metabolites, these have not correlated with the biologic differences. For dieldrin, the primary metabolite in the rat and mouse is pentachloroketone (PCK). Baldwin et al. (37) found that pretreatment in the rat increased the proportion of PCK excreted in the urine, but this did not occur in the mouse. From the studies of Hutson (38), the main differences between these species appear to be a more rapid metabolism of dieldrin in rats, with a greater production of PCK, and the appearance of small amounts of polar urinary metabolites in mice. It was concluded that the hydroxylation of HEOD is less efficient in mice and that the formation of the glucuronide of 9-hydroxy HEOD results from the action of the hepatic microsomal monooxygenase and uridine diphosphoglucuronyl transferase.

For rats, mice, rhesus monkeys, and chimpanzees (39), the major metabolite of dieldrin is the 9-hydroxy derivative, which is found in the feces or as a conjugation product in the urine. In man this metabolite is found in the feces of industrially exposed workers (40). The overall conclusion has been that the pattern of metabolites does not in itself explain the species differences in response to dieldrin or the other OCI.

Information on the metabolism of DDT was summarized by Hayes (35). As mentioned earlier, there is bioassay data to indicate that one metabolite, DDE, may induce mouse liver tumors. However, this metabolite is not unique to the mouse and does not provide an explanation of the species differences. Gingell and Wallcave (41) showed that DDT does not stimulate (or may slightly decrease) its own metabolism in mice but does so in the hamster. Chhabra and Fouts (42) examined the effects of adrenalectomy or castration on the drug-metabolizing enzyme activity in animals exposed to DDT and concluded that rats and mice may differ in their hormonal regulation of basal levels of enzymes and response to enzyme inducers.

Many of the organochlorine insecticides are only slowly metabolized, but the liver may respond with visible changes within hours of exposure. While this may suggest that the parent compound itself either is the active moiety or is capable of initiating a chain of indirect events leading to tumor formation, small amounts of metabolites can be found within 15 min of intravenous injection in the rat (43).

While metabolism has not been shown to be involved directly in the chain of causation, there are hints that it may be relevant. For instance, as cited above, with some OCI, there are indications of some covalent binding to cell organelles.

Ploidy

Several studies have indicated changes associated with exposure to organochlorine insecticides. Abraham et al. (44) have reported increased ploidy with mirex. van Ravenzwaay and Kunz (45) examined the effect of dieldrin on ploidy in the CF1 mouse. They found that in the untreated controls, there was a linear increase in octaploid nuclei with age. Age-dependent changes occurred earlier in mice exposed to dieldrin, and it was suggested that dieldrin might advance the biologic age of the liver. The relevance of ploidy to the carcinogenic process is questioned by other investigators such as Danielsen et al. (46).

Effects of Cell Organelles, Enzymes, and Dynamics

It has already been indicated that the organochlorine insecticides induce the production of smooth endoplasmic reticulum and induce the monooxygenases of the P-450 system. Wright et al. (24,25) found that the response of the smooth endoplasmic reticulum was not so pronounced in the mouse as in the dog and rat. Wright also reported that in the monkey, there was less enzyme induction and proliferation of the endoplasmic reticulum (7). In man there has not been any evidence of enzyme induction in workers exposed to aldrin or dieldrin (47), although this does occur with exposure to endrin or exposure to phenobarbitone.

The effect of dieldrin on cell enzymes has been discussed in several reports, including those of Ruebner et al. (48), who also discussed the formation of Mallory bodies in the livers and tumors of mice exposed to dieldrin. This implies some interference with the spatial relationships within the cell system.

Organochlorine Insecticides as Promoters

Many investigators have suggested that nongenotoxic carcinogenic agents act by promoting previously initiated cells. This raises the issue of what causes the initiation. Tennekes et al. (17) have examined the potential effects of feed and bedding on the response of CF1 mice to dieldrin. Using a semisynthetic diet and filter paper for bedding for comparison, they found that both a conventional diet and sawdust bedding induced the liver monooxygenase system, although this effect was less pronounced than with exposure to 10 ppm dieldrin. There was no evidence in these studies that there was a difference in the incidence of liver tumors due to the type of food or bedding or that there was an interaction between dieldrin and the feed or bedding. Thus, it does not appear that initiation arises from the presence of initiators in feed or bedding.

An alternative hypothesis is that an intrinsic factor is associated with initiation. Buchmann et al. (49) have just reported a preliminary investigation of mutations in codon 61 of the Ha-*ras* proto-oncogene in CF1 mice and have found that the mutations are similar to those found in the $B6C3F_1$ mouse. They have identified mutations in two tumors induced by aflatoxin and two induced by phenobarbitone but no such mutations so far in dieldrin-induced mouse liver tumors. Because of the limited nature of the study, clearly the potential role of oncogenes is still an open question.

Several studies in rats show that OCI induce lipid peroxidation in the liver. Junqueira et al. (50) reported a dose-related increase in superoxide anion production in rats dosed with lindane, with a similar increase in lipid peroxidation. Goel et al. (51) found an approximately threefold increase in lipid peroxidation in rats given 26 mg/kg dieldrin. No OCI studies have been identified in other species or with dosing over a longer time frame and at lower doses, although hexachlorobenzene and the polychlorinated biphenyls have been shown to induce liver lipid peroxidation during more prolonged exposure of rats. Given the higher rate of oxygen utilization by the mouse and its lower capacity to deal with free radicals, a higher rate of damage could be anticipated.

Junqueira et al. (52) suggested that in rats there may be two components to lindane-induced oxidative stress on the liver following acute dosing. The early component relates to the reductive metabolism of lindane, and the second was associated with the induction of cytochrome P-450. This supports the hypothesis that both the metabolism of OCI and their ability to increase free radical production may be important in the induction of mouse liver tumors.

Intercellular Communication

As discussed extensively in this book, many studies have shown that the organochlorine insecticides can inhibit intercellular communication. Ruch and Klaunig (53) and Klaunig and Ruch (54) reported a good correlation with the liver effects in several mouse strains and the F344 rat with phenobarbital, DDT, lindane, and dieldrin. Unfortunately, only males were examined, so that the interesting sex difference in response seen in the $B6C3F_1$ mouse could not be correlated. Other studies by Suter et al. (55) provide excellent descriptions of the events seen in these systems.

Among the possible explanations for interference with intercellular communication is the generation of endogenous genotoxic products such as free radicals. The fact that antioxidants inhibit some parts of the process, such as the effects of intercellular communication, suggests that this aspect requires serious consideration. For instance, Ruch and Klaunig (56) reported that superoxide dismutase, *N, N'*-diphenyl-*p*-phenylenediamine, and vitamin E prevented the inhibition of intercellular communication induced by lindane and DDT. Both singlet oxygen and lipid peroxides have been proposed as possible common inducers of the liver monooxygenase system by Paine (57).

Loch-Caruso and Trosko (58) have discussed the underlying processes that can lead to interrupted cell communication. These include changes in the composition of the intracellular fluid such as the concentration of calcium and cAMP, pH changes, free radical production, and substances altering key receptors, as well as events that alter the phenotypic and genotypic expression of cells. There is evidence that more than one of these changes could be occurring. However, there is little evidence that these events are necessarily unique to the mouse liver. Klaunig and Ruch (59) suggested that cAMP might mediate the blocking effects of DDT and phenobarbital.

Calcium and Other Ions

The target of the OCI is the nervous system of insects. In mammals, the central nervous system is also the major site of toxicity. Although there are some differences between individual OCI, as a class they are potent inhibitors of the Na, K, or Mg ATPases, which maintain the concentration of cellular cations that are key to nerve function. The OCI are also active inhibitors of similar enzymes in other tissues. Wang and Matsumara (60) showed that in vitro, dieldrin altered the flux of sodium and calcium in rat liver cells and speculated that these changes might be related to the proliferation of the smooth endoplasmic reticulum. Certainly, calcium is an important element in cell regulation (61). Pounds and Rosen have also published an excellent review of the role of calcium in intracellular processes and the potential for short-term effects via calmodulin and longer-term effects via protein kinase C (62).

Mehrotra et al. (63) showed that subchronic exposure of rats to dieldrin reduced brain calmodulin and its regulation of calcium-ATPase activity. Other tissues were not examined.

Klaunig and Ruch (59) demonstrated that both phenobarbital and DDT decrease the levels of cAMP in cultured mouse hepatocytes. They also showed that caffeine and dibutyryl cAMP increased the levels of cAMP and prevented the interference in intercellular communication caused by phenobarbital or DDT. These authors have also shown that antioxidants prevent this interference and suggested a relationship between free radical increases caused by phenobarbital and DDT, decreases in cAMP, and interference with intercellular communication.

Warngard et al. (64) investigated the effects of DDT on calmodulin and concluded that it bound to the protein in a manner similar to established inhibitors. However, the latter did not inhibit intercellular communication.

Receptors

While it is well established that some stable organochlorine compounds, such as the dioxins, have important interactions with receptors such as the Ah receptor, OCI have not been investigated extensively for the same kinds of activity. Compounds such as DDT are known to have estrogen-like activity; thus it would not be surprising to find that receptor binding was part of their processes of toxicity.

Discussion

There is not a unique event or process that can be correlated directly with mouse liver tumors, other than increased liver cell division. While there are several processes that may have a plausible role in the outcome, in most cases, other organs or species may be cited in which the outcome is not carcinogenesis and no increase in cell division has been seen. It would seem, therefore, necessary to consider either quantitative differences between tissues or combinations of events (the "silver shotgun" rather than the "silver bullet").

The outlined working hypothesis proposed is that the OCI may have several possible parallel, but interacting mechanisms, including the following.

- The OCI interfere with ion transport and can alter the calcium intracellular concentration, and this in turn modulates other cell processes.
- Combination with a surface or cytosolic receptor leads to enzyme induction. This increases the rate of production of free radicals that are not completely neutralized before intracellular targets are hit, so that cell damage is caused.
- Protein kinase C may be switched on by the radicals, and in turn this alters arachidonic acid metabolism and the production of biologically active kinins.

• Oxidative damage may also occur in the mitochondria, affecting the fidelity of replication of cell proteins, membranes and enzymes.
• These processes result in the production of growth factors or changes in (onco)gene structure or expression, with the result that the rate of cell division increases or cannot be switched off.

Interference with or absence of one or more of these processes may prevent the tumorigenic outcome. Tumorigenesis requires sustained effects, although effects on the nervous system could be anticipated after only brief periods of ionic change. Other environmental conditions might exert their own influence through similar mechanisms. For example, caloric intake would be expected to be related to oxygen consumption and radical production, and changes in protein intake could change the level of activity of oxygenase enzymes.

It may be postulated that the unique aspect of the mouse is that it is at one extreme of the range of oxygen utilization but does not need as high a degree of protection against oxygen damage because under normal conditions of life, its survival to reproduce is limited by predation and disease rather than by its ability to protect itself against the gradual accumulation of tissue damage. On the other hand, the human has a prolonged period of growth and reproduction and requires a much higher degree of protection against the cumulative effects of tissue damage. Substances such as the OCI appear to dissect these apparent species and tissue specificities by highlighting the large differences in biologic outcomes that may result from quantitative differences in enzyme and other cell systems.

Open Forum

Dr. Ernest McConnell: Has anybody ever correlated enzyme induction with tumor response? The level (magnitude) of enzyme induction is well below the level of tumor induction. Maybe the mechanisms are similar, and it is merely a matter of dose-response.

Dr. Don Stevenson: Alan Wright looked at dieldrin in the CF_1 mouse and found that enzyme induction may not be seen initially at 1 ppm, and at that stage we were still getting an increase in tumors. However, after some weeks, enzyme induction may be measurable.

Dr. McConnell: Did you observe tumor increase at levels when you saw no increase in enzyme induction?

Dr. Stevenson: Yes, no initial enzyme induction—although this may occur as exposure continues. We have the interesting situation that endrin, which is negative in the mouse, causes enzyme induction in man, while dieldrin does not, as far as we can tell after looking for it with several indicators in people who have had relatively high exposure.

Dr. McConnell: So you think enzyme induction has nothing to do with tumor formation?

Dr. Stevenson: I don't think it necessarily has nothing to do with it, but

I don't think it has everything to do with it. As Dennis Parke indicated, I think the increased oxygen cycling may be part of this, but it is not the entire answer.

Dr. Michael McClain: Enzyme induction per se does not cause tumors. I think the important thing to recognize is that you need enzyme induction in combination with a defect in the liver of these sensitive mouse strains. In both sensitive and resistant mouse strains, the level of enzyme induction and other responses in the liver appear to be similar. However, the sensitive strains show a tumor response, and the resistant strains do not. Thus, it is really enzyme induction in combination with a defect in the liver that leads to tumors. In humans treated with phenobarbital there is enzyme induction, but no evidence of tumors. One perspective on this difference is that humans do not have the same type of liver defect as the C3H/He or the $B6C3F_1$ mouse, so tumors are unlikely to occur in humans even when you do have enzyme induction.

Dr. McConnell: The fascinating observation is that you have tumor formation in the absence of enzyme induction with an enzyme inducer.

Dr. McClain: I have not really seen data on that. No one has yet determined that with phenobarbital there is a tumor response at non-enzyme-inducing doses. The doses that have been studied that produce tumors are much higher than the dose required for enzyme induction. Again, I would say that it is not enzyme induction per se that is causing tumors. It could be the cellular events that are secondary to enzyme induction or adaptive responses of the hepatocyte to enzyme induction. The mitogenic effect observed with enzyme inducers may be more important than the level of enzyme induction. I think that considerable work still needs to be done with phenobarbital and chlorinated hydrocarbons. We still have a long way to go in understanding what some of these cellular responses are and what the specific mechanisms are.

Dr. Dennis Parke: I would like to comment first on the phenobarbitone; 500 mg/kg is a very high dose of phenobarbitone for rats, and I imagine it has the same effect on mice. These animals are almost in a state of narcosis.

Dr. McClain: Let me correct one thing. The dose of phenobarbital is 500 mg/kg of diet. Most of the phenobarbital mouse studies are done at 0.05% (500 ppm) in the diet. This concentration is equivalent to a dose of around 75 mg/kg body weight; 1,000 ppm is equivalent to about 150 mg/kg. Barbital is less potent than phenobarbital in terms of enzyme induction and central nervous system (CNS) effects. The dose in the barbital study was 500 mg/kg body weight. In our ongoing studies at doses of phenobarbital up to 150 mg/kg/day, we see no evidence of CNS depression in treated CD-1 or $B6C3F_1$ mice.

Dr. Parke: Well, they may not be narcosed, but they are certainly less active, even at that dose. I doubt whether they are eating enough. Phenobarbitone does various things. If it produces enzyme induction, it increases the rate of protein turnover enormously and escalates the rate of protein synthesis several-fold. That requires the dietary protein to be directed toward protein synthesis in the cell. That, in itself, depletes intracellular glutathione. The sulfur amino acids are simply not available. We have unpublished results that demonstrate this effect. The second thing is that phenobarbitone causes a depletion of folate. This effect has been shown in humans, rats, and mice. I do not know what the overall effect of that depletion of folate will be, but it eventually starts inhibiting protein synthesis. Although the enzyme induction rises for the first few months, it then comes down and reaches zero simply because there is not enough folate. If you give those animals folate, then the enzyme induction goes up again. This must have a disastrous effect on the intracellular metabolism, because it inhibits protein synthesis.

The other thing is that these levels of phenobarbitone produce high levels of futile cycling in the rat and probably 20 times that level of futile cycling in the mouse. So you are producing a very high rate of hydroxyl radicals and you have tied up the intracellular glutathione. No wonder you get malignancy in the mouse. I think the same is true of the chlorinated chemicals. These often require glutathione for their metabolism. So I would say that the situation you are looking at in the mouse liver tumors is one of oxygen radical induction of tumors, which is the major cause of spontaneous tumors in the mouse and the mouse liver. In other words, you are looking at nonfindings.

Dr. McClain: I think there are other possible mechanisms besides that.

Dr. Parke: I am not denying that, but that mechanism has been completely ignored.

Dr. James Trosko: Dr. Stevenson, you raised the possibility that the initiation-promotion-progression model might not be an adequate paradigm to understand and explain carcinogenesis. Did you offer an alternative paradigm, and why did you raise the question?

Dr. Stevenson: I really feel that some of these materials are having dramatic cytotoxic or other effects on the cells; it may well be that those effects are key in the ultimate process rather than simply something that has happened in terms of initiation or promotion. I have information that makes me wonder whether cells that become initiated are not in fact already promoted by some of the things we are talking about. So I am really looking for an alternative explanation. I don't want to rule anything out. I don't think we should necessarily classify these things all into neat boxes.

Dr. Michael Gallo: I think the question of cytotoxicity and its role in promotion or progression may be nothing more than a simple selection pressure. This goes back to some of the wonderful experiments that Gary Williams and his group have done over the years. We are just not looking at cytotoxicity in the proper way. It is very difficult to think of a series of progression steps without cytotoxicity. One thing I would like to see come out of this meeting is some deep thought on this multistep process. I don't think we should pigeon-hole substances as initiators, promoters, and progressors, but I think we have to keep that process in mind.

Dr. Stanley Goldfarb: Dr. Stevenson, I was intrigued with the Druckery curves that you showed. In a 1985 paper, Stan Vesselinovitch did a similar plot of his data and showed quite different slopes for foci vs. carcinomas. On this basis he concluded that the number of hits necessary to cause transformation of a normal cell to a focus was different than the number of hits necessary to cause cancer. They were also qualitatively different. Your projection appears to support a quite opposite interpretation. Is that correct?

Dr. Stevenson: Yes. I think that may relate to the nature of the two materials we were working with. We used dieldrin; he was working with nitrosamines.

Dr. Goldfarb: What is emerging here is the idea that we should view benign and malignant mouse liver tumors as part of a continuum, rather than separate them sharply into discrete entities as we have in the past.

Dr. Stevenson: One thing that really intrigues me is that the slope for each dose group is parallel. Intuitively, I would not have thought that would be the case. The other implication is that the population is actually homogeneous and it is a matter of chance as to which animals develop a tumor first. If you had a mixed population, then clearly that slope would suddenly change, so I think plotting this way out may have some additional advantages.

References

1. U.S. Food and Drug Adminstration, Division of Pharmacology. *Appraisal of the Safety of Foods, Drugs and Cosmetics*. Washington, DC: U.S. FDA, 1959.
2. Davis KJ, Fitzhugh OG. Tumorigenic potential of aldrin and dieldrin in mice. *Toxicol Appl Pharmacol* 4:187–189, 1962.
3. Annau E. Estimation of pentose nuclei acid and desoxynucleic acid in the liver and brain tissue of mice following the feeding of the insecticide aldrin. *Can J Biochem Physiol* 32:178, 1953.
4. Cabral JRP, Raitano F, Mollner T, Bronczyk S, Shubik P. A carcinogenicity study of the pesticide dieldrin in hamsters. *Cancer Lett* 6:241–146, 1979.
5. Cabral JRP, Hall RK, Rossi L, Bronczyk SA, Shubik P. Lack of carcinogenicity of DDT in hamsters. *Tumori* 68:5–10, 1982.
6. Adamson RH, Sieber SM. Chemical carcinogenesis studies in nonhuman primates. In Langenbach R, Nesnow S, Rice JM (eds): *Organ and Species Specificity in Chemical Carcinogenesis*. New York: Plenum Publishing, 1983, pp 129–159.
7. Wright AS, Donninger C, Greenland RD, Stemmer KL, Zavon MR. The effects of prolonged ingestion of dieldrin on the livers of male rhesus monkeys. *Ecotoxicol Environ Safety* 1:477–502, 1978.
8. Higginson J. DDT: Epidemiological evidence. *IARC Sci Publ* 65:107–117, 1985.
9. Ribbons PH. Mortality study of industrial workers exposed to aldrin, dieldrin and endrin. *Int Arch Occup Environ Health* 56:75–79, 1985.
10. Warnick SL, Carter JE. Some findings in a study of workers occupationally exposed to pesticides. *Arch Environ Health* 25:265–270, 1972.
11. Morgan DP, Roan CC. Liver function in workers having high tissue stores of chlorinated hydrocarbon pesticides. *Arch Environ Health* 29:14–17, 1974.
12. Deichman WB, Keplinger M, Sala F, Glass E. Synergism among oral carcinogens. IV. The simultaneous feeding of four tumorigens to rats. *Toxicol Appl Pharmacol* 11:88–103, 1967.
13. Walker AIT, Thorpe E, Stevenson DE. The toxicology of dieldrin (HEOD). I. Long-term oral toxicity studies in mice. *Food Cosmet Toxicol* 11:415–432, 1972.
14. Thorpe E, Walker AIT. The toxicology of dieldrin (HEOD). II. Comparative long-term oral toxicity studies in mice with dieldrin, phenobarbitone, beta-BHC,

gamma-BHC. *Food Cosmet Toxicol* 11:433–442, 1973.
15. Ashby J, Tennant RW. Chemical structure, Salmonella mutagenicity and extent of carcinogenicity as indicators of genotoxic carcinogenesis among 222 chemicals tested in rodents by the U.S. NCI/NTP. *Mutat Res* 204:17–115, 1988.
16. Jones G, Butler WH. Morphology of spontaneous and induced neoplasia. In Butler WJ, Newberne PM (eds): *Mouse Hepatic Neoplasia.* Amsterdam: Elsevier/North Holland, 1975, pp 21–57.
17. Tennekes HA, Wright AS, Dix KM, Koeman JH. Effects of dieldrin, diet and bedding on enzyme induction and tumour incidence in livers of male CF-1 mice. *Cancer Res* 41:3615–3620, 1981.
18. Stevenson DE, Walker AIT. Hepatic lesions produced by dieldrin and other enzyme inducing compounds. *Eur J Toxicol* 2:83–85, 1969.
19. Dix KM. An Investigation into Some Aspects of the Pathogenesis of Rodent Liver Cell Neoplasms. Doctoral Thesis, University of Surrey, Guildford, England, 1981.
20. Tennekes HA, Edler L, Kunz HW. Dose-response analysis of the enhancement of liver tumour formation in CF-1 mice by dieldrin. *Carcinogenesis* 3:941–945, 1982.
21. Kunz HW, Tennekes HA, Port RE, Schwartz M, Lorke D, Schaude G. Quantitative aspects of chemical carcinogenesis and tumor promotion in liver. *Environ Health Perspect* 50:113–122, 1983.
22. Druckery H. Quantitative aspects in chemical carcinogenesis. *UICC Monograph Series* 7:60–78, 1967.
23. Tennekes HA, van Ravenzwaay B, Kunz HW. Quantitative aspects of enhanced liver tumour formation in CF-1 mice by dieldrin. *Carcinogenesis* 6:1457–1462, 1985.
24. Wright AS, Potter D, Wooder MF, Donninger C, Greenland RD. The effects of dieldrin on the subcellular structure and function of mammalian liver cells. *Food Cosmet Toxicol* 10:311–332, 1972.
25. Wright AS, Akintowa DAA, Wooder MF. Studies on the interactions of dieldrin with mammalian liver cells at the subcellular level. *Ecotoxicol Environ Safety* 1:7–16, 1977.
26. Miller EC, Miller JA. The presence and significance of bound aminoazo dyes in the livers of rats fed p-dimethylaminoazobenzene. *Cancer Res* 7:468–480, 1947.
27. Maslansky CJ, Williams GM. Evidence for an epigenetic mode of action in organochlorine pesticide hepatocarcinogenicity: A lack of genotoxicity in rat, mouse and hamster hepatocytes. *J Toxicol Environ Health* 8:121–130, 1981.
28. Ashwood-Smith MJ. The genetic toxicology of aldrin and dieldrin. *Mutat Res* 86: 137–154, 1981.
29. Upton AC, Clayson DB, Jansen JD, Rosenkranz HS, Williams GM. Report of ICPEMC Task Group 5 on the differentiation between genotoxic and non-genotoxic carcinogens. *Mutat Res* 133:1–49, 1984.
30. Kupfer D, Bulger WH, Nanni FJ. Characteristics of the active oxygen in covalent binding of the pesticide methoxychlor to hepatic microsomal protein. *Biochem Pharmacol* 35:2775–2780, 1986.
31. Charles A, Rosenbaum DP, Ashok L, Abraham R. Uptake and disposition of mirex in hepatocytes and subcellular fractions in the CD1 mouse liver. *J Toxicol Environ Health* 15:395–403, 1985.
32. Saul JW, Gee P, Ames BN. *Modern Biological Theories of Aging.* New York: Raven Press, 1987.
33. Busser M-T, Lutz WK. Stimulation of DNA synthesis in rat and mouse liver by various tumor promoters. *Carcinogenesis* 8:1433–1437, 1987.

34. Moolgavkar SH, Dewanji A, Venon DJ. A stochastic two-stage model for cancer risk assessment. I. The hazard function and the probability of tumor. *Risk Anal* 8:383–392, 1988.
35. National Institute of Occupational Safety and Health (NIOSH). *Special Occupational Hazard Review for Aldrin/Dieldrin.* DHEW (NIOSH) Publication No. 78-201. Rockville, MD: NIOSH, 1978.
36. Hayes WJ. *Pesticides Studied in Man.* Baltimore: Williams & Wilkins, 1982.
37. Baldwin MK, Robinson JR, Parke DV. A comparison of the metabolism of HEOD (dieldrin) in the CF1 mouse with that in the CFE rat. *Food Cosmet Toxicol* 10:333–351, 1972.
38. Hutson DH. Comparative metabolism of dieldrin in the rat (CFE) and two strains of mice (CF1 and LACG). *Food Cosmet Toxicol* 14:577–591, 1976.
39. Mueller W, Nohynek G, Woods G, Korte F, Coulston F. Comparative metabolism of dieldrin-C14 in mouse, rat, rabbit, Rhesus monkey and chimpanzee. *Chemosphere* 4:89–92, 1975.
40. Richardson A, Robinson JR. The identification of a major metabolite of HEOD (dieldrin) in human faeces. *Xenobiotica* 1:213–219, 1971.
41. Gingell R, Wallcave L. Species differences in the acute toxicity and tissue distribution of DDT in mice and hamsters. *Toxicol Appl Pharmacol* 28:385–394, 1974.
42. Chhabra RS, Fouts JR. Stimulation of hepatic drug metabolizing enzymes by DDT polycyclic hydrocarbons or phenobarbital in adrenalectomized or castrated mice. *Toxicol Appl Pharmacol* 28:465–476, 1974.
43. Moersdorf K, Ludwig G, Vogel J, Korte F. Die Ausscheidung von Aldrin ^{14}C und Dieldrin ^{14}C sowie ihrer Metaboliten durch die Galle. *Med Exp* 8:90–95, 1963.
44. Abraham R, Benitz KF, Mankes R. Ploidy patterns in hepatic tumors induced by Mirex. *Exp Mol Pathol* 38:271–282, 1983.
45. van Ravenzwaay B, Kunz W. Quantitative aspects of accelerated nuclear polyploidization and tumour formation in dieldrin treated CF-1 mouse liver. *Br J Cancer* 58:52–56, 1988.
46. Danielsen HE, Steen HB, Lindmo T, Reith A. Ploidy distribution in experimental liver carcinogenesis in mice. *Carcinogenesis* 9:59–63, 1988.
47. Jager KW. *Aldrin, Dieldrin, Endrin and Telodrin.* Amsterdam: Elsevier/North Holland, 1970.
48. Ruebner BH, Gershwin ME, French SW, Meierhenry E, Dunn P, Hsieh LS. Mouse hepatic neoplasia: Differences among strains and carcinogens. In Popp JA (ed): *Mouse Liver Neoplasia: Current Perspectives.* Washington, DC: Hemisphere, 1984, pp 115–143.
49. Buchmann A, Bauer-Hofmann R, Schwarz M, Balmain A. Detection of mutations in the Ha-ras proto-oncogene in the liver tumors of the CF1 mouse (Abstract). In Spandidos EA (ed): *Ras-Oncogenes.* New York: Plenum Publishing, NATO ASI Series, in press, 1989.
50. Junqueira VBC, Simizu K, Videla LA, Barros SB de M. Dose-dependent study of the effects of acute lindane adminstration on rat liver superoxide anion production, antioxidant enzyme activities and lipid peroxidation. *Toxicology* 41:193–204, 1986.
51. Goel MR, Shara MA, Stohs SJ. Induction of lipid peroxidation by hexachlorocyclohexane, dieldrin, TCDD, carbon tetrachloride and hexachlorobenzene in rats. *Bull Environ Contam Toxicol* 40:255–262, 1988.
52. Junqueira VBC, Simizu K, van Halsema L, Koch OR, Barros SBM, Videla LA. Lindane-induced oxidative stress. I. Time course of changes in hepatic microsomal

parameters, antioxidant enzymes, lipid peroxidative indices and morphological characteristics. *Xenobiotica* 18:1297–1304, 1988.

53. Ruch RJ, Klaunig JE. Effects of tumor promoters, genotoxic carcinogens and hepatocytotoxins on mouse hepatocyte intercellular communication. *Cell Biol Toxicol* 2:469–483, 1986.
54. Klaunig JE, Ruch RJ. Strain and species effects on the inhibition of hepatocyte intercellular communication by liver tumor promoters. *Cancer Lett* 36:161–168, 1987.
55. Suter S, Trosko JE, El-Fouly MH, Lockwood LR, Koestner A. Dieldrin inhibition of gap junctional intercellular communication in rat glial cells as measured by the fluorescence photobleaching and scrape loading/dye transfer assays. *Fundam Appl Toxicol* 9:785–794, 1987.
56. Ruch RJ, Klaunig JE. Antioxidant prevention of tumor promoter induced inhibition of mouse hepatocyte intercellular communication. *Cancer Lett* 33:137–150, 1986.
57. Paine AJ. Excited states of oxygen in biology: Their possible involvement in cytochrome p450 linked oxidations as well as in the induction of the p450 system by many diverse compounds. *Biochem Pharmacol* 27:1805–1815, 1973.
58. Loch-Caruso R, Trosko JE. Inhibited intercellular communication as a mechanistic link between teratogenesis and carcinogenesis. *CRC Crit Rev Toxicol* 16:157–183, 1985.
59. Klaunig JE, Ruch RJ. Role of cyclic AMP in the inhibition of mouse hepatocyte intercellular communication by liver tumor promoters. *Toxicol Appl Pharmacol* 91:159–170, 1987.
60. Wang CM, Matsumara F. Dieldrin: Effect on the ion transport activities in liver tissues. *Bull Environ Contam Toxicol* 4:144–151, 1969.
61. Trump B, Berezesky IK. Ion regulation, cell injury and carcinogenesis. *Carcinogenesis* 8:1027–1031, 1987.
62. Pounds JG, Rosen JF. Cellular Ca^{2+} homeostasis and Ca^{2+}-mediated cell processes as critical targets for toxicant action: Conceptual and methodological pitfalls. *Toxicol Appl Pharmacol* 94:331–341, 1988.
63. Mehrotra BD, Ravichandra-Reddy S, Desaiah D. Effect of subchronic dieldrin treatment on calmodulin-regulated Ca^{2+} pump activity in rat brain. *J Toxicol Environ Health* 25:461–469, 1988.
64. Warngard L, Fransson R, Drakenberg TB, Flodstrom S, Ahlborg UG. Calmodulin involvement in TPA and DDT induced inhibition of intercellular communication. *Chem Biol Interact* 65:41–49, 1988.

Mouse Liver Carcinogenesis: Mechanisms
and Species Comparisons, pages 385–393

Acute Hyperplasia and Peroxisome Proliferation Induced by Methylclofenapate: A Species Comparison and Implications for Liver Carcinogenesis

J. A. Styles, M. D. Kelly, N. R. Pritchard, and C. R. Elcombe
ICI Central Toxicology Laboratory, Cheshire, UK SK11 8AH

The refinement of testing methods and strategies for the detection of genotoxic carcinogens is revealing a growing number of nongenotoxic mouse and rat liver–specific carcinogens. These chemicals are nonmutagenic in standard assays both in vivo and in vitro, and although they have no discernible structural commonality, they are similar in biologic activity. Typically, the chemicals have low toxicity and induce liver cell hypertrophy through peroxisome and smooth endoplasmic reticulum proliferation and hyperplasia by DNA replication and cell division at dose levels that do not cause liver necrosis. Examples are the hypolipidemic drugs methylclofenapate (MCP), nafenopin, and clofibrate and pesticides such as dieldrin.

Species Differences in Liver Hyperplasia and Peroxisome Proliferation

It has been shown that rodent species exhibit a range of hyperplastic responsiveness to chemicals that induce peroxisome proliferation (1–21).

Comparison of the hyperplastic responses elicited by MCP has revealed that within a species, the peak of S-phase activity is directly related to dose, while duration is inversely related to dose (Fig. 1). Estimation of the areas under the response curves indicates that they are similar, suggesting that a limited number of cells is capable of responding to MCP. Comparison of the frequency of S-phase activity at a dose of 25 mg/kg/day in four species (Fig. 2) showed that the duration of responses was similar but the peak levels differed. Responses in the mouse and rat were high, in the hamster there was a slight response, and the guinea pig was unresponsive. This parallels the extent of peroxisome proliferation in the four species.

Cytologic analysis of the cells in S phase after MCP administration revealed that the majority of responding cells were 2 × 2N (Fig. 3). Furthermore, the analysis showed that the proportion of 2 × 2N cells reduced during the wave

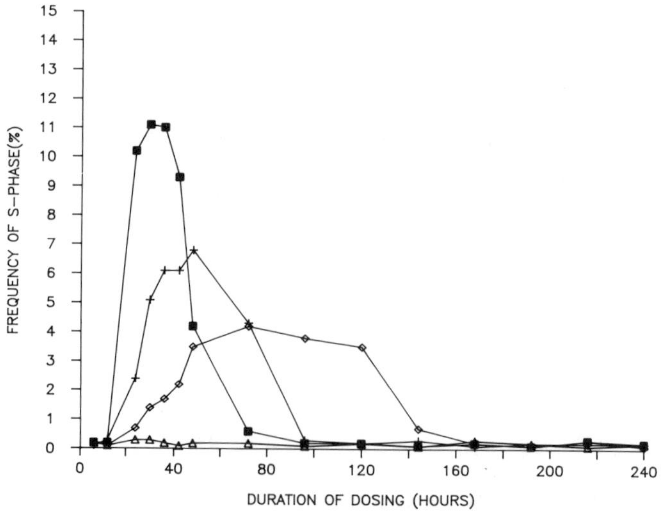

Figure 1. Frequency of S phase in hepatocytes from mice administered MCP at three dose levels by gavage.

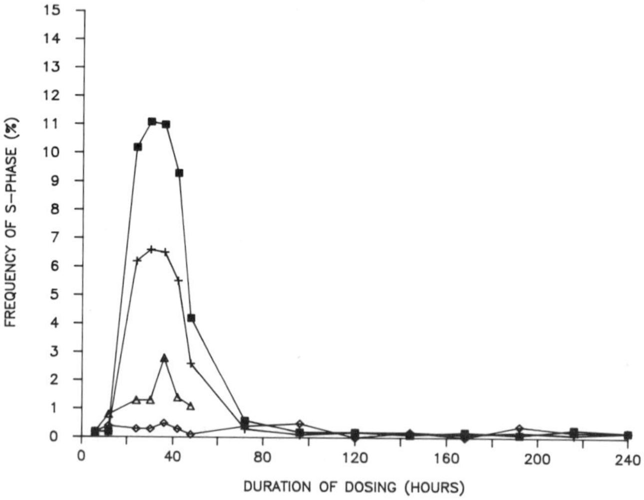

Figure 2. Percentage of hepatocytes undergoing DNA synthesis in four rodent species treated with 25 mg/kg MCP administered daily by gavage. Each point is the mean from three animals; ~3×10^4 cells were analyzed from each animal.

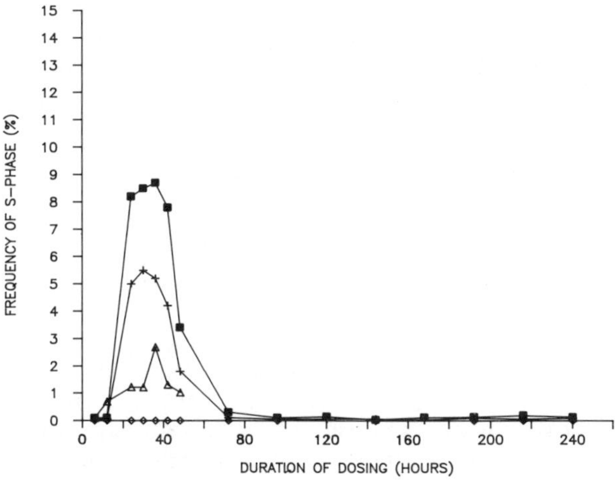

Figure 3. Percentage of 2 × 2N hepatocytes undergoing DNA synthesis in four rodent species treated with 25 mg/kg MCP administered daily by gavage.

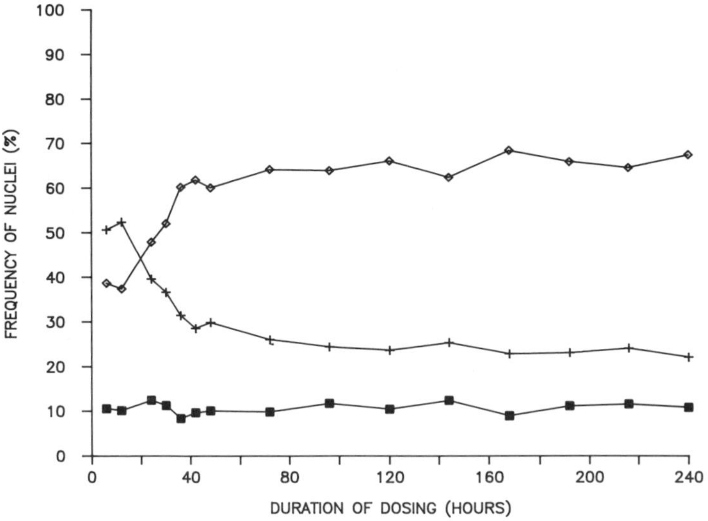

Figure 4. Nuclearity analysis of hepatocytes from mice treated with 25 mg/kg of MCP daily by gavage.

of hyperplasia; there was a concomitant increase in the fraction of 4N cells in the hepatocyte population (Fig. 4). The results given in Figure 5 show that the proportion of hepatocytes undergoing S phase (labeling interval ~6 h) is much greater than the fraction of hepatocytes entering mitosis (colchicine blockade interval ~4 h).

Mechanisms of Cell Division in Rodent Hepatocytes

The presence of hyperplasia in most organs can be detected in three ways (Fig. 6). The first is by the detection of DNA replication using radiolabeled thymidine and autoradiography or, more recently, using bromodeoxyuridine as a label and detecting its presence in DNA with a monoclonal antibody. This measurement is the labeling index (LI). The second method of detecting hyperplasia is by estimating mitotic activity after the blockade of mitosis by a spindle poison (most commonly colchicine). This is the mitotic index (MI). Finally, the end result of cell division can be estimated by counting the increased cell number (N). In practice, this is only possible in cultured cells.

The assumption made when the above methods are employed is that the various compartments of the cell cycle always follow in sequence, such that LI = MI = N.

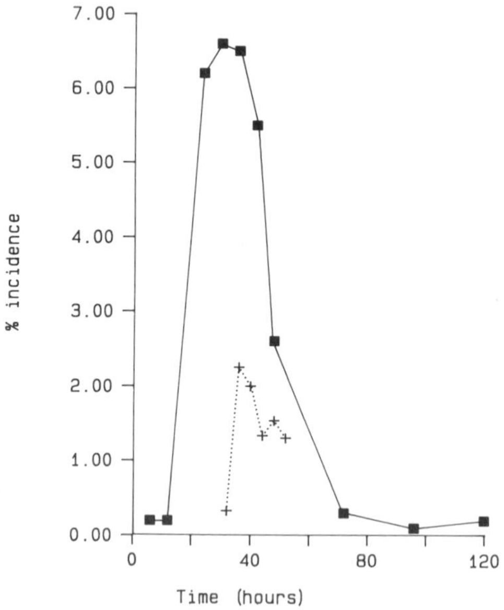

Figure 5. Comparison of percentage of hepatocytes undergoing DNA synthesis in rat with frequency of mitosis after dosing with 25 mg/kg MCP.

In the rodent liver, however, there appear to be abridged cell cycles that cause the formation of 2 × 2N cells by the suppression of cytokinesis (Fig. 7). Subsequently, the 2 × 2N cells appear to enter a cycle in which mitosis is suppressed and give rise to two tetraploid cells (Fig. 8). It has been suggested that the 2 × 2N cells undergo mitotic fusion (22). While this has not been shown directly, it does not appear to be the mechanism operating in hyperplasia, since the MI would not then be less than the LI. James (22) has suggested that mitotic fusion of 2 × 2N cells occurs in the rodent liver following 2/3 partial hepatectomy (PH). This would require fusion of the centrioles. Studies on regeneration after 2/3 PH have shown that all nuclear and ploidy classes of hepatocytes undergo labeling, and there is a general increase in ploidy without a change in the proportion of 2 × 2N cells (2), suggesting that the normal polyploidization process is accelerated.

Cytologic Responses of Hepatocytes to Genotoxins and Mitogens

Studies on the cytologic response to genotoxic hepatocarcinogens (2,23–31) indicate that the response is quite different from that observed after regeneration or during induced liver growth. The early response to DNA-reactive carcinogens appears to be the conversion of some 2 × 2N cells to 2N by cytokinesis induction. This change appears to be followed by proliferation of 2N cells by unabridged cell cycles to give rise eventually to tumors through the acquisition of a malignant phenotype. A comparison of the responses of the

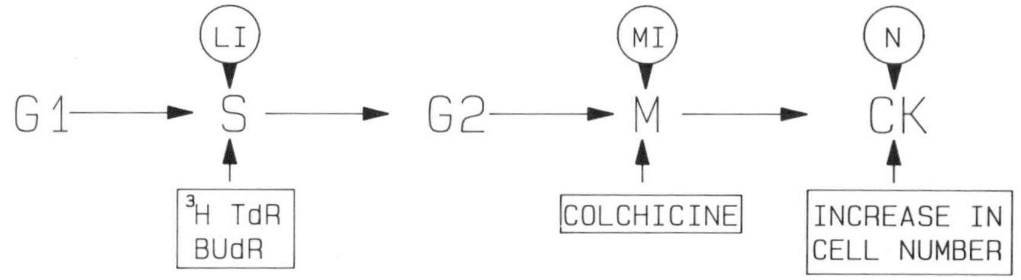

Figure 6. Schematic of normal cell cycle and the various methods for measuring cell replication.

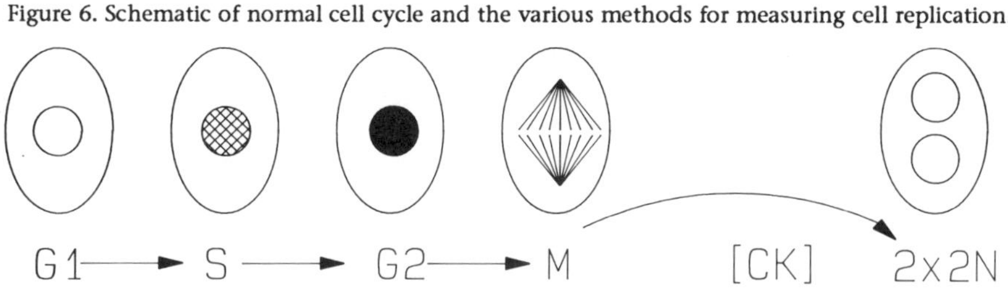

Figure 7. Schematic of cell cycle in rodent liver showing the suppression of cytokinesis giving rise to 2 × 2N cell.

rodent liver to liver growth–inducing agents and to DNA-reactive carcinogens is given in Figure 9. The response of hepatocytes to the noncarcinogenic mitogen 4-acetylaminofluorene appears to resemble that following 2/3 PH. S phase occurs in all nuclear classes, and after hyperplasia occurs, there is no change in the proportions of nuclear classes (Styles et al., in preparation).

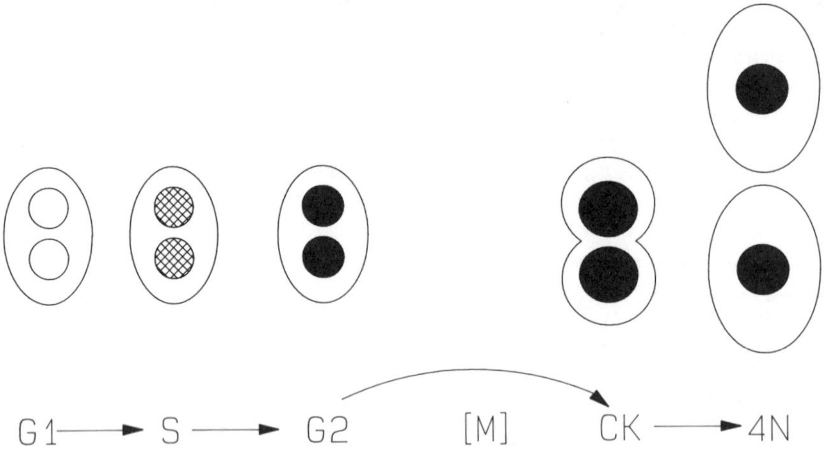

Figure 8. Schematic of cell cycle of 2 × 2N cell in rodent liver showing the suppression of mitosis giving rise to 4N cell.

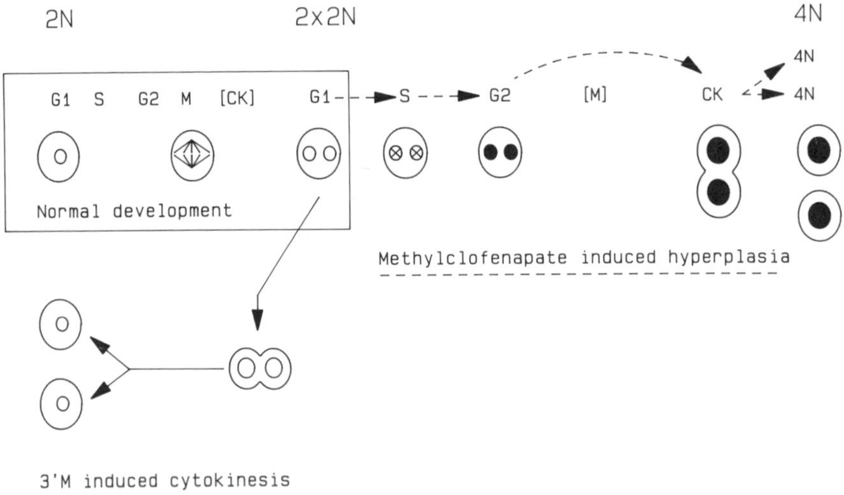

Figure 9. Schematic of cytologic mechanisms in rodent liver in response to the liver growth–inducing hepatocarcinogen MCP and the genotoxic carcinogen 3′-methyl-4-dimethylaminoazobenzene (3′M).

It is unclear, in both kinds of response, what the connection is between early cytologic changes and the subsequent appearance of tumors. As outlined above, the DNA-reactive carcinogens are assumed to induce mutation in all cell types, but the expression of such changes is most likely to be manifest in 2N cells rather than in 2 × 2N or higher ploidy hepatocytes, as increased gene copies are likely to suppress phenotypic changes. Clonal expansion of 2 × 2N cells into a tumor would be obvious upon histopathologic examination, and cytometry of liver tumors usually shows a 2N origin. In the case of tumors induced after liver growth, the link with the acute effects is even more obscure, since it is even more unlikely that the tumors would arise from cells other than 2N. Recent experiments (Styles et al., in preparation) have shown that after acute hyperplasia and subsequent cessation of MCP dosing, the liver recovers its ability to respond to a second challenge by MCP by 10 weeks, the fraction of 2 × 2N cells having been restored to normal levels in that period. This suggests that during continuous dosing, the acute hyperplastic wave may be followed by a chronic low-level hyperplasia as nonresponsive 2 × 2N cells evolve into responsive forms, which then undergo S phase and conversion to 4N cells. Over a prolonged period, the control of the 2N phenotype may change if this stem cell population is continually stimulated to replace 2 × 2N cells; some 2N cells may acquire constitutive cycling ability. Such cells could undergo clonal expansion, and presumably, some might evolve by further phenotypic changes and give rise to tumors.

Our species comparison study showed that the liver hyperplasia induced by MCP depended largely on the presence of a sensitive subpopulation of 2 × 2N hepatocytes. All the rodents examined had 2 × 2N cells, but the fraction of responsive cells varied among species. Unfortunately, the peroxisomal proliferation paralleled the hyperplasia; thus it was not possible to assess the relative significance of the two phenomena with respect to liver cancer. To separate the phenomena, the effect of MCP on the Snell dwarf mouse was studied (Styles et al., in preparation). This strain of mouse is a pituitary dwarf caused by a simple Mendelian recessive that produces a syndrome of developmental abnormalities in the homozygotes, including a failure of the liver to polyploidize. The liver of the dwarf mouse, like that of the guinea pig, possesses 2 × 2N cells, but the proportion is about 10% of hepatocytes compared with about 50% in the heterozygous wild-type Snell mice. After administration of 25 mg/kg/day by gavage of MCP, hyperplasia was detected in the wild-type mice in a response curve similar to that seen in other strains, but no hyperplasia occurred in the dwarf animals. In contrast, peroxisome proliferation occurred to the same extent in both phenotypes, increasing from ~2% to 25% of cytoplasmic volume. Thus it appears that hyperplasia is independent of peroxisome proliferation. Whether MCP is carcinogenic to the dwarf mouse is unknown, but a bioassay in both dwarf and wild-type Snell mice is now under way, and its outcome could define the significance of peroxisome proliferation and hyperplasia in the hepatocarcinogenicity of liver growth–inducing agents such as MCP. Thus if tumors are induced in both phenotypes, preceding events, such as hyperplasia and nuclearity changes, will become but secondary phenomena. However, should the dwarf mice fail to produce tumors while the wild-type mice prove sensitive to carcinogenesis, the reverse will apply; peroxisome induction will recede into insignificance as an indicator of rodent liver carcinogenicity.

References

1. Styles JA, Kelly M, Pritchard NR, Elcombe CR. A species comparison of acute hyperplasia induced by the peroxisome proliferator methylclofenapate: Involvement of the binucleated hepatocyte. *Carcinogenesis* 9:1647–1655, 1988.
2. Styles JA, Kelly M, Elcombe CR. A cytological comparison between regeneration, hyperplasia and early neoplasia in the rat liver. *Carcinogenesis* 8:391–399, 1987.
3. Cohen AJ, Grasso P. Review of the hepatic response to hypolipidaemic drugs in rodents and assessment of its toxicological significance to man. *Food Cosmet Toxicol* 19:585–605, 1981.
4. Svoboda D, Grady H, Azarnoff D. Microbodies in experimentally altered cells. *J Cell Biol* 35:127–152, 1967.
5. Rodricks JV, Turnbull D. Interspecies differences in peroxisomes and peroxisome proliferation. *Toxicol Ind Health* 3:197–212, 1987.
6. Gray RH, de la Iglesia FA. Quantitative microscopy comparison of peroxisome proliferation by the lipid-regulating agent gemfibrozil in several species. *Hepatology* 4:520–530, 1984.
7. Lake BG, Gray TJB, Foster JR, Stubberfield CR, Gangolli SD. Species difference in di-(2-ethylhexyl) phthalate-induced hepatic peroxisome proliferation. *Dev Toxicol Environ Sci* 11:595–598, 1984.
8. Lake BG, Gray TJB, Foster JR, Stubberfield CR, Gangolli SD. Comparative studies on di-(2-ethylhexyl) phthalate-induced hepatic peroxisome proliferation in the rat and hamster. *Toxicol Appl Pharmacol* 72:46–60, 1984.
9. Mitchell AM, Joseph K, Elcombe CR. Species difference in hepatic peroxisome proliferation. *Hum Toxicol* 4:552, 1985.
10. Reddy MK, Lalwani ND, Qureshi SA, Reddy JK. Induction of hamster hepatic peroxisomal beta-oxidation and peroxisome proliferation-associated 80,000 mol. wt. polypeptide by hypolipidemic drugs. *Hum Toxicol* 1:135–147, 1982.
11. Blane GF, Pinaroli F. Fenofibrate: Etudes de toxicologie animale en rapport avec les effets secondaires chez les malades. *Nouv Presse Med* 9:3737–3746, 1980.
12. Holloway BR, Bentley M, Thorp JM. Species differences in the effects of ICI 55,897 on plasma lipids and hepatic peroxisomes. *Ann NY Acad Sci* 386:439–442, 1982.
13. Orton TC, Adam HK, Bentley M, Holloway B, Tucker MJ. Clobuzarit: Species differences in the morphological and biochemical response of the liver following chronic administration. *Toxicol Appl Pharmacol* 73:138–151, 1984.
14. Holloway BR, Thorp JM, Smith GD, Peters TS. Analytical subcellular fractionation and enzymic analysis of liver homogenates from control and clofibrate treated rats, mice and monkeys with reference to fatty acid oxidizing enzymes. *Ann NY Acad Sci* 386:435–455, 1982.
15. Osumi T, Hashimoto T. Enhancement of fatty acyl-CoA oxidizing activity in rat liver peroxisomes by di-(2-ethylhexyl)phthalate. *J Biochem* 83:1361–1365, 1978.
16. Elcombe CR, Batten PL, Jackson SJ, Pratt IS, Rhodes C. Comparative toxicity of di-(2-ethylhexyl)phthalate (DEHP) in rats and marmosets (Abstract). *The Toxicologist* 3:507, 1983.

17. Elcombe CR, Mitchell AM. Peroxisome proliferation due to di(2-ethylhexyl) phthalate (DEHP): Species differences and possible mechanisms. *Environ Health Perspect* 70:211–219, 1986.
18. Reddy JK, Rao MS. Xenobiotic-induced peroxisome proliferation: Role of tissue specificity and species differences in response in the evaluation of the implications for human health. *Arch Toxicol [Suppl]* 10:43–53, 1987.
19. Rao MS, Reddy JK. Peroxisome proliferation and hepatocarcinogenesis. *Carcinogenesis* 8:631–636, 1987.
20. Reddy JK, Lalwani ND. Carcinogenesis by hepatic peroxisome proliferators: Evaluation of the risk of hypolipidemic drugs and industrial plasticizers to humans. *CRC Crit Rev Toxicol* 12:1–58, 1983.
21. Reddy JK, Lalwani ND, Reddy MK, Qureshi SA. Excessive accumulation of autofluorescent lipofuscin in the liver during hepatocarcinogenesis by methylclofenapate and other hypolipidemic peroxisome proliferators. *Cancer Res* 42:259–266, 1982.
22. James J. The genesis of polyploidy in rat liver parenchymal cells. *Eur J Cell Biol* 15:410–419, 1977.
23. Styles J, Elliott BM, Lefevre PA, Robinson M, Pritchard N, Hart D, Ashby J. Irreversible depression in the ratio of tetraploid:diploid liver nuclei in rats treated with 3′-methyl-4-dimethylaminoazobenzene (3′M). *Carcinogenesis* 6:21–28, 1985.
24. Holzner JH, Barka T, Popper H. Changes in deoxyribonucleic acid content of rat liver cells during ethionine intoxication. *JNCI* 23:1215–1225, 1959.
25. Laird AK, Barton AD. Cell proliferation in precancerous liver: Relation to presence and dose of carcinogen. *JNCI* 27:827–839, 1961.
26. Simard A, Cousineau G, Daoust R. Variations in the cell cycle during azo dye hepatocarcinogenesis. *JNCI* 41:1257–1263, 1968.
27. Neal GE, Judah DJ, Butler WH. Some effects of acute and chronic dosing with aflatoxin B1 on rat liver nuclei. *Cancer Res* 36:1771–1778, 1976.
28. Irving CC, Roszell JA, Fredi JL. Effects of chronic feeding of 2-acetylaminofluorene on nuclear populations in rat liver. *Adv Enzyme Regul* 16:365–377, 1977.
29. Schwarze PE, Pettersen EO, Shoaib C, Seglen PO. Emergence of a population of small, diploid hepatocytes during hepatocarcinogenesis. *Carcinogenesis* 5:1267–1275, 1984.
30. Seglen PO, Schwarze PE, Saeter G. Changes in cellular ploidy and autophagic responsiveness during rat liver carcinogenesis. *Toxicol Pathol* 14:342–348, 1986.
31. Deleener A, Castelain P, Preat V, De Gerlache J, Alexandre H, Kirsch-Volders M. Changes in nucleolar transcriptional activity and nuclear DNA content during the first steps of rat hepatocarcinogenesis. *Carcinogenesis* 8:195–201, 1987.

Mouse Liver Carcinogenesis: Mechanisms
and Species Comparisons, pages 395–408

Interspecies Comparisons of In Vitro Hepatocarcinogenesis

Michael M. Lipsky, Katherine E. Cole,*† Ih-Chang Hsu,* Myong W. Kahng,* Thomas W. Jones,* and Benjamin F. Trump*‡*
**Department of Pathology, University of Maryland School of Medicine, Baltimore, MD 21201; †Laboratory of Human Carcinogenesis, National Cancer Institute, Bethesda, MD 20892; ‡Maryland Institute for Emergency Medical Services System, University of Maryland Medical Systems, Baltimore, MD 21201*

Species differences in susceptibility to the carcinogenic potential of xenobiotics present problems for reliable interspecies extrapolation, particularly with respect to risk assessment. Our laboratory has been developing strategies to explore interspecies variations in toxicologic/carcinogenic reactions to xenobiotic exposure, using a correlated in vitro–in vivo approach (1–3). In this paradigm, direct comparisons of cellular and subcellular reactions to xenobiotic exposure can be made between different species in vitro for a given target tissue, such as the liver. At the in vitro level, both experimental animal and human hepatocytes can be studied. Comparisons can also be made within an individual experimental animal species between in vitro experiments and in vivo tumorigenicity for different chemical carcinogens. In this experimental design, therefore, a scientific basis for extrapolating to in vivo human risk can be established (Fig. 1).

We have used this approach with primary cultures of liver cells to investigate the potential carcinogenicity of aflatoxin B_1 (AFB) and several other hepatocarcinogens (3–7). We have attempted to correlate xenobiotic metabolism, covalent binding to DNA, mutagenicity, and the induction of specific ultrastructural alterations in hepatocytes in vitro from species that are highly sensitive to AFB carcinogenicity to those parameters for species less sensitive to AFB carcinogenicity and to humans, in whom definitive evidence for AFB carcinogenicity is lacking. This chapter summarizes our research in this area and attempts to demonstrate the rationale and strengths of our basic scientific approach. Our data show that primary cultures of human and experimental animal hepatocytes can be used to evaluate species variations in sensitivity to chemical toxins and carcinogens. This can be particularly useful when a compound's toxic/carcinogenic action is hypothesized to be due to a pattern of metabolism/detoxification unique to a particular species.

Methods and Experimental Design

For the studies described herein, we used either primary monolayer cultures of differentiated hepatocytes or the cytosolic and microsomal fractions of homogenized liver. All experiments on isolated hepatocytes were performed during the first 24–48 h of primary culture.

Animals and Animal Care

Male Fischer-344 and Sprague-Dawley rats (125–150 g) and CD-1 and BALB/c mice (35–40 g) were obtained from Charles River Laboratories (Wilmington, MA) and held for 2–3 weeks prior to use in the University of Maryland School of Medicine Central Animal Facility. All animals used for cell isolation or tissue sampling were anesthetized with sodium pentobarbital (35–50 mg/kg, intraperitoneally) prior to testing.

Cell Isolation (Rats and Mice)

Hepatocytes from experimental animals were isolated using a modification of the method of Moldeus et al. (8) as previously described (5,6). The cell suspensions were filtered through a 10-micron mesh filter with a resultant variability that was routinely greater than 90% as measured by trypan blue exclusion.

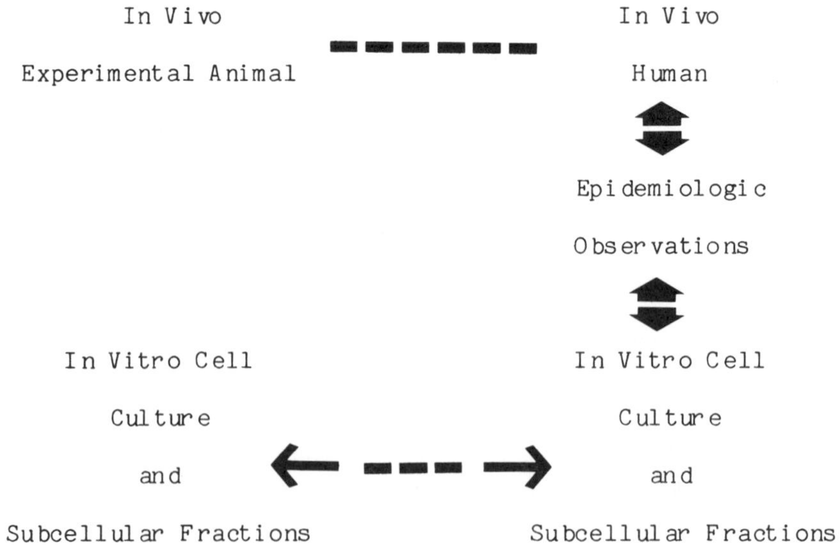

Figure 1. Experimental design for interspecies extrapolation. Adapted from refs. 1 and 2.

Cell Isolation (Human)

Human hepatocytes were isolated by a method previously developed in our laboratory (4). The cell suspensions were filtered through a 10-micron mesh filter; resultant viability was 65%–90% (mean, 81.7% ± 5.5%).

Hepatocytes were isolated from donor livers obtained at immediate autopsy (IA), i.e., 30–90 min postmortem. The liver tissue was obtained at autopsy in strict compliance with the laws of the state of Maryland and the regulations of the University of Maryland Human Volunteers Committee. The IA program maintains strict confidentiality of all medical records, and no identification of individual autopsy patients by name is available to the investigator. IAs were performed on two types of donors: those selected for organ donation whose livers were not used and accident victims who died of acute shock and whose conditions (e.g., trauma or lacerations) made the liver unacceptable for organ transplantation. Patients who tested positive for human immunodeficiency virus type I (HIV-I) or hepatitis assays were excluded from the donor pool, as were those whose deaths were caused by a drug or other toxic, xenobiotic overdose. Smoking, general medical, and alcohol and drug abuse history data were available on the liver donors. Table 1 describes a typical group used for the research described in this report.

Table 1. Representative Liver Donors

Case no.	Sex	Age	Cause of death	Source of tissue
HL-1	F	76	MVA	IA
HL-2	M	25	Head injury	IA
HL-3	M	26	Head injury	RA
HL-4	M	52	Head injury	IA
HL-5	M	37	Chest injury	RA
HL-6	M	21	MVA	IA
HL-7	M	36	MVA	IA
HL-8	M	38	Unknown	RA
HL-9	M	18	MVA	IA
HL-10	M	25	Head injury	IA
HL-11	F	40	MVA	IA
HL-12	M	35	MVA	IA
HL-13	F	29	MVA	RA

Abbreviations: MVA, motor vehicle accident; IA, immediate autopsy (30–90 min postmortem tissue collection); RA, routine autopsy (6–12 h postmortem tissue collection).

Primary Culture of Isolated Hepatocytes

Hepatocytes from all species were cultured in an identical fashion. The final cell pellet was resuspended in Waymouth's medium containing 10% fetal bovine serum, insulin (1 µg/ml), and gentamycin (50 µg/ml) and seeded onto collagen-coated dishes (9). The dishes were maintained in a humidified incubator at 37°C for a 4-h attachment period. Four hours after plating, the medium was replaced with fresh medium containing the chemicals to be studied or the appropriate solvent. All exposures were carried out for 24 h. The cell culture system was used to collect data on basic ultrastructural alterations induced by exposure to AFB, dimethylnitrosamine (DMN), and actinomycin D; the covalent binding of AFB and DMN to DNA; and the mutagenicity of AFB and DMN as determined by a cell-mediated mutagenesis assay (3–7).

Isolation of Microsomes and Cytosol

The microsome isolation was done by the method of Fleischer and Kervina (10). Pieces of human liver were thawed in cold rinsing solution. Keeping all materials on ice, thin slices were made, rinsed several times to remove blood, and finely minced with scissors. The minced liver was homogenized in 10 vol (10 ml/g tissue) of homogenizing medium (0.25 M sucrose buffered with 10 mM HEPES, pH 7.5) using a Polytron (Brinkman Instruments, Westbury, NY).

Pooled livers from rats or mice were minced and homogenized in a Potter-Elvejhem homogenizer. All subsequent procedures were the same for human, rat, and mouse tissue. The homogenates were filtered through several layers of cheesecloth. The filtered homogenate was centrifuged at 25,000 × *g* for 10 min to remove large particulate matter. The resulting supernatant was centrifuged at 124,000 × *g* for 1 h in an L2-65B Beckman ultracentrifuge. The pellet was suspended in the homogenizing medium and recentrifuged at 124,000 × *g* for 30 min. The resulting microsomal pellet was suspended in homogenizing medium to a concentration of 1 ml/g original tissue and stored at –70°C until used. The supernatant from the first ultracentrifugation was designated as the cytosol fraction and was stored at –70°C. Storage at –70°C for 4–6 weeks showed no loss of activity according to comparison studies. The protein content of microsomes and cytosol was assayed by the method of Lowry et al. (11).

Incubation for AFB Binding to Exogenous DNA

The in vitro binding of AFB to calf thymus DNA mediated by liver microsomes was done by the method of Lotlikar et al. (12). The incubation medium contained 0.1 M Na-phosphate buffer, pH 7.4, 0.1 M sucrose, 2 mM NADPH, 2 µM [^{3}H]AFB (specific activity, 125 Ci/mol), 0.1 mg calf thymus DNA, and liver microsomes containing 0.5–1.0 mg protein in a total volume of 1.0 ml. After a 2-min preincubation, microsomes were added and incubated for 30 min at 37°C in a shaking water bath. Triplicate assays were done for each sample.

After incubation, 0.9 mg calf thymus DNA in 1 ml of 1 mM NaOH was added as a carrier. DNA with bound [^{3}H]AFB was isolated by the method of Croy et al. (13). Isolated DNA was centrifuged, washed twice with cold 95% ethanol, and finally dissolved in 1 ml of 10 mM HEPES (pH 7.4) containing 1 mM EDTA. Tritium radioactivity was counted on 100 μl aliquots mixed with 10 ml of scintillation fluid (MP, Beckman) in a liquid scintillation counter (Beckman LS-7800). The results are expressed as nanomoles of AFB bound per milligram of microsomal protein.

NADH Cytochrome b5 Reductase

NADH-dependent cytochrome b5 reductase activity was assayed in the microsomes on the day of isolation using the method of Prough et al. (14). The increase in absorbance at 550 nm due to the reduction of cytochrome c was measured, and the enzyme activity was calculated using the equation $\varepsilon_{550} = 21$ mM^{-1}. The enzyme activity is expressed as moles of cytochrome c reduced per minute per milligram of protein.

Results

AFB Binding to DNA in the Microsomal System

Human liver microsome-mediated binding of [^{3}H]AFB to calf thymus DNA is shown in Table 2. The extent of AFB binding was highly variable between cases, ranging from 0.9 to 14.6 nmol of [^{3}H]AFB/mg microsomal protein. This result demonstrated at least a 16-fold interindividual variability for normal human liver to metabolize AFB to reactants that bind to DNA. NADH-dependent cytochrome b5 reductase activity for each case (Table 2) ranged from 0.27–1.0 units/mg protein. The activity of this enzyme in the 13 human liver samples did not correlate with the magnitude of AFB–DNA binding when compared on a milligram protein basis.

A cytosol fraction (5 mg protein) from the same human liver was added to each microsome preparation in DNA-binding experiments to determine the effects of cytosolic detoxification on AFB–DNA binding. The addition of 5 mM glutathione (GSH) in the presence of cytosol was also studied. These results are shown in Table 3. In 5 of 13 cases, addition of cytosol had no effect on the amount of AFB–DNA binding when compared with binding using microsomes alone. The remaining 8 cases showed decreases in DNA binding of 13%–37% compared with binding using microsomes alone. When 5 mM GSH was added to the microsome and cytosol preparation, the binding pattern was quite different. In 4 cases (HL-2, -3, -5, and -13), AFB–DNA binding increased to 116%–346% of the binding using microsomes alone; in 4 cases (HL-4, -6, -8, and -9) there were no changes, and the remaining 5 cases showed decreases of up to 37%. There was no correlation between the effect of adding cytosol alone and cytosol plus GSH in individual cases.

Table 2. [^{3}H]AFB Binding to Calf Thymus DNA Mediated by Human Liver Microsomes and Cytochrome b_5 Activity

Case no.	[^{3}H]AFB bound (nmol/mg protein)	NADH-cytochrome b_5 reductase (μmol/min mg protein)
HL-1	14.58 ± 0.21 (3)*	1.01†
HL-2	12.47 ± 0.65 (3)	0.69
HL-3	11.55†	0.38
HL-4	9.52 ± 0.37 (3)	0.39
HL-5	5.36 ± 0.31 (3)	0.27
HL-6	3.73 ± 0.35 (3)	0.43
HL-7	3.13	0.20
HL-8	2.97	0.53
HL-9	2.13 ± 0.19 (3)	0.50
HL-10	2.01 ± 0.13 (3)	0.54
HL-11	1.29 ± 0.03 (3)	0.89
HL-12	1.27 ± 0.25 (4)	0.92
HL-13	0.87 ± 0.24 (3)	0.52

*Mean ± S.D.; (number of replicate assays).
†Average of duplicate assays.

Table 3. Effects of Cytosol and GSH on AFB Binding to DNA

	[^{3}H]AFB bound (nmol/mg protein)*				
Case No.	Ms	(Ms+Cyt)	% of Ms	(Ms+Cyt+GSH)†	% of Ms
HL-1	14.58	14.58	100	10.21	70
HL-2	12.47	10.28	82	14.51	116
HL-3	11.55	9.82	85	13.47	117
HL-4	9.52	8.31	87	9.87	104
HL-5	5.36	3.93	73	10.09	188
HL-6	3.73	3.56	95	3.70	100
HL-7	3.13	2.50	80	2.58	82
HL-8	2.97	3.05	102	2.96	100
HL-9	2.13	2.13	100	2.07	97
HL-10	2.01	1.68	83	1.31	65
HL-11	1.29	1.28	99	1.03	80
HL-12	1.27	0.81	64	0.80	63
HL-13	0.87	0.51	63	3.01	346

*Average binding of microsome preparation repeated from Table 2 for comparison.
†Microsome + cytosol + 5 mM GSH.
Abbreviations: Ms, microsomes; Cyt, cytosol.

Microsome-mediated [^{3}H]AFB binding to DNA using rat and mouse liver was also studied for comparison to the human data (Tables 4 and 5). Rat liver is known to be highly sensitive to AFB-induced carcinogenesis. Addition of cytosol had little effect on the binding of AFB to DNA, and the addition of 5 mM GSH increased binding to 181% of that using microsomes alone. Mouse liver is considered to be less sensitive to AFB-induced carcinogenesis. The binding of AFB to DNA decreased 68% when mouse liver cytosol was added to microsomal preparations; this was partially restored when GSH was present.

Comparison of AFB and DMN Binding in the Culture System Using Human, Rat, and Mouse Liver Cells

The binding of [^{3}H]AFB to DNA from rat, mouse, and human liver cells exposed to 2×10^{-7} M [^{3}H]AFB (specific activity, 2.6 μCi/nmol) is given in Table 6. The rat liver cells bound the highest amount of AFB (203 pmol/mg DNA), while the mouse liver cells bound the lowest amount of AFB (1.4 pmol/mg DNA). Cells from three human livers bound 53 pmol/mg (male), 69 pmol/mg (female), and 25 pmol/mg DNA (female) (6). Table 5 also details the binding of DMN to DNA of liver cells exposed to [^{14}C]DMN (0.5 mM; specific activity, 1 μCi/ml). Both rat and mouse liver cells effectively formed covalent bonds between

Table 4. [^{3}H]AFB Binding to Calf Thymus DNA Mediated by Microsomes from Rat and Mouse and Effects of Cytosol and GSH

	[^{3}H]AFB-Bound (nmol/mg protein)				
Animal	Ms	(Ms+Cyt)	% of Ms	(Ms+Cyt+GSH)	% of Ms
Rat*	3.14 ± 2.15 (3)	2.90 ± 1.15 (3)	92.4	5.69 ± 2.16 (3)	181.2
Mouse†	4.68 ± 1.68 (4)	1.98 ± 0.72 (4)	42.3	2.99 ± 1.31 (4)	63.9

*Livers from three rats used for assay.
†Livers from mice were pooled to obtain 10 g (~6). Assays were repeated four times; each assay was performed in triplicate.
Abbreviations: MS, microsomes; Cyt, cytosol.

Table 5. NADH-Cytochrome b_5 Reductase Activity

Animal	NADH cytochrome b_5 reductase (μmol/min/mg protein)*
Rat	1.12 ± 0.37 (3)
Mouse	0.62 ± 0.16 (3)

*Data are expressed as the mean ± 1 SD. The number of experiments is shown in parentheses.

Table 6. Binding of Chemical Carcinogens to Hepatocyte DNA in Cultured Cells*

	pmol Bound/mg DNA		
Chemical carcinogen	Human	Rat	Mouse
AFB	53,69,25	203	1.4
DMN	6.5×10^3, 12×10^3	8.9×10^3	9.4×10^3

*Each value for human liver cells represents a different individual. All experiments were performed in duplicate or triplicate for human cells. Values from rats and mice were from three animals of each species. Data taken from Cole et al. (6) and Hsu et al. (7).

Table 7. Nucleolar Segregation in Hepatocytes Exposed to AFB

Species	AFB (μg/ml)	Segregation of nucleoli (%)*
Rat	0.01	12
	0.1	47
	1.0	82
Mouse	1.0	0
	10.0	0
Human	0.1	13
	1.0	62

$$*\% = \frac{\#\text{ of cells with nucleolar segregation} \times 100}{\#\text{ of total cells examined}}.$$

Table 8. Nucleolar Segregation Induced by AFB, DMN, and Actinomycin D

	% of Cells with Nucleolar Segregation*		
Species	AFB	DMN	Actinomycin D
Human	60 (38–82)	31	86 (65–100)
Rat	84	38	98
Mouse	0	79	55

*AFB, 32×10^{-6} M; DMN, 5×10^{-2}M; actinomycin D, 7.9×10^{-8} M. Standard deviations for most data points were <10%, except for human cells exposed to AFB and actinomycin D. For those, two numbers in parentheses represent the range of values seen in five human livers evaluated.

DMN and DNA (8.9×10^3 pmol/mg DNA and 9.4×10^3 pmol/mg DNA, respectively). Cells from two human livers (both male) bound 6.5×10^3 and 12×10^3 pmol/mg DNA (3).

Comparison of V-79 Cell-Mediated Mutagenesis of AFB and DMN Using Human, Rat, and Mouse Liver Cells

Hepatocytes from all three species evaluated effectively increased the mutation frequencies of cocultivated V-79 cells. At a concentration of 0.2 mM DMN in the media, the number of ouabain-resistant mutants/10^6 surviving cells was 8 for mouse hepatocytes, 7.5 for rat hepatocytes, and 4.5 for human hepatocytes (3). Species differences were noted in this assay for AFB mutagenicity. The number of mutants/106 survivors was 13 for rat hepatocytes but did not go above the control level of 1 or 2 for mouse hepatocytes. The liver cells from humans produced a range of results: two cases had only the background level of mutants (1 or 2), while one case was positive, producing five mutants (3).

Comparison of Nucleolar Segregation Induced in Human, Rat, and Mouse Hepatocytes by Chemical Carcinogens

Macrosegregation of nucleolar components has been observed in hepatocytes of animals exposed to a number of chemical carcinogens. Studies in our laboratory have determined that isolated hepatocytes in primary culture also develop nucleolar segregation as a response to chemical carcinogen exposure (5,7). Initial studies of the effects of AFB on human, rat, and mouse hepatocytes in vitro demonstrated that the pattern of nucleolar segregation correlated with the in vivo tumorigenic sensitivity of a given species (7). Rat hepatocytes developed nucleolar segregation in a dose-dependent manner to AFB exposure, varying from 12% at 0.01 μg/ml of AFB in the media to 86% at 1.0 μg/ml (Table 7). Mouse hepatocytes failed to develop nucleolar segregation even when exposed to AFB at a concentration of 10 μg/ml. The human hepatocytes displayed nucleolar segregation but required a higher dose of AFB to elicit a response, ranging from 13% at 0.1 μg/ml to 68% at 1.0 μg/ml (Table 7).

In these initial studies, however, it could not be ruled out that the effects we observed were related to some peculiarity in the culture system, such as the selection of a uniquely sensitive cell type. Additional experiments were conducted to evaluate the induction of nucleolar segregation as an indicator of sensitivity to a chemical carcinogen. In addition to AFB, we exposed cells to the hepatocarcinogen DMN and actinomycin D, a compound that intercalates directly between deoxyguanosine residues within double-stranded DNA (15–17). Table 8 provides the summary data from this study. Hepatocytes from humans, rats, and mice developed nucleolar segregation (86%, 98%, and 55%, respectively) when treated with actinomycin D at 7.9×10^{-8} M. As in our previous experiments, 84% of rat hepatocytes displayed nucleolar segregation after exposure to AFB (3.2×10^{-6} M), while mouse hepatocytes did not develop a response.

The human hepatocytes in these experiments developed segregation (60%) at the 3.2×10^{-6} M concentration of AFB in the media. These studies validated our initial work in that the pattern of in vitro nucleolar segregation paralleled the in vivo tumorigenic sensitivity of a species to AFB. DMN, a potent hepatocarcinogen in many species, induced nucleolar segregation in human (31%), rat (38%), and mouse (79%) hepatocytes at a concentration of 5×10^{-2} M in the media.

Discussion

The realistic and valid extrapolation of toxicology and carcinogenesis data between species, particularly between experimental animals and humans, has been difficult. It is, however, extremely important to reliable evaluation of chemical compounds with respect to quantitative risk assessment. Our laboratory has used a correlated in vivo–in vitro approach to investigate interspecies differences in toxicologic and carcinogenic responses and sensitivity to xenobiotics (1–7). Primary cultures of hepatocytes from different species, including teleost fishes, rats, mice, and humans have proved effective for comparing chemical metabolism, covalent binding to DNA, mutagenicity, and the induction of ultrastructural alterations. While primary cultures of rodent hepatocytes have been used in toxicologic research for many years (18–24), it has only been recently that primary cultures of human hepatocytes have been utilized in toxicologic investigations (3–7,25–30). The limited availability of suitable human livers for cell isolation has generally restricted the use of human hepatocytes as an investigative model system. Recent advances in primary cell culture methods and the recognition that viable cells can be obtained from livers of a variety of donors, including those patients dying from acute trauma, have resulted in an increasing number of laboratories using human hepatocytes for research. Primary cultures of human hepatocytes are particularly useful in the experimental approach we employ. They provide a valuable link in investigations on interspecies differences in mechanisms of action of xenobiotic toxicity as a basis for determining differential sensitivity/susceptibility to carcinogenic activity. We have used this basic approach to investigate species differences in susceptibility to AFB-induced carcinogenesis.

The formation of AFB–DNA adducts by humans has been indirectly demonstrated by the identification of the AFB–Gua I adduct in urine samples collected from people in different regions of the world that have high incidences of liver cancer (31,32). These data have been used as indirect proof of the in vivo activation of AFB to reactive metabolites in humans (31). Metabolism of AFB in human liver tissue obtained at autopsy has been reported (33,34). Aflatoxin Q_1 was the major metabolite detected in two studies using an isolated microsomal preparation. Aflatoxin 8,9-dihydrodiol and aflatoxin M_1 and P_1 were also detected (33,35). There was, however, little evidence for the production of glutathione conjugates when human liver cytosol was added to this microsomal system (33).

Our research demonstrates that AFB is activated by human liver microsomes to metabolites that covalently bind to exogenous calf thymus DNA. There was a 16-fold interindividual variation in the levels of AFB–DNA adducts formed. The mean binding level (5.45 nmol/mg protein) of AFB produced by human liver microsomal preparations was in the same range as for both the rat and mouse liver microsomal preparations. It is interesting that two of the three liver preparations from females produced the lowest binding data (HL-11 and -13). This observation is consistent with the report of Moss and Neal (33), in which male human liver metabolized AFB two to three times faster than female liver. That study, however, was based only on three cases, two male and one female. Female rats have also been shown to be less susceptible to the toxic and carcinogenic effects of AFB than male rats (36). One liver preparation from a female in our study (HL-1), however, showed a high degree of microsome-mediated AFB–DNA binding. This liver was from a 76-year-old woman. This would seem to deviate from many reports that show decreased xenobiotic metabolism with aging, including metabolism of AFB in aged rats (37). The reason for the high level of binding in this case is not known.

Lotlikar et al. (12) showed a good correlation between species sensitivity to AFB hepatocarcinogenicity and the ability of cytosolic GSH S-transferases to detoxify AFB metabolites. In the current study, the effects of cytosolic detoxifying enzymes on microsome-mediated AFB–DNA binding were evaluated in human liver preparations and compared with data on the rat, a species highly sensitive to AFB hepatocarcinogenicity, and the mouse, a species weakly sensitive to AFB hepatocarcinogenicity. In the rat, the addition of cytosol had no effect on the level of microsome-mediated AFB–DNA binding. In the mouse, the addition of cytosol decreased the level of microsome-mediated AFB–DNA binding. The human liver preparations gave mixed results. There were decreases in microsome-mediated AFB–DNA binding by cytosol in eight cases, ranging from 13% to 37%, and no significant changes in the remaining five cases. The effects of cytosol on the mouse liver are similar to those obtained by Lotlikar et al. (12) in another resistant species, the guinea pig.

Using isolated liver cells, we demonstrated a good correlation between the level of total AFB–DNA binding and species sensitivity to AFB carcinogenesis (6). In these studies, human cells were capable of producing AFB–DNA adducts, but at a lower level than the rat liver cells. Studies from other laboratories have shown that the formation of AFB-glutathione conjugates (12,38) and active DNA repair (39) also correlate with decreased sensitivity of the mouse to AFB carcinogenesis. It is the balance, therefore, between activation, detoxification, and DNA repair that probably accounts for species differences in sensitivity to AFB carcinogenesis (6,7).

An interesting observation followed up in our laboratory was the relationship between AFB–DNA binding and the induction of nucleolar segregation (7). Both rat and human hepatocytes displayed nucleolar segregation after exposure to AFB. The level of AFB–DNA binding appeared to correlate with the degree of segregation (percent of cells). It has been hypothesized by Simard and Bernhard (40) that nucleolar segregation may be a morphologic manifestation of carcinogen-DNA binding. Our studies on DMN and actinomycin D demonstrate that all three species we tested are capable of nucleolar segregation. The in vitro induction of nucleolar segregation correlated well with the in vivo carcinogenic susceptibility to AFB (7).

The in vitro primary cell culture system using experimental animal and human liver cells is useful in evaluating susceptibility to xenobiotic toxicity and carcinogenicity. The intact cells allow the evaluation of the dynamic interplay between activation, detoxification, and repair pathways. The system also allows direct comparisons of important cellular and molecular mechanisms of toxic response to xenobiotic exposure between species, including humans. This experimental design can provide important data to be included in a total-weight-of-evidence approach to human risk assessment.

Acknowledgments

This research was supported in part by Environmental Protection Agency grants R-809835-01-1 and R-809599010, Department of Energy contract DE-AS05-83ER60158, and National Cancer Institute contract N01-CP-51000.

References

1. Trump BF, Harris CC. Human tissues in biomedical research. *Hum Pathol* 10:245–248, 1979.
2. Harris CC, Trump BF. Human tissues and cells in biomedical research. *Surv Synth Pathol Res* 1:165–171, 1983.
3. Hsu IC, Harris CC, Lipsky MM, Snyder S, Trump BF. Cell and species differences in metabolic activation of chemical carcinogens. *Mutat Res* 177:1–7, 1987.
4. Hsu IC, Lipsky MM, Cole KE, Su CH, Trump BF. Isolation and culture of hepatocytes from human liver of immediate autopsy. *In Vitro Cell Dev Biol* 21:154–160, 1985.
5. Cole KE, Hsu IC, Trump BF. Comparative ultrastructural effects of aflatoxin B_1 on mouse, rat, and human hepatocytes in primary culture. *Cancer Res* 46:1290–1296, 1986.
6. Cole KE, Jones TW, Lipsky MM, Trump BF, Hsu IC. In vitro binding of aflatoxin B_1 and 2-acetylaminofluorene to rat, mouse, and human hepatocyte DNA: The relationship of DNA binding to carcinogenicity. *Carcinogenesis* 9:711–716, 1988.
7. Cole KE, Jones TW, Lipsky MM, Trump BF, Hsu IC. Comparative effects of three carcinogens on human, rat and mouse hepatocytes. *Carcinogenesis* 10:139–143, 1989.
8. Moldeus P, Hogberg J, Orrenius S. Isolation and use of liver cells. *Methods Enzymol* 52:60–71, 1978.
9. Michalopoulos G, Pitot HC. Primary cultures of parenchymal liver cells on collagen membranes. *Exp Cell Res* 94:70–73, 1975.

10. Fleisher S, Kervina M. Long-term preservation of liver for subcellular fractionation. *Methods Enzymol* 31:3–6, 1974.
11. Lowry OH, Rosebrough HJ, Farr AL, Randall RJ. Protein measurements with the folin phenol reagent. *J Biol Chem* 193:265-275, 1951.
12. Lotlikar PD, Jhee EC, Insetta SM, Clearfield MS. Modulation of microsome-mediated aflatoxin B_1 binding to exogenous and endogenous DNA by cytosolic glutathione S-transferases in rat and hamster livers. *Carcinogenesis* 5:269–276, 1984.
13. Croy RG, Essigmann JM, Reinhold VN, Wogan GN. Identification of the principal aflatoxin B_1-DNA adduct formed in vivo in rat liver. *Proc Natl Acad Sci USA* 75: 1745–1749, 1978.
14. Prough RA, Imblum RL, Kouri RA. NADH-cytochrome c-reductase activity in cultured human lymphocytes. *Arch Biochem Biophys* 176:119–126, 1976.
15. Goldberg IH, Rabinowitz M, Reich E. Basis of actinomycin action. I. DNA binding and inhibition of RNA-polymerase synthesis reactions by actinomycin. *Proc Natl Acad Sci USA* 48:2094–2101, 1962.
16. Gellert M, Smith CE, Neville D, Felsenfeld G. Actinomycin binding to DNA: Mechanism and specificity. *J Mol Biol* 11:445–457, 1965.
17. Sobell HM. The stereochemistry of actinomycin binding to DNA. *Cancer Chemother Rep* 58:101–116, 1974.
18. Acosta P, Sorensen EM. Role of calcium in cytotoxic injury of cultured hepatocytes. *Ann NY Acad Sci* 407:78–92, 1983.
19. McQueen CA, Williams GM. The use of cells from rat, mouse, hamster and rabbit in the hepatocyte primary culture/DNA repair test. *Ann NY Acad Sci* 407:119–130, 1983.
20. Williams GM. Detection of chemical carcinogens by unscheduled DNA synthesis in rat liver primary cell cultures. *Cancer Res* 37:1845–1851, 1977.
21. Borek C, Williams GM (eds). Differentiation and carcinogenesis in liver cell cultures. *Ann NY Acad Sci* Vol. 349, 1980.
22. Steward RA, Dannan GA, Guzelian PS, Guengerich FB. Changes in the concentration of seven forms of cytochrome P-450 in primary cultures of rat hepatocytes. *Mol Pharmacol* 27:125–132, 1985.
23. Grisham JW. Use of hepatic cell cultures to detect and evaluate the mechanisms of action of toxic chemicals. *Int Rev Exp Pathol* 20:123–210, 1979.
24. Grisham JW, Charlton RK, Kaufman DG. In vitro assay of cytotoxicity with cultured liver: Accomplishments and possibilities. *Environ Health Perspect* 25:161–171, 1978.
25. Begue JM, LeBigot JF, Guguen-Guillouzo C, Kieckel JR, Guillouzo A. Cultured human adult hepatocytes: A new model for drug metabolism studies. *Biochem Pharmacol* 32:1643–1646, 1983.
26. Strom SC, Jirtle RL, Michalopoulos G. Genotoxic effects of 2-acetylaminofluorene on rat and human hepatocytes. *Environ Health Perspect* 49:165–170, 1983.
27. Lisa B, Tel G, Seddan J, Boobis AB, Davies DS. Drug metabolizing activity of freshly isolated human hepatocytes. *Br J Pharmacol* 19:279–294, 1985.
28. Byard J, Dougherty KK. Comparative metabolism and toxicity of chemical carcinogens in primary cultures of hepatocytes. *In Vitro Cell Dev Biol* 21:489–494, 1985.
29. Moore CJ, Gould MN. Metabolism of benzo[a]pyrene by cultured human hepatocytes from multiple donors. *Carcinogenesis* 5:1577–1582, 1984.
30. Rijntües PJM, Moshage HJ, Van Gemert PJL, DeWaal R, Yap SH. Cryopreservation

of adult human hepatocytes: The influence of deep freezing storage on the viability, cell seeding, survival, fine structures and albumin synthesis in primary cultures. *J Hepatol* 3:7–18, 1986.
31. Autrup H, Bradley KA, Shamsuddin AK, Wakhisi J, Wasunna A. Detection of putative adduct with fluorescence characteristics identical to 2,3-dihydro-2-(7'-guanyl)-2-hydroxy-aflatoxin B_1 in human urine collected in Murang'a district, Kenya. *Carcinogenesis* 4:1193–1195, 1983.
32. Groopman JD, Donahue PR, Zhu J, Chen J, Wogan GN. Aflatoxin metabolism in humans: Detection of metabolites and nucleic acid adducts in urine by affinity chromatography. *Proc Natl Acad Sci USA* 82:6492–6496, 1985.
33. Moss EJ, Neal GE. The metabolism of aflatoxin B_1 by human liver. *Biochem Pharmacol* 34:3193–3197, 1985.
34. Buchi GH, Muller PM, Roebuck BD, Wogan GN. Aflatoxin Q_1: A major metabolite of aflatoxin B_1 produced by human liver. *Res Commun Chem Pathol Pharmacol* 8:585–592, 1974.
35. Roebuck BD, Wogan GN. Species comparison of in vitro metabolism of aflatoxin B_1. *Cancer Res* 37:1649–1656, 1971.
36. Gurtoo HL, Motycka L. Effect of sex difference on the in vitro and in vivo metabolism of aflatoxin B_1 by the rat. *Cancer Res* 36:4663–4671, 1976.
37. Jayaraj A, Hardwick JP, Diller TW, Richardson AG. Metabolism, covalent binding, and mutagenicity of aflatoxin B_1 by liver extracts from rats of various ages. *JNCI* 74:95–103, 1985.
38. O'Brien K, Moss E, Judah D, Neal G. Metabolic basis of the species difference to aflatoxin B1 induced hepatotoxicity. *Biochem Biophys Res Commun* 114:813–821, 1983.
39. McQueen CA, Kreiser DM, Williams GM. The hepatocyte primary culture/DNA repair assay using mouse or hamster hepatocytes. *Environ Mutagen* 5:1–8, 1983.
40. Simard R, Bernhard W. Le phenomene de la segregation nucleolaire, specificité d'action de certains antimetabolites. *Int J Cancer* 1:463–479, 1966.

Mouse Liver Carcinogenesis: Mechanisms
and Species Comparisons, pages 409–414

Research Opportunities and Reducing Uncertainty

William H. Farland
Office of Health and Environmental Assessment, Environmental Protection Agency, Washington, DC 20460

In preparing for the conference this book is based on, I spoke with the session chairpersons about the research to be presented and the speakers they had chosen. I decided to come to the conference, listen, and try to sum up from my perspective current research needs related to risk assessment.

First, it is easy to see from the information presented that we can talk about events in mouse liver carcinogenesis at a number of levels. For example, there was extensive discussion about the molecular events leading to initiation, promotion, and progression. At a different level, we heard about the utility of mouse liver carcinogens in studies describing some cellular events leading to changes in cell-to-cell communication and cellular architecture. In addition, we heard a general consensus that tumors produced in the mouse liver appear to represent a continuum between those traditionally described as benign and those deemed malignant in other systems; that was an important observation. Finally, data on mouse liver carcinogenesis were generated via many different protocols. Clearly, this was one reason that we needed to evaluate the unifying principles of mouse liver carcinogenesis.

Why are these topics of particular interest to the Environmental Protection Agency (EPA) with regard to risk assessment? In the EPA, risk assessment is central to the way that science is both developed and used to support regulatory decisions. The risk assessor has the responsibility not only to help develop the science but also to feed information into the risk management process. While more than risk information is considered in risk management (economic, social and political considerations also are weighed, for example), it is important. The risk assessor uses the research and data collection that goes on within the federal government, university labs, industry, and elsewhere. At the same time, as we do risk assessments, we gain a good perspective on types of information that need to be developed to improve our risk assessment capabilities. The discussions at this conference addressed many of these issues.

A fairly simple paradigm that we follow is: carcinogenic risk = exposure × hazard. An important product of this assessment process is risk characterization. In other words, we are ultimately trying to make sense of all the data available to characterize risks and to determine whether it is possible to develop a qualitative or quantitative appreciation of the potential likelihood of carcinogenicity in an exposed human population. This approach leads us to suggest that some data may be particularly useful in defining a carcinogenic hazard, a judgment that does not depend on exposure or dose. In other words, when we make a statement about carcinogenic hazard, we are talking about the likelihood that a response in one species will be seen in another species at some dose—not at any dose, not at environmentally relevant doses, but at some particular dose.

To put this hazard in perspective from a quantitative standpoint, then, we look at the magnitude of the risk at various doses that might be associated with, or at least relevant to, human exposure. It is important to understand that, when we do a quantitative risk assessment, the current procedures we use do not provide us with a "best estimate" of the risk. Often, they use the animal data and allow us to produce what we euphemistically call a "plausible upper bound" on the risk, based on a 95% confidence limit on the data and, in most cases, an assumption that the slope of the dose-response curve is not any greater at low doses than in the observable range. We expect that the true risk at those low doses would probably be less, and we make statements that those responses may even be zero; that is, that the range of the risk goes from the plausible upper bound down to zero.

To hypothesize a zero risk at some finite dose, we must assume the existence of a threshold. At extremely low doses, this assumption depends greatly on whether the mechanism of carcinogenic action is additive to or independent of background processes. We heard from Chris Kemp on androgen receptors and Marshall Anderson on N-nitrosodiethylamine some suggestions that, in these cases, mechanism of action may be additive to background. Mike McClain talked about the work coming out of Germany now with dieldrin and the fact that some of the processes under study appear to be additive to background. Many questions remain regarding the existence of thresholds in carcinogenesis.

Where do I see things going? First, I think there is a feeling that the way the agency deals with the two major components of risk assessment—exposure and hazard is to look only at an applied dose (i.e., a dose that is going to be applied to an animal or that is assumed to affect a human) and tumor production. Our process has become much more sophisticated than that. What I am suggesting here is that there is currently a trend toward trying to understand the relationship between exposure and dose and toward understanding mechanisms of carcinogenic hazard that we would like to continue.

The basic assumption made is that we have the ability to describe dose as something more specific, and more useful, than point-of-contact exposure. Here we begin talking about the use of physiologically based pharmacokinetic modeling. Dr. Parke discussed the impacts of such an approach at this conference. Drs. McClain and Diwan also talked about the importance of enzyme induction in terms of dealing with this issue. In the recent risk assessment for methylene chloride, for instance, human "dose" was reduced by approximately an order of magnitude based on pharmacokinetic differences between high and low dose considerations in rodents and humans (Dr. Parke suggested it might even be as much as a factor of 40). The goal of such efforts is to get closer to what we call the biologically effective dose. We are looking toward research that will take us closer to this goal. For the time being, however, we are concentrating on better measures of exposure or biomarkers of exposure. Some of the data recorded by Miriam Poirier and reported herein suggest that DNA adducts may be particularly useful biomarkers of exposure and may lead us toward something approaching a biologically effective dose. However, we still have a lot of research to do to determine how those adducts are going to be used in exposure or dose assessment.

To describe carcinogenic hazard, one can look at tumor production in animals and correlate this with the expected response in humans. Several speakers at this conference described that type of work and its impact on the weight of evidence for human response, particularly for tumors arising in multiple species or strains. They described a variety of approaches that could be taken toward developing correlations using modern computer technology and extensive data bases.

However, as discussed at this meeting in the evening session and in Gary Williams' talk, what many investigators are focussing on is mechanisms of carcinogenic action. This will help us understand not only tumor production in the experimental animal but also its relevance to human risk assessment, supporting inferences regarding similar mechanisms of action, and the site of response. These activities become the basis for an argument that correlations are not enough; that one really needs to begin to understand the underlying mechanisms for those correlations. This point is illustrated by the types of questions that Mike McClain raised in his presentation. He asked whether a positive response in the rat thyroid and mouse liver really is an indication of a robust, multi-species response or only an indication of a particular underlying mechanism that one ought to be concerned about when extrapolating data to humans, particularly when using a quantitative approach to risk assessment. Those are the kinds of answers we are looking for. Judging from presentations made at this conference, genetically manipulated strains will be particularly useful for this type of research. We heard discussions about implications related to identification of stem cells, use of genetically susceptible hybrids and, certainly, oncogene work.

Finally, I would like to focus on the dose-response issue in risk assessment. This important topic was well represented at this conference. Figure 1, used in Jim Trosko's presentation, is a simple representation of a two-stage model for carcinogenesis that has a very long history. Suffice it to say it is probably most recently adapted from the Moolgavkar-Knudson-Venzon discussions on retinoblastoma. A point to keep in mind is that the model of initiation, promotion, and progression is becoming more, rather than less, complex. That was very well addressed here and leads to some points regarding what types of research we would like to see.

First, I think it is particularly important for us to look at the issue of the identity of the stem cell. I thought Dr. Fausto's discussion regarding oval cells vs. hepatocytes as stem cells in the mouse liver were particularly tantalizing. I think the issue is really whether to look for stem cell responses or whether, at high enough doses, one might be recruiting cells that are further down the line toward differentiation and somehow causing cellular events that bring about

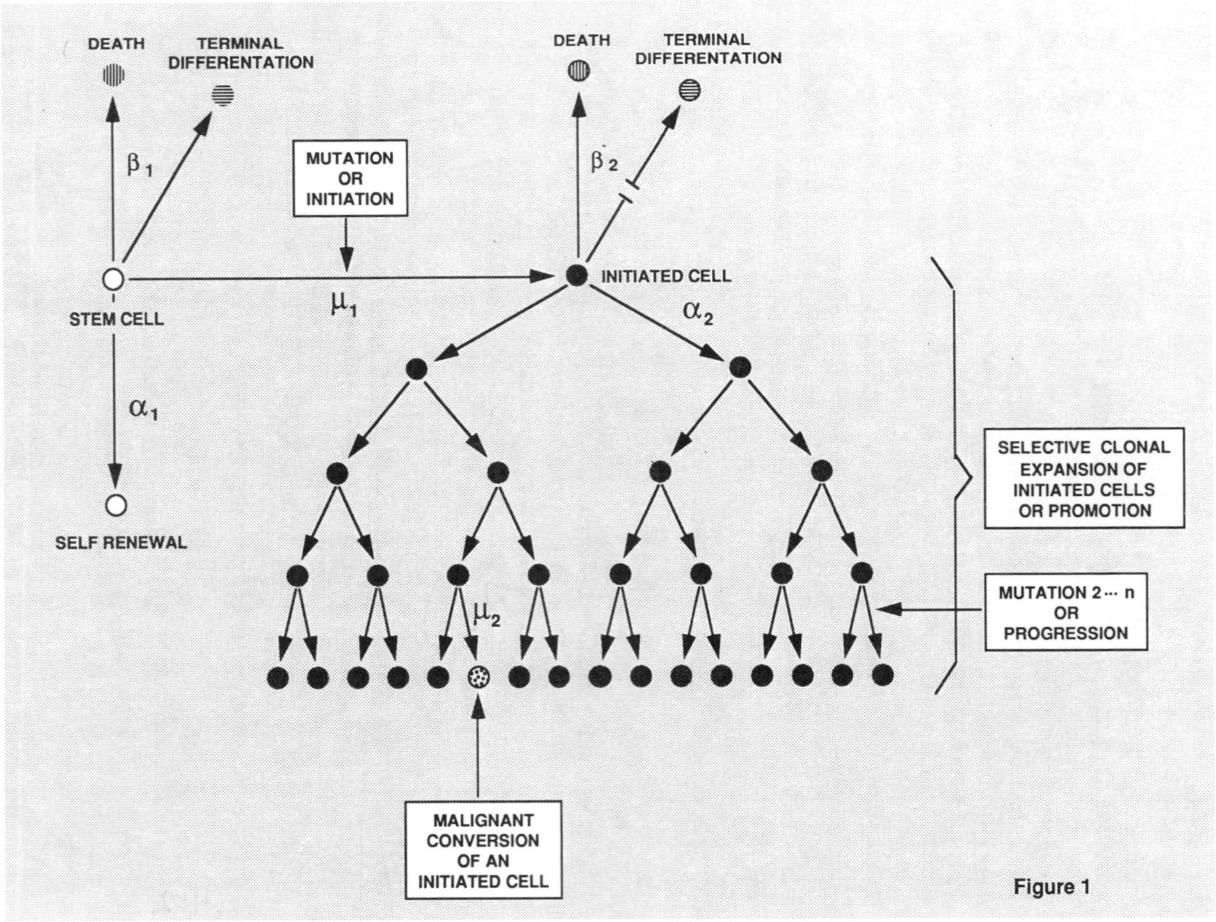

Figure 1. Two-stage model of carcinogenesis.

either a dedifferentiation response or produce a responsive cell. Certainly, we are learning there is more complexity here than was originally considered in terms of the potential fate of the stem cell. One particularly important idea is that, among stem cells, there are fates other than self-renewal and death. Differentiation may reduce the population of cells susceptible to initiation. This becomes important as we look at our concept of initiated cells. As illustrated in Figure 1, initiated cells are drawn from the stem cell population by a mutation or series of mutations at a finite rate, u1. This generally considered to be the start of a sequence of events leading to carcinogenicity: an initiation response occurs that provides an initiated cell and there is a proliferative response, cell promotion, that essentially increases the probability of some additional mutations or irreversible events leading to malignant transformation.

This conference and proceedings volume suggest that under certain conditions, which may be represented by an infant mouse single-dose protocol, one may be able to go from the first step, initiation, to the second step, progression or transformation, very rapidly, without invoking an increase in probability based on promotion but by then having tumor formation occur because of rapid proliferation associated with certain malignant cells. I think it was particularly interesting that we could begin to look at cell kinetics with regard to shell/core growth theories, because if we are going to build a biologically based carcinogenesis model and actually evaluate dose-response with this type of model, we have to know something about the kinetics of cell replication or proliferation as they relate to that response. Dr. Goldfarb's data, particularly, might fit into that type of model development. He and others also made a strong point that we cannot look at this solely from the standpoint of increased cell division—that we have to look at the dynamics between proliferation and cell death, as well. Several researchers suggested the role that cell death may play in oncogenic or carcinogenic expression. Tumor growth may not be simply a change in the rate of cell division but may represent a change in the rates of cell death within a transformed population. One suggestion that has been made, but I think is yet untested, is that the pool of initiated cells is somehow prevented from going down the road toward terminal differentiation, as is shown in Figure 1. The suggestion made earlier about Dr. Fausto's work relating to stem cell conversion may apply equally here. Whether tissues contain cells with the potential to give rise to tumors after limited numbers of subsequent events may be determined by whether initiated cells are able to move toward differentiation.

Dr. Ward talked about the effect of nutrition on the development of mouse liver tumors under experimental conditions. Here, again, is a case when the type of laboratory model or protocol that one employs may have profound effects on the way that tumors develop. What we are really asking, however, is whether the biological mechanisms underlying these tumors are species-specific. Several researchers have suggested that, as one learns more about the mouse liver tumor response, many of the underlying mechanisms are likely to be found in all levels of biologic organization and are not specifically related to the mouse. If we can determine where chemicals work in this process in the mouse, we will have some advantage in trying to extrapolate, both qualitatively and quantitatively to the human.

As we follow Figure 1 down through selective clonal expansion of initiated cells, or promotion, we can reflect on the cell-to-cell communication research and the sequelae to changes in cell communications findings presented by several speakers during Dr. Trosko's session. This is a particularly important area, ripe for additional research simply because of the parallels that can be drawn between normal cell development and the apparent aberrant development in individual cells that have been turned down the road toward carcinogenicity.

The conference provided provocative discussions by Drs. Mirsalis, Styles, and Goodman, particularly concerning the role of ploidy in facilitating an increase in cell proliferation that may be associated with the carcinogenic response. This also is an area of active research interest. Of particular interest with regard to our current thinking on chemical carcinogenesis is alternative mechanisms to begin a process to produce a particular response that would be perpetuated even in the absence of the chemical and in the absence of genetic damage as described traditionally. This describes the heritable epigenetic response that Dr. Goodman discussed. This issue becomes particularly important when one considers the question of reversibility of the promotion response once the chemical has been removed. Methylation, gene expression, and oncogenes all may be important in this type of response. These may affect the final outcome of the tumor response when a burst of proliferation or long-term proliferation occurs.

Several speakers focused on receptor binding, including Mike Gallo. A consensus is building among many investigators that receptor binding is part of the carcinogenic process for a number of chemicals. The parallels between epidermal growth factor and the underlying mechanism of action in many carcinogens argue that we need to do more not only to understand that process but also to model it as something that may be directly relevant to quantitative risk assessment. While many inferences may be drawn from mouse liver responses, knowledge of such processes may allow us to develop a better quantitative model for carcinogenic processes in general.

The EPA will consider many of the issues discussed at the conference and in this volume as we evaluate our risk assessment guidelines. In addition, such discussions will help guide our research agenda and will help us build quantitative and qualitative models for carcinogenicity that will allow us to represent the best science possible in the field of carcinogenicity risk assessment.

Mouse Liver Carcinogenesis: Mechanisms
and Species Comparisons, pages 415–427

General Discussion

Dr. Gary Williams: I am wondering about one thing: if cancer were a disease that occurred only in the mouse liver, would we all be here tonight? I assume the answer is no—that we're interested in mouse liver tumors because we think they have something to do with a more important disease, human cancer. I think we need to leave here with at least some approximation of what mouse liver cancer has to do with human cancer. I happen to think that we could probably, if the public cooperated, prevent a large portion of human cancer without understanding anything about mouse liver tumors. Anyone want to take issue with that? If we all agree, that leaves us with the question, What is it we want to learn from the mouse? I think that the mechanism of carcinogenesis in the liver is vitally important.

Let me put forth another proposition: the mechanism may not be the same in all strains of mice. We've seen data suggesting that the C3H mouse and its progeny, the B6C3F_1, are endowed with some kind of responsiveness and hereditary predisposition that doesn't exist in all other strains of mice. Yet we are using the B6C3F_1 for testing chemical carcinogens. I have insight into that evolution because I was at the National Cancer Institute at the time they were developing their bioassay program. At that point, the F-344 rat had been selected largely on the basis of its susceptibility to aromatic amine carcinogenesis. The problem with the C3H mouse was that it had such a high background of spontaneous tumors that there was some doubt as to whether it would accurately detect carcinogens. Contracts were let to look at hybrids between the C3H and low liver tumor incidence strains of mouse, and out of that came the B6C3F_1. I remember Stan Vesselinovitch figured prominently in this.

That project represented an implicit acceptance that a strain of mouse susceptible to liver tumor development was desirable, although an effort was made to dampen that susceptibility by creating the F_1 hybrid.

Today, we can introduce animals that are even more cancer-prone. I think Tom Slaga was involved in developing the SENCAR mouse, and now, of course, we have the transgenic mice, into which activated oncogenes have been introduced. There is another intriguing question—if we were to pick a bioassay mouse today, would we want one of these transgenic strains? And if that answer is no, to what extent does that decision relate to the B6C3F_1? Does anyone want to take that up?

Dr. Norman Drinkwater: Two comments. One, the notion that there's a distinction between hazard identification and risk assessment. I would propose that the C3H mouse or the B6C3F_1 is a reasonable test animal to use for hazard identification because the mouse's genetic background in some sense acts as an amplifier of any applied chemical's carcinogenic activity. That genetic background increases the animals' response to a variety of classic genotoxic DNA-damaging carcinogens, at least as measured in the preweanling mouse assay. I realize there's a problem with translating that assay to the traditionally used chronic bioassay. But the data also indicate that the genetic background acts synergistically with agents that are tumor promoters in the mouse liver. For at least two compounds that we've studied—hexabromobiphenyl and testosterone, an endogenous promoter—the response is similar in both the less susceptible C57BL/6 mouse and the more susceptible C3H mouse. That is, the fold-increase in tumor response is the same for both genetic backgrounds, but the tumor yield is obviously much higher in the C3H mouse.

The problem is not whether that mouse is an appropriate model for hazard identification, but rather how you use the data that comes out of a positive result in that assay. What kind of information does that give you in terms of risk assessment? It's quite possible that the level required to achieve that response would never be encountered in a human population—especially for an agent that acts as a tumor promoter in that model system, for example, a drug or some environmental agent. If that is the case, using that mouse is an efficient way to identify chemicals that require risk assessment. I'm sure that a lot of people will argue with that.

Dr. Robert LeBoeuf: I think that the distinction between hazard identification and risk assessment is very neat and tidy on paper, but in practice, with the current risk assessment models, once you've identified a material as a carcinogenic hazard, it's very difficult to draw a distinction between a real hazard and a nonhazard. The virtual safe doses are so low that it's very difficult to distinguish whether it's a theoretical problem or a real human risk.

Dr. Bernie Daniel: Nevertheless, for hazard identification, you want something that avoids false negatives, so I think that Norm's point is well taken. For hazard identification, sensitive strains are needed. Bob's point is also true—we may have to live with uncomfortable political problems when you subsequently don't see cancer in the whole animal. However, from the viewpoint of human safety, this is the price we must pay.

Dr. Williams: Then you're saying, regardless of the rate of false positives, you would like to eliminate false negatives in hazard identification.

Dr. Don Stevenson: If you have a positive in the mouse liver, and you have very little noise elsewhere, what additional information would you need to better downgrade that hazard identification step to one where you would be prepared to go to other end points? There will clearly be other end points—neurotoxicity, liver damage, behavioral depression, or whatever.

One feeling I get from this meeting is that we may not know the entire mechanism within the next 5 years. The cell proliferation that may be seen in one species and not in another may be one of the real things we need to look at rather seriously on a much broader basis.

Dr. Williams: Do you think it's important whether the compound reacts with DNA? Since I have the floor here, let me go over the history of the terms genotoxic and epigenetic. To the best of my knowledge, the term genotoxic was coined by Herman Druckrey in 1971, at a meeting in Sweden. At that time, the working group I was a member of was looking for a term to describe the ability of chemicals to alter genetic material. So a rather general description of genotoxicity was put forward—something like "the ability of a chemical to interact with and alter both chromosomal and extranuclear DNA." Druckrey was rather infatuated with the work of Gaffi in East Germany, suggesting that mitochondrial DNA might be a target for carcinogenesis, so he wanted that "extranuclear DNA" in there.

That report was buried away in a Swedish journal for a few years until John Weisburger and I became interested in whether there were agents that did not have the property of reacting with DNA, carcinogens specifically. We appropriated the term genotoxic to describe chemicals that essentially, as Miller and Miller described, had the ability to function as electrophilic reactants, interacting with DNA. We then adopted the term epigenetic, which interestingly had also been used by Miller and Miller. They used "epigenetic" specifically to refer to chemicals that did not function as electrophilic reactants. So we formalized, if you will, the dichotomy between those chemicals that react with DNA and those that do not.

I still think that is a meaningful distinction. On several occasions, we have analyzed the properties of human carcinogens. Someone at this meeting also showed a slide indicating that 80%–90% of chemicals shown to be associated with cancer in humans clearly can react covalently with DNA. Certainly there are a number of chemicals that do not—agents such as asbestos, estrogen, and oral contraceptives—so we are not implying that nongenotoxic carcinogens or epigenetic carcinogens should be dismissed. Our feeling is that compounds that are clearly DNA-reactive represent a distinct kind of hazard. Certainly, this is the case if you look at the carcinogenicity studies that have been done with DNA-reactive agents, for example, the ED_{01} study with AAF (2-acetylaminofluorene) at the National Center for Toxicology Research (NCTR) or the BIBRA (British Industrial Biological Research Association) study on diethylnitrosamine/dimethylnitrosamine. While there was a threshold in the AAF study, it was only for the bladder, and it was low, as Miriam Poirier has shown. On the other hand, dose-response studies done on the agents that do not have the capacity to react with DNA, generally speaking, show much higher thresholds. I must immediately exclude agents like TCDD that interact with receptors. I think that as we understand the molecular mechanisms and the fact that agents that bind to receptors can initiate a cascade of reactions that amplify the chemicals' interaction with the cell, we see that those chemicals should be presumed to have very low thresholds. However, other chemicals such as saccharin, nitrilotriacetic acid, or a host of others that are inert, have thresholds for biological effects that are below the levels that elicit carcinogenicity in the experimental animal. These chemicals have not been associated with cancer in humans, although there certainly has been widespread usage and exposure to these chemicals.

So I stand by our position that this is one piece of information helpful in the risk assessment. Does anyone want to take issue with that?

Dr. James Popp: If you're suggesting that information on genotoxicity be used in risk assessment, how do you propose that it be incorporated into the risk assessment process? That's the crux of the issue. I suspect you would have strong support, but implementation is the problem.

Dr. Williams: Our feeling is that for a DNA-reactive carcinogen, there is no way to ascertain the safe level of exposure for humans. Therefore, we suggest that they be regarded as qualitative hazards. If whoever is concerned with the chemical can devise some way to demonstrate that that chemical does not represent a qualitative hazard to humans, then they should put that forward. Basically, we think risk assessment is confined to the epigenetic category. I believe that different types of risk assessment are needed for different types of agents.

I don't think that any of the present mathematical models provide a "best estimate" of human risk. They may be useful in representing a societal consensus on how such chemicals should be handled, but biologically, those models don't appeal to me for a number of reasons. Start with the fact that in human exposure we're dealing with a heterogeneous population. Some proportion is exposed to hazards that others are not. We have cigarette smokers and non-smokers, we have people who eat high-fat diets and vegetarians. It's not like a well-controlled animal study. I don't think any model can factor in what we know about differences in human susceptibility.

It has proved to be basically sound to establish a no-observed-effect level and then factor in a safety factor of 100. That sounds very simplistic, but in fact, that has been the basis of product safety in many countries for the past 40 or more years, and is based on accumulated experience in the food toxicology area. This approximation has served us well. There have been, as far as I know, few if any noxious substances that have been involved in human toxicology when kept below those levels.

Dr. Bob Sielken: As a biostatistician, I seem to be on the hot spot at the moment. Let me respond to a few of the things you mentioned, without touching on the social value of safety factors or their past performance from a societal point of view. The models that have been used, particularly in the early years of risk assessment, and are still used in some circles, are primarily curve-fitting models, and are not biologically reflective at all. Therefore there is a lot of uncertainty, disbelief, and dissatisfaction with those models.

Modelers are trying to develop more biologically reflective models. To make overly simplistic models more reflective, there is a push in the direction of pharmacokinetics. We are also trying to find ways to incorporate things like cell proliferation. At this meeting I heard that there are a lot of covariants that could influence the dose-response relationship. There are oncogene activation levels, proliferation levels, enzymatic levels, proneness to certain things; if we could identify the important cofactors, they could be incorporated into the mathematical models better than they are now.

Similarly, if the growth portion of the process were deemed important, the statistical models could be designed to reflect that growth—the actual growth and not proliferation as an enhancing event for later mutations.

I think there's been dissatisfaction with the old curve-fitting models, as well as a hope that conferences like this would identify mechanisms or physical phenomena that we could try to model. I would say that the models, as they've been used, can provide point estimates of phenomena within the species and protocols where the experimental data have been collected. If you want to get the best information from the data that you have, you can get more information from the point estimates than you can from, for example, an upper bound or a lower bound. The information is there and can be utilized.

There's information on time-to-response, intermittent exposure, and on timing and age susceptibility that is not being used in the models—some of us are trying to incorporate that information.

Another thing you've mentioned is the diversity, or heterogeneity, of the human population. That phenomenon could be incorporated if we compiled data on, for example, the distribution of metabolic rates. If we could establish the distribution of a metabolic rate for a particular carcinogen, we could incorporate that into the risk assessment—not just the metabolic rate for a most-, least-, or average-susceptibility person, but for a population distribution of metabolic rates. Such incorporation could give risk probabilities on a percentile basis for a population. I hope the dissatisfaction with quantitative risk assessment won't lead to anything but its improvement.

Dr. James Trosko: As a geneticist, I'd like to make two responses. First, I agree that we have to shoot for a biologically based risk assessment system. That means understanding mechanisms. I realize I'm the only one who still believes that cell communication has anything to do with cancer, but assume for the moment that it does play a role. I want to give an example of how chemical interaction may give us some predictive values, because I think we're starting to understand some of the mechanisms affecting cell-to-cell communication. For example, there are at least five different mechanisms described in the literature that have been ascribed to downregulating gap-junctional communication: activating the protein kinases, increasing calcium, changing pH, activating the cAMP protein kinase, and free radicals. There may be others. Take TPA (12-*O*-tetradecanoylphorbol-13-acetate), for example, a documented tumor promoter in the SENCAR mouse. TPA downregulates gap junctions very efficiently, presumably by activating protein kinase C (PKC), its receptor. It has now been shown by Ahlborg's group in Sweden and at Michigan State by Aylsworth that if you add TPA in vitro to cells that respond to it, at a very low level, to just tickle the PKC, the gap junctions are barely inhibited. If you add DDT to the same cells, at a level low enough to inhibit the gap junction slightly, there is no effect on PKC. DDT is another tumor promoter—at least to the liver, and not the skin, to my knowledge. DDT doesn't act through PKC, presumably it acts by raising intracellular calcium levels.

Now if you add those low levels of DDT and TPA together, you get a synergistic effect. In this case, the chemicals strongly block cell communication. I think we understand the mechanism for this effect. Protein kinase C is a phospholipid-sensitive, calcium-dependent enzyme. When DDT raises intracellular calcium, and TPA is the phospholipid ligand for PKC, the two interact and activate the enzyme. To my knowledge, no one has done that experiment in vivo to see if there will be synergism. But I predict that it probably would not work in the animal because of the pharmacokinetic information—that TPA works on the skin, DDT on the liver, and never the twain will meet in vivo.

Ahlborg's group then asked, What would happen if we interacted DDT with dieldrin (because presumably they work by the same mechanism)? When they did the same kind of experiment, they got additivity rather than synergism. Finally they asked, What if we take quercetin, a PKC inhibitor, and add it to DDT? They got antagonism. All these experiments were done in vitro. None have been done in vivo. At least in regard to promotion, and also with initiation mechanisms, we've got to understand all of these interactions. We're a long way from there, but we've got to at least start thinking that way.

I would also like to address the question of which mouse or animal model we should use for hazard identification, as opposed to risk assessment. I think that you're going to find genetic predisposition, as well as genetic resistance, to both the initiation and promotion phases of cancer. We have human examples of that. Maybe, instead of designing genetic animal model systems, we should start to think about studying genetically predisposed human syndromes to understand mechanisms of predisposition to the initiation and promotion phases of human cancer. Otherwise, we would have to design lots of genetic models for the animals because the mice may be genetically predisposed to skin, liver, kidney, or some other cancer. Something that's positive in one of these strains may be risk or hazard identification for that chemical, but it may also be a hazard to another organ in the human, which you don't have an animal screen for.

In fact, I started out my research in Xeroderma pigmentosum. Jim Cleaver beat me in the race to identify this cancer-prone syndrome. We were thinking the same thing, that there must be human beings out there who are sensitive to UV light. It turns out there are three or four human syndromes marked by sensitivity to UV light. He was lucky: he picked Xeroderma. I picked something else, and I lost this race. I've decided that there are "human SENCAR and $B6C3F_1$ mice" out there now, in fact right in this room. For example, using the Knudsen model, or the retinoblastoma model, how many of us—epidemiologists or biostatisticians—have climbed out of the ivory tower to start looking at hazard chemical identification and risk assessment in these human beings. They're rare, but they are there. People with Xeroderma develop plenty of skin cancers. However, all of their cells, not only their skin cells, lack DNA repair. Why don't they get liver cancer or other internal cancers? They eat the same chemicals as the SENCAR mice and Fischer rats. Xeroderma appears to be an initiator-prone syndrome. They lack repair, and therefore they have a higher mutation frequency. Veronica Maher and Justin McCormick were first to demonstrate this. Those with Down's syndrome, on the other hand, are prone to leukemias. Their DNA repair is perfectly normal, so that's probably a promoter-prone syndrome, not a initiator-prone syndrome. And retinoblastoma, the classic model that the Moolgavkar model is based on—how many people are looking at these individuals? More and more are surviving today because of surgery and radiation treatment. But now they're coming down with osteosarcomas and other kinds of cancers. Why aren't we looking at these "human SENCAR mice," so to speak? These are the kinds of genetic syndromes for which we already have model systems that can be used to identify potential hazards and maybe even make more realistic risk assessments.

Dr. Sielken: There are several interesting challenges in the things that have just been said, not only to statisticians, but to everyone in this room. I might add a couple of other challenges that go along with those you mentioned. It is clear that in the case of mixtures, or with the possibility of interaction, understanding mechanisms becomes even more important. There won't be a default situation or a guideline that necessarily provides an upper bound on risk, because you don't know what the default risk is for an interaction—the risk could go up or down or stay the same. Therefore it will be harder to decide what to do when there aren't any defaults to fall back on.

The other thing that will be interesting is the conduct of risk assessment and risk management if the statisticians and mathematicians develop different models based on different perceptions of carcinogenic mechanism. If we develop models that reflect cell-to-cell communication, oncogene activation, or cell proliferation, then we may have two or three different models for a particular cancer risk. We can develop multiple quantitative risk assessments, each one based on one of two or three different models, that reflect our best estimates of mechanism. How much of this scientific background should you pass on to the people who have to make risk management decisions? You've suggested multiple mechanisms, and there is probably no possibility of absolute proof of one over another. Given that situation, society will ask the scientific community to put value judgments on the relevance of the different mechanisms and indicate the strength of evidence for particular models. That means that some day you all are going to be asked to quantify how much you really think the evidence supports this mechanism vs. that mechanism—and that will be a difficult decision to make, but a necessary one.

Dr. Stevenson: I'd like to go back to Jim's statement. We did the in vivo experiment about 20 years ago. DDT and dieldrin are pretty well additive in the mouse, but Deitman reported an experiment in the rat using a cocktail of four compounds, including aldrin and DDT. If anything, the level of tumors went down. Can you expand on the intriguing observation that antitoxins block these compounds' effect on intercellular communication?

Dr. Trosko: I want to respond in general to work that Gary and Dr. Ito and others conducted; I think Mike Gallo even made reference to it earlier. It's the "good news/bad news" story about these chemicals. For almost every example—phenobarbital, PBB, TCDD, DDT—that I know of, you can say "this chemical is bad news in one species for one organ," then someone else will say, "in another species, in another sex, or another condition, the chemical protects against cancer." It gets to the point that you cannot determine the goodness or badness of the particular chemical from the structure alone. You have to put it in a biological context. I think Gary has a whole list of these kinds of chemicals. Retinoic acid is a good example: it seems to have reversible effects. We need retinoic acid for life; however, we now know it is a human teratogen. A woman only has to be exposed to it at a critical time during the fetus' gestation for it to have an irreversible teratogenic effect on the fetus. It has also been shown to be an anti-tumor promoter of skin cancers under one set of conditions, yet it can be a tumor-promoter under another set of conditions. Many of these chemicals that have biological effects also have thresholds. In some cases they have positive effects, in other cases they are negative, in some cases the effects are reversible, and in other cases they're not. Again, it depends on the biological context. If we believe we're going to be able to identify chemicals that will do just one thing, I think that's an illusion.

Dr. Drinkwater: It's certainly clear that there are individuals who are predisposed to tumor development. If they do develop tumors, in terms of appropriate human models, the difficulty is in identifying those agents that were responsible. As for appropriate animal models—in spite of having worked with it for a while, I have no particular fondness for the B6C3F_1 mouse. But were you to pick any other inbred mouse strain or F_1 hybrid you would be faced with a similar problem. Maybe someone here can identify an inbred mouse strain that does not show a high susceptibility to some specific organ for both spontaneous carcinogenesis and chemically induced carcinogenesis. But I think that if you look at the pattern of sensitivities, every inbred mouse strain has its own complement of organs for which it is highly susceptible to tumor induction. Also, if the rat were better studied, I think the situation would be similar in inbred rat strains. So then we would be faced with exactly the same problem. If the B6C3F_1 is an inappropriate model, what would be better?

Dr. Dennis Parke: Another species.

Dr. Drinkwater: Which species? You've got a whole variety of rodents—guinea pigs, hamsters. Hamsters are probably more susceptible than mice or guinea pigs.

Dr. Parke: Well, primates are an unacceptable animal to use for such testing both from the cost point of view and the societal point of view these days. It seems to me that this issue is not simply restricted to scientific questions. There's the consideration of societal values in risk estimation, and so forth. I will make the following foolish assertion, one that will doubtless get me in trouble. Were you to pick another species, develop appropriate inbred strains, and look at these inbred strains' susceptibility, you would identify for each of them a particular susceptibility to tumor induction. The issue is not whether a better one is available but whether we can use the information that we have to make the ones that we're using more effective.

Dr. Debra Kaden: What about an outbred strain?

Dr. Drinkwater: Festing has dealt with that issue extensively, and the difficulty with an outbred strain is that the same patterns of variation are there, but you limit the population size that would be susceptible to any particular tumor site. So, in effect, you are reducing the size of your test by using an outbred strain. That argument is made fairly cogently in several reviews that Festing has written over the last 10 years or so.

Dr. Michael Gallo: That is, in fact, one of the problems with the primates. We did a great deal of work with drug-metabolizing enzymes in rhesus and speciosa monkeys and in gorillas; they aren't any better predictors of the human response than are many of the rodents species. Alexander Pope, 300 years ago, said the best study of man is man in his essays on humanity. That may sound glib, but we're still in the same boat. I think we have the same problems when we go to other animals for risk-assessment models. I guess if I had to really look for the one thing to understand before I try to look at risk assessment, I would really want to know about the delivered dose to a receptor site. I'm not talking about administered dose or even the dose delivered to a tissue. Gene McConnell said yesterday that when we understand that mechanism at a pharmacokinetic level, we'll have come a long way. That's an area in basic research that is still wide open. The oncogenes and everything else are important, but we have to understand the dose—and that's where tissue culture work comes in handy, regardless of the source of the cells.

Dr. Stanley Goldfarb: I'd like to go back for a moment to the original statement—the idea of a high spontaneous background incidence. That incidence is a function of the kind of end point you use. When 2-year studies in mice show that there are one or two grossly visible tumors per mouse at the end of that 2-year period, it is said that there is a high background in the control animals. From this approach, we have an estimate of about 1,600 neoplasms after 28 weeks. We consider these neoplasms, already estimated at 28 weeks with 5 mg of diethylnitrosamine/kg. This approach is taken not only by me but others as well, at least in the single-dose mouse model. If you think in terms of the time, if there are large numbers of tumors forming, some of which have variable growth rates, this means that the control animals are showing a low background rate—a very small number of slow-growing tumors that are growing out after a 2-year period. In the Fisher rat it's a similar situation: if you wait 3 years with Fisher 344 rats, the literature says that you'll get a noticeable background rate.

Comment: Yes, 1% or so.

Dr. Kaden: Is it only 1% at 3 years? In control animals? You have to think about when you are examining the animals. You have to quantitate and use the tumor multiplicity parameters that are measured and that go into the risk assessment equation. Viewed that way, the control animals in mouse studies have a relatively low background.

Dr. Williams: That's an interesting perspective—that the mice have a low background. I mean, 20% lifetime rates for liver cancer certainly don't reflect the human experience.

Comment: That's in the liver. What about the prostate in the male?

Dr. Eugene McConnell: The issue of animal tumor incidence vs. human cancer incidence deserves clarification. Every species and strain of animal has its own unique tumor type in which incidence will be high, approaching 80% or more, and that includes humans. The problem of tumor incidence in humans is that most tumor data is based on mortality data. If you evaluated human cancer incidence the way it is done in the mouse, the incidence would be much higher than the reported figures. For example, a wart (squamous papilloma), mole, or basal cell tumor that many of us have taken off during our lifetime would be counted as a tumor. Most of those benign tumors that we survive would have been called cancer in animals because they are allowed to develop to a lethal stage. If the lesions we call cancer in the rat prostate gland were looked for in humans, over 75% of us, if we reached 75 years of age, would show up in the cancer column for this lesion alone. There are also fibromas that occur in the uterus at high rates. There are pituitary lesions similar to those scored as tumors in rodents that occur in over 20% of elderly people. Therefore, it is fair to suggest that the rodents used in carcinogenesis studies are not relevant because they have a high incidence of a particular type of tumor. The pattern in rodents just happens to be different from humans, but the overall incidence is not that much different.

Dr. Phil Leder: As I hear all the high-powered biological and mathematical arguments, I'm forced to play the reductionist here. I want to expand on what you said at the start, Gary. Why this conference? Is it because mouse liver tumors have something to do with human cancer? I think it could be taken one step further: it's because animal tumors, including mouse liver tumors, are used to project human cancer risk in a nonthreshold manner. I think "nonthreshold" is why we're here—because if we were considering mouse liver tumors as we do, necrosis, or a nonneoplastic change in these rodents, we would add safety factors. Everyone in industry could live with that, I'm sure. I think it's this nonthreshold idea that really binds us.

Two questions. In the case of a genotoxic liver carcinogen, there's at least a theoretical basis in saying there is no threshold. That's based on probability. Is that a given, and is that something that people wish to concede? Aside from the theory, does the biology say that no-threshold risk assessments should be performed on such agents? And then when you consider the epigenetic agents, is there a reason to consider them anything other than a threshold type of phenomenon? Any comments?

Dr. Jon Mirsalis: Actually, Phil, I think that's a very good point—but it brings up something that is becoming our new millstone. Until quite recently, many of us, not all of us, were in the "mutagens are carcinogens" and vice-versa school. Well, now it's been handed down that there are genotoxic, or mutagenic or DNA-damaging, agents and then there are nongenotoxic, or epigenetic, agents. I'm concerned that now people are going to the other extreme and classifying things as genotoxic or nongenotoxic on some very shaky data. I don't think anyone here would disagree with me that aflatoxin is a genotoxic hepatocarcinogen and dioxin is not genotoxic—but lately I'm seeing compounds that cause tumors referred to as positive as genotoxic carcinogens based on a positive Ames test—and that's insanity. Here's a good example that's been pretty extensively studied: a pair of hair dyes, HC blue #1 and #2. We still don't quite understand HC blue #1, but if you look at the in vitro GENETOX data, it looks like aflatoxin. From the bioassay data, it looks like phenobarbital. In the bioassay, it produces mouse liver and rat thyroid tumors at doses of 20,000 ppm in the diet. When you put it into in vitro UDS (unscheduled DNA synthesis), or ^{32}P-postlabeling with isolated hepatocytes, it turns the cells bright blue; it's a flaming genotoxic agent—DNA adducts, DNA repair, strand breakage, covalent binding, whatever you want to do. Then, in vivo, it also turns the liver bright blue, and when you isolate the cells and spin down the pellet, the cells are bright blue. But there are no adducts, no DNA repair, no strand breakage, although there is a lot of S-phase synthesis. John Ashby, somewhat tongue-in-cheek, calls these "genotoxic/nongenotoxic" carcinogens. They're genotoxic in vitro but they're probably producing tumors by a mechanism that has absolutely nothing to do with genotoxicity. There are still people who will say, "Oh, well, it's genotoxic in vitro, and it produces tumors, therefore it's a genotoxic hepatocarcinogen." I think that's wrong. We ought to be very careful about how we use the terminology. The list of compounds like that is going to get bigger and bigger as we start studying.

Another example is 4,4′-methylenedianiline, which is one of the 12 most potent carcinogens the NTP (National Toxicology Program) has ever identified. I believe it produces primarily liver tumors in the rat and mouse, but I don't recall what the other tumor sites are. Again, it's a screaming positive in in vitro tests, but negative in every in vivo system it's been looked at. We're studying S-phase synthesis in rat liver, and it's so strong we've made it our new positive control. So again, here's a compound that's genotoxic in vitro, and not a shred of evidence that it's genotoxic in vivo. It's probably operating under a different mechanism. We ought to keep these things in mind before we immediately rush out and make two columns, genotoxic and nongenotoxic. I'm sure in a few years, we're going to find a third column that's "other," and maybe four or five other columns.

Question: What's the name of the blue dye?

Dr. Mirsalis: The chemical name for HC blue #1 is 2,2-([4-(methylamino)-3-nitrophenyl]imino)*bis*(ethanol). HC blue #2 is a structurally similar analogue that is almost equally genotoxic and is noncarcinogenic—or right on the line. I'd call it an inconclusive or equivocal. Just to complicate matters, it was recently discovered that there was an impurity in the original HC blue #1 compound that was used for the bioassay, and it is only about 97% pure. If you do an ethanol purification, the pure HC blue #1 is Ames-negative. However, when it was rebioassayed, the compound still produced mouse liver tumors. Anyway, I think this is a nongenotoxic hepatocarcinogen compound that happens to be genotoxic in an in vitro test.

Dr. Williams: I have been dismayed myself at how carelessly these distinctions are applied sometimes. One of the reasons that, for example, we chose the term epigenetic for the nongenotoxins is that nongenotoxicity is not a mechanism; to put a chemical in that category, you have to have ample evidence from structure, in vitro, and in vivo studies showing lack of interaction with DNA, as well as a plausible mechanism for the actual epigenetic mode of action.

The specific chemical I want to talk about, because it has been mentioned here several times, is methapyrilene. That's one I think the jury is still out on. The genetic toxicology data is mixed, largely negative, as your findings, Jon, and ours show, but we were confronted with a very perplexing observation when we did a promoting study with methapyrilene. We gave 2-acetylaminofluorine first, followed by methapyrilene, and we saw magnificent enhancement of liver tumors. But also in that study, we gave the reverse administration, where methapyrilene was given first, followed by an interval off the drug, and then AAF. We got the same enhancement. Given that, we cannot conclude that methapyrilene is operating only by a promoting, or cytotoxic, or a cell proliferative mechanism. Where actual in vivo data raises the possibility of some subtle, poorly understood effect, before that chemical can be categorized as a type other than genotoxic, a lot more effort is required.

Dr. Parke: I'd like to address a few points that the previous speakers made about the species-specific target-organ carcinogenicity, and also kinetics and metabolism. I'd also like to tell you very briefly about an incident most of you probably don't know about.

About 10 years ago, there was a major world crisis in the safety evaluation of the progestins. They were showing carcinogenicity in the beagle, in the breast. The Australians very unwisely decided that they wanted all progestational agents to be evaluated for potential carcinogenicity in the dog, unopposed by estrogens. Anyone could have predicted the dire results. After 4 years, one or two of the beagle dogs started showing tumors of the breast. Histologically, the tumors had a well-defined margin and were pronounced benign—very common in elderly dogs, and these dogs were about 6 or 7 years old. Unfortunately, the next year, in the histology, the tumors showed penetration of the basement membrane, and the following year, some of them even metastasized. So we had to change our diagnosis. Even then the veterinarians, the real dog experts, were telling us, "If you keep beagle dogs to about 11 years of age, they will all get carcinoma of the breast." Even so, we had to take these studies seriously and called all the regulatory agencies around the world, including the FDA, to London for a meeting similar to this to decide what we were going to do. Solly Zuckerman said, of course, the dog is an anomalous animal—it is bound to have progestational receptors dominating in the breast, rather than estrogen receptors. The consensus at that meeting was to do receptor studies, to find the particular progestogen concentrations in the beagle breast fluid and further to compare them with humans. The companies managed to get this done—I think half a dozen pharmaceutical companies worked together. They determined the receptor concentration in breast tissue of women and dogs; they determined the chemicals' affinities, and those of their metabolites, for that breast tissue. Finally, they determined those chemicals' concentrations in the breast fluid of women and dogs. The result was that there was about a 10^6 safety factor. I think the concentration of progestational receptors in the breast was about 1,000 times less in homo sapiens than dogs; the chemical's concentration in the breast was about 1,000 times less, and the affinity for the receptor was many times less. We concluded that we had a safety factor of between 10^6 and 10^8. Even the FDA accepted that it was a reasonable margin of safety, so the incident was never made public. It became a little six-line issue from the DHHS (Department of Health and Human Services), that the beagle is a very sensitive animal for evaluating progestins. I think many of us were still a little unhappy, but one of the pharmaceutical companies had very wisely decided to do a similar study in the same species of dogs on natural progesterone, a few months after they started the studies on the progestational agents. They were able to tell us that at about a tenth of the dose of the synthetic progestational agent progesterone produced metastasizing carcinoma in the same species of dog, and we were all well assured.

One of the things that puzzles me about chlorinated chemicals' carcinogenicity in the mouse is, Why are you making such heavy weather over it? This problem with the beagle, which affected at least 30 million women who would have been told they'd been taking carcinogens for years, was solved within a year. This debate has been going on for over 20 years—the question of how we decide the safety of chemicals when they give rise to liver tumors in the mouse. Why don't we make a decision? I would say the decision is that the mouse is an unsuitable species.

Dr. Stevenson: I suggest that we do the same—and look at the receptor affinities in the two species and so forth.

Dr. Mirsalis: I want to respond very quickly to that. I agree with your first statement, that to call something nongenotoxic you must have pretty good evidence. On the other hand, I think I stand on our call on methapyrilene. The compound has been studied to death, and it is a very unusual compound—it clearly doesn't fall in the same class as a lot of the other, more traditional tumor promoters. But, if you look at the target organ, which is rat liver, it doesn't covalently bind to DNA, but it does bind to protein. It doesn't induce adducts by ^{32}P-postlabeling. It doesn't induce strand breakage or DNA repair—it's really pretty clean. I'm familiar with your initiation/promotion study, and I'd be happier if you'd promote it with something other than AAF, which, of course, is very genotoxic. Try the same thing initiating with methapyrilene and promoting with partial hepatectomy or carbon tetrachloride or another nongenotoxic agent. If you get the same result, then I'll concede the point.

Methapyrilene does two things: it causes very significant mitochondrial proliferation and induces significant hypomethylation. Willie Lijinski has suggested that the mitochondrial proliferation, which is significant and rather unusual, may be related to gene regulation or other effects.

In the second case, Dan Casciano and Willie Lijinski have both come out quite recently with data showing that methapyrilene induces hypomethylation. That goes back to what Jay Goodman was saying earlier, that this may be affecting gene regulation. I think that the data on methapyrilene are very strong, that it's not working by a genotoxic mechanism in rat liver, the target organ.

Index